The motif is the Sanscrit for "Particle." This ancient root of Indo-European languages serves to illustrate the antiquity of particulate technology as an art.

Particulate Science and Technology

by

JOHN KEITH BEDDOW

Professor of Chemical and Materials Engineering
University of Iowa

Chemical Publishing Co., Inc.
New York, N.Y.

ISBN 0-8206-0254-x

Printed in the United States of America

ACKNOWLEDGMENTS

Most of those who helped me with this book are mentioned by name in references and footnates. Special thanks are offered to Professor Thomas Farrell, Professor of Engineering Communications in the University of Iowa who really helped me to say what I wanted to say. I would also like to thank Kay Chambers for so skillfully making the drawings. For typing assistance I would like to thank Rosanne Huff, Rachel Levy and Kathy Walters.

I am grateful to the University of Iowa for a semester leave of absence during which the text was written. Finally, I would like to thank Earla Marshall for sending out literally hundreds of letters asking for permission to use many of the diagrams in this book.

The author wishes to thank the following sources for permission to use the items mentioned below:

Academic Press and C. Billings, Tables 5.2 and 5.3, Fig. 5.2; Academic Press and C. Mercer, Tables 5.11, 5.12, 5.13, 5.14, 5.15, 5.16, 5.17, 5.18, Figs. 6.5, 7.23, 7.12, Table 7.8; Academic Press and K.J. Whitby and B.H.Y. Liu, Figs. 5.9 and 7.11; Academic Press and L.A. Zadeh, Figs. 6.49 and 6.53; Academic Press and B. Hodkinson, Fig. 7.7 and Table 7.5; Academic Press, Fig. 7.25; Academic Press and L. Waldmann, Table 7.9; Academic Press and M. Corn, Table 7.13; Academic Press and C.N. Davies, Tables 7.19, 7.16, 7.18, 7.15; Academic Press and P. Goldsmith, Figs. 7.13a and 7.13b. Accounts of Chemical Research (jo.) and G.A. Samorjai, Fig. 3.60; Addison Wesley Publishing Co., Fig. 2.5 and Fig. 5.1 based on an original of theirs; Addison-Wesley, Tables 7.3, 7.3, Figs. 7.29, 7.1, 7.3, 7.26, Tables 7.20, 7.21, Figs. 7.32, 7.33, 7.34, 7.35, 7.36; Air Pollution Control Association Journal, Figs. 3.52, 3.53 and Table 3.10; Allis-Chalmers Corp., Fig. 3.24; Ambac Industries, Inc., Michigan Dynamics Division, Dynapore®, Fig. 4.35; American Ceramic Society, Table 7.10; American Industrial Hygiene Association, Tables 8.5, 8.6, 8.16; American Institute of Chemical Engineers and W.R. Marshall, Jr., Figs. 3.5, 3.6, 3.9, 3.10 and 3.11; American Institute of Chemical Engineers, Figs. 4.51, 4.52, 4.53, 4.54, 6.15, 7.24; American Laboratory (jo.), Fig. 6.11; American Scientist (jo.), Figs. 7.4 and 7.8, Table 7.7; American Society for Metals,

Fig. 2.37; American Society for Testing and Measurement, Figs. 2.9, 4.47, Table 6.2; Ann Arbor Science Publishers, Inc., Figs. 2.1, 2.7, 2.12, 2.14, 2.16, 2.17, 2.18, 2.23, Tables 2.3, 2.4, 2.5, 2.6; Annals of Chemistry (jo.), Fig. 3.46; Teoman Ariman, Editor, Univ. of Notre Dame, Fig. 7.22; Teoman Ariman, Editor, Univ. of Notre Dame and D.C. Drehmel, Table 7.23; Associated Book Publishers, Ltd., Figs. 6.2 and 6.3; Associated Book Publishers, Ltd., Figs. 8.1, 8.2, 8.3 and Table 8.2 are based on originals of theirs; Aufbereitungs-Technik (publisher) Processing Preparation, Fig. 4.4; The British Ceramic Society, Fig. 6.1; British Journal of Applied Physics, Institute of Physics and D.H. Lucas, H.J. Lowe, Table 5.4; British Journal of Applied Physics, Institute of Physics and P.W. Howe, Fig. 6.7; British Medical Journal and P.C.Elmes, Fig. 8.15; British Medical Journal and R.F. Hounam, Tables 8.17 and 8.18; British Occupational Hygiene Society and Pergamon Press, Fig. 5.8; Buffalo Forge Co., Fig. 5.7; Buhler-Miag, Inc., Fig. 4.28; Chapman and Hall and K.N. Palmer, Figs. 8.1, 8.2, 8.3 and Table 8.2; Chemical and Engineering News, Fig. 3.55; Chemical Engineering (jo.), Table 3.9; Chemical Engineering (Japan), Fig. 4.46; Chemical Engineering Progress (jo.) and W.R. Marshall, Jr., Figs. 3.14, 3.15, 3.16, 3.17, 3.18, 3.19, 3.20, 3.21, 3.22; Chemical Engineering Progress (jo.), Figs. 4.1, 4.33, Table 4.11; Chemical Physics Letters and Yu A. Borisov, Figs. 3.56 and 3.57; Chemical Reviews (jo.), Table 7.17; The Chemical Society, London, Tables 6.4 and 8.9; The Clean Air Society of Australia and New Zealand, Department of Health, Fig. 7.19; Consolidated Engineering Co., Fig. 4.26; Constable Publishers and H.E. Rose, Figs. 2.6, 3.30, 3.36 and 3.37; Craven Fawcett (Wakefield) Ltd., England, Fig. 3.27; CRC Press, Inc., Figs. 2.8, 6.30, 6.38, 6.50, 6.51, 6.52; CRC Press and G.C. Lowrison, Figs. 3.23, 3.28, 3.29, 3.38, Tables 3.2, 3.3, 3.4, 3.5; Department of the Air Force, Table 2.2; Dynapac Manufacturing Co., Fig. 4.19; Educagroep, Educa International, Netherlands, Tables 6.6, 6.7; Edward Arnold, Ltd. (publ.), Fig. 2.24; Ellis Horwood Ltd., Publishers, Fig. 6.8 and Tables 6.1, 6.5, 6.8; Elsevier Scientific Publishing Co., Table 4.13, 4.14; Elsevier Sequoia, S.A., Powder Technology (jo.), Figs. 3.48, 3.50, 3.51, 4.44, 4.57, 4.48, 4.15, 4.18, 4.39, 4.43, 6.17, 6.18, 6.41, 6.47, 6.48, 6.4, 6.31, 6.33, 6.34, 6.35, 6.36, 6.10, 6.43, 6.44, 6.45, 6.46, 6.3, 6.9, and Tables 4.6, 4.7, 4.22, 6.18; Engineering Experimental Station, Georgia Tech., Table 7.14; Engineering Experimental Station, University of Utah, Figs. 4.2, 4.6, 4.12, 4.13; Fearon-Pitman Publishers, Table 5.9; Gordon and Breach Science Publishers, Inc., Fig. 2.30 and Table 2.8; Nick Gothard, Gothard Industries, Corp., Table 4.19; Heyden and Son, Inc., Figs. 6.61, 6.62, 6.63, 6.55, 6.56, 6.57, Table 6.22; Industrial and Engineering Chemistry Process Design and Development, American Chemical Society, Figs. 7.21, 7.38, 4.55, Table 5.8, Industrial and Engg. Chemistry. American Chemical Society and W.R. Lane, Fig. 3.12; – and D.W. Fuerstenau, Fig. 3.45; – and James H. Gary, Fig. 4.38, – and A.B. Metzner, Fig. 4.24; – and J.M. Smith, Fig. 4.29 and Tables 4.9, 4.10; Industrial Research/Development, Fig. 6.59; Institute of Mining and Metallurgy, Table 3.6; Institution of Mechanical Engineers, Fig. 3.34 and 3.35; Interdisciplinary Science Review (jo.), Heyden and Son, Ltd., Table 8.8; International Journal of Powder Metallurgy, Figs. 4.58 and 4.61; IPC Science and Technology Press, Ltd., Fig. 4.17; Japan Society of Powder and Powder Metallurgy, Table 4.12; John Wiley and Sons, Figs. 3.47, 3.49; John Wiley and Sons and H.C. van de Hulst, Fig. 7.23; Journal of Applied Physics and F.E. Luborsky, Fig. 7.31; Journal of Chemical Physics and R.C. Baetzold, Fig. 3.58; Journal of Colloid and Interface Science and B.H.Y. Liu, Fig. 3.54; Journal of Geology (Univ. of Chicago Press), Table 4.19; Journal of Metals, Figs. 3.3, 3.13, 3.25, 3.41 and Table 3.1; Journal of Physics D: Applied Physics, Table 2.7; Journal of Research Association of Powder Technology, Japan, Figs. 6.26, 6.35 and Tables 6.13, 6.14; Journal of Sedimentary Petrology, Fig. 5.17 and Table 6.10; Journal of the Oil and Colour Chemists' Association, Fig. 2.3; H. Kane, Fig. 4.60; Koppers Co., Inc., Figs. 3.31, 3.32, 3.33; Homer Lewis, Los Alamos Scientific Laboratory, Tables 5.5, 5.6, 5.7, 5.10, 5.20, 5.21, 5.22, and Figs. 5.10, 5.11, 5.12, 5.13, 5.21, 5.22, 5.23, 5.24, 5.25, 5.26, 5.27, 5.28, 5.29; Macmillan Publishing Co., Figs. 4.31, 4.45, 4.4; Jeffrey Steinfeld, Massachussetts Inst. of Technology, Fig. 3.59; McGraw Hill Book Co., Figs. 3.44, 5.5, 5.6; Metal Powder Industries Federation, Fig. 4.64; Methuen & Co., Ltd. and R.A. Bagnold, Fig. 8.9a based on an original of theirs; National Physical Laboratory (HMSO), England, Fig. 2.11 and Table 3.8; Nature (jo.), Fig. 8.13; Newell-Dunford Engineering Ltd., England, Fig. 3.23; Oxford University Press, Fig. 4.63; Pergamon Press, Inc., Figs. 2.27, 2.28, 2.34, 5.8, 6.6, – and P.M. Heertjis, Fig. 4.32, – and M. Ashby, Figs. 4.65, 4.66, Table 4.26, – and Journal of the Franklin Institute, Figs. 5.14, 5.15; Physikalische Zeitschright (jo.), S. Hirzel Verlag, Table 7.22; Pion Ltd., Table 7.24; Pipeline and Gas Journal, Fig. 4.22; Planseeberichte fur Pulvermetallurgie (jo.), Figs. 6.22, 6.23, 6.24, 6.18, 6.19, 6.20, 6.21 and Table 6.11; Plenum Publishing Co., Tables 5.19, 8.13, 8.14, 8.15 and Figs. 5.18, 5.19, 8.14; Powder Metallurgy International, Verlag Schmid GMBH, Table 4.20, Figs. 6.13, 6.40; Praktische Metallographie, Dr. Riederer Verlag GMBH, Fig. 6.58, Tables 6.20, 6.21; Prentice Hall, Figs. 7.27, 7.28, 7.30 are based on originals of theirs; Process Engineering, Table 4.8 and Fig. 4.37; Processing (formerly British Chemical Engineering), Fig. 3.40; Quarry Management and Products, Fig. 6.16, Research, Figs. 4.34a, 4.34b; Scanning Electron Microscopy, Fig. 2.41; SRI International, Table 2.1, Science (jo.), American Association for the Advancement of Science, Fig. 6.12; Scientific Publishing Co., Fig. 2.10; Sedimentology (jo.) and Blackwell Scientific Publications, Ltd., Fig. 6.17; Solids Flow Control Corp., Fig. 4.16a; Stanford Research Institute, Table 8.4; Syracuse University Press, Figs. 3.2, 3.4; Transactions Institute of Chemical Engineers, Figs. 4.42, 7.16, Tables 6.6, 8.3, – and Paul Eisenklam, Fig. 5.30; Trans Tech Publications, Table 4.1 and Figs. 4.5, 4.14; U.S. Department of Commerce, Figs. 8.10, 8.11; U.S. Energy Commission, Fig. 2.14c; University of Pittsburgh, Office of Continuing Education, Figs. 4.21, 4.23, 4.25, Table 4.3; H.C. van de Hulst, Fig. 7.5; Van Nostrand Publishers, Fig. 8.6 and Table 8.4; Vibra Screw, Inc., Fig. 4.16b; Western Publishing Co.,

Figs. 8.4, 8.5; Wiley-Interscience, John Wiley and Sons, Inc., Table 4.15; The Young Industries, Fig. 4.28; Z. Physik chem. Neue Folge, Figs. 2.33, 2.36.

I would be glad to be informed of errors (omissions) in the above list so changes can be made in the next edition.

CONTENTS

PREFACE

Unity in Diversity

An essay to integrate the field of particulate science and technology seems opportune. Concern with specialized interest areas results in slow and usually steady progress and is a comfortable situation to work in. However, particularity of interest constrains ideas, whereas a sound unifying concept knows no natural boundary. Our field is currently asking searching questions and seeking definitive answers to broaden our outlook. We are ready to see what the other person is up to, to learn from the observations of others, and profit by interrelating their ideas.

However, in a broader and more fundamental sense, most of us who work in the field of finely divided matter would assert that, collectively, our present knowledge is only a promise of what it will be. Further, we individually know so little of what we need to know and, to add to our troubles, we realize that new knowledge is being accumulated at a pace faster than we can injest on a regular diet of reading and study.

A rewarding approach to the problems inherent in mastering this diversity is to develop an appreciation for the wholeness of the field. Necessary specialized excursions can then be related to the whole-field view and fitted into a rational scheme of scientific explanation. This book is an attempt to provide a methodical whole-field view of fine particle science and technology. It in no way pretends to be exhaustive. There is a rich abundance of searching treatises on particular subjects available now to the specialist, so a general view both unifying the good work already accomplished and providing a comprehensive structure for guiding what is to come seems appropriate at this juncture.

Particulate science and technology is a fascinating field of study. I don't know anyone working in it who does not find it somewhat exciting. It is an ancient art and a baffling science simultaneously. I count myself among the lucky ones who wonder why matter in finely divided form behaves as it does.

INTRODUCTION

"There is a tide. . ."

UNDERDEVELOPED TECHNOLOGY

The science of the nature and behavior of fine particles, like an underdeveloped country surrounded on all sides by highly structured industrial states, looks wistfully at developed entities like physics, chemistry, mathematics, biology, material science, engineering sciences, and computer science, to mention but a few well-structured disciplines. As with all cases of underdevelopment, a great deal of "aid" from these technically advanced neighbors is needed if progress in fine particle science is to materialize. However, the drive to progress must be derived from the urgencies sensed by those already working in the field who have the wit and foresight to develop and implement the creative necessities.

In any field of human technical endeavor where progress is made there is an advantageous juxtaposition of three elements: industrial and commercial need, intellectual challenge, and the possibility of success. For the study of particulate matter all three elements now coexist in good measure, but they have only just come together. The efforts of many have made this conjunction possible and although it is unfair to mention some and omit others who have made their contributions, still it seems appropriate to mention some major contributors.[1–19]

In this century the sciences have been developing new knowledge at a frenetic pace in a direction predetermined by the discovery of electricity. The physicist has become interested in the very small and, with the engineer, in the macroscopic to the spectacularly large, and the chemist is concerned with structures ranging up to molecular sized units. This pattern of study has left the great middle ground of fine particles in a relatively unexplored condition. During the past 200 years, we have seen the armies of science in their inexorable advance on the relationships between the structure and properties of solids, liquids, and gases. It is now the turn of matter in finely divided form to be comprehended and exploited. We can only guess why it has taken so long to reach this stage, but one can imagine that it may have something to do with the extreme difficulty of theoretical and experimental work in the area, the staggering diversity of particulate matter in finely divided form, and the glamor of the older established disciplines which enjoy a hard won prestige.

Within the last 20 years or so there have been signs of a gradually developing awareness and interaction among those concerned with particulate matter. Although we still have the ceramicists and the powder metallurgists, the pharmacists and the chemical engineers are still ploughing their own furrow, though they interact more with each other than ever before. This seems to be true on campus, in the meeting hall, and especially in industry. Perhaps the most public manifestation is in the start-up of the journal "Powder Technology" in the 1960's; in the change of name of the "International Journal of Powder Metallurgy" to that of "International Journal of Powder Metallurgy and Powder Technology" in the 1970's. Other evidence includes: the publishing of Orr's book on *Particulate Technology* by Macmillan in 1966, the holding of a workshop on "Fine Particle Technology" in 1975 (National Science Foundation)*, the publication of various series of texts on different aspects of particulate matter on both sides of the Atlantic, the founding of the Fine Particle Society and its aspiration to correlate and serve the needs of its members in this field, in Japan the foundation of the Research Association for Powder Technology, the establishment of the International Powder Institute, the formation of the International Fine Particle Research Institute, and the services of the Powders Advisory Center. We can conclude, therefore, that there are serious efforts being made to move in the direction of integration so that we can better focus on our needs and coordinate our energies to meet them.

The key to success lies in our educating ourselves. In Europe a number of research centers like Karlsruhe Erlangen, Bradford, and Loughborough and in Canada, Laurentian University, are studying and teaching. But in one major industrial enterprise of this world, the United States of America, there is no single institution devoted to the study of matter in finely divided form, although there are numerous individual students scattered throughout the country. Some have said that there is a pressing need for establishing an American institution, many others have nodded their assent, but to date none has been organized. Apparently, the greatest single obstacle to this development is not the lack of funds, but the absence of a developed

---

*Program Director, Dr. Morris Ojalvo

consensus of the whole field view and a realization of its potential long-term benefits to us all. This situation exists because of the lack of serious discussion about the problems inherent in formulating policy, developing structure, and implementing the former through the latter. *In essence, just how do we conjoin the academic with the industrial, and reward the components in such a diverse interdisciplinary field as particulate matter and at the same time avoid the production of a new breed of specialist who will fragment the very cohesion that is so vital and so assiduously being developed.*

RESEARCH AND DEVELOPMENT NEEDS

Our technological requirements for comprehending the state and behavior of particulate matter during the next decade have been admirably summed up in the NSF Workshop Report.[20] They fall into four main categories.

Particulate Properties and Structure

The description of particle shape by means of morphological analysis of particle profiles offers a promising way of expressing visual images mathematically in a variety of functional forms. The functions can then be related to the properties of the particulate matter. Fundamental investigations of particle surface phenomena are required to establish the basic geometric properties of particulate matter so that they may then be related to their physico-chemical properties. Coordinately, establishment of the relationships between surface, subsurface, and interior structures and particle properties must be another of the goals of fine particle research and development. There is also a major experimental challenge to learn more about the electrical, optical, thermal, and magnetic characteristics of fine particles. Further, as more and more industrial processing becomes automated, there is an increasing need for the on-stream testing-cum-monitoring of powder technology operations. This demand raises a further need for study in this area because we do not know enough about what technology there is available.

Particle-Fluid Separation

There is an urgent need for developing dynamic concepts about the methods for the separation of very fine particles (less than 0.5 micron) from gases. More direct approaches to the study of the structure of flocs and of agglomerates generally are necessary. In particular, there is a need to establish the pattern of relationships between the development and form of floc structures and the behavior of sedimentation, thickening, and dewatering processes at all of their stages. Moreover, developing methods for studying the behavior of individual particles in rheologically complex fluids as well as in their deposition on collecting surfaces is a prerequisite to understanding the mechanisms and improving the technology. The transmission of interparticle forces in changing conditions (e.g., the alteration of particle-fluid drag or the change of applied pressure level, etc.) and the way in which the particles become rearranged must be directly observed under carefully controlled experimental conditions if we are to master the complexities of the motion of particle sets in dynamic environments.

Particle Handling

The following quotation from Zenz is an eloquent indicator of the formidable nature of our problems in this area:[21] "As mundane as it might first appear, there is today no universally guaranteed design procedure for assuring that a bin filled with any powder will be able to be emptied whenever desired." A new approach to the problems of mixing and blending must take into account the complex properties of the particualte matter being handled. When particles are being moved, usually entrained in a fluid, their level of interaction with each other and with the carrier fluid and also with the system surfaces and measuring devices should be related to the basic properties of the particles, the fluid, and the overall system. Phenomena to be studied include bridging, flow, choking, saltation, resilience and elutriation, to name but a few.

Controls

Minimizing the deleterious effects of applying powder technology requires controls. In general these systems problems are extremely involved. Dealing with only single phase liquids (for example) is a sufficiently complex system control problem, but effectively controlling in detail the behavior of particulate matter in different situations (as outlined in the immediately preceding sections of this introduction) raises the difficulty of our task to a high level. In brief, we have to adopt an overall systems approach which will take into account technical, economic, health, social, and political factors inherent in a given situation. Just a glance at some of the industrial hygiene and related journals reveals the extent and seriousness of the problems of minimizing unfavorable effects from the industrial use of fine particles.

ABOUT THIS BOOK

The general concept offered in this book and some of the more important conclusions and observations that I have developed during the course of its writing are stated and examined in Chapter 1. Chapter 2 emphasizes the primacy of the individual particle in particulate science and technology. Chapters 3 and 4 may be said to deal with PARTICULATE TECHNOLOGY; they describe the ways in which particles are formed, the various production methods associated with each, followed by a descriptive treatment of some of the standard processing steps for dealing with finely divided matter. The next three chapters deal with what might be appropriately termed PARTICULATE SCIENCE. In Chapter 5 there is a discussion of a number of descriptors with particular emphasis on fine particle statistical descriptors. Chapter 6 reviews some of the more recent work on particle shape characterization and particle size analysis. Chapter 7 examines the nongeometric properties of fine particles including optical, electrical, magnetic, thermal, and chemical properties, and relates them to various aspects of particle settling and diffusion.

The essential message of the final Chapter 8 is vital to all of those working in the fine particle field. Particulate hazards are insidious: if they don't blow you up or irreparably damage your health they may just ruin your landholding quietly and steadily. Some of what must be learned and some presently known means for protecting us against folly in the use of fine particles are explored. Fine particles should be treated with the greatest of respect.

THE APPROACH

The individual particle may range in size from the vanishingly small to the visually obvious and may be simply shaped or multifaceted in outline. Regardless, we have to be able to identify and collect the shape-size information, and to condense it without losing the pattern that the information represents. This is no easy task and we have yet to solve it. Chapter 2, entitled "The Single Particle," is presented early in the book in order to focus attention on the fact that the single particle is indeed the fulcrum of our study. We have to develop a precise method for collecting the size-shape information and handling it for the single particle. Then we can confidently describe sets of particles with a corresponding general description (coupled with their chemical and physical properties), and then be able to relate particle properties to their behavior so that we can make predictions of their behavior in known conditions and thereby control them.

Chapters 5, 6, and 7 can be read following Chapter 2 without disturbing the continuity. Although particle size measurement is discussed, it is treated generally because there are already very informative specialist texts on the subject. Particle morphology analysis is treated at greater length because there is no other current source on this topic (outside of the journal literature) and also because this subject is important to the development of fine particle science. The approach taken in Chapter 6 might, at first, seem complicated, but in light of the overall strategy it is not. Thus, to obtain the necessary information and to analyze the morphology of a particle (and then many more) require sophisticated image analysis capability and the ability to handle and condense large amounts of information. At the present time, there are instruments available with the requisite image analyzing capabilities. By using these in conjunction with an appropriate set of information handling techniques (pattern recognition, orthogonal functions, fuzzy sets, etc.) advances are being made in the morphological analysis of particles. A crucual part of this program is the development of a physically meaningful set of morphological descriptors which can be culled from the information gathered about the particle. Chapter 7 discusses some of the physical and chemical properties of fine particles including electric, magnetic, optical, and thermal properties as well as settling. Nowhere in this chapter is there a statement about the effect of morphology upon particle physical-chemical properties,* because their interrelations are unknown. At the present time, therefore, we investigate

*With one exception, Chapter 7, section

and study the sets of characteristics discussed in chapters 6 and 7 respectively. We are not yet able to describe one particle in terms of its associated size-shape/chemistry-physics continuum. This dichotomy is real and profound and the literature presents it this way. Clearly it must be eliminated in the course of time so that we can better describe the reality that is, and build our science on that.

Many advances have been made and continue to be made in particle production, in the understanding of mechanisms of particle formation, and in particle handling and processing. For those familiar with these aspects of finely divided matter, Chapters 3 and 4 may be overlooked or read cursorily. The beginning student will find these two chapters essential. The treatment is not surfeit with references but there are significant ones so that the reader can find the original sources and related items. The chapter on hazards is last in order but not in importance; it should not be overlooked.

Chapter 1

IN THE SCHEME OF THINGS

1.1 PARTICULATE SCIENCE AND TECHNOLOGY

The objective of this book is to establish the concept that the behavior of fine particles when a specific force is applied to them in a particular medium is determined by their structure; by their geometric, chemical, and physical properties; and by their interactions among themselves with the applied forces and the medium. Fine particle science describes the macroscopic, microscopic, and atomic level structures of individual particles and their sets in terms of the fundamentals of physics, chemistry, and mechanics, and interrelates these three levels with the behavior of a set (powder) and its component particles to permit the development of technology.

The study of particulate science and the consequent technology is a field of considerable intellectual, economic, and industrial magnitude. It is not an exaggeration to say that our whole industrial civilization is based upon the processing of particulate matter, yet until recently it has been largely uninvestigated by the scientist and engineer. Despite the fact that particulates have become important during the last half century, studies are still relatively unorganized, uncoordinated, and unrecognized. Neither the industrial nor the academic communities have yet perceived the predominating importance of particulates in the affairs of mankind. Our industrial and commercial progress is becoming more and more dependent upon our understanding and our mastery of the complex and broad science and technology of fine particles. Consequently, more work is urgently needed in the particulate field. It is also imperative that those of us who are working in the field make every effort to disseminate and correlate information concerning the knowledge that we have gained to date, and, in addition, to proselytize others to join in the exciting and revolutionary advances that are going to be made in the next twenty years or so.

Industrial applications have long been made, but new demands arise continually. The aspirin tablet and the powder metallurgy gear are produced by powder compaction in a die. Plastics granules are thermomechanically molded. Pottery is fired in an oven; bread is baked in an oven; powder metallurgy parts are sintered in a furnace; sinter cake is produced for blast furnace burden; some liquid phase sintering occurs during the production of cement clinker in a rotary kiln. Most mineral processing activity includes a great expenditure of effort and resources to separate the valued mineral from the gangue; effluent from industrial plants must be cleansed of noxious gases and liquids (if they are present) and especially particulate matter, which is almost always present.

Matter in finely divided form is often much more reactive than the same substance in solid form and this instability can lead to fires, explosions, and, in the case of protein material, to spoilage. An important fraction of the world's electric power is consumed in the crushing and grinding of macroscopically continuous solids in order to produce finely divided particles. The list of processes for treating particulate matter is so large that it would appear to be never ending: food technology; minerals processing; ceramics and refractories production; paints, inks, and dyes manufacture; cement, hard metal, welding rod, and building product industries.

Earth moving and soil mechanics are generally important provinces of fine particle science and technology, as well as many agricultural operations. Chemical engineering processing deals with large amounts of finely divided matter. Metallurgy deals with continuous solids but upon closer examination it can be observed that the continuous phases contain dispersions of other phases, precipitates, and inclusions. Geology looks to particles to explain earth movements and to trace the results of forces operating on and in our planet.

The behavior of finely divided matter in the more mobile gaseous and liquid continua is an important area of study in the engineering disciplines. In pharmacy, the production of solid dosage forms is a sophisticated branch of particulate technology which is complicated even more by the interactions within the human system that the medication may undergo.

No matter what their individual processes, material, or end product, all of those persons involved in fine particle technology require adequate samples of what they process and appropriate concepts, methods, and tools with which to characterize these samples. It is possible to obtain data of this type that is reproducible and that can be related in a descriptive way to the subsequent events that the particulate matter is to pursue. However, with our present methods we are not able to do much more than establish general patterns of relationships. The reasons for this limitation are not hard to fathom. For example, consider the more or less standard requirement to determine the particle size. Such determinations are linked to the property of particle shape. Therefore when a "size" is quoted, in a very fundamental sense we only have a small item of needed information with which we can characterize the complex population of particles. It is for this reason that there is currently a push to develop useful ways to measure particle morphology.

However, a set of particles represents a total surface area that is large in comparison with that of the same mass in the form of a single solid. Methods are available to determine surface area and it is therefore convenient to relate this measure to some property of the set of particles that is of concern (for example, the reactivity). However, the viewpoint that such properties of a rather small portion of solid are determined by a thin, two-dimensional external layer is too limited because we clearly have to consider the material directly underneath this extremum. As in the case of particle shape, this area of structure too is in the melting pot.

When we can interrelate particle size, morphology, chemistry, and physics we will have a fine particle science capable of generating new technologies so badly needed in this field. Within such a diversity of industrial activity the implementation of new technologies will likewise have broadside benefits for society. We are not yet in this happy condition. However, the encouraging developments in particle morphology analysis, in structural analysis, and in related areas ensures our eventual success.

1.2 OUR REALM

Matter, whether in a solid, a liquid, or a gaseous state, can exist in the form of finely divided particles. Familiar examples are dust and powder, both solid particles in a gaseous medium. Mist and spray are examples of fine liquid droplets in a gaseous medium. Bubbles in a liquid are gas particles, and perhaps pores in a solid are examples of finely divided particles called cavities in continuous media. The schematic diagram in Fig. 1.1 illustrates the range of fine particle systems that are encountered.

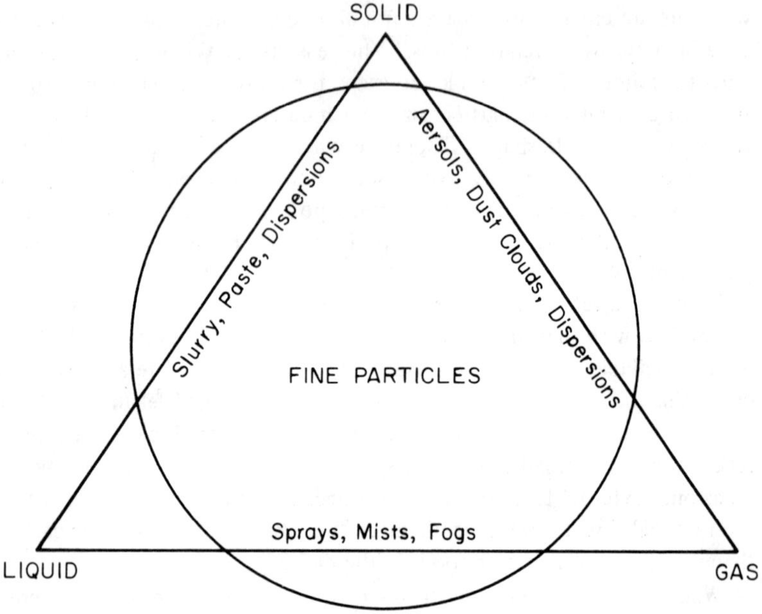

Fig. 1.1

Fig. 1.1 serves to emphasize two major points that should be kept in mind:

1. For the most part, particulate systems involve sets of particles interacting with each other and with the enveloping medium, all within one or more force fields.
2. An understanding of fine particle systems necessitates corresponding understanding of solids, liquids, and gases as states of matter, all of which can be divided into fine particles and all of which are media for particles of another state.

These two considerations alone tend to call for better models than those available to date, as well as removing the deficiencies due to the morphology/chemical-physical dichotomy already alluded to. The basic reason for calling particulate science and technology interdisciplinary is simply that a cohesive discipline must be constructed from many diverse fields. Perhaps another and equally valid reason is that the complexity of the problems inherent in this field defy the aspirations of individuals working separately to solve them. When we work together, sharing disciplines and exchanging information, we develop an interdisciplinary science with which to continue to advance our knowledge and our technology.

In the following chapters, an examination of some of the more basic aspects of the subject serves to illustrate the great advances in particulate science and technology that are being initiated. In summary, it is a wide open, dynamic field that challenges fruitful investigation.

REFERENCES

Note: The interested reader may obtain an introduction to the work of these authors. The items listed below are usually conveniently available in technical libraries.

1. Bagnold, R. *The Physics of Blown Sands and Desert Dunes.* London: Methuen & Co., 1941.
2. Dalla Valle, *Micromeritics.* New York: Pitman, 1948.
3. Derjaguin, in *Powders in Industry.* SCI monograph 14, London, 1961, pp. 102–113.
4. Exner, H. E. *New Theoretical Approaches to the Sintering Mechanisms.* paper presented at seminar on "Solid Particles." Arad, Israel, 1975.
5. Fuerstenau, D. W. *Powder Technology* 1, 1967, pp. 174–182 (with D. S. Cahn).
6. Gaudin, A. M. *Principles of Mineral Dressing.* New York: McGraw-Hill, 1955.
7. Hausner, H. H. *Handbook of Powder Metallurgy.* New York: Chemical Publishing Co., 1973.
8. Kaye, B. H. is soon to publish a book on Fine Particle Characterization.
9. Kingery, W. D. *J. App. Phys.* 30 (1959): 301.
10. Kuczinsky, G. C. *Sintering and Related Phenomena.* New York: Gordon & Breach, 1967 (with N. A. Hooton and C. F. Gibbon).
11. Lawley, A. in *Powder Metallurgy for High Performance Applications.* Ed. Burke and Weiss, Syracuse University Press, 1972, pp. 1–23.
12. Lenel, F. V. ibid, pp. 119–137.
13. Meloy, T. P. *Powder Technology.* 16, 2 (1977) 233–253.
14. Orr, C. *Particulate Technology.* New York: Macmillan, 1966.
15. Rose, H. E. *Ball, Tube and Rod Mills.* London: Constable, 1958 (with Sullivan).
16. Rumpf, H. in *Agglomeration.* Ed. by Knepper, New York: Wiley, 1962, p. 399.
17. Tiller, F. M. *Chem. Eng. Prog.* 49 (1953) 467–479.
18. Underwood, E. E. *Quantitative Stereology.* Reading, Mass.: Addison-Wesley, 1970.
19. Williams, J. Editor of *Powder Technology.* Numerous publications therein dealing with mixing and segregation.
20. Wasan, D. *Powder Technology Research Directions.* NSF, Washington, 1977.
21. Zenz, F. ibid.

To this list many other names could be added, especially H. H. Heywood who was responsible for much of the pioneering work in particle shape analysis.

Chapter 2

THE SINGLE PARTICLE

2.1 THE PRIMACY OF THE SINGLE PARTICLE

To deal practicably with materials in finely divided form, it is necessary to begin with accurate knowledge of the single particle, not only of its chemistry and shape, but of its total characteristics seen in a whole-field view. The gains for both science and technology, for students and for industry, are potentially great for understanding, prediction, and control of materials in finely divided form. When an adequate level of sophistication in the measurement of particle size and shape is attained, it will then be possible to map the internal details of the particle chemistry, defect content, and other physical properties within the established morphological framework. This internal detail has been termed texture.* Specifications dealing with particulate matter now lack detailed and precise information concerning the characteristics of the particles of which they are composed. This inadequacy of information inhibits the development of accurate criteria for establishing specifications. Practically complete descriptions of individual particles enables one to describe sets of particles and leads directly to an increased level of precision in specification development. This level of precision can be adjusted to mutually agreed gradations.

At the same time, the development of a more precise characterization of individual particles enable us to hypothesize and test more detailed models of particle formation and production processes. This insight will raise these respective technologies to a much higher level of understanding and control. Simultaneously, the behavior of sets of particles will be more rigorously related to the fundamental characteristics of their individual members. This will open up particle processing technology to a new era of exploitation in which science will increasingly provide understanding of design problems and ultimately guide the formulation of solutions to these

*Professor Julius Tou, personal communication, August 1977.

problems and the technology to deal with them.

The above exposition is just another way of restating the importance of our taking the whole-field view of particle science and technology. One must not, however, infer that this is a panacea. For example, it has been observed that, in order to select an appropriate water pump for pumping water up a pipe, it is unnecessary to know the molecular weight of water. Similarly, many problems in processing technology are practically solved at the macroscopic level.* The truth of this statement can not be challenged today but, with the rapid development of Particulate Science and Technology, tomorrow may tell another story. In the interim, the level of our understanding will improve as the whole-field view is adopted.

Information Needed

The whole-field view of matter in finely divided form necessitates the ability to analyze and describe the size-shape/chemical-physical continuum of a single particle with accuracy and completeness. At the present time we are not able to carry out this task because we do not have the analytical techniques and the information-handling capability that is required.

Once the data for a single particle can be obtained, data for the many particles of a suitable sample will then be required. This data must then be appropriately manipulated so that it can be condensed or compressed, while retaining the original pattern. The amount and diversity of information to be handled is large. It is believed that the techniques are available for dealing with this problem. The question is which techniques are appropriate and when should they be used in order to optimize the application of the whole-field view of finely divided matter.

One initial problem is that the particle is quite unique—unless perfectly spherical any specific particle is not exactly like its neighbor. And yet the differences may be miniscule. Clearly the means for describing the morphological characteristics of even one particle must be sophisticated and potentially capable of a high degree of discrimination. However, a discriminatory ability which forces us to distinguish among all the particles encountered is not required. Such a level of differentiation would yield an almost infinite variety, which would be useless. To achieve practicability through identifying manageable yet sufficiently discriminating detail, something analogous to the chemist's atomic series is required. Analytical tools that enable us to classify, codify, seek, and discover the order that is (we

*Dr. John C. Williams, personal communication, August 1977.

believe) inherent in nature and awaiting our happy discovery is needed.

Developing Necessary Methods

Morphological analysis of fine particles is a central, essential, and fundamental prerequisite to the development of a science of fine particles. It is well known that, for anything but the most regular particle, the measurement and indication of its size by a single "diameter" is a very approximate means of characterizing the particle, ignoring as it does the whole intricacy of its shape. Clearly, methods must interrelate the size-shape characteristics of the particle. For measuring individual particles of finely divided matter, there are numerous methods available based upon a variety of physical principles. These are reviewed in Chapter 6. The problem with many of these methods is that they physically classify the particles according to their response to the applied forces of the testing system being utilized. An arbitrary scale is then imposed upon this set of measured responses. The set of particles (corresponding to the sample which in turn corresponds to the parent particulate matter) is then assigned to a particular "size" with appropriate mean, standard deviation, and so on. This type of size analysis yields a representation of the size of the finely divided matter (as a group) but does not report the size of any one particle. For quality control purposes this may be useful but it is not for basic scientific investigation.

Chemical analysis of the surface of matter has taken great steps forward during the last decade and, although there has been a tendency to experiment on flat surfaces, for reasons of experimental convenience, there appear to be no very unusual problems associated with the examination of the surface of fine particles.*

The morphology and surface chemistry of a particle have been seen as important characteristics. However, the importance of the interior structure and chemistry (including defect content) is less recognized and so they have been correspondingly less investigated. Alternatively, the properties of continuous solids are well characterized in comparison. One would expect that the physical properties of fine particles are much affected by the interiors. This information awaits intensive investigation, particularly with respect to the relationships between these properties and the particle morphology. Physical property measurements on individual particles in the case of electrical, magnetic, optical, and other properties are clearly de-

*J. Kottke, personal communication, July, 1977. However, see J.T. Armstrong in SEM '78, paper 73, "Methods of Quantitative Analysis of Individual Microparticles with Electron Beam Instruments."

pendent upon variations in morphology, chemistry, and structure. The ways in which these interrelate are not understood. They have not been subject to intensive investigation. It is clear that we are a long, long way from developing adequate information for a whole-field view of matter in finely divided form. If this single fact were to be reiterated at appropriate points throughout this book, it would doubtless become monotonous and annoying to the reader. Therefore, a policy of reviewing at the beginning of each chapter some of the advantages of the whole-field view and some of the problems to be overcome in developing such a view shall be adopted.

In the remainder of this chapter, the ways in which others rate the important characteristics of single particles, describe how they analyze and measure some of them, and discuss the developments necessary for improving our methods of obtaining information for taking a whole-field view of a particle are discussed.

To summarize, if the concept of a whole-field view for finely divided matter is proselytized here, it is necessary to be able to describe with accuracy the size-shape/chemistry-physics continuum for a single particle. Once this can be done, there are data handling methods available for dealing with sets of particles or methods can be developed. Modes of particle formation and the behavior of finely divided matter can then be related to the now fully characterized particles. In this way science can be developed to radically influence the generation of new technology.

2.2 CONCEPT AND DEFINITION OF A PARTICLE

In order to facilitate subsequent discussions concerning the properties and behaviors of particulate assemblies, it is necessary to enumerate and discuss the various characteristics of individual particles of matter. A particle is variously described as:[1]

1. a small part, portion, or division of a whole,
2. a very minute portion or quantity of matter, and
3. the smallest sensible component part of an aggregation or mass.

Although it has been claimed that the characteristics of a single particle seldom interest us[2] there is surely an equally substantial counterclaim that the basis of Fine Particle Science is rooted in the characteristics of the individual particle.

Table 2.1

CHARACTERISTICS OF PARTICLES AND PARTICLE DISPERSOIDS

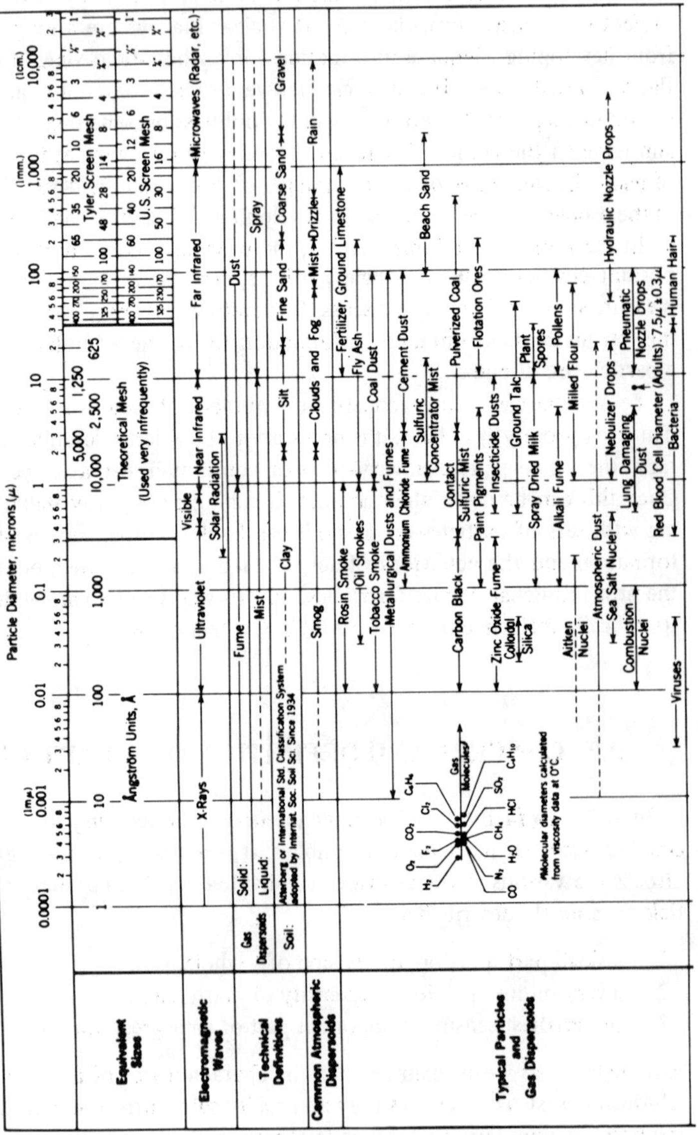

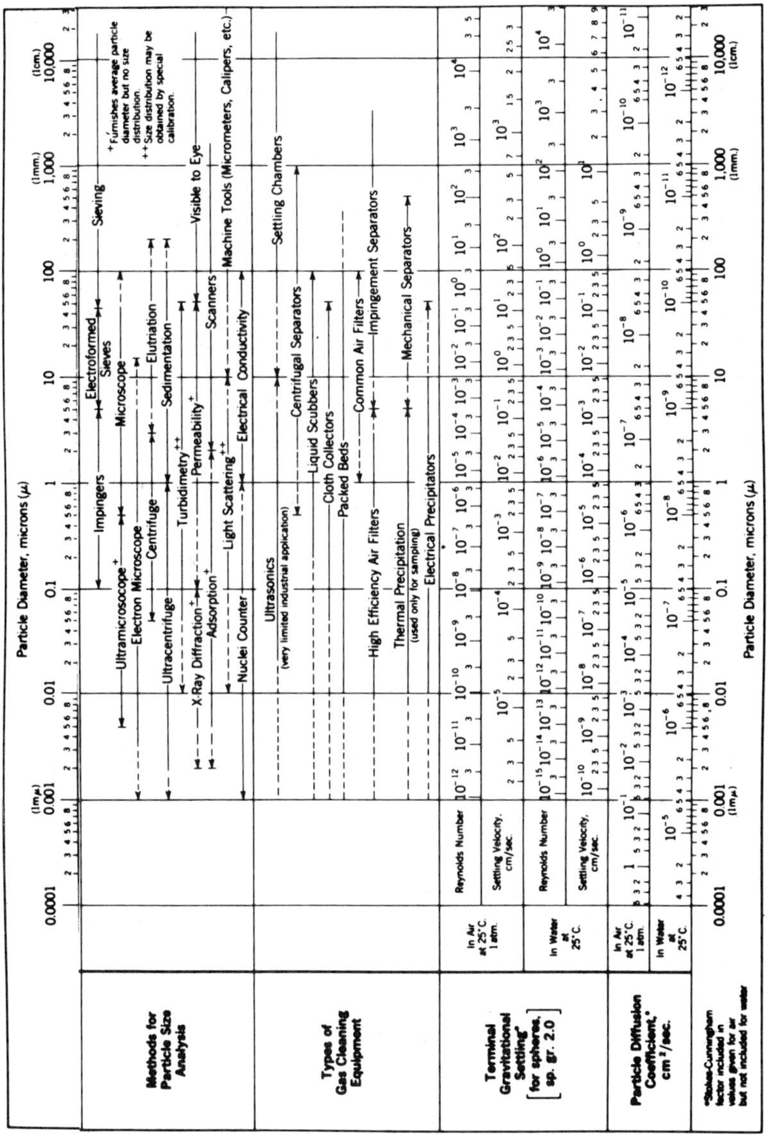

Reprinted from Stanford Research Institute Journal, Third Quarter, 1961.
Single copies 8 1/2 by 11 inches free. Multiple copies or 20 by 26 inch wall chart available for sale.

Table 2.1[3] gives a comprehensive picture of the size ranges of a wide variety of "typical particles" and of gas dispersoids too. This diagram serves to emphasize that when we use the word particle, the word evokes an interest in its size in addition to its substance. The common usage of the word "dust" corresponds to a size range somewhere between approximately 100 microns all the way down to 0.01μ and below. The word "fume" usually describes particles in the small end of this range and words such as "drops," "granules," and "pulverized" tend to be used to indicate the coarser end of this range. Particles coarser than one micron are usually manufactured powders and those smaller than this are considered unwelcome by-products from human activity. But there are many exceptions. For example, paint pigments, spray dried products, and carbon black are manufactured and are smaller than one micron; whereas spores, bacteria, and even red blood cells are coarser. Particles which consist essentially of liquid are termed droplets. These are not considered specifically in this book. This discussion is confined to solid particles.

Before moving on, two terms often confused are briefly discussed: aggregate and agglomerate. The main difference between the two is that an aggregate consists of a particle (or an assembly of particles) held together by strong bonds whereas the agglomerate is held together by weak (i.e., relatively weak) bonds.

Regions of a Particle

The structural regions of a particle which can be considered separately from each other comprise: the surface, the subsurface, and the bulk interior. The difficulties of analyzing and describing these regions are considerable, and are discussed in the following section of this chapter. In addition to the above structural regions there are numerous other characteristics which are of variable importance depending on the particular circumstance. A fairly comprehensive list is shown below[4]:

Materials Characteristics

Structure
Theoretical density
Melting point
Plasticity

Elasticity
Purity (impurities)

Characteristics Due to the Process of Fabrication

Particle size (particle diameter)
Particle shape
Density (porosity)
Surface conditions
Microstructure (crystal grain structure)
Type and amount of lattice defects
Gas content within a particle
Adsorbed gas layer
Amount of surface oxide
Reactivity

With such a plethora of characteristics and regions it is always useful to a ascertain what scientists and engineers consider to be the most important specifics. One such correlation is available for ceramic particulates and is reproduced in part in Table 2.2.[5]

Table 2.2

A SURVEY AMONG 130 PERSONS IN
CERAMICS MANUFACTURING AND RESEARCH[5]

Item	*Ranking*		*Percentage Representation*		*Number Respondents Listing Item*	
	M	R	M	R	M	R
Size distribution	1	2	20	15	26	19
Surface area	4	1	11.5	18.4	15	24
Characteristic size	3	4	12.3	11.0	16	14
Shape	2	3	15	13	20	17

Note: M indicates respondent whose interest is primarily in manufacturing; R indicates a respondent whose interest is primarily in research.

Other items listed in the original report but not included in Table 2.2 include: surface reactivity, adsorbed gases, impurities, surface energy.

In the area of ceramics it can be concluded that the individuals presumably best qualified to comment observed that the three most important characteristics of the particle are:

size
surface area
shape

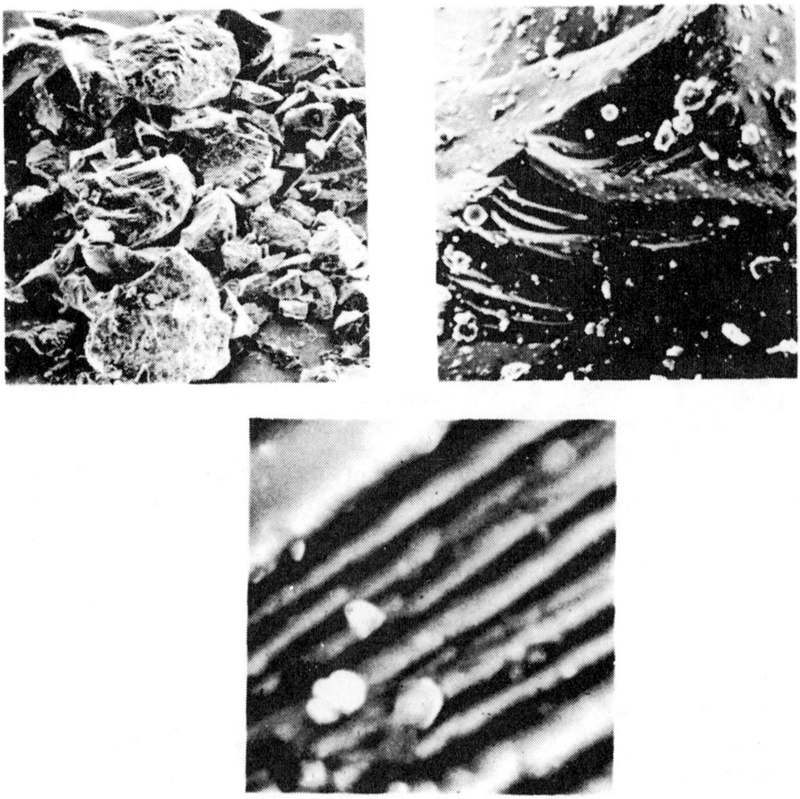

Fig. 2.1 Milk Powder: sharp edged lactose particles cemented together with casein. Fracture surface is conchoidal. Left to right 100 μm; 10 μm; 1 μm. *Particle Atlas,* Ed. 2, Vol. 3, Plate 328, p. 697.

The Occurrence of Fine Particles and Their Constitution

To write a list of materials which occur in particle form would be a mammoth task. Nevertheless, it is important to note the immense range of operations involving particles. Therefore, a short list is given below:

Geological/agricultural.
Ceramic/powder metallurgical manufacturing of components.
Chemicals/drugs/dyes/pigments and fillers in plastics, etc.
Cement/food manufacturing.
Biological systems interaction/oceanography/air pollution.
Coal/oil shale/wood and paper products.

A number of examples of various types of particles from a variety of sources is shown in Fig. 2.1-2.22. Comments are included, where appropriate, as a footnote to each figure.

There are many ways in which particles can be observed. These include: optical microscopy, transmission electron microscopy, scanning electron microscopy (this is particularly good for surface details), electron probe microanalysis, and several other methods which are reviewed later in this

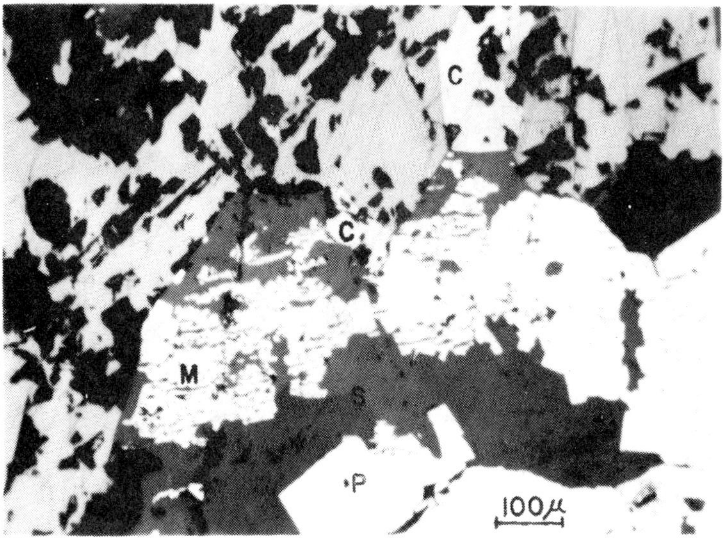

Fig. 2.2 Las Lamas mine ore. Euhedral pyrite (P), marcasite (M) replacing pyrite, chalcopyrite (C), sphalerite (S) and gangue. Parallel nicols 100 X.[80]

discussion. Results of some of these methods of observation are exemplified in Fig. 2.1–2.22.

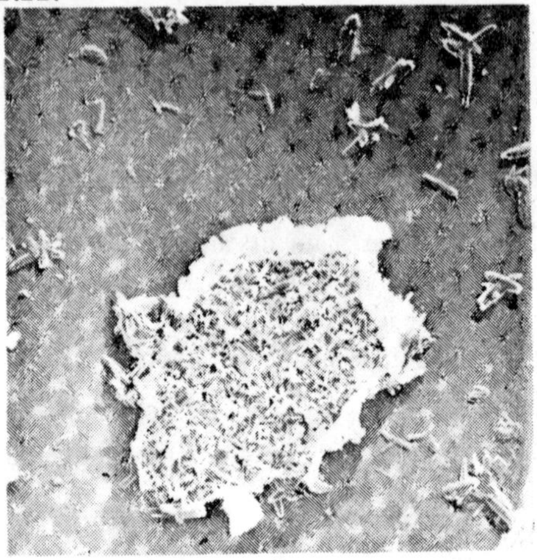

Fig. 2.3 This is an electron micograph of a surface replica of a pigment particle (yellow iron oxide) 6250 X magnification. The large particle is clearly an agglomeration of many smaller ones.
(Reprinted with permission of Jo. of the Oil & Colour Chemist Association).

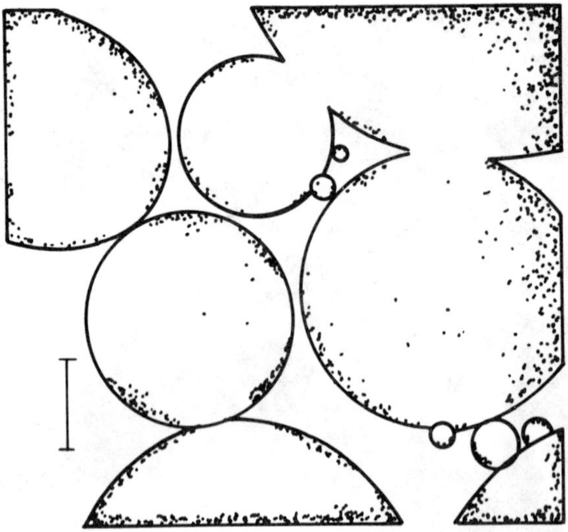

Fig. 2.4 Alpha ferric oxide particles made from a sulfate feed salt. Average particle size is 1.5 microns and the average crystallite size is more than 0.2 microns.

Fig. 2.5 Particles of clay (clay crystals of kaolinite-hydrated alumino silicate) 33,000 X magnification. EM.[71]

Identification of Individual Particles

There are many characteristics of individual particles that can be used to identify them and many tests that can be carried out (see Table 2.3). The sensitivities of the methods which can be used in identification procedures vis-a-vis the size and weights of particles that are necessary for proper identification are surprisingly gross as is shown in Table 2.4. The sensitivity in terms of size and weight limitations is steadily being improved so that most particles can be identified as single nanogram particles and many are also identifiable at the picogram level.[6]

The source book for morphological analysis of individual particles is the "Particle Atlas." Using the techniques and information presented therein makes it possible to readily identify many particles by simple morphological analysis. The exact features being observed are dependent upon whether the optical or electron microscope is being used as shown in Table 2.5.

The Atlas uses a six digit code system which is based on the basic six classification characteristics as follows:

transparency
color (transmitted)

color (reflected)
birefringence
refractive index (relative to medium)
shape

Table 2.3

CHARACTERIZATION AND IDENTIFICATION OF SINGLE PARTICLES

Light Microscope
- size (three principal diameters)
- shape (crystal form, habit, interfacial angles, axial ratios, histological features)
- cleavage
- surface characteristics (reflected light for opaque particles)
- homogeneity (characterize all phases)
- color (reflected and transmitted light)
- transparency, translucency or opacity
- refractive indices, dispersion (dispersion staining data, temperature coefficient of refractive index)
- birefringence, extinction angles, dispersion
- sign of elongation
- optic sign, optic axial angle, dispersion
- correlation of optics and morphology
- photomicrograph

Physical and Chemical Properties
- density
- magnetic susceptibility
- dielectric constant
- solubility data
- melting temperature
- polymorphism

Microchemical Tests

Electron Microscopy
- micrographs
- shape
- surface (by replication)
- scattering of electrons
- selected area diffraction

Table 2.3 *(cont'd.)*

X-Ray Diffraction
cell dimensions
powder pattern (*d* spacings, intensities, hkl's)
density
formula weight
formula weights per cell
Electron Probe Analysis
qualitative
quantitative
electron fluorescence
chemical composition

Particle Atlas, Ed. 2, Vol. 1, Table 1, page 309.

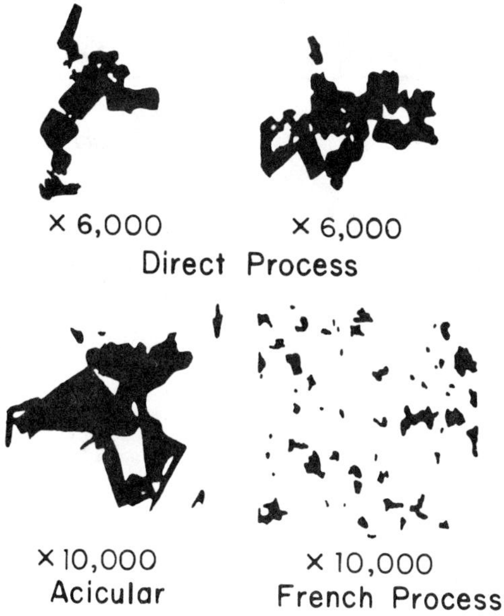

Fig. 2.6 Electron micrograph silhouettes of zinc oxide pigment particles made by three processes. The acicular particles give the better paint job because failure of the film occurs by exfoliation, which necessitates less preparation for repainting than either of the other two examples (these tend to crack and chip off).[72]

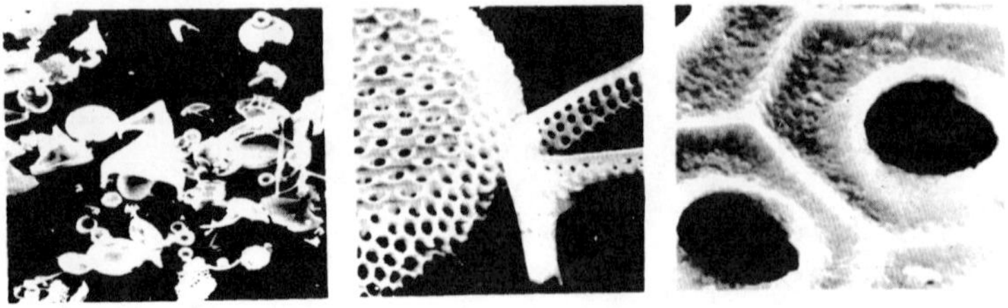

Fig. 2.7 Diatoms, essentially siliceous unicellular (or colonial) aquatic plants (80 X magnification). (*Particle Atlas,* Ed. 2., Vol. 3, Plate 4, p. 589).

Table 2.4

PHYSICAL PROPERTIES OF SINGLE PARTICLES

	*Minimum Particle Size and Weight** (microns)	(grams)
Melting point	2	10^{-11}
Solubility	1	10^{-12}
Density	20	10^{-6}
Microchemical tests	1	10^{-12}
Hardness	10	10^{-9}
Cleavage	10	10^{-9}
Molecular weight	20	10^{-8}
Refractive indices	5	10^{-10}
Refractive index	0.5	10^{-13}
Fluorescence, U.V. or electron	1	10^{-12}
Magnetism	1	10^{-12}
Selected area diffraction	0.02	10^{-17}
X-ray diffraction	10	10^{-9}
Electron microprobe	1	10^{-12}

*These values assume a density of one.

Particle Atlas, Ed. 1, Table XXXI, p. 99.

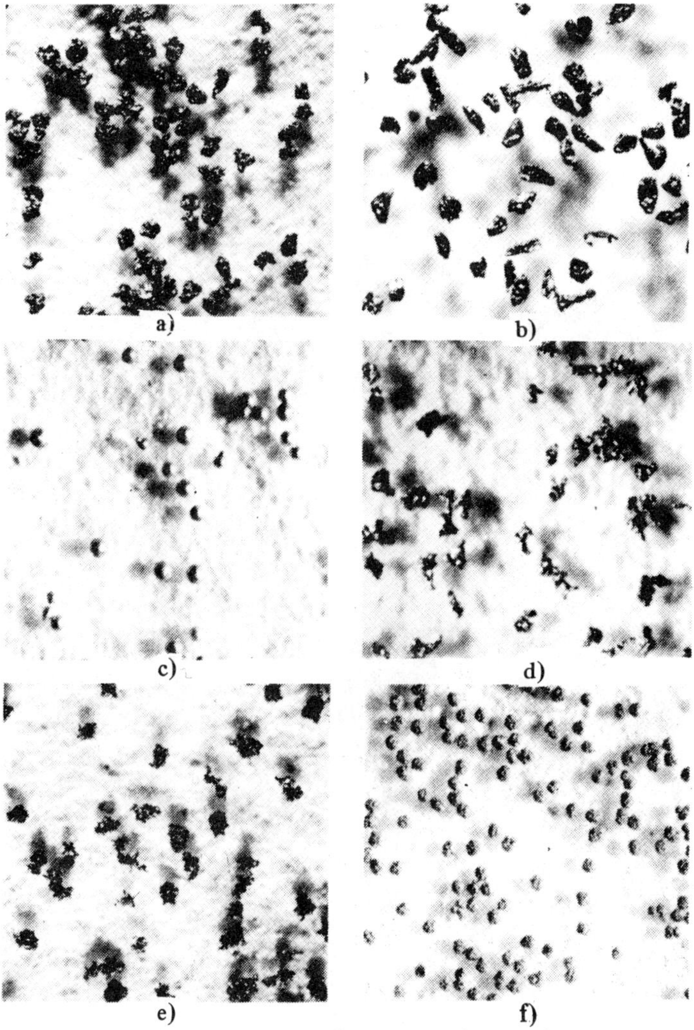

Fig. 2.8 Six different types of metal powder particles produced by different manufacturing processes: (a) sponge iron made by reduction of oxide, (b) electrolytic iron made by grinding of electroplate, (c) atomized brass made by gas atomization of a stream of liquid metal, (d) atomized stainless steel (as for (c), (e) reduced tungsten powder made by a process in which the ore is digested and subsequent chemical processing steps lead to the precipitation of the particles from solution, (f) an example of nickel powder also made by chemically induced precipitation in solution under special conditions (Sherrit–Gordon Mines).[74]

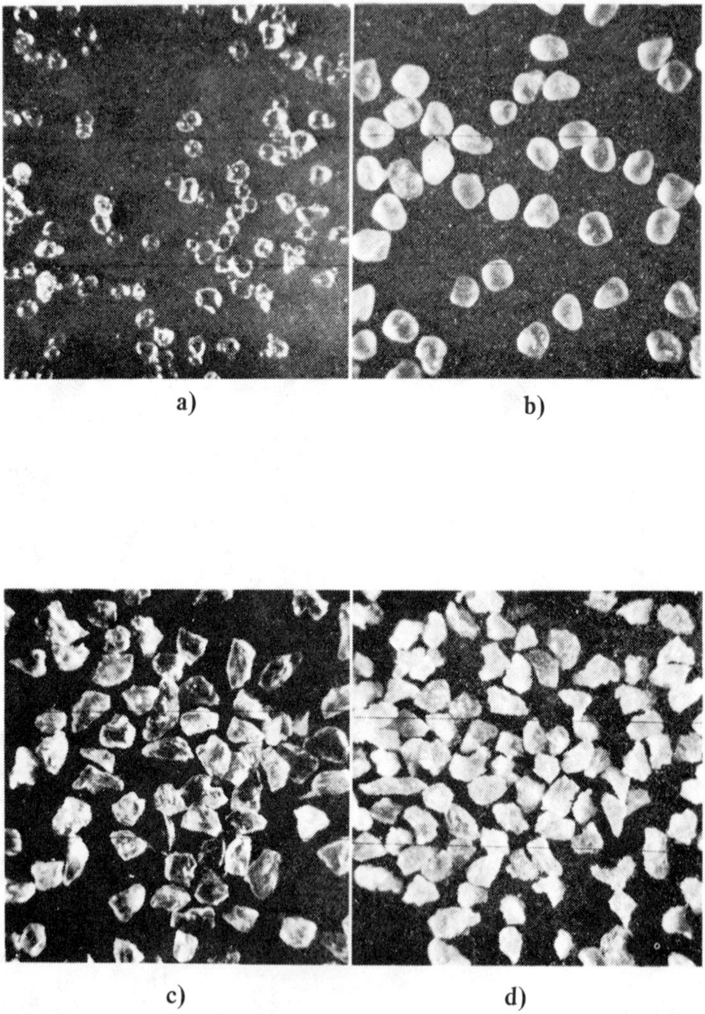

Fig. 2.9 An example of four types of particulate materials that have been used in powders researches by Whitby (Mechanics of Fine Sieving ASTM Symposium on Particle Size Measurement 1958, pp. 3–24): (a) glass beads, (b) St. Peter sand, (c) flint, and (d) wheat middlings.[75]

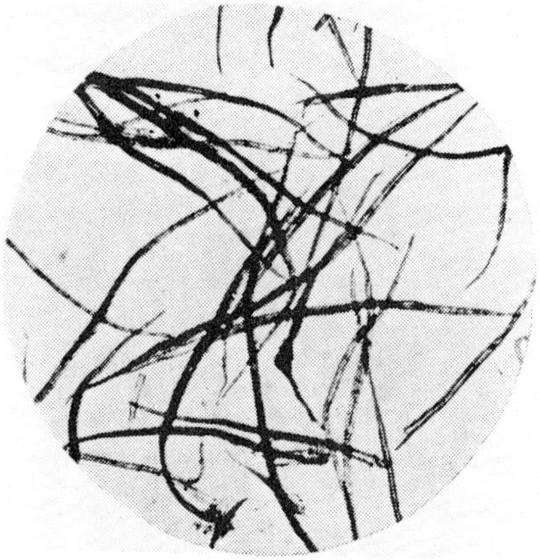

Fig. 2.10 Birch fibers, 110 X magnification.[7]

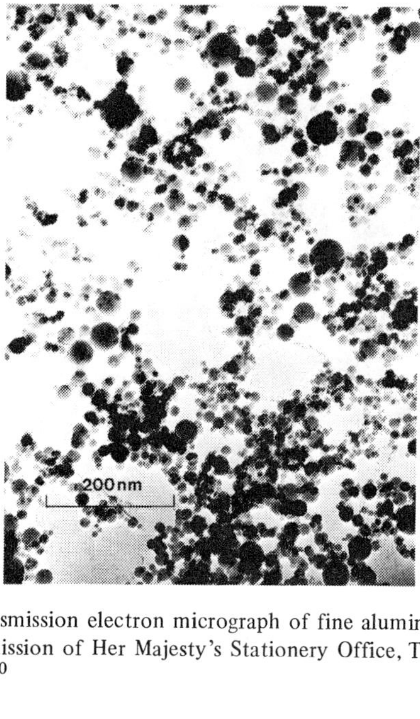

Fig. 2.11 A transmission electron micrograph of fine alumina particles (Reprinted with permission of Her Majesty's Stationery Office, Teddington, Middlesex, England).[20]

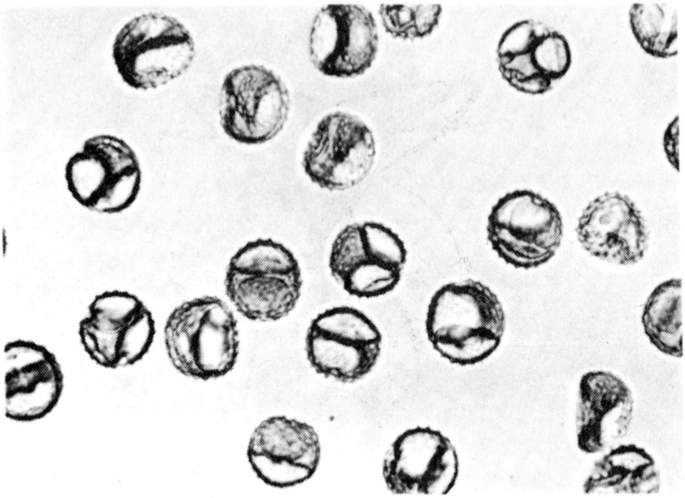

Fig. 2.12 Ragweed pollen is one of most frequently counted items in pollen counts, 500 X magnification (*Particle Atlas,* Ed. 2, Vol. 4, Plate 22A, p. 339).

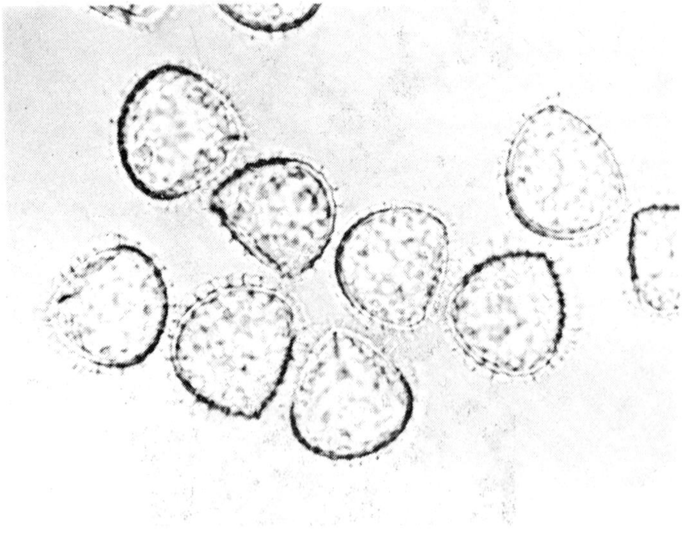

Fig. 2.13 Lycopodium spores. These are the spores of a club moss, 100 X magnification (*Particle Atlas,* Ed. 2, Vol. 4, Plate 38, p. 345).

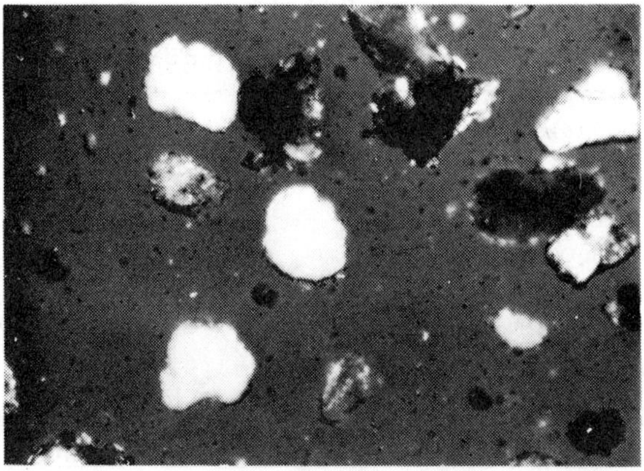

Fig. 2.14 A street air sample taken from an industrial area after a complaint was received. The particulates include: asphalt (dark), quartz, sodium carbonate spheres, and many others, 40 X magnification (*Particle Atlas,* Ed. 2, Vol. 4, Plate 606, p. 563).

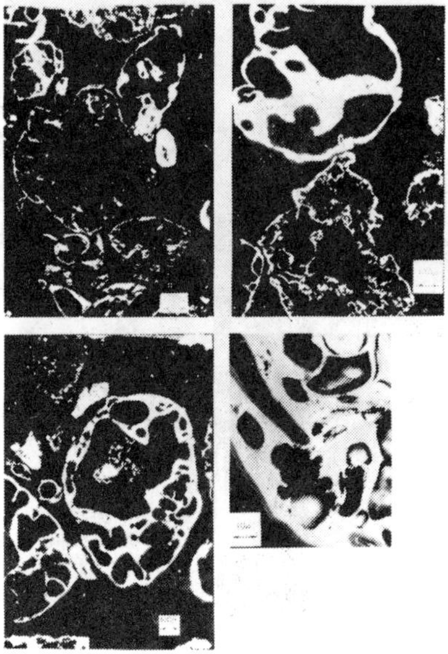

Fig. 2.15 Optical micrograph of high volatile bituminous coal particle after second stage hydrogasification (heating between 1200–1800°F in a steam/hydrogen/hydrocarbon gaseous mixture).[79]

Table 2.5

MORPHOLOGICAL PROPERTIES*

Light microscope	*Electron microscope*
shape	shape
surface	surface
homogeneity	homogeneity
transparency	scattering
color	selected area diffraction
refractive indices	
birefringence	

*Sensitivity: Light microscope—1μ, 10,000 A, 1 pg, 10^{-12}g
Electron microscope—0.01μ, 100 A, 0.01 pg, 10^{-18}g

Particle Atlas, Ed. 2, Vol. 4, Table XLIX, page 275.

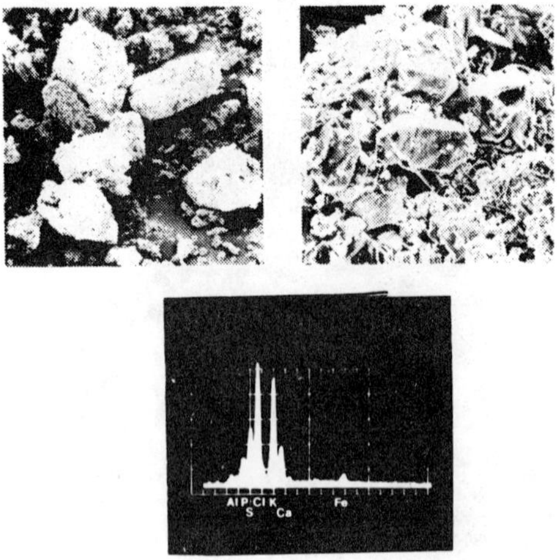

Fig. 2.16 5-10-30 Fertilizer. Physical mixture of potassium chloride and triple superphosphate. EXDRA results on left (energy dispersive x-ray analyzer). From left to right 100 μm, 10 μm, 1 μm (*Particle Atlas,* Ed. 2, Vol. 2, Plate 292, p. 685).

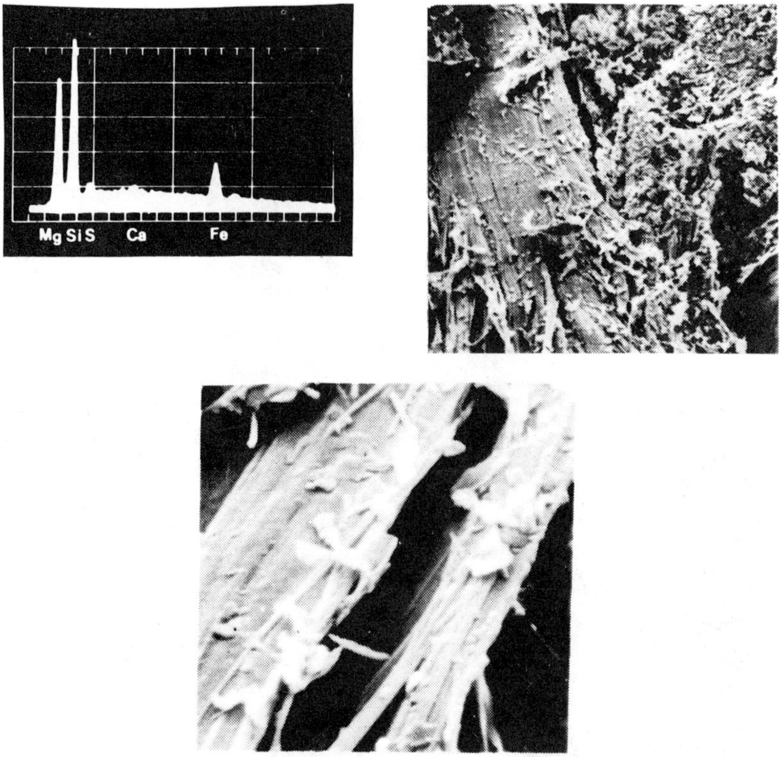

Fig. 2.17 Asbestos Mortar Insulation–Chrysotile asbestos and gypsum mortar. EXDRA results show main components. From left to right 100 μm, 10 μm, 1 μm (*Particle Atlas,* Ed. 2, Vol. 2, Plate 245, p. 669).

The scheme developed for the morphological analysis using the six digit code and upon which the Atlas is based is shown in Fig. 2.23.

In the case of an unknown particle for which identification is sought, the investigator first classifies it according to the characteristics itemized above. Consider the following example taken from the Atlas 1973 edition:

The particle is examined in the microscope and is found to be

transparent	colored	isotropic	high index	equant	

the corresponding code digits are

0	1	0	1	00	or 20:010100

Fig. 2.18 Mercerized cotton from left to right 100 μm, 10 μm, 1 μm (*Particle Atlas,* Ed. 2, Vol. 2, Plate 60, p. 607).

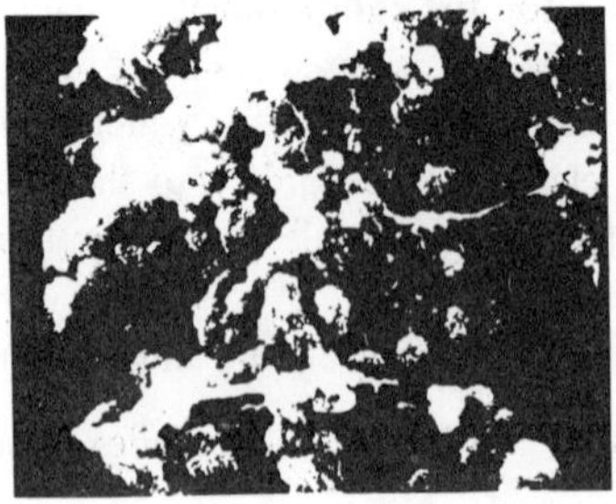

Fig. 2.19 Portland Cement (stub)–Crystal Bridging, 5800 X, 33 days.[78]

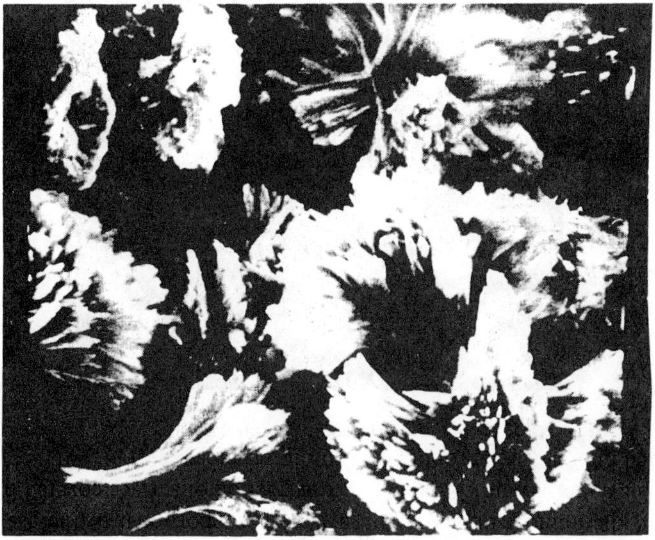

Fig. 2.20 Mix I (tub)–"Sheaf of Wheat," 5000 X, 1 day.[78]

Fig. 2.21 Portland Cement (tub)–Flower Structure, 1050 X, 10 days.[78]

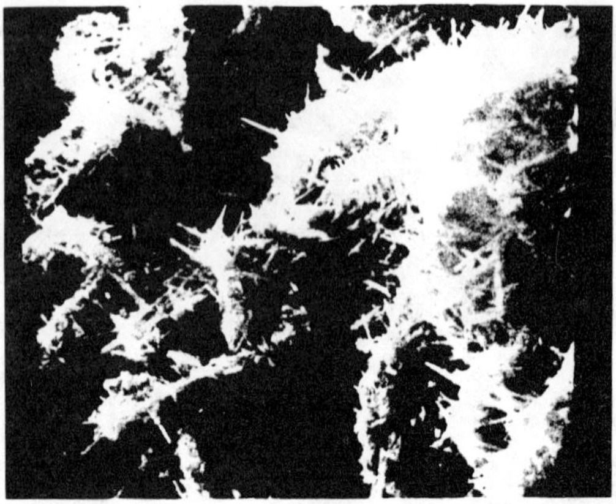

Fig. 2.22 C_4AF (tub)–Needle Structure, 4900 X, 12 days.[78]

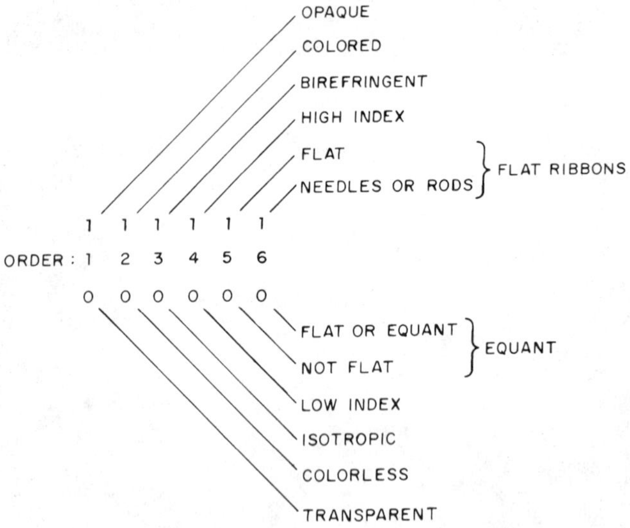

Fig. 2.23 Definition of the six-digit code (*Particle Atlas,* Ed. 2, Vol. 4, Figure p. 805).

A table of additional information corresponding to this code is reproduced in Table 2.6. This gives the investigator a very narrow field of possibilities from which to make the final identification. The Atlas can also be useful in describing a known particulate. In these cases a direct comparison with a sample of the known substance can be obtained.

Table 2.6

ADDITIONAL IDENTIFYING CHARACTERISTICS FOR PARTICLES DESCRIBED MORPHOLOGICALLY AS 20:010100

Substance	*Atlas Nos.*	*Density*	*Refractive Indices*	*Additional Data*
Coal, bituminous	560	ca. 1.1	> 1.66	thin edges, brown
Sphalerite, ZnS	196	3.9–4.1	ca. 2.40	yellow to yellow green
Slag (chrome steel)	418	ca. 2.8	> 1.66	yellow, green, blue
Thoria	429	9.7	ca. 2.2	thin edges, red to brown
Electric furnace dust	381	ca. 2.4	ca. 1.66	yellow to brown
Wood sawdust	546	ca. 1.3	< and > 1.66	amber yellow

Based on Table 20:010100, *Particle Atlas,* Ed. 2, Vol. 4, pages 828–29.

2.3 THE PARTICLE SURFACE

An interface is considered to be a boundary at which two dissimilar bulk portions of matter meet. The interface therefore marks the line of discontinuity between the two portions. The characteristics of the discontinuity may correspond to a change in composition, a change in structure, a change of orientation, or a change in phase. The term surface is associated with an interface between a solid and a fluid, particularly when the fluid is a gas. It is a definition which has its origins in the macroscopic workings of the senses. Similarly, many of the measurement techniques available for assessing the behaviors of surfaces (as compared with techniques which permit us to observe surfaces) are comparatively gross in sensitivity, whereas much of the theoretical approach considers atomistic and/or molecular characteristics of surfaces.

It is clear from the various representations in Fig. 2.1–2.22 that the surfaces of particles are very far from smooth. In fact all surfaces are rough except for cleavage surfaces (mica) and they are invariably covered with adsorbed molecules.[7] Even the smoothest surface has irregularities of 100–1000 Angstroms and normally, multimicron-sized features are present in all but the most carefully prepared surfaces.[8] The view that a surface is a two-dimensional entity is an oversimplification. As illustrated in Fig. 2.24, the properties of the solid gradually change as the surface is neared. Therefore, in order to be specific the term surface region should be used in order to indicate that the surface has structure.

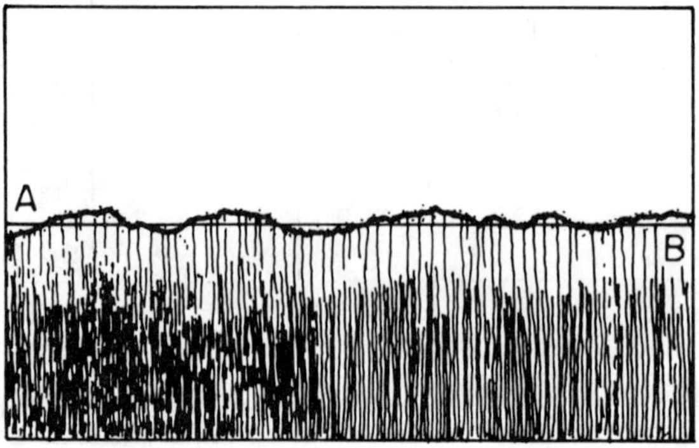

Fig. 2.24 Cross section of a typical solid surface. The solid is shaded, and the gradation of shading indicates gradual change of properties on nearing the interface with air (white). The dots are adsorbed molecules.[23]

In general, the reversible work necessary to creat a unit area of a planar interface is defined as

$$\gamma dA = dE - TdS - \sum_i \mu_i dn_i$$

where

dA = area

E = surface excess of internal energy

S = surface excess of entropy

μ_i = excess of chemical potential

n_i = excess number of i atoms in interface

Now

$$F = E - TS$$

or

$$E = F + TS$$

Therefore

$$dE = dF + TdS + SdT$$

i.e.,

$$\gamma dA = dF + TdS + SdT - TdS - \sum_i \mu_i dn_i$$

and so

$$\gamma dA = dF + SdT - \sum_i \mu_i dn_i$$

Therefore,

$$\gamma = \left(\frac{\partial F}{\partial A}\right)_{n_i, T, V} \tag{2.1}$$

in which

F is the Helmholtz Free Energy,
A is the area of the interface, and
n_i is the excess number of atoms of type i at the interface.

Two alternative forms of Equation 2.1 are as follows

$$\gamma = f - \sum_i \mu_i \Gamma_i \tag{2.2}$$

or

$$d\gamma = -SdT - \sum_i \Gamma_i d\mu_i \tag{2.3}$$

in which $f = F/A$, $s = S/A$, $\Gamma_i = n_i/A$, and μ_i is the excess of chemical potential of i type atoms at the interface.

For convenience, the position of the interface can be arbitrarily defined so that Γ_i is zero. The units of γ defined in Eqs. (2.1) - (2.3) are those of energy per unit area. γ is a scalar quantity variously described as surface energy, surface free energy, or surface tension.

Because the surface represents a discontinuity it is in a state of tangential stress σ. In the case of solids this level of surface stress varies with direction in any surface, i.e., it is orientation dependent. Surface energy and surface stress are not equal. If we consider the isothermal stretching of a surface (one component), equating the work done to the energy increase

$$\sigma dA = d(\gamma A) = \gamma dA + A d\gamma$$

$$\therefore \sigma = \gamma + A \frac{d\gamma}{dA} \qquad (2.4)$$

For example, in the case of gold, extrapolation of high temperature values of surface tension, obtained from the zero creep-type, experimental method[9], gives a value for γ of 1800 ergs/cm^2 down to 50°C. Measurements of surface stress on gold give a value of 1175 ± 200 dynes/cm at 50°C[10]. This description is that in general surface stress and surface energy are not equal for solids. However, if the temperature is raised, stress relaxation may more readily occur in the presence of defects.

When the characteristics of the surface down at the atomic level are considered they are no longer uniform simply because the potential field along the surface varies depending on whether the measurement is taken immediately above a surface atom or between two, three, or four surface atoms. Because the potential energy varies according to location, it is possible to conceive of the potential field contours in a direction parallel to the atomic surface as shown in Fig. 2.25. This contour potential map is for a square lattice. The dots are for maxima and the crosses are for minima of potential energy. Because the surface so described is not energetically uniform and consists of a series of peaks and vallies, but is nevertheless regular, it is more accurate to use a specific term which emphasizes this regularity. The word homotattic has been coined consisting of the Greek homos (same) and tasso (arrange).[11]

The process of attraction of an incoming atom occurs by means of secondary bond forces and the potential energy varies with location of the adsorbent and also with interatomic distance (i.e., the distance between the adsorbate and adsorbent species). The relationship has the form:

$$U = -a/r^6 + b/r^{12} \qquad (2.5)$$

Eq. (2.5) represents the potential energy which is variously described as the 6–12 potential or the Lennard-Jones potential.[12] The r^6 term is the attractive term and the r^{12} term is the repulsive term. a and b are constants. An adsorbing species experiences a different potential field depending upon the position on the surface as shown in Fig. 2.26. Curve A in Fig. 2.26 corresponds to the variation of potential as the atom approaches point A in Fig. 2.25. Curve B corresponds to point B and Curve C corresponds to an approach to point C.

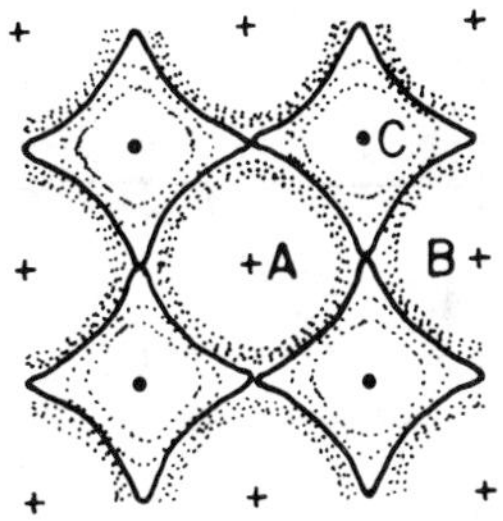

Fig. 2.25 Potential countours of surface of solid dots are maxima and crosses are minima of potential energies (for A, B, C see 2.26).[76]

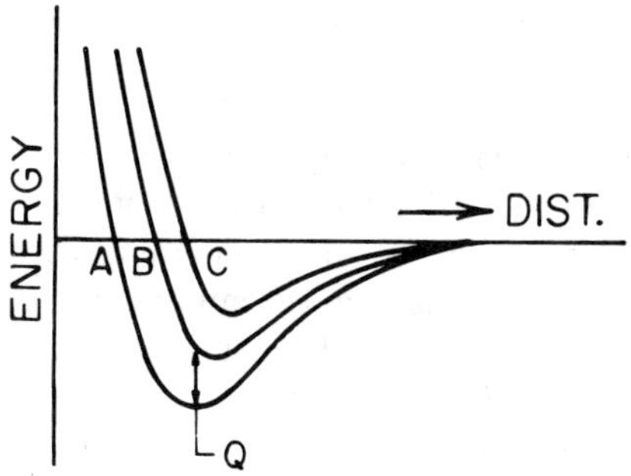

Fig. 2.26 Adsorbing species experience a different potential field depending upon where it approaches the surface. Curve A indicates approach to point A in Fig. 2.25, and so on.[76]

This Lennard-Jones model of the surface atomic structure of a solid leads to two important observations:

1. The equilibrium position for the adsorbed atom is at distance d_o from the surface and it has a binding energy of $\Delta\epsilon_o$ as shown in Fig. 2.27. It is possible to differentiate between physisorption (mainly due to Van der Waals type forces) and chemisorption (mainly due to

normal chemical bond formation). The heats of adsorption are approximately 0.25 eV and 5 eV, respectively. Clearly, the binding energy of a particular adsorbent depends upon position on the surface (Fig. 2.26).

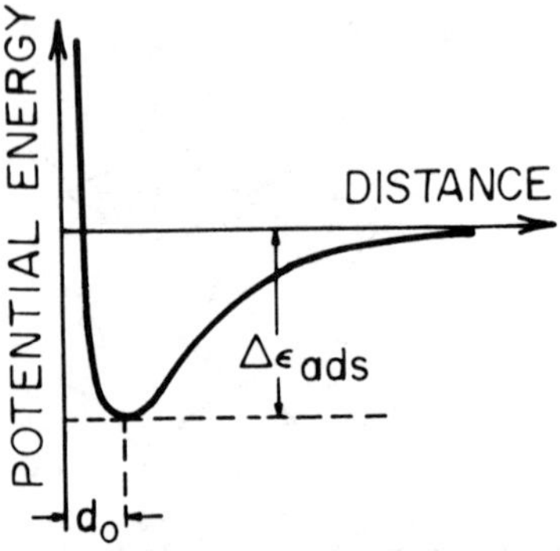

Fig. 2.27 Schematic potential energy versus distance curve for an atom near a planar crystal surface. The minimum corresponds to the equilibrium position for the adsorbed atom at distance d_o and with binding energy $\Delta\epsilon_{ads}$. With a real surface the potential energy distance curve depends on the coordinates in the plane of the surface.[16]

2. The process of surface diffusion may be considered to occur by activation (due to thermal agitation) to an amount Q (see Fig. 2.26). This permits the adsorbed atom to move across the surface and locate on one of the equipotential contours. The atom can then move along this contour for a distance until an interaction occurs that causes it to locate in another "hollow" on the surface. Some work on diffusion of atomic species on W surfaces is reported in Table 2.7.[13] Clearly, the diffusivity is very orientation dependent. At elevated temperatures, surface diffusion can be seen as analogous to a two-dimensional ideal gas. In this case the variations in the field are more or less smoothed out as the temperature rises.

Features of the Surface of a Solid

Since an ideal surface is described as homotattic in the case of a crystalline material, for a noncrystal it might be expected that energy uniformity is more nearly realized. The structure of a liquid surface should closely

Table 2.7*

DATA ON THE DIFFUSION OF ADATOMS ON TUNGSTEN SURFACES FROM FIELD ION MICROSCOPY STUDIES[13]

	Ta	Mo	W	Re	Ir	Pt
ΔH_σ^m	0.78	–	0.87	1.04	0.78	≈ 0.6
(110) W D_{σ_0}	4.4×10^{-2}	–	2.1×10^{-3}	1.5×10^{-2}	8.9×10^{-5}	$\approx 10^{-4}$
ΔH_σ^m *	0.49	0.56	0.76	0.86	0.58	–
(211) W D_{σ_0}	0.9×10^{-7}	2.4×10^{-6}	3.0×10^{-4}	2.2×10^{-3}	2.7×10^{-5}	–
ΔH_σ^m	0.67	–	0.84	0.88	–	–
(321) W D_{σ_0}	1.9×10^{-5}	–	1.2×10^{-3}	4.8×10^{-4}	–	–

ΔH_σ^m in eV. D_{σ_0} in cm^2/s.

The diffusion coefficient for adsorbed atoms has been expressed as $D_\sigma = D_{\sigma_0} \exp(-\Delta H_\sigma^m / kT)$ where D_{σ_0} is the pre-exponential or frequency factor and ΔH_σ^m the enthalpy change associated with its motion. The latter will essentially be the barrier height $V_0/2$.

(From data given by D. W. Bassett and M. J. Parsley, *J. Phys. D.: Appl Phys.* 3, 707 (1970).)
(Courtesy of the Institute of Physics.)

* Some values in error due to incorrect data analysis or due to effect of lateral adatom–adatom interactions.

agree with this idea of an energetically uniform surface (and so should glass, but it is far too surface reactive due to contamination for a useful study). In reality, a solid surface almost always contains ledges, terraces, kinks, cracks, dislocations, adsorbed impurities and other discontinuities. Even the most practically attainable perfect surface consists only of homotattic patches of various types, one type of which will be most numerous.[14]

One can analytically distinguish between three specific types of surface of a crystalline solid:[15]

1. Singular surfaces have the highest atomic packing factor and therefore correspond to close packed planes with simple indices (see Fig. 2.28).
2. Vicinal surfaces which are slightly inclined at an angle to the singular surfaces and consist of terraces are shown in Fig. 2.29.
3. General surfaces for which the terrace-ledge-kink model (TLK Model) is appropriate as shown in Fig. 2.30. The coordination of the various types of surface atoms varies with its position in the TLK model as shown in Table 2.8.

Table 2.8

COORDINATION OF VARIOUS TYPES OF SURFACE ATOMS

Position of atom	*Key to Fig. 2.30*	*No. of bonds*
In terrace	*a*	9
At vacancy	*b*	8
In ledge	*c*	7
At kink	*d*	6
At ledge	*e*	5
Adatom	*f*	3

Because the origin of surface energy is due to the atoms at the surface being less tightly bound than those in the interior, the energy state associated with a loosely bound atom is higher than that of a more tightly bound

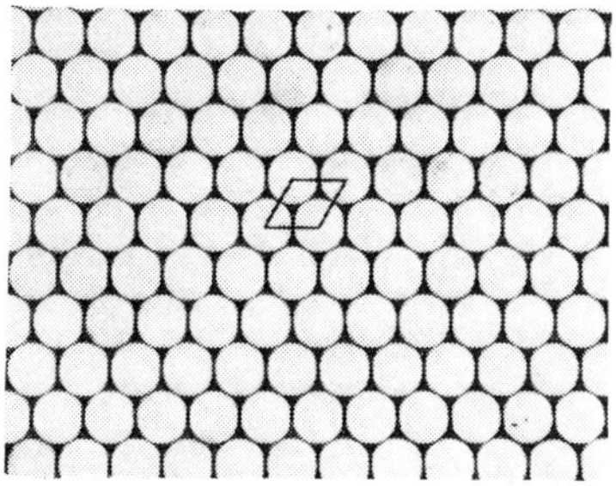

Fig. 2.28 An example of a singular surface. A hard sphere model of an ideal (111) surface of a face centered cubic monatomic crystal. Primitive unit mesh is as indicated.[16]

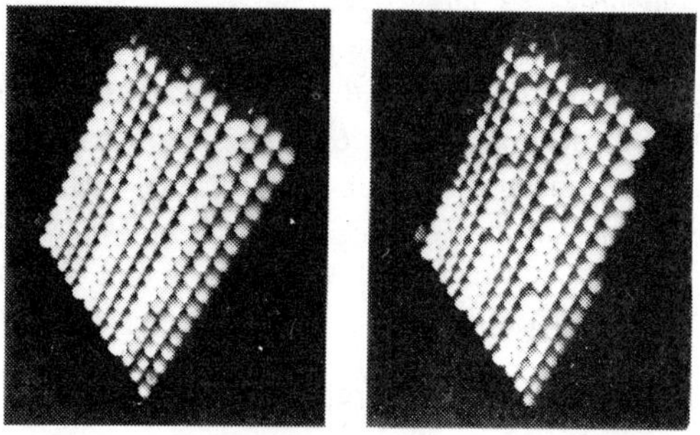

Fig. 2.29 Two examples of vicinal surfaces. A hard sphere model of a (a) FCC monatomic crystal vicinal to (100) plane, steps are evident, and (b) a FCC monatomic crystal vicinal to (100) plane showing steps with kinks.[16]

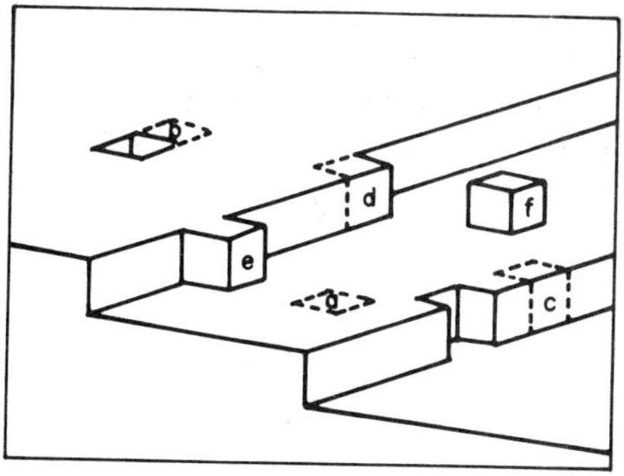

Fig. 2.30 Schematic diagram of a vicinal surface showing various types of surface defects.[73]

atom. This means that the energy state of the atoms in Table 2.8 increases as the number of bonds decreases. One important consequence of this variation in energy state is that the process of absorption indicated consists of these steps to start with:

1. atom approaches potential field of surface and becomes an adatom,
2. adatom diffuses on surface and becomes an atom at a ledge, and
3. atom moves along ledge and becomes an atom at the kink.

This sequence is logical because it represents a series of steps by which the original adatom seeks lower and lower energy states.

One interesting aspect of surface defect formation is that the energy of formation of a vacancy is small (e.g., 0.25 eV for gold surface at 1000°K) which means that there is a large fractional concentration of vacancies on gold at this temperature (0.1). Because these interact, the energies of formation become concentration dependent and a catastrophic increase in vacancy concentration occurs. This is observed microscopically as a surface roughening.[16] Other mechanisms can be postulated such as the formation of large numbers of kinks at temperatures well below the melting point. There is also experimental evidence to show that other surface defects tend to interact with each other.[17]

Surface-Fluid Interactions (Gas/Solid)

Langmuir derived a relationship between the surface coverage by a monolayer of adsorbent and the gas pressure by equating the rate of adsorption with the rate of desorption at equilibrium. If n_a is the number of molecules arriving and f_s is the fraction sticking, where F is the fraction of the surface covered by the (assumed) gas monolayer then $(1 - F)f_s n_a$ is the rate of buildup of the monolayer. At equilibrium this equals the rate of molecules leaving the surface which is Fr_1, therefore

$$(1 - F)f_s n_a = Fr_1$$

or

$$F = \frac{f_s n_a}{f_s n_a + r_1}$$

because the number of molecules arriving is proportional to the gas pressure, this equation indicates that F, the fraction of surface covered by the adsorbed monolayer is proportional to the gas pressure or

$$F = \frac{kP}{1 + kP} \tag{2.6}$$

Eq. (2.6) is Langmuir's isotherm and is shown graphically in Fig. 2.31.[18] The assumptions involved in this isotherm are:

1. each elementary space on the surface is capable of adsorbing one molecule, which is localized on that site,
2. each site on the surface is equivalent,
3. there is no interaction between adsorbed molecules,
4. adsorption is restricted to a monolayer.

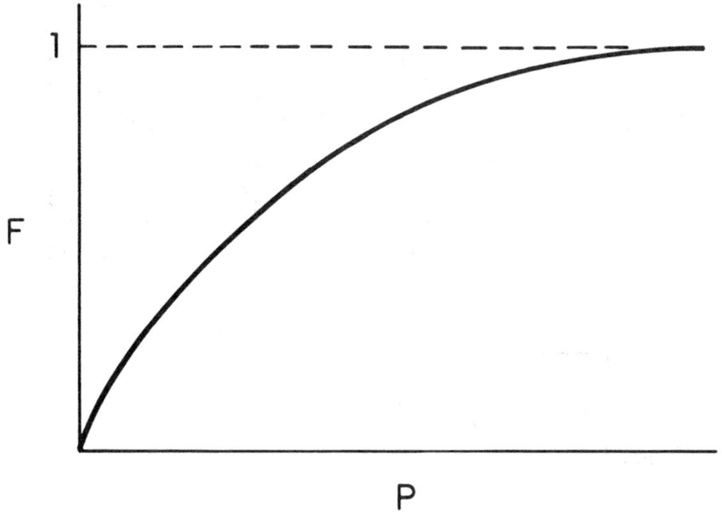

Fig. 2.31 Langmuir's Isotherm.[18]

The BET theory of multilayer adsorption was developed some 20 years after Langmuir's theory.[19] Five general types of curves were postulated in this theory and the vast majority of gas adsorption isotherms plotted follow one or the other of the BET isotherms.[20] The five main types of isotherm are shown in Fig. 2.32. The equation which represents the BET isotherm is

$$V_{ads} = \frac{kp/p_o}{(1 - p/p_o)(1 - p/p_o + kp/p_o)} = \frac{kx}{(1 - x)(1 - x + kx)} \tag{2.7}$$

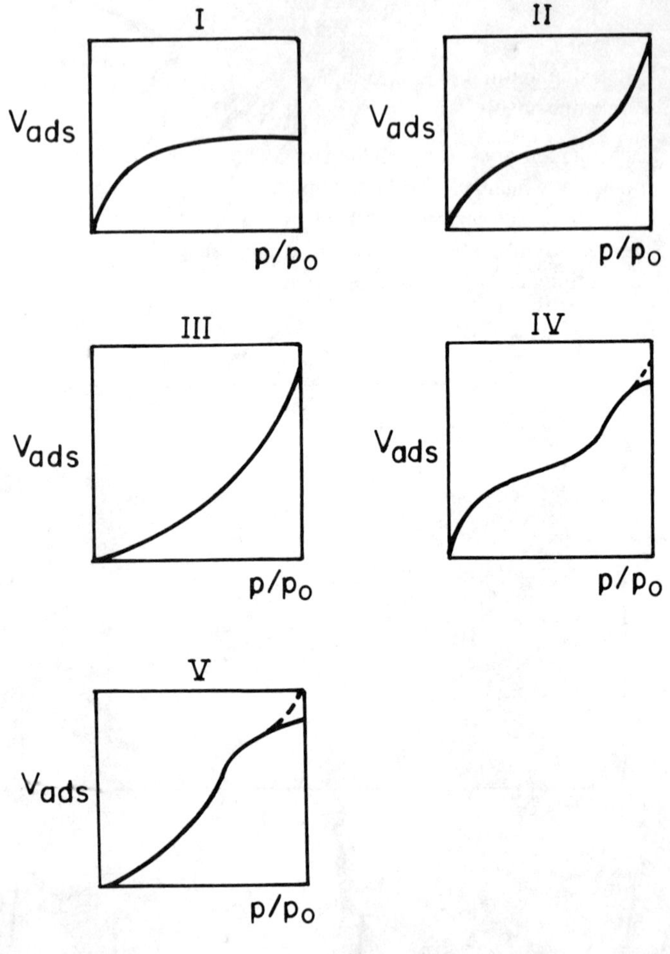

Fig. 2.32 The main types of gas adsorption isotherm.[20]

where p_o is the saturation vapor pressure of the adsorbate at temperature and V_{ads} is the STP volume of gas adsorbed at p/p_o.

In the case of types IV and V (BET isotherms), at high values of p/p_o there are variations in the shapes of the curves obtained when there are pores present below the surface. In the case of types II and IV isotherms, the sharpness of the knee at low p/p_o indicates that some localized adsorption or chemisorption may be occurring. The BET method is used to obtain information about surface area and size of particulates.

It has been observed that the BET theory accepts all of Langmuir's assumptions except for the one concerned with the monolayer of adsorbed gas molecules.[14] At least one scientist has commented that the BET theory never assumed that adsorption occurs on a uniform surface and also never assumed any interactions between adsorbate molecules. It is claimed that these two effects probably counteract each other as the fraction of surface coverage increases.[21]

Surface-Fluid Interaction (Liquid/Solid)

It has been shown that in the case of a slight displacement of a gas–liquid interface (see Fig. 2.33), the changes in area of the gas–liquid interface; the solid–liquid interface and the volume of the interfacial region are related:[22]

$$dA^{gl} = -J dV^{\ell} + \cos\theta \, dA^{gs} \tag{2.8}$$

where J is the mean curvature and equals $-(1/R_1 + 1/R_2)$ in which case R_1 and R_2 are the radii of curvature. θ is the wetting angle.

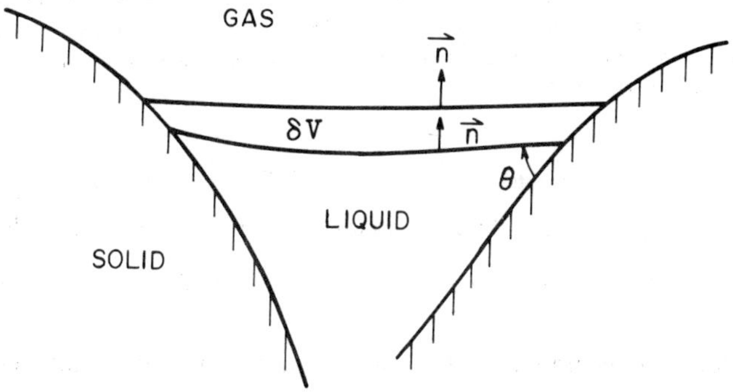

Fig. 2.33 Displacement of Interface.[22]

Note that if the value of the angle is 90°, cos 90 is zero and the relationship becomes a well recognized one in capillarity where the situation is restricted to one in which fluid–fluid interfaces are present, namely:

$$-J dV = dA \tag{2.9}$$

Consider the interface between two regions—alpha and beta as shown in

Fig. 2.34. *DS* is a mathematically defined interface which coincides approximately with the transition region between alpha and beta as indicated. This model permits us to write the free energy for the entire system F_{total} in terms of F^{α}, F^{β}, and F^{s}. The latter quantity is the surface excess value of F: excess quantity (+ or −) in the actual surface phase over the amount that would have been there if both bulk phases had remained homogeneous right up to the dividing surface *DS*.

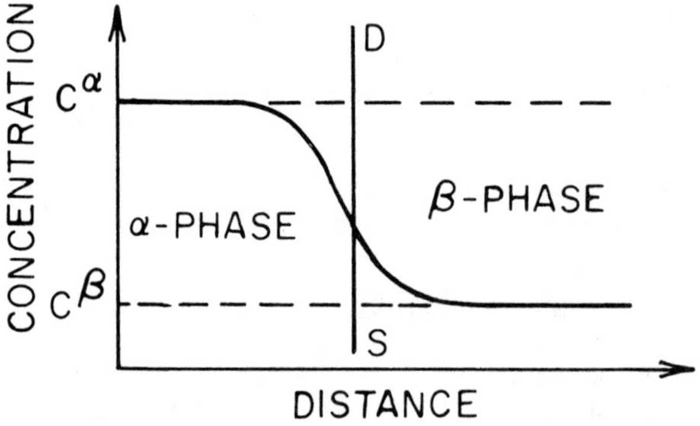

Fig. 2.34 Variation of the concentration of a particular component across the interface between two phases α and β. C^{α} and C^{β} are the concentrations of that component in the two phases at large distances from the interface. *DS* is the dividing surface.[16]

$$F_{\text{total}} = F^{\alpha} + F^{\beta} + F^{s} \tag{2.10}$$

For a multicomponent system a change in the Helmholtz free energy is

$$dF = -SdT - PdV + \gamma dA + \sum_i \mu_i dN_i \tag{2.11}$$

so that for a one component system at constant temperature and volume

$$dF = \gamma dA \tag{2.12}$$

Consider the system shown in Fig. 2.33 in which the interface $g\ell$ advances a small increment; it follows from Eq. 2.10 that

$$F_{\text{total}} = F^{g} + F^{\ell} + F^{s} \quad \text{or} \quad F^{s} = F_{\text{total}} - (F^{g} + F^{\ell})$$

and so

$$dF^{s} = dF_{\text{total}} - (dF^{g} + dF^{\ell}) \tag{2.13}$$

Now

$$dF^g = -P^g dV^g \tag{2.14}$$

and

$$dF^\ell = -P^\ell dV^\ell \tag{2.15}$$

Because the $g\ell$ interface has moved there is:

a change in V^g of dV^g
a change in V^ℓ of dV^ℓ
where $dV^g = -dV^\ell$
a change in $A^{\ell s}$ of $dA^{\ell s}$
a change in A^{gs} of dA^{gs}
where $dA^{\ell s} = -dA^{gs}$
a change in $A^{g\ell}$ of $dA^{g\ell}$.

Combining this information with Eq. (2.13) - (2.15)

$$dF^s = \gamma^{g\ell} dA^{g\ell} + \gamma^{\ell s} dA^{\ell s} + \gamma^{gs} dA^{gs} - (-p^g dV^g - p^\ell dV^\ell)$$

so

$$dF^s = \gamma^{g\ell} dA^{g\ell} - \gamma^{\ell s} dA^{gs} + \gamma^{gs} dA^{gs} + p^g dV^g - p^\ell dV^g$$

at equilibrium

$$dF^s = 0$$

Therefore

$$(p^g - p^\ell) dV^g + \gamma^{g\ell} dA^{g\ell} + dA^{gs}(\gamma^{gs} - \gamma^{\ell s}) = 0$$

From Eq. (2.8)

$$dA^g = -J dV^\ell + \cos\theta \, dA^{gs}$$

So

$$dA^{g\ell} = J dV^g + \cos\theta \, dA^{gs}$$

Therefore

$$[(p^g - p^\ell) + \gamma^{g\ell} J] \, dV^g + [\gamma^{g\ell} \cos\theta + \gamma^{gs} - \gamma^{\ell s}] \, dA^{gs} = 0 \tag{2.16}$$

Because dV^g and dA^{gs} are independent variables, each term in the square bracket must equal zero. Consequently

$$(p^g - p^\ell) = -J\gamma^{g\ell} \tag{2.17}$$

and

$$\gamma^{g\ell} \cos\theta + \gamma^{gs} - \gamma^{\ell s} = 0 \tag{2.18}$$

Eq. (2.17) is a form of the Laplace equation (also called Gibbs-Kelvin equation). Eq. (2.18) is a form of Young's equation.

Experimental Observations of Surfaces

There are many ways of examining the surfaces of solids in order to determine both surface structure and surface chemical composition. Some of the more popular ones are exemplified in Fig. 2.1-2.22. Rather than elaborate on each individual technique, they are reviewed in tabular form in Table 2.9. In this table, references are given in the last column so that the interested reader may obtain fairly up-to-date sources on the methods listed.

2.4 THE SUBSURFACE REGION

We earlier identified three regions or features of a particle: the surface, the interior, and the subsurface region. The surface can be briefly defined as the exterior boundary between fluid and solid but the definition of the subsurface region is more arbitrary and involved.

The sketch shown in Fig. 2.35 attempts to give some indication of the assorted characteristics which a subsurface region of a particle might exhibit at any one time. In general, one can conjecture that there are two types of characteristics: structural and compositional.

The external surface of the particle is defined in such a way that the phase on one side of the surface is all fluid. On the other side of the interface, the particle may contain surface fissures, capillaries or pores reaching into the interior. In this case if one were to slice away the subsurface region it would be found to consist of a mixture of the solid and fluid phases concerned in proportion depending upon the extent of fissuring and/or capillary volume. The solid phase in the slice may contain a variety of stress and defect gradients and in addition it could conceivably contain one or more of internal boundaries such as grain boundaries, sharply demarcated boundaries between different solid phases or more diffused

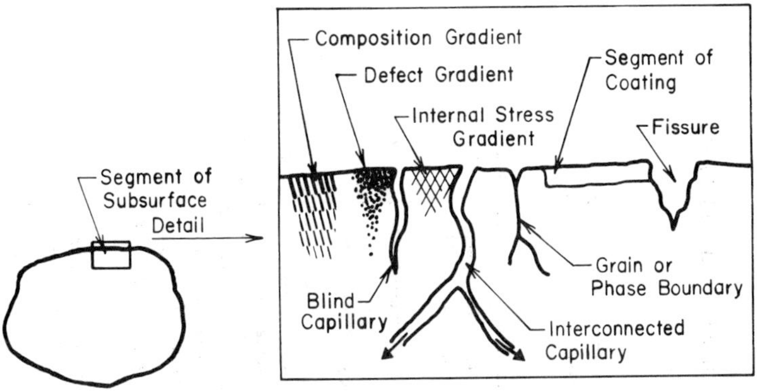

Fig. 2.35 Features of the subsurface region of a particle.

boundaries such as those formed by composition gradient between the surface and the particle interior. Examples of chemical or compositional features may be obtained in coatings, diffusion layers, and in encapsulations.

Fissures and Capillaries

A particle with a surface containing fissures and/or capillaries is endowed with a larger surface area per unit volume than a smooth featureless surface. Although these contributions are classified as surface, they differ from the true external surface of the particle in at least two aspects:

1. The surfaces of fissures and particularly the surfaces of capillaries are less accessible than the external surface. Because of these variations in accessibility, it is to be expected that different areas of the surface of a particle possess differing proclivities for interaction with the particle environment. In particular, one would expect solid or liquid reaction products to hinder reactions in the subsurface region.
2. Fissures and also capillaries have their concavity in common and it is known that the vacancy concentration beneath a surface depends upon radius of curvature according to the relationship:[40]

$$\frac{X_v - X_v^o}{X_v^o} = -\frac{\gamma V}{RT}\left(\frac{1}{r_1} + \frac{1}{r_2}\right) \tag{2.19}$$

where

r_1 and r_2 are the two principal radii of curvature,

Table 2.9

METHODS OF OBSERVING SURFACE STRUCTURE AND CHEMICAL COMPOSITION

Surface Structure

Method	*Magnification* X	*Resolution*		*Depth of Focus*	*Mode of Operation*	*Reference*
		Horizontal	Vertical			
Optical Microscopy Phase contrast		0.25 micron			Contrast due to uneven reflections. For low absorption (poor contrast) materials.	23
2 beam interferometry	2,000	1 micron	300 A		Similar to Michelson Interferometer.	24
Multiple beam interferometry		> 1 micron	5 A		Development of Fabry-Perot interferometer	
Transmission Electron Microscope	300,000	A few A		0.15 micron	Electrons scattered and fluoresce the screen differently, giving contrast.	25
Scanning Electron Microscope		150 A		1 mm	Electron beam at angle to surface. Can also analyze emitted x-rays to show element distribution in surface.	26
Field Ion Microscopy	3,500,000	1 A			Possible to use in "atom probe" type of work.	27
Mirror Electron Microscope		750 A			Can reveal variations in topography, work function and magnetic domains at and near the surface.	28

Table 2.9 *(cont'd.)*

Surface Structure

Method	*Magnification* ×	*Resolution*		*Depth of Focus*	*Mode of Operation*	*Reference*
		Horizontal	Vertical			
Field Emission Microscopy	c.f. FIM	20 A			Needle tip (1000 A radius) in high intensity field, electron emission excites screen. Tip radius r and fluorescent screen radius R gives magnification R/r.	29
Low Energy Electron diffraction					100 eV beam of electrons diffracted from surface. Shows surface structures.	30

Surface Composition

Group I methods	Based on electronic energy levels	31
Photoelectron and secondary electron spectroscopy		32
X-ray appearance potential measurements		
Electron probe microanalysis	Very often used. Chemical analysis in a 1 micron diameter volume.	33
Auger electron spectroscopy		34
Group II methods	Based upon mass.	
Mass spectrometer identification based upon either: 1. Thermal desorption, 2. Ion bombardment, or 3. Field desorption.		35
Energy loss of inert gas ions backscattered from the surface.		
Others include: Work function measurements		36
Surface electrical properties		37
Paramagnetism		38, 39

X_v^o = vacancy fraction on a flat surface,

X_v = vacancy fraction,

V = molar volume,

γ = surface tension (assumed isotropic).

Eq. (2.19) indicates a higher concentration of vacancies beneath a concave surface than under either a flat one or convex one.

It is possible for a fluid to pass through or into the subsurface region by means of capillary action. Consider the arbitrary capillary shown in Fig. 2.36. In this case, the area is B, the perimeter is L and the height of the fluid is h. Consider a small increment in height of the fluid up the duct dh.

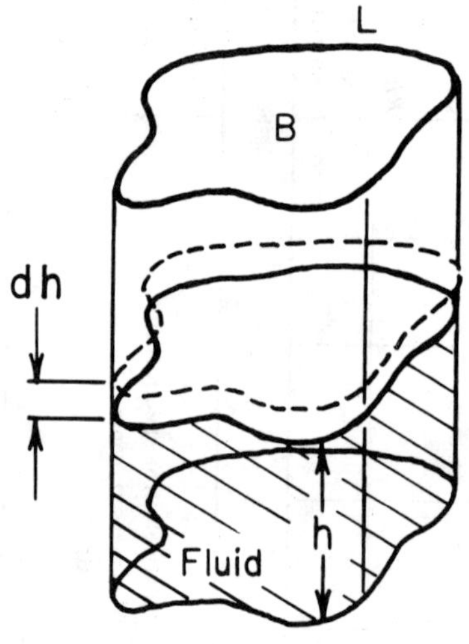

Fig. 2.36 Capillary of arbitrary cross section.[22]

From Eq. (2.8)

$$dA = -JdV + \cos\theta \, dA'.$$

In this case, A does not change so dA is zero. dA' is Ldh and dV is Bdh. Therefore, substituting these in Eq. (2.8),

$$-JBdh + \cos\theta \, Ldh = 0$$

so

$$J = \frac{L}{B} \cos\theta \tag{2.20}$$

From Eq. (2.17)

$$\Delta P = -\gamma J$$

or in terms of density

$$\Delta \rho g h = -\gamma J$$

Substitution from Eq. (2.20) and transposing gives

$$h = \frac{\gamma \cos\theta}{\Delta\rho g} \frac{1}{B/L} \tag{2.21}$$

Eq. (2.21) is a general equation representing the height of capillary rise in a duct of arbitrary section. Note the ratio B/L may be used as a shape factor.[41]

Internal Boundaries

There are numerous types of internal boundaries in crystalline solids including (in approximate order of increasing energy) twin boundaries, small angle grain boundaries, coincidence boundaries, stacking faults, in addition to phase boundaries and large angle grain boundaries. As shown in Fig. 2.37, grain boundaries, for example, act as "high diffusivity paths."[42] In this example it is to be noted that the grain boundaries make their contribution at lower temperatures.

Internal boundaries that intersect with the free surface of the particle can effectively contribute to reduction of the stress required for crack propagation. The Orowan modification of the Griffith's crack theory is represented by

$$\sigma_{crit} = \left[\frac{2E(\gamma + p)}{\pi c}\right]^{1/2} \tag{2.22}$$

where σ_{crit} is the stress level for brittle fracture to occur, E is the elastic modulus and p is the plastic work term. In brittle materials p is small and the influence of the surface energy term γ becomes significant. In this situation the present of either

1. an adsorbed layer can reduce the energy term and hence reduce

and make brittle fracture more easy, or

2. a grain or phase boundary coupled with an adsorbed layer or surrounding fluid reduces the energy term in two ways. If the boundary angle with the surface is ϕ, then at equilibrium

$$2\gamma^{s\ell} \cos \phi/2 = \gamma_{gb} \tag{2.23}$$

where γ_{gb} is the grain boundary surface energy.

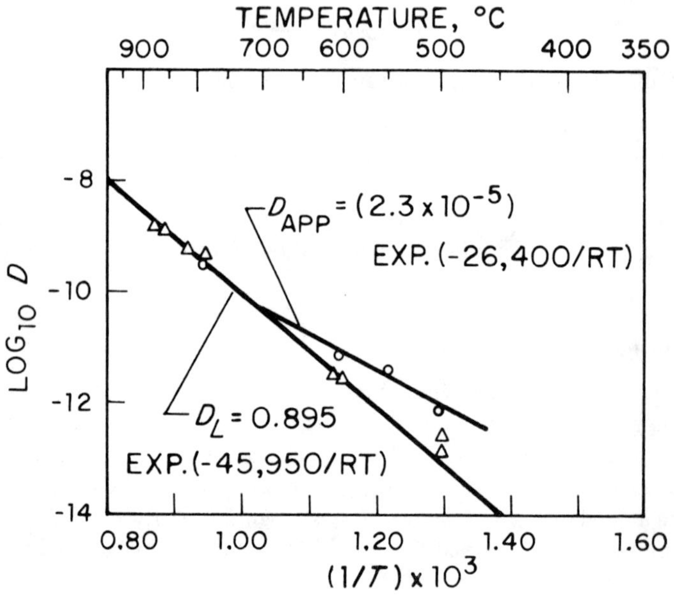

Fig. 2.37 Values of the self-diffusion coefficient obtained for silver using single-crystal and polycrystal samples (After D. Turnbull, in *Atom Movement.* p. 129, ASM, Cleveland, 1951).

Now let us note the various surface energies as follows:

transgranular crack in gaseous fluid, energy is $2\gamma^{gs}$. . . . I

transgranular crack in liquid, energy is $2\gamma^{s\ell}$ II

grain boundary crack in gas, energy is $2\gamma^{gs} - \gamma_{gb}$ . . . III

grain boundary crack in liquid, energy is $2\gamma^{s\ell} - \gamma_{gb}$. IV

Because these energy values are I $>$ II $>$ III $>$ IV, it follows from Eq. (2.22) that the critical stress level for brittle fracture has the same sequence. This is to say that the presence of these grain boundaries (or phase boundaries) tends to reduce the fracture strength of the particles, especially if they are made of inherently brittle material.

Defects in Subsurface Regions of Ionic Solids

In the case of metallic and homopolar bonded materials, the interatomic interactions are of short range, but in the case of ionic solids the interionic potentials are of very long range and in consequence the defect distribution may be disturbed for appreciable distances below the surface. One result of this is that in this subsurface region and also in the vicinity of internal boundaries and dislocations, there is a possiblity that a nonequilibrium concentration of charges occurs (remember, in ionic solids the point defects are charged). So space charge regions occur near lattice discontinuities. The presence of space charges may be of importance in such phenomena as sintering, surface conductivity, and film formation on surfaces.[16]

Diffusion Layers and Coatings

Variations on a theme of coatings and diffusion layers are almost endless so the following remarks are restricted to comments concerning some well understood phenomena.

One commonly practiced process is the alteration of carbon content of a ferrous alloy either by carburization or decarburization. In either case the relationship between depth of the diffusion layer (either carbon rich or carbon depleted) is

$$x = \beta(Dt)^{1/2} \tag{2.24}$$

where beta is a constant, D is the diffusion coefficient and t is time. Oxide layer build-up on metals and alloys occurs by diffusion of ions and electrons through the oxide layer. In one case the diffusion of electrons is the rate controlling factor (for example the reactions of halogen gases with metals) and in the other case the flow of ions is limited (for example, oxidation of Fe in oxygen). In both of the above examples, the build-up of a layer on the substrate effectively creates a subsurface region. However, in some cases such as that of internal oxidation and also during annealing in hydrogen atmospheres, true subsurface reactions occur. Nitriding (of steel) is yet another example of this type of phenomena. These phenomena are not discussed in detail here and the reader is referred to a suitable review.[43]

The above diffusion layers are bonded to the interior of the particle by the continuity of the lattice via primary chemical bonding. However, many layers on substrates are actually physically deposited (by fluid flow, by condensation, or by plastic flow for example) under such conditions in which primary bonding is hindered or because the materials are dissimilar. In the case of dissimilar materials, adhesion between the coating and the particle interior is dominated by London dispersion forces which result from momentary dipoles resulting from rapid fluctuations in electron density, and the interaction of these dipoles with the dipoles in nearby volume elements. Because each volume element can interact with all of the other, the total interaction is the sum of all the interacting pairs. This leads to a capability for the development of appreciable, long range, attractive forces over hundreds of angstrom units.[44]

Another way in which adhesion between dissimilar materials can be increased is by the formation of electrical double layers at interfaces of dissimilar materials which store enough energy to give considerable adhesive strength.[45] The formation of electrical double layers is important when considering the behavior of particulates in gaseous and liquid media and is discussed in later chapters.

2.5 INTERIOR OF THE PARTICLE

The interior of the particle consists essentially of three elements:

1. The material(s) substance(s),
2. Porosity which may be discrete or continuous or both, and
3. Voids.

One, two, or all three of these features may be present in the same particle simultaneously and the distribution of pores and cavities may be uneven. In addition, the material substance contains various defects (point, line, or surface) and these also may be distributed unevenly. A study of porosity is more in the province of research in the properties particulate assemblies and is discussed at that time (note that methods for determining the true density of a particle[46] and the size distribution of micropores in a particle[47] have been proposed). The reader is referred to texts on materials science for further information regarding the nature of defects occurring in solids.

The existence of impurities in particles is revealed by appropriate chemical analysis and is also observed by microscopic means in cases where distinct impurity phases exist. The presence of stress inhomogeneities in the form of microstrains has been shown by the well-known, x-ray line broadening experiment.[48] Some effects of these microstrains include: increased chemical reactivity of pure copper, increased rate of solution of LiF in water, increased tendency to transform austenite to martensite in a stainless steel.[49] The origin of the microstrains may be thermal or due to a natural deformation.[50]

2.6 PARTICLE SIZE

Concept of Dimension

The concept of size is concerned with the bulk or dimensions of anything. In the case of an exact sphere, it is immediately apparent that its size can be represented by one dimension only—its diameter. However, as the shape of the particle becomes more complex, the size becomes increasingly more difficult to represent. This means that the size of a particle and the shape of a particle are intimately related.

Particle Dimensions—Diameters

For a nonspherical particle the degree of difficulty in merely choosing an appropriate diameter is higher than for a sphere. One frequently quoted dimension in these cases is the projected area diameter (D_p). In this case, the observer fits the outline of the particle into a circle of the same area (see Fig. 2.38). The measurement can be carried out with a planimeter or with a graticule.* The former method is tedious and the latter is dependent upon the observational skill of the experimenter or observer (i.e., this is open to the potential criticism of being a subjective method). The personal error in this method has been discussed briefly elsewhere.[51] One way of avoiding the deficiency of the equivalent area diameter is to use the so-called Feret diameter, D_F.[52] This diameter is the distance between parallel tangents and is illustrated in Fig. 2.38. It is clear from this diagram that different measures of D_F can be obtained from the particle unless the observer specifies the direction of measurement. For this reason, the direction is specified usually in a left-to-right direction in the field of view.

Yet another diameter is called the Martin's diameter, D_M. In this measure, the dimension taken is the length of the chord through the particle

* I am omitting sophisticated instruments from this discussion (Quantimet, Bausch and Lombe, Leitz).

which also bisects the particle, i.e., through the centroid.[53] As in the case of Feret's diameter, it is necessary to specify the direction of measurement.

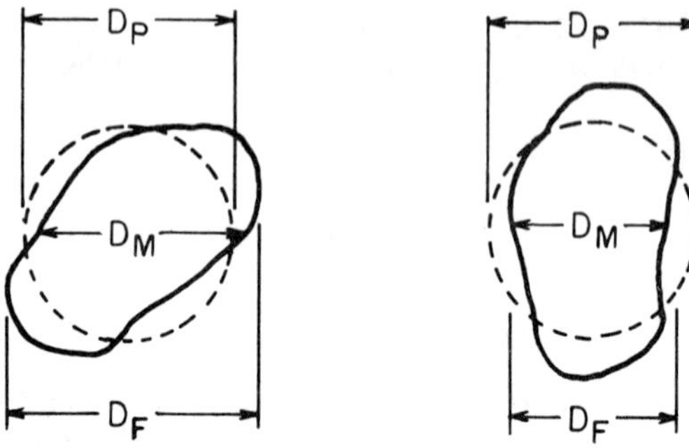

Fig. 2.38 Geometric diameters of an irregularly shaped particle. D_p is the diameter of dashed circle.[77]

Because the observer is usually interested in the sizes of particles in a given sample, the diameters of interest are more often statistical diameters. For example, Feret's statistical diameter is the distance between parallel tangents for a number of particles. Specifically

$$\bar{D}_F = \frac{P}{\pi} \qquad (2.25)$$

where P is the perimeter of one particle and $\bar{D}_F$ is therefore the Feret diameter for a single particle averaged over all possible directions. For a group sample, D_F for a number of particles is the average perimeter divided by pi. Similarly, the value of Martin's diameter for a number of particles is the mean of chords through the centroids of a group of particles. Yet another diameter is the mean length of chords drawn at random through the outline of a particle. This measure is known as Croften's diameter and for a single particle[54]

$$\bar{c} = \frac{\pi A}{P} \qquad (2.26)$$

For a group sample, C for a number of particles is the mean of the ratio of their area to perimeter times pi.

Size Classification Strategies

The measurement of particle size is normally carried out in order to relate particle size to some behavior(s) of the population to which the sample of measured particles belong. Furthermore, because in a given population of particles there is hardly ever a monsize, the observer is concerned with statistical representations of the measures taken relating to size. Both of these factors underpin the general principles upon which the strategy of particle size measurement is based.

To size means to classify or arrange according to size. The process of classification necessitates the establishment of a convenient number of classes into which the various sizes of particles can be grouped. The distribution of the population of particles among the groups can then be represented statistically in terms of mean and variance, etc. This strategy which is really a sophisticated application of the fuzzy set[55] coupled with normal statistical techniques, permits us to carry out particle size analysis in which the information per expenditure of time and effort is small (see Chapters 5 and 6 for further details of methods). Once this step has been taken and the individual is willing to utilize the classifications and statistics, the choice of the experimental sizing technique may involve a variety of considerations including economics, expedience, and the purpose to which the information will be put once it is obtained. In general, apart from microscopic methods, the sample of powder particles is forced to react in a carefully regulated environment, for example, mechanical sieving, sedimentation in gaseous or liquid fluids, precipitation methods, filtration methods, etc. Differently sized groups of particles react differently to the same environmental forces and this separation based upon behavior in the stated conditions permits the observer to carry out the appropriate classification and statistical representation. Often the choice of the experimental environment in the test is made with a view to its close relationship to the expected environment which the powder faces later on. All of the foregoing serves to underline the fact that in the measurement of the dimension(s) of particle size there is no one absolute or set of absolutes. All is relative. One can take the viewpoint that the observers utilize fundamentals in order to carry out size measurements and deduce meaningful results, but the very plethora of dimensions needed to specify the "size" of a population of powder particles indicates that mathematically the measurement of particle size is approximate.

2.7 CONCEPTION AND DEFINITIONS OF SHAPE

It appears to be a basic human habit to assign to objects a property which we call shape. When an observer is presented with an object, his eyes scan it very rapidly. The more familiar he is with such objects, the more he has learned as a result of similar exercises in the past, the more rapidly is the scan completed and the shape identified. For example, it is a very simple matter for us to identify the shape of a cup. More complex and less familiar shapes can be identified with greater expenditure of effort.

A natural consequence of scientific interest in shape analysis is the development of shape measurement instrumentation. One can appreciate the difficulty involved in the instrument measurement of shape when it is learned that a mechanical system which analogs the human scanning approach mentioned above took 11 minutes to finally recognize a cup correctly.[56] If it takes 11 minutes for a comparatively sophisticated instrument to measure the shape of a simple thing like a cup, one might conjecture that, in the case of the rich variety of shapes encountered in particulates, the complexity would defy satisfactory measurements being obtained at an acceptable cost. Nevertheless, the formidable nature of the problem has not prevented the formulation of a whole variety of shape assessment schemes. Many of these are ingenious, and some of them have been and remain extremely useful (see for example the work of Heywood).[57] These are reviewed and classified in Chapter 6.

There is a number of definitions of shape in any standard dictionary. Those judged most pertinent are listed below:

1. external appearance as determined by outlines or contours[58]
2. that which has form or figure[58]
3. a pattern to be followed[58]
4. that quality of a material object (or geometrical figure) which depends upon constant relations of position and proportionate distance among all the points composing its outline or its external surface.[59]

One may summarize the foregoing by saying that the shape of a particle is the recognized pattern of relationships among all of the points which constitute the external surface. From this definition, certain corollaries may be deduced as follows.

Shape is a property; it is an intrinsic characteristic of a material system. Shape is not an extrinsic characteristic: shape 1 cannot be added to shape 2 to make shape 3 which has more shape than either 1 or 2. However, shape

1 and 2 may be combined in some way in order to produce shape 3 which may or may not be different from its components 1 and 2. In the case of a particle of material, its size tells us about the quantity of matter in it, and its shape tells us about the pattern in which the overal quantity is fitted together.

Shape can be a property of an abstraction (e.g., a geometric figure) and it can also be a property of a piece of concrete material, because shape is formed in the human consciousness through a process of pattern recognition. Shape is composed of every point on a surface. Because the surface of even a small object can be conceived as consisting of a myriad of points, theoretically the assessment of its shape involves the observer in making a myriad of observations. Doubtless it is this daunting prospect that forces us to scan the object, sampling from the myriad of points on the surface. In some cases, the scientific means for measuring shape involve arbitrary sampling methods, raising serious doubts concerning the accuracy of these methods.

Shape is concerned exclusively with the outline of the external surface of the object. Consequently, the shape characterization of a portion of matter which contains pores, for example, does not provide any information pertaining to them if the pores are completely internal. Only if they intercept the outline of the external surface of the object does the shape description give any information about the pores. And even in this case, the description does not express the porosity, which is an internal condition of the object, but only the effect of the craters on the pattern of points in the outline. Finally, although both two and three dimensional objects possess the property of shape, three dimensional shape is often projected as a two dimensional outline in order to reduce the complexity of shape assessment.

Differences between two shapes are perceived by noting differences between the pattern of the relationships among the point coordinates of each shape. There is an infinite variety of shapes ranging from those which might be described as having monotonic symmetry to those with monotonic assymmetry.

Measuring Particles so as to Describe Shape Meaningfully*

A shape assessment scheme may be viewed as consisting of a series of

*The various schemes which have been suggested, developed, and utilized to measure particle shape are reviewed in Chapter 6.

three operations, starting with data collection, followed by data processing, and ending with the judgement procedure. Depending upon the scheme being used, the type of data collected can vary all the way from taking one datum point per particle sampled to taking many data points per particle sampled. The data processing step can vary all the way from a simple tabulation of a result to an involved statistical manipulation, depending upon the particular scheme used. In the final step entitled judgement procedure, the formulated package of information is allotted a category which, by definition is unique to the particular scheme invoked and identifies the shape of the particle.

In general all investigators have been forced to consider the tradeoffs among the following parameters:

The final, useful information obtained. This is usually denoted as the shape of the particle(s) being studied.

The data collected and processed.

The effort expended.

The precision of description.

The final information obtained should permit the observer to unequivocally identify the shape of the particle(s) in question. Thus, it is necessary to establish a set of rules for choosing reference points which are mutually acceptable to the majority of observers. Once established, this set of rules would permit the development of a scale of measurement of particle shape.

What of the relationship between the initial measurements, the data processed, and the final information obtained (i.e., the assessed shape)? At one extreme, if no data existed, then no shape measurement could be formulated. At the other extreme, with an infinite number of data points, complete information regarding the particle shape could be deduced. Between these two extremes it is not clear what the general relationship between data points chosen and the shape assessed is. But in many cases investigators have assumed the form of the curve drawn in Fig. 2.39. This curve suggests that beyond a certain level of data collection, the utility of additional data decreases. The law of diminishing returns appears to have been one of the underpinning factors in the work of previous researchers, but it is an important question whether or not this, for the most part unstated inherent assumption, is justified. The answer seems to be that it is not. Basically, the assumption overlooks the fact that a shape assessment scheme is designed: not only can the investigator choose the number of data points to be collected, but he can also choose the type of data with which he will deal. Shape does not have to be assessed as a result of a singular type of measurement; it can be assessed (and it often is so assessed) as a result of two or more different types of measurement such that the shape is established as a set of numbers, or of words, or of both. Finally, it is clearly a matter of choice as to how many data points are required to locate the assessed shape of a particle on a scale of measurement, simply because this is what the investigator decides is necessary when he originally sets up a particular scheme.

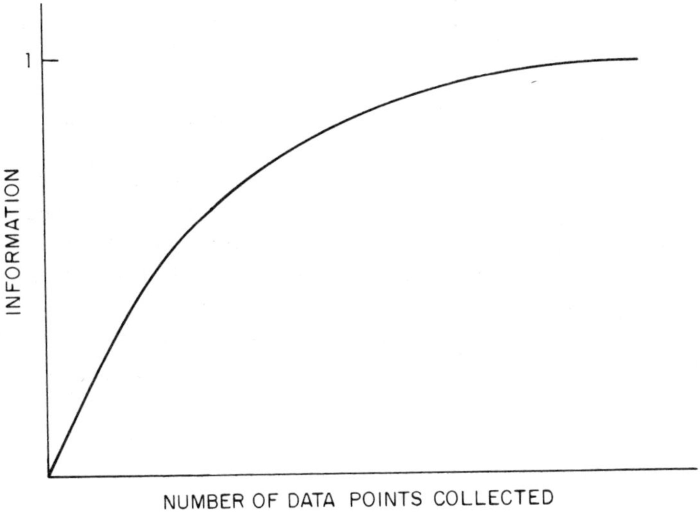

Fig. 2.39 Illustrating a fallacy.

When estimating the degree of completeness of observation necessary for accurately assessing particle shape, it is reasonable to consider the problem in some respects as analogous to many other problems of measurement. For example, in the case of precision in measuring an object's weight, it costs a nominal amount to estimate its order of magnitude–the object weights between 1 and 10 lb; it costs a small amount more to weigh the object to the nearest lb; it costs more again to find out the weight to the nearest decimal point; and still more to the nearest two decimal places. Finally, to have a degree of precision to the nearest ten or hundred atoms would be costly and probably prohibitively expensive. The lesson to be learned here is that it becomes progressively more and more expensive to increase the degree of precision required of a measurement or a set of measurements.

In particle shape measurements there appear to be three specific areas of concern with respect to precision and accuracy. These refer to:

1. the individual measurements themselves,
2. the method of treating those measurements, and
3. the conclusions drawn from the processed data.

In delineating these areas, it is established that the measurement which produces the individual data point is not necessarily the same as the operation which identifies the shape. For example, in the Church scheme, the operation which produces the individual data points consists of the technique of measuring both Martin's and Feret's diameter of a particle.[60] The subsequent data processing operation which combines all of these measurements in the form of a ratio of the two diameters (this ratio constitutes a shape factor) involves the use of statistical methods to handle the data.

Inherent in the concept of precision is the resolving power of the scheme. For example, in the case of the measurement of a particle diameter through a microscope, the resolution obtainable is a function of the system used. In this case the resolving power is an expression of the distance which must separate two points before they can be distinguished from each other through the instrument. The degree of resolving power of a scheme describing shape determines the smallest differences between two particles that can be assessed and therefore assigned to two different shape categories. (Once again, this presupposes that a scale of shape assessment exists.) Furthermore, because the concept of shape deals with all of the points of the surface, the resolving power of the scheme is closely related to the data point sampling procedure. For example, consider the scheme of curve generation using Fourier series (see class 3 in Fig. 2.40). In this case one is able to vary the number of data points per particle. Let us say that 100 points per inch of outline are selected in a particular run. It may be found later that the shape can still be recognized even if only ten points per inch are sampled. This means that the resolution of the scheme is more powerful than is necessary.

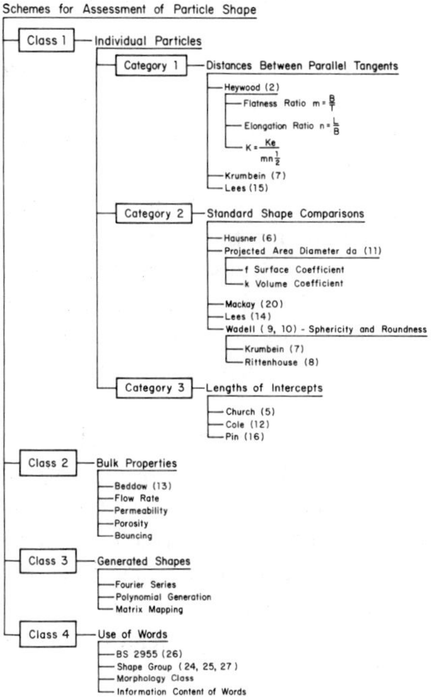

Fig. 2.40 A proposed classification of schemes of shape analysis.

Errors in measurement vary in kind and potential magnitude from one class of shape assessment schemes to another and also vary within classes. Therefore, these errors are as numerous in type as are the measurement techniques themselves. Sources of potential errors also include:

1. insufficiency of individual data points with which to work,
2. contrasymmetrical sampling procedure of data points (this would be especially important in the methods of shape generation), and
3. an inadequately developed judgement procedure.

It is possible to discern that two extremely different points of view regarding the objective of assessing shape have governed the development of the scheme for measuring shape. One point of view is that, when the points have been measured, the data collected, processed, and formulated into a shape measurement, the investigator should be able to recreate the original particle shape. Hausner put this viewpoint forward by indicating that the shape measurement should help the scientist to visualize what the particle looks like.[61] Because he has insisted that the final shape measurement should permit the reproduction of the original shape as closely as possible, this demand will be referred to as the Hausner Requirement. One may conjecture that the merit of this point of view lies in the fact that we appreciate shape mainly by its visual impact.

The contrasting point of view may be represented by saying that it is not particularly important for the scientist to visualize what the original particle looks like from the final shape measurements obtained. The objective here is to obtain a numerical shape measurement which can be used to establish a ranking by and for comparison with other numerical shape measurements. This point of view has the merit that because the data processing step produces a generalized composite picture of the particle shape, in all probability there are a few of the actual shapes that correspond to this general picture. This implies that the contrary argument is somewhat artificial. However, on the debit side the presumption in this second point of view is that because in reality the measurement and analysis of all the points of the particle surface cannot be done, it should not be attempted.

Briefly, the aim of one extreme is the reproduction of the original particle shape and the aim of the other extreme is the production of a number (or of a set of numbers) which would constitute shape factors, and be conveniently manipulable in mathematical formulae.

The various schemes used or proposed for assessing particle shape are reviewed in Chapter 6 and are conveniently classified in Fig. 2.40. The way in which the schemes have been classified is described below. Four general classes of schemes for particle shape assessment have been delineated, each class being further divided into categories.

Class 1. In this class, shape is assessed on an individual particle basis. There are three categories. In the first one the particle shape is related to distances measured between tangents parallel to the particle contour. In the second category, a variety of standard shape outlines is used as a shape comparator. In the third category, the lengths of specific types of intercepts are measured and used to characterize particle shape.

Class 2. Schemes belonging to this class involve taking measurements of bulk properties of powders. The use of the ratio of tap density to apparent density, the use of the bulk flow rate of the powder, and so on, are examples.

Class 3. Those schemes in which particle shapes are generated by various mathematical techniques including Fourier Series analysis and matrix mapping are included in this class.

Class 4. This class may be conveniently divided into four categories corresponding to: the use of shape group, the use of essentially separate verbal definitions, the analysis of the information content of words, and the use of morphology classes. It should be noted that the first three classes deal with means of observation, whereas the fourth class deals with means of description. This explains why class 4 schemes are often used as an adjunct to the other three.

Methods based on the Hausner Requirement are generally those in which multiple data points are obtained for each particle examined, and the resulting description is compared with a constructed prototype. Examples illustrating this may be taken from two of the four classes identified in Fig. 2.40, namely Class 1.2 and Class 3 which are titled Standard Shape Comparisons and Generated Shapes respectively. Keeping in mind that shape is concerned with all of the points of the surface, it is clearly necessary to sample the surface points in some way in order to reduce the effort expended, and this is done in Polynomial and in Fourier generation by selecting one coordinate point as representing a given area or length of a particle surface (to date these two methods have only been attempted with two dimensional models). Standard shape analysis methods, including those developed by Hausner,[61] Krumbein,[62] Rittenhouse,[63] Wadel,[64, 65] and those using surface coefficient (f) and volume coefficient (k),[66] are all sample points and make a comparison in some way. However, all do not give equal weight to the Hausner Requirement. For example, the use of f and k involves comparison of the sampled points of the particle under observation with an equivalent sphere. Hausner used an equivalent rectangle. Rittenhouse developed a scale of shape variation (sphericity) for visual identification of pebbles. In all of these standard shape analysis schemes there is a considerable approximation involved in the comparison of selected points with the standard shape, and this is the same as saying that the level of discrimination, or resolving power, is reduced in these schemes as contrasted with the ideal expressed by the Hausner Requirement.

At the other end of the spectrum from the classes of shape assessment discussed above are those corresponding to Class 1.3 and Class 2 respectively, which involve simple measurements on either individual particles or bulks of particles. As was mentioned before, the scheme proposed by Church utilizes statistically projected values of Feret's and Martin's diameters in the form of a ratio to represent the shape of a monosize assembly of particles (elliptical). Cole's scheme uses high speed intercept, perimeter, and area counting techniques to obtain data for large numbers of particles.[67] The data can then be handled in a convenient way to produce a variety of factors which can be used as shape indicators. Bulk schemes such as the one suggested by Beddow (tap/apparent density),[68] the bounce scheme, and the flow rate scheme are of a fundamentally different kind than all the others in that the observer assesses the shape of the particles of which the powder is composed by making a standardized observation of the behavior of the powder in bulk. In one sense this class (Class 2) has a built-in statistical processing operation in that the behaviors result from interactions in which all of the particles of the sample participate. In addition the schemes are, for the most part, easily applied, that is to say, of all the four classes discussed in this paper, the bulk schemes of Class 2 involve the observer in the least effort.

In conclusion, some of the schemes such as those of Class 1.2 and Class 3 make a concerted effort to try to comply with the Hausner Requirement. Other schemes such as those of Class 1.1, 1.3, and 2 produce numerical factors which can be used in conjunction with Class 4 schemes to attempt to more closely approach the requirement. The strategy that should be pursued in the future involves the development of schemes which fulfill the Hausner Requirement to the maximum extent while simultaneously producing representative numerical measurements of particle shape. Such representative measurements could more safely be used in mathematical treatments relating shape to behavior than is possible with the many inferential measurements available at the present time. As has been observed, in some cases only a few data points (one or more) per particle are collected, and this forces the reporting of particle shape in more or less symbolic terms, usually in the form of a shape factor in which little or no actual representation is incorporated. In other cases, many data points per particle are collected and this permits the reporting of the shape more in accordance with the Hausner Requirement.

Finally, in the outline of Table 2.10 is presented a scheme by scheme preliminary assessment of the presently available particle shape measurement schemes against the criteria stated earlier in this section. This outline should be used only as a guide because the individual comparisons may change, depending upon the specific particulates being examined.

The Beginning of a Whole Field Approach?

Recent work in multicompositional particle characterization using the

Table 2.10

PRELIMINARY ASSESSMENT OF SCHEMES FOR MEASURING PARTICLE SHAPE

Scheme	*Experimental Criteria*			*Theoretical Criteria*			
	Simplicity	Rated Effort	Controlled Selection	Simplicity	Visualization	Resolution	Mathematics
Class 1							
Category 1	*a*	*b*	*u*	*a*	*b*	*a*	*a*
Category 2	*a*	*b*	*u*	*a*	*a* - *b*	*a*	*a*
Category 3							
Church	*b*	*b*	*u*	*a*	b^{ϕ}	*a*	*b*
Cole	*a**	*a**	*a* - *b*	*a*	*b*	*a*	*a*
Pin	*b*	*b*	*u*	*b*	*b*	*b*	*b*
Class 2							
All categories	*a*	*a*	*u*	*a*	*b*	*u*	*b*
Class 3							
Fourier series	a^x	$-^x$	u^x	a^x	a^x	a^x	a^x
Polynomial	*a*	*u*	*u*	*a*	*u*	–	–
Matrix mapping	–	–	–	*a*	–	–	–
Class 4							
Shape group	*a*	*a*	*u*	*a*	*a* - *b*	*a*	*a*
B S 2955	*a*	*a*	*u*	*a*	*a* - *b*	*a*	–

KEY

– Unknown or no basis for judgement
a Acceptable, meets criterion
b Does not meet criterion, needs further development to do so
x In process of development
u Uncertain as to acceptability
* Equipment is expensive
ϕ Only usable for elliptical particles at present

semimicroprobe for determining sizes and compositions of complex particles has been reported.[70] Fig. 2.41 shows some results of an analysis of a particle of quartz-chalcopyrite composition. The top image shows the raster image obtained with back scatter electrons and the bottom image is a digital map of the image. The regions of quartz and chalcopyrite are clearly shown. The computed particle parameters are given in the accompanying data table. If this chemical analysis with the morphological anlaysis were interrelated, this would represent a significant move in the direction of the whole-field view of fine particles.

Fig. 2.41 (a) Raster image of a quartz-chalcopyrite composite particle obtained with back-scattered electrons. *PW* = 50 μm. (b) Digital map of image (a). 1 = quartz, 2 = chalcopyrite. 1 character space = 1 μm. In the computer array, only the coordinates of each change of position are stored.[70]

Computed Particle Parameters*

Sample	Bougainville	7-73-106
Frame		1
Particle		8
Rectangle		
Location	X	210
Location	Y	320
Dimension	X	50
Dimension	Y	44
No. of Chords		39
Max X Chord		46
X Min		3
X Max		49
Y Min		3
Y Max		41
Centroid X		26.6
Centroid Y		22.5
Perimeter		142.5
Area		1267
Subarea Type 2		67

*For particle shown in Fig. 2.41. Lengths in μm, areas in square μm.

Recent Developments

More recently, there has been a concerted effort to analyze particle profiles by digitizing the profile, converting the data to a continuous function and transforming this to a set of (for example) Fourier coefficients. This approach is discussed more fully in Chapter 6.[69]

REFERENCES

1. "Oxford English Dictionary," Oxford, 1961.
2. Orr, C. Jr., and Dallavalle, J. M. *Fine Particle Measurement.* New York: MacMillan, 1959.
3. Lapple, C. E. *J. Stanford Research Institute.* 5 (1961): 95.
4. Hausner, H. H. *Planseeber.* 14, 2 (1966): 75–84.
5. Stover, E. R. "Technical Report AFML-TR-67-56." *Wright-Patterson AFB.* Ohio, 1967.
6. McCrone, W.C., Delly, John Gustav. *Particle Atlas.* 2nd ed. Ann Arbor Science Publishers, Ann Arbor, MI, 1973, 1979.
7. Holliday, L. *Composite Materials.* New York: Elsevier, 1966.
8. Bikerman, J. J. *Science of Adhesive Joints.* New York: Academic Press, 1961.
9. Hayward, E. R., and Greenough, A. P. *J. Inst. Metals* 88 (1960): 217.
10. Mays, C. W., Vermaak, J. S., Kuhlmann-Wilsdolf, D. *Surface Sol.* 12 (1968): 134.
11. Sandford, C., and Ross, S. *J. Phys. Chem.* 48 (1954): 288.
12. Lennard-Jones, J. E., *Trans Faraday Soc.* 28 (1932): 347. *Proc. Roy. Soc. A* 158 (1937): 244.
13. Basset, D. W., and Parsley, M. J. *J. Phys. D: App. Phys.* 3 (1970): 707.
14. Ross, S. *Surfaces and Interfaces.* edited by Burke, J. J. Syracuse Univ. Press, 1967.
15. McClean, M. *Problems in Mat. Sci.* vol. 1, edited by Marchant, H. G. Gordon & Breach, New York, 1972.
16. Blakely, J. M. *Crystal Surfaces.* Oxford: Pergamen Press, 1973.
17. Blakely, J. M., and Schwoebel, R. F. *Surface Sci.* 26 (1971): 321.
18. Langmuir, I. *J. Am. Chem. Soc.* 40 (1918): 1361.
19. Brunauer, S., Emmet, P. H., and Teller, E. *J. Am. Chem. Soc.* 60 (1938): 309.
20. Veale, C. R. *Fine Powders.* New York: Wiley, 1972. (Applied Science Publishers Ltd., Barking, Essex, England) 1972 Crown Copyright.
21. Brunauer, S. *Solid Surfaces.* Am. Chem. Soc., Washington, 1961.
22. Hwang, S. "The Gauss Equation in Capillarity." *Zeitschrift fur Physikalische Chemie Neue Folge,* Bd. 105, S. 225–235 (1977) by Akademische Verlagsgesellschaft, Weisbaden, W. Germany.
23. Chalmers, B. "Physical Examination of Metals." vol. 1, *Optical Methods.* London: Edward Arnold, 1944.

24. Tolansky, S. *An Introduction to Interferometry.* New York: Longmanns, 1955.
25. Beck, J. A., and Davies, A. L. *Electron Microscopy and Microanalysis of Metals.* New York: Elsevier, 1968.
26. Thornton, P. R. *Scanning Electron Microscopy.* London: Chapman & Hall, 1968.
27. Hochman, R. F., et al. *Applications of Field Ion Microscopy in Physical Metallurgy and Corrosion.* Atlanta: Georgia Tech. Press, 1969.
28. Bok, A.B. "A Mirror Electron Microscope." Delft, The Netherlands, 1968.
29. Hren, J. J. *Surface Sci.* 23, 1 (1970).
30. May, J. W. *Advances in Catalysis.* 21, 151 (1970).
31. Siegbahn, K. *Nov. Reg. Soc. Sci. Vp, Ser.* 4, 20 (1967).
32. Parl, R. L., et al. *Rev Sci Inst.* 41 (1970): 1810.
33. Birks, L. S. *Electron Probe Microanalysis.* New York: Wiley, 1963.
34. Taylor, N. J. *Techniques of Metals Research.* vol. VII, New York: Wiley-Interscience, 1971.
35. Smith, D. P. *Surface Sci.* 25, 171 (1971).
36. Riviere, J. C. *Solid State Surface Sci.* 1, 179 (1969).
37. Frankel, D. R. *Electrical Properties of Semiconductor Surfaces.* London: Pergamon Press, 1967.
38. Selwood, P. W. *Adsorption and Collective Paramagnetism.* New York: Academic Press, 1962.
39. Flood, E. A. Solid–Gas Interface. vol. II, 1967.
40. Kuzynski, G. C. *Powder Metallurgy for High Performance Applications.* eds. Burke & Weiss, Syracuse U. Press, 1971.
41. Hwang, S. T. personal communication.
42. Shewmon, P. G. *Diffusion in Solids.* New York: McGraw Hill, 1963.
43. Bernard, J. *Met Rev.* 9, 473 (1964).
44. Fowkes, F. M. *Surfaces and Interfaces.* ed. Burke et al., Syracuse U. Press, 1967.
45. Derjagvin, B. V., and Aleinakova, I. N. *Research in Surface Forces.* ed. D. V. Dergajuin, trans by Consultants Bureau, 1963.
46. Blum, J. F., and Bear, E. J. USAEC Report NCLO. 677, 1957.
47. Yarnton, D., and Simpson, G. R. *Powder Metallurgy.* 8 (1961): 42–64.
48. Gillies, D. C., and Lewis, D. *J. Less Common Metals.* 13 (1967): 179–185.
49. Lewis, D., and Northwood, D. O. *Strain.* vol. 4, 3, July 1968.
50. Hirschorn, J. S. P. Tech. 4 (1970–71): 1–8.

51. Davies, C. N. *Nature.* vol. 195 (August 25, 1962): 768–770.
52. Feret, L. R. *Assoc Int Pour l'Essai des Mat.* 2D, Zurich, 428, 469 (1932).
53. Martin, G., Blyth, C., and Tongue, H. Trans Ceramic Soc. 23,60 (1923).
54. Croften, M. W. *Encyclopedia Brittanica.* 9, 19, 768 (1885).
55. Zadeh, L. A. *Inf. & Control.* 8 (1965): 338-353.
56. This was illustrated in a P.B.S. Film, "The Human Mind," Spring 1973. The work was reported as part of a research program on pattern recognition at The University of Edinburgh, Scotland. Note that the cup is equivalent to an interconnected phase as described by DeHoff and as such he claims that concepts of shape are inadequate. Instead, topological concepts have to be introduced in order to describe such systems first of all qualitatively and later on quantitatively. See *Quantitative Microscopy* by DeHoff, R. T., and Rhines, R. N. pp. 291–325. New York: McGraw-Hill, 1968.
57. Heywood, H. "Symposium on Particle Size Analysis." *Trans. Inst. Chem. Engrs.* 25 (1947): 14–24. See also the report by R. T. DeHoff in *Quantitative Microscopy* by DeHoff, R. T., and Rhines, F. N., pp. 128–148. New York: McGraw-Hill, 1968.
58. Consolidated Webster Encyclopedia Reference Dictionary, 1945.
59. Oxford English Dictionary, 1972.
60. Church, T. *Powder Technology.* 2, (1968-69): 27–31.
61. Hausner, H. H. *Planseeber.* 14, 2 (1966): 75–84. This concept has been further developed as a dynamic shape factor of particles. See Medalia, Avrom I. *Powder Technology.* 4 (1970–71) 117–138.
62. Krumbein, W. C. *Journal of Sedimentary Petrology.* 11, 2 (1941) 64–72.
63. Rittenhouse, G. *Journal of Sedimentary Petrology.* 13, 2 (1943): 79–81.
64. Wadel, H. *Journal of Geology.* 40 (1932): 443–457.
65. Wadel, H. *Journal of Geology.* 43 (1935): 250–280.
66. Allen, T. *Particle Size Measurement.* pp. 18–21, London: Chapman & Hall, 1968.
67. Cole, M. *American Laboratory.* pp. 19–28, June 1971. See also Fischmeister, H. F. "Automatic Measuring and Scanning Devices in Stereology." in H. Elias, ed. *Stereology.* New York: Springer-Verlag OHG, 1967; and Elias, H. (ed.), "Proceedings of the 2d International Congress for Stereology, Chicago, 1967," New York: Springer-Verlag OHG, 1967.
68. Kostelnick, M., and Beddow, J. K. *Advances in Powder Metal-*

lurgy, Vol. IV, Processes. edited by Hausner, H. H., pp. 29–48, New York: Plenum, 1971.

69. Beddow, J. K. "On Schemes for Shape Analysis of Individual Particles" Report A390 ChME 74-007, University of Iowa, Engineering College, Iowa City, Iowa 53242, February 1974.
    Beddow, J. K., and Philip, G. C. *Planseeberichte für Pulvermetallurgie.* 23, 1 (1975): 3–14.
    Meloy, T. P. *Powder Technology.* 16, 2 (1977): 233–253.
    Ehrlich, R., and Weinberg, B. *J. Sed. Pet.* 40, 1 (1970): 205–212.
70. Grant, G., Hall, J.S., Reid, A.F., and Zuiderwyk, M. "Multicompositional Particle Characterization Using the SEM-Microprobe" 1976, Scanning Electron Microscopy/1976 (Part III), ITT Research Institute, Chicago, 401–408.
71. Van Vlack, L. H. *Materials Science for Engineers.* Reading: Addison-Wesley, 1970.
72. Rose, H. E. *The Measurement of Particle Size in Very Fine Powders.* London: Constable & Co., p. 19, 1958.
73. *Problems in Materials Science I.* vol. I, ed. by Merchant, Harish D., New York: Gordon & Breach, 1972.
74. Sands, R. L., and Shakespeare, C. R. *Powder Metallurgy.* Cleveland: CRC Press, 1966.
75. Prof. K. T. Whitby, "Mechanics of Fine Sieving" from *Particle Size Measurement.* ASTM Special Technical Publication No. 234, 1958.
76. Trans. Faraday Society, 28, (1932) 347.
77. Mercer, T. *Aerosol Technology in Hazard Evaulation.* New York: Academic Press, 1973.
78. Hasbrouck, Jerald E., "A Scanning Electron Microscope Study of Hydrating Cement and Its Constituents." M.S. Thesis, University of Iowa, Civil Engineering, 1973.
79. Mason and Shora, "Coal and Char Transformation in Hydrogasification" in *Fuel Gasification,* ed. Robert F. Gould. Washington, D.C.: American Chemical Society, 1967.
80. Roldán-Quintana, Jaime. M.S. Thesis, "The Geology and Mineralization of the San Felipe Area, East-Central Sonora, Mexico," Department of Geology, University of Iowa, Iowa City, IA, July 1976.

Chapter 3

FORMATION AND PRODUCTION OF PARTICULATES

3.1 THE SEVERAL PROCESSES

In this chapter the various processes for the production and handling of particulate matter are described and discussed. Theoretical considerations are left for later treatment in the book. We have already made the bold claim that an immense quantity of *particulate* materials processing is conducted in our industrial/technological civilization. Various percentages have been quoted—all the way from 50 percent through 90 percent of the total material processed as being particulate matter. (The actual proportion may in reality be somewhere in the middle of this range.) If we process such an enormous mass of particulate matter, where does it all come from?

1. Large quantities of particulates are produced according to (preplanned) genetic designs by the agricultural industries.
2. Geological formations serve as a major source of particulate matter in the form of minerals, ores, fossil resources, etc.
3. The chemical and fuel industries produce large quantities of particulates, as well as processing many.
4. Many of the multifarious human activities lead to the production of particulates. Examples of these include products of combustion, dusts created and/or raised by activities including agriculture, horticulture, transportation, and particle processing in general. To this list naturally occurring particle production from processes such as droughts, volcanic activity, prevailing winds, water erosion, forest fires, etc., can be added. Of course, many of these interact with the activities of humans to augment the effect experienced.

In concentrating on the ways in which material is derived in particulate form, it quickly becomes apparent that there are only a few ways in which

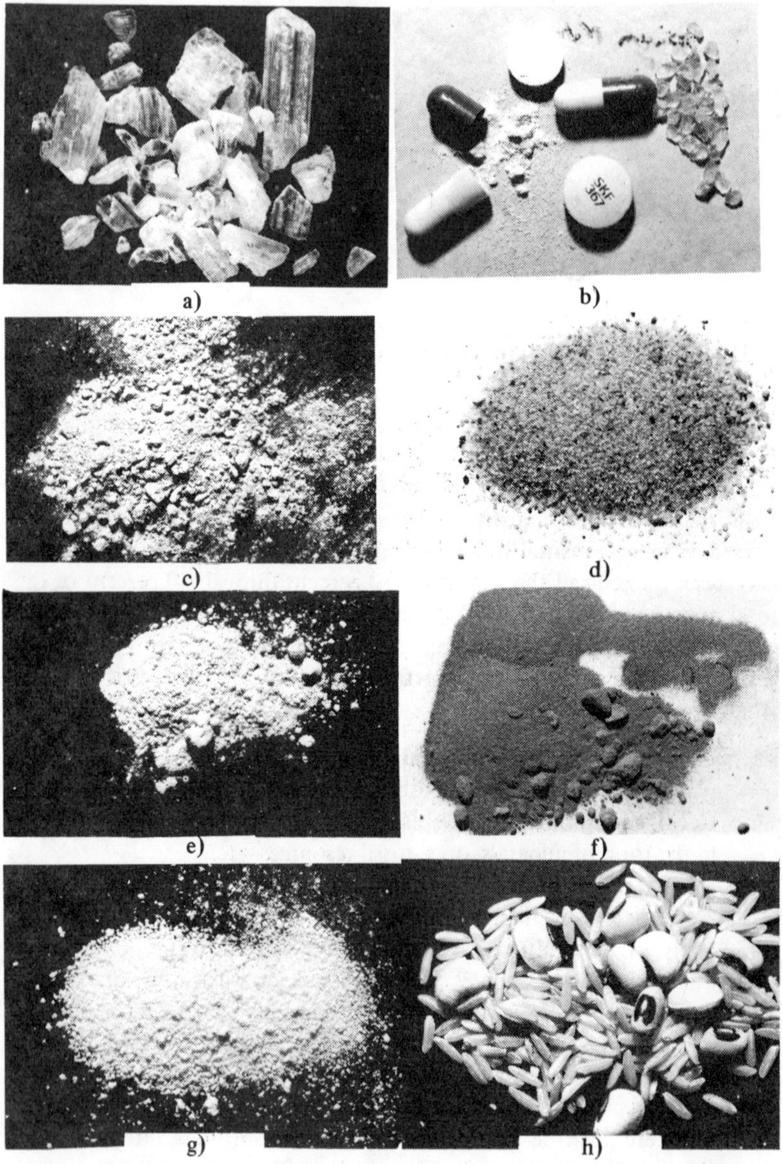

Fig. 3.1 Some Examples of Particulate Materials. a) Potassium nitrate crystals. b) Pill, panmycin, aspirin, erythromycin. c) Clay powder. d) Sand. e) Cement. f) Zinc powder. g) Household detergent. h) Legumes and rice.

the particles can be produced. The natural processes of growth and the process of combustion have already been mentioned. To these we can add: atomizing, spraying, comminution, precipitation, and electroplating. In present practice, atomizing processes usually utilize a second fluid in order to break up the subject fluid into droplets. These droplets subsequently freeze and form solid particles. The process of atomizing is popular in the metal powder industry. Spraying may or may not utilize a second fluid to break up the subject material. It is particularly popular in the chemical and related industries. Comminution is, of course, so widely practiced that it cannot be related to any one industrial group. However, the cement industry carries out a great deal of comminution as does the mineral processing industry. Nucleation and growth in solution is widely practiced in the chemical and related industries including fine chemicals, food technology, pharmaceuticals, etc. Some examples of particulate materials are shown in Fig. 3.1.

3.2 ATOMIZING OF METAL POWDERS[2]

The bulk of atomized metal powders is produced by the two fluid method of atomizing.[2] In this method the liquid to be atomized is broken up by the impact of another stream of fluid, either gas or liquid (the latter usually water). The general features of nozzle design are shown in Fig. 3.2. The flow of the liquid metal stream and that of the atomizing fluid are controlled quite independently. The metal usually flows into the atomizing zone under gravity and the atomizing fluid is introduced under pressure. Any number of jets of atomizing fluid may be used. The actual process occurring may be quite complicated, but this simple example illustrates the basic operations.[2]

When one jet alone is used, it may be in the form of a conical, annular, concentric jet through which the liquid metal stream flows. The actual atomizing occurs at the junction of the liquid metal and atomizing fluid streams. This is termed the primary stage of atomization. It is here that the molten metal stream is accelerated by the atomizing fluid. It is also cooled by the same jet, until freezing occurs. In this way, the metal stream is separated into discrete droplets which when deposited solidify to form small pieces of solid granular or powdered metal.

Air is often used in gas atomizing, although this tends to promote oxidation of the particle material. Water atomization produces a much more

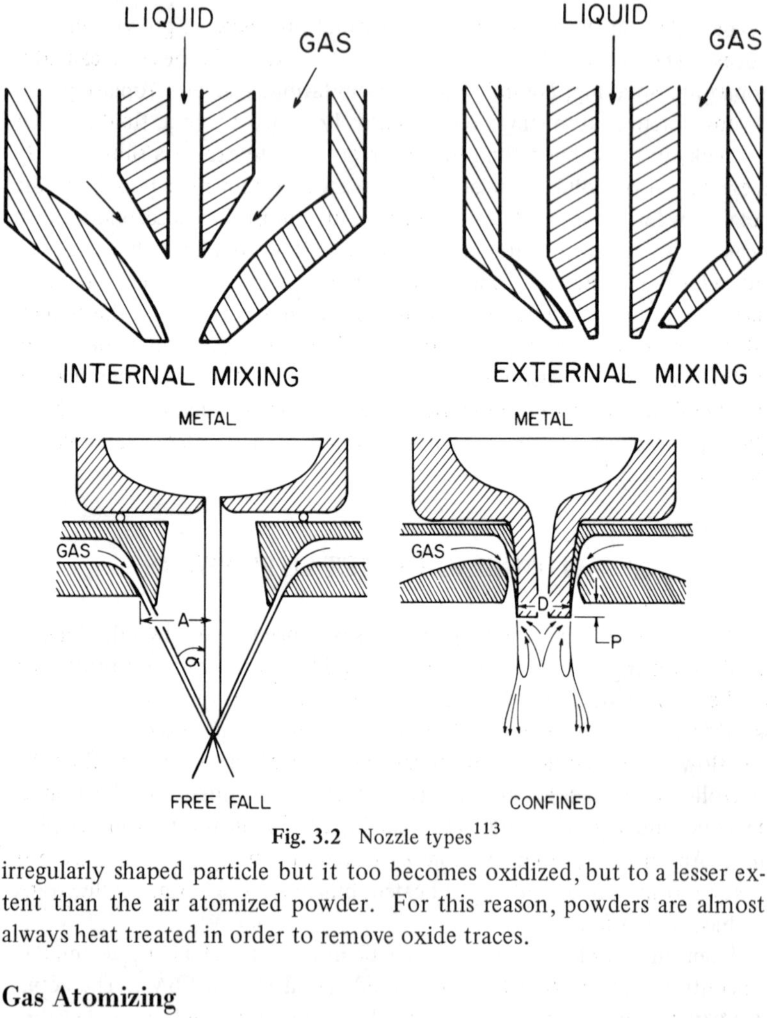

Fig. 3.2 Nozzle types[113]

irregularly shaped particle but it too becomes oxidized, but to a lesser extent than the air atomized powder. For this reason, powders are almost always heat treated in order to remove oxide traces.

Gas Atomizing

A multijet configuration has been used to model the atomization process.[3] This model predicted that at the intersection of the jet streams, powder particle diameter varies inversely with the jet velocity.[3] The graph of the data is shown in Fig. 3.3. A study of the influences of gas pressure, velocity, flow rate, jet length, and metal superheat (i.e., degrees above melting point) upon the metal particle shapes of cast iron, as characterized by a simple aspect ratio, showed that the two most important variables are the jet distance and the jet pressure.[4] Thus, increasing the jet distance or reducing the jet pressure or both promoted the production of more irregularly shaped particles. In other alloys that were atomized these changes did not produce a shape variation as they did cast iron. These other materials included: 1020 steel, 4620 steel, stainless steel, Inconel, and Cu.[4] These observations confirm an earlier hypothesis that above a certain critical level of surface tension, metal droplets spherodise into solid particles. Below this critical surface tension, irregular, solid particles may be formed.

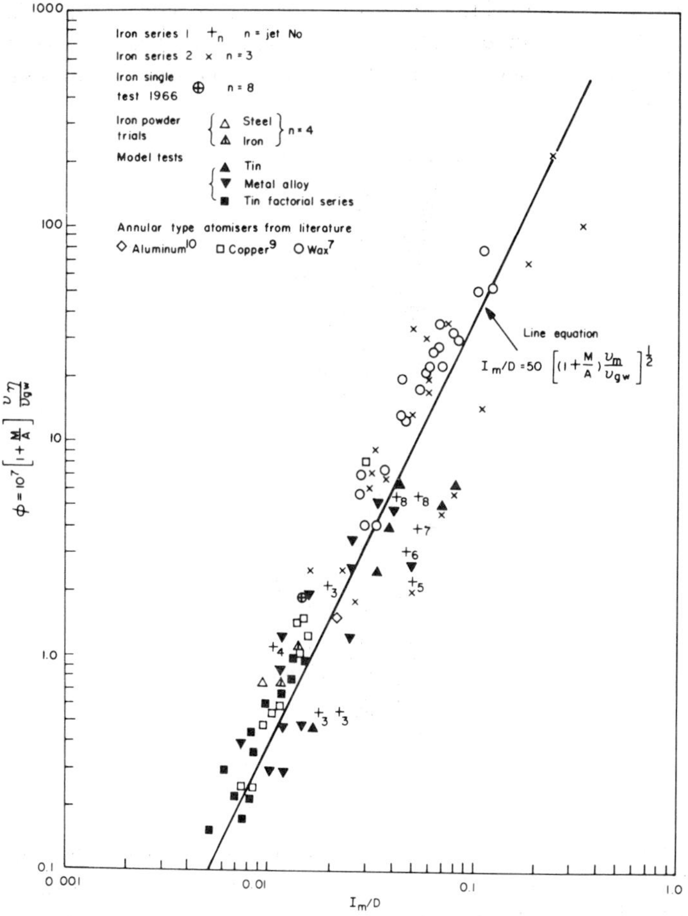

Fig. 3.3 Correlation of atomization data for liquid metals.[3]

Water Atomizing

The most common method for water atomization of metal powders is that which guides the flow of a liquid metal stream into the apex of two flat water jets. In the water atomization process as applied to liquid metals certain factors affect particle size and particle shape variously.[5] Finer particle sizes in the finished product were said to be favored by a short metal stream, a short jet length, and high water flow rate. A low metal flow rate also favored small particle sizes. Superheat of the liquid metal (and low metal viscosity) favored fine powder production. Spherical particle shapes were said to be produced by superheat, large apex angles, and narrow metal melting ranges. A reproduction of the flat jet stream arrangement is shown in Fig. 3.4.

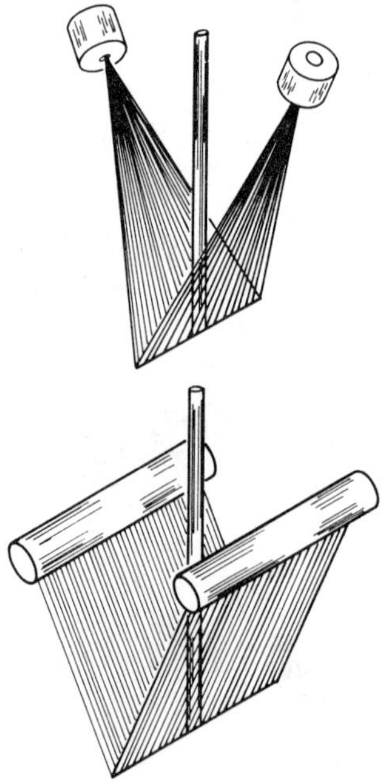

Fig. 3.4 Flatstream V-jets.[5]

A study of the water atomization of a number of liquid metals including 4620 steel, 304 stainless steel, copper, bismuth, alloy, and cast iron was made.[6] The general observations mainly dealt with particle size but some study of the factors affecting particle shape was also made. The most important variable in controlling particle size is water jet velocity. An equation is presented which relates these two as follows:

$$d = \frac{200{,}000}{V}\left(\frac{1}{\rho K}\right)^{1/3} \tag{3.1}$$

in which

d is the mass median particle size,
V is the water jet velocity,
ρ is the solid metal density, and
K is the energy transfer constant.

It was also observed that high water jet velocity and low liquid metal surface tension independently favor the formation of irregular particle shapes in the finished powder.

3.3 SPRAYING AND ATOMIZING

These processes have been used in various fields of application for many years but to a limited extent only. Spray drying became extensive after World War I, but spray atomizing did not arouse widespread interest until after World War II. There are three main types of atomizers: pressure nozzles, spinning discs, and pneumatic nozzles. These are described following a discussion of the principles of jet break up.

Jet Break Up

There are three fundamental characteristics of a fluid which markedly influence droplet formation density, viscosity, and surface tension. The most important single consequence of droplet formation is that it greatly increases the relative surface area of the fluid, particularly at small droplet sizes.

The first examination of jet disintegration assumed nonturbulence and no viscosity effect; that led to the conclusion that overcoming surface tension alone was responsible for break-up.[9] It was assumed that a lateral disturbance (see Fig. 3.5) would produce a droplet when its amplitude α became $D/2$, in which D equalled the diameter of the jet initially. This disturbance was believed to be in the form

$$\alpha = \alpha_0 e^{qt} \tag{3.2}$$

in which

α_0 = initial amplitude of the disturbance, and

q = a function of fluid tension, density, initial drop size, and wavelength λ of the disturbance.

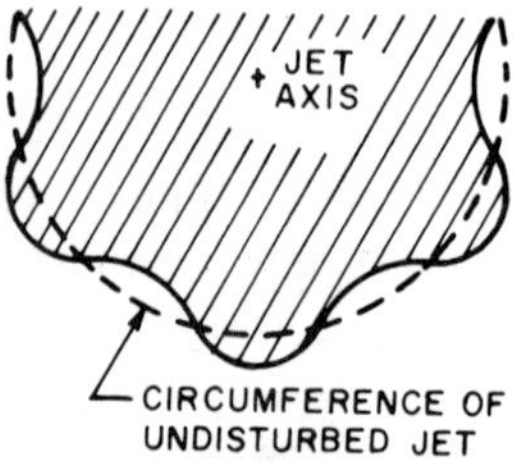

Fig. 3.5[7]

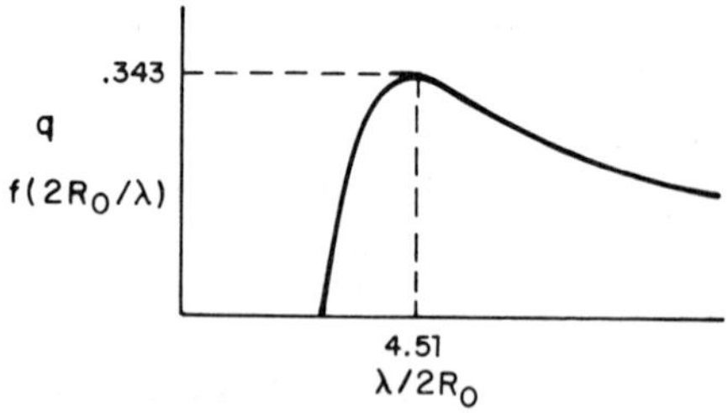

Fig. 3.6[7]

A plot of q versus λ/D in Fig. 3.6, shows that the disturbance is at a maximum at $\lambda/D = 4\frac{1}{2}$. In the case of water the critical λ/D ratio was shown to be 4.38,[9] and later it was shown to be 4.69.[10]

By equating the volume of the droplet formed, the mechanism of Ray-

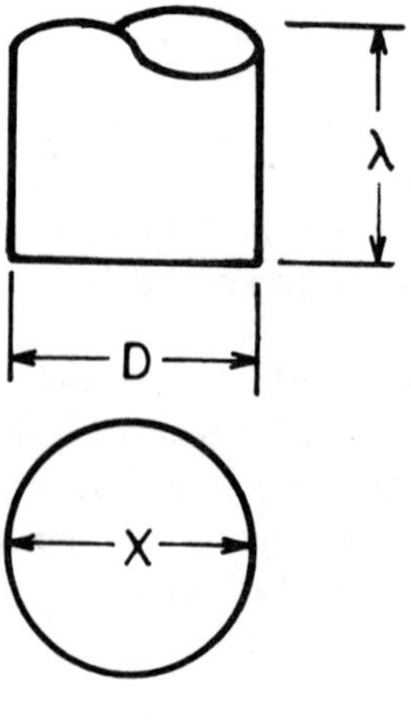

Fig. 3.7

leigh disintegration predicts an individual droplet diameter of approximately twice the diameter of the tube from which the liquid issues forth. The estimate is as follows (refer to Fig. 3.7):

Rayleigh disintegration predicts droplet diameter (X) to be

$$4.5D \frac{\pi D^2}{4} = \frac{4}{3} \frac{\pi X^3}{8}$$

$$X^3 = \frac{4.5D^3 \cdot 3}{2} = 6.75D^3$$

Therefore X is $\sim 1.9D$. (3.3)

Filaments and Fine Droplets

The original Rayleigh prediction is that the jet will break up into large diameter drops. The mechanism is one in which the droplets form by liquid flow away from the intervening regions. Experimental observation has since shown that a small filament of fluid is left behind by the flow and this filamentary material subsequently breaks up into small satellite drops.[11] Gerrard's photomicrograph is shown in Fig. 3.8 to illustrate this phenomenon. This observation explains the presence of very fine droplets which are otherwise inexplicable.

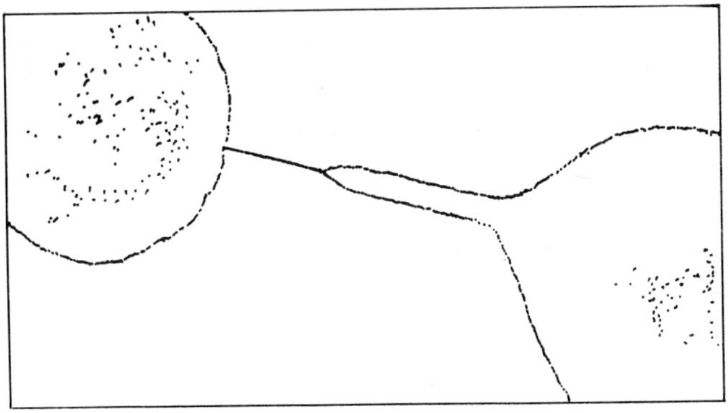

Fig. 3.8 Breakup. Filament (1–2 μ) forms between drops as liquid jet nears the breakup point.[21]

The effect of fluid viscosity has been calculated according to Weber[12] as shown:

$$\lambda/D = \pi(2)^{1/2}[1 + 3\mu/(\ell_L \sigma d_o)^{1/2}]^{1/2} \quad (3.4)$$

in which

λ is the fluid viscosity,
σ is the surface tension, and
d_o is the droplet diameter.

The Type of Disintegration of the Fluid

By the use of dimensional analysis it has been demonstrated and experimentally confirmed,[13] that jet disintegration may occur in one of three regimes:

1. Rayleigh or varicose disintegration under conditions where both the Reynolds number and Weber numbers are low,
2. true atomization in conditions where both numbers are high, or
3. a transition zone of sinuous disintegration at intermediate levels of Weber and Reynolds number.

The details of the dimensional analysis and Ohnesorge's chart of results confirming the hypothesis are given below:

Dimensional analysis of jet disintegration

$\ell \overset{\partial}{=} ML^{-3}$	density
$\mu \overset{\partial}{=} ML^{-1}T^{-1}$	viscosity
$\sigma \overset{\partial}{=} MT^{-2}$	surface tension
$\nu \overset{\partial}{=} LT^{-1}$	velocity
$D \overset{\partial}{=} L$	diameter

where $\overset{\partial}{=}$ indicates "dimensionally equal to."

Following the method of Bingham

$$F(\ell\mu\sigma\nu D) = 0$$

$$M^\circ L^\circ T^\circ \overset{\partial}{=} (ML^{-3})^{a_1} (ML^{-1}T^{-1})^{a_2} (MT^{-2})^{a_3} (LT^{-1})^{a_4} (L)^{a_5}$$

Collecting terms of like dimension

$$M \qquad a_1 + a_2 + a_3 = 0$$

$$L \qquad -3a_1 - a_2 + a_4 + a_5 = 0$$

$$T \qquad -a_2 - 2a_3 - a_4 = 0$$

Solving for a_1, a_4, a_5

$$a_1 = -a_2 - a_3$$
$$a_4 = -a_2 - 2a_3$$
$$a_5 = -a_2 - a_3$$

$$\ell^{-a_2-a_3} \mu^{a_2} \sigma^{a_3} v^{-a_2-2a_3} D^{-a_2-a_3} \quad (\mu/\ell v D)^{a_2} (\sigma/\ell v^2 D)^{a_3}$$

Thus dimensional analysis shows that two ratios are important in jet break up: the Reynolds number and the Weber number

$$We = \ell_c v^2 D/\sigma \tag{3.5}$$

$$Re = vD\ell/\mu \tag{3.6}$$

The ratio of these two numbers, termed Z by Ohnesorge, is plotted against Reynold's number in Fig. 3.9. Zone I has an axially symmetric disturbance and Zone II has a helically symmetrical disturbance.

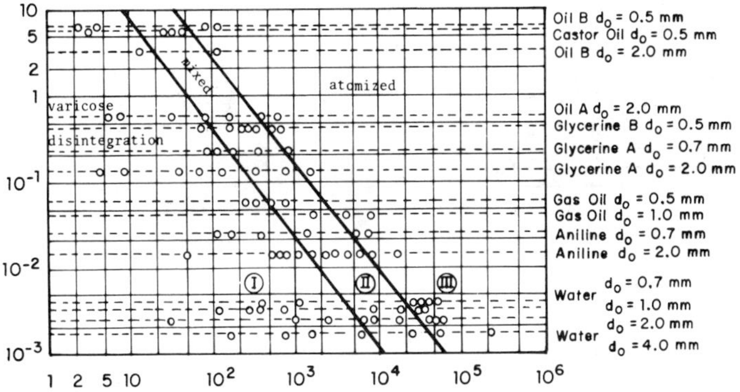

Fig. 3.9 Chart due to Ohnesorge showing jet breakup as a function of Reynolds number and liquid properties. Region I corresponds to Rayleigh breakup.[7]

Efficiency

The efficiency of droplet production by spray atomization is very low, not only because the liquid viscosity is not negligible but also because the droplets have to be separated from each other so that they do not collide or otherwise recombine in some way (thus defeating the purpose of the procedure). It has also been calculated that to produce particle sizes of less than 10 microns by spray atomization would require a very large and prohibitive amount of energy.[14] It is for this reason that small particle sizes are manufactured with pneumatic atomization. Even there, the efficiency is low, and it has recently been observed that the process of atomization is expensive in the case of metal powder production and that new, cheaper, and better production processes are needed.[15]

Swirl Nozzles

A variety of designs of swirl nozzles are utilized (e.g., Fig. 3.10 and Fig. 3.11) .

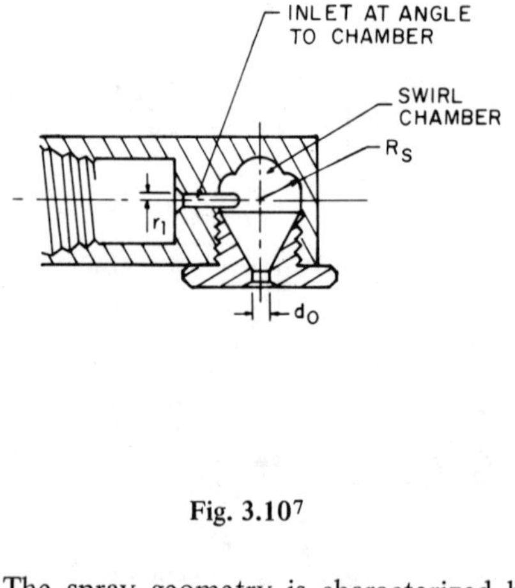

Fig. 3.10[7]

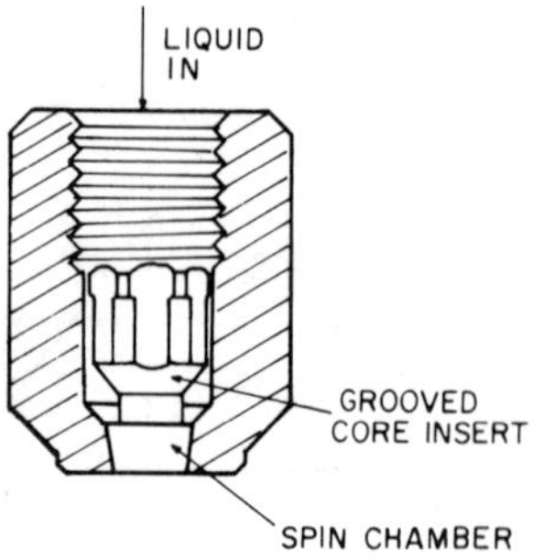

Fig. 3.11[7]

The spray geometry is characterized by a conical sheet of fluid, with its apex at the orifice which contains a cone of air. For a free vortex the torque = d/dt (MVR) = 0; therefore $v \alpha 1/r$, i.e., as $r \to 0$, $v \to \infty$ which is clearly impossible; therefore, the vortex opens up to form a hollow cone as observed. Experimental observations, which are confirmed by theoretical considerations, show that the air core is initiated as a boundary layer in the swirl chamber and nozzle prior to issuing forth from the orifice. A detailed review is presented elsewhere.[7] As the viscosity of the fluid is increased, to achieve atomization the pressure has to be increased in order to increase the velocity of flow.

Spinning Disc Atomizers

In this method the liquid is accelerated by centrifugal means, rather than by pressure as in spray atomizing. The rate of atomization can be altered, within limits, without changing other factors whereas in spray atomizing the pressure or orifice size must be changed. The spray normally appears as an open umbrella. Theoretical calculations for droplet formation have not been particularly useful because of disturbances caused by the pumping action of the disc; and the drag between the disc and fluid, the fluid and air, respectively. Experimental work has revealed three distinct mechanisms of jet formation.[16] These are: drop, ligament, and film formation. Drop and ligament formation probably correspond to 1 and 2 of the Ohnesorge diagram, Fig. 3.9. It was found that as the speed of rotation is increased, the film begins to break nearer and nearer to the disc perimeter. It was postulated that film fracture occurs at less than a certain minimum thickness.

At low feed rate, size is inversely proportional to centrifugal force; at high feed rate, size is inversely proportional to peripheral speed.[17]

Pneumatic Atomization

This method is used in paint spraying, metal powder atomizing, carburetors and some small scale drying applications. It is particularly useful for producing fine droplet sizes ($< 50\mu$). The mechanism of droplet formation is that the friction or drag at the liquid-air interface tears the descending jet of liquid into filaments or ligaments. These break up into small droplets as in Rayleigh break-up.[18,19] Theoretical calculations of droplet size fail to predict the production of the really fine droplets occurring in this method,[20] because the mechanism of jet disintegration is Rayleigh-like break-up. Therefore, a small drop of say 5μ (the minimum size predicted in references 19 and 20) fractures to produce 1 μ size drops and even smaller. The disintegration mechanism is shown in Fig. 3.12 where a small isolated drop in a high velocity air stream dishes and blows up into a bag which subsequently collapses into a bracelet of fine droplets.

This Rayleigh type of break-up was predicted some 40 years ago[21] by a simple utilization of Lord Rayleigh's argument. But the flash photography techniques were insufficiently sensitive at that time to confirm the observation. This is shown in Fig. 3.12.

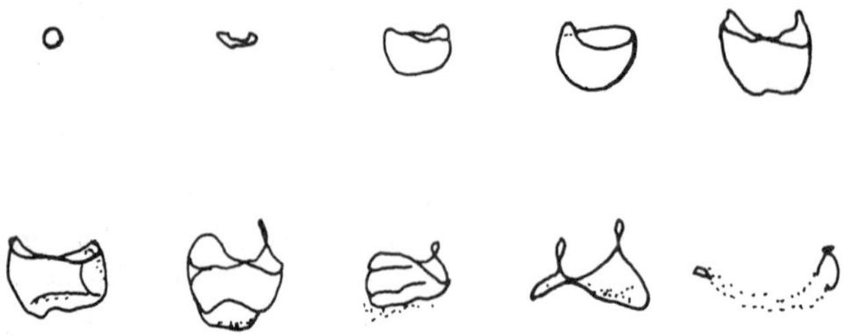

Fig. 3.12 High speed photographs by Lane[20] showing the break-up mechanism of a drop suddenly impacted by a high-velocity jet of air.

From Eq. (3.3) it is expected that if the droplet diameter is small, around 5 microns, then the ligament is approximately 10^{-3} cm. Substituting this value as the value of α in Eq. (3.2), and assuming that α_o is of the order of an Angstrom (i.e., at $t = 0$, amplitude of disturbance is zero), and q_{max} is 0.345 $(8\sigma/\rho d^3)^{1/2}$, this value of q_{max} corresponds to t_{max} which is the total length of time that the filament exists. By substituting the figures quoted, the length of time that a filament exists is of the order of 10^{-4} seconds. This period of time is so short that it allows quenching rates high enough to affect the dendritic spacing[22] but not high enough to produce metastable structures that occur in splat cooling.[23]

Splat Cooling

In splat cooling, cooling rates in excess of 10^6 °C per second can be achieved. An early example of a device for achieving this operation is shown in Fig. 3.13. Some examples of the metastable structures that have been obtained are given in Table 3.1. The affects of such high rates of cooling include: the alteration of the normal solid solubility relationship,[24] the stabilization of a new structure for use as a superconductor, etc.;[25] and the production of alloys in true amorphous form (e.g., Te–Ge alloys which are amorphous and stable at room temperature but crystallize when heated by a dendritic method of growth). In a recent book, Martin and Doherty have observed that the properties of metastable phases must be studied and understood if the potential of these materials is to be realized commercially in the future.[26]

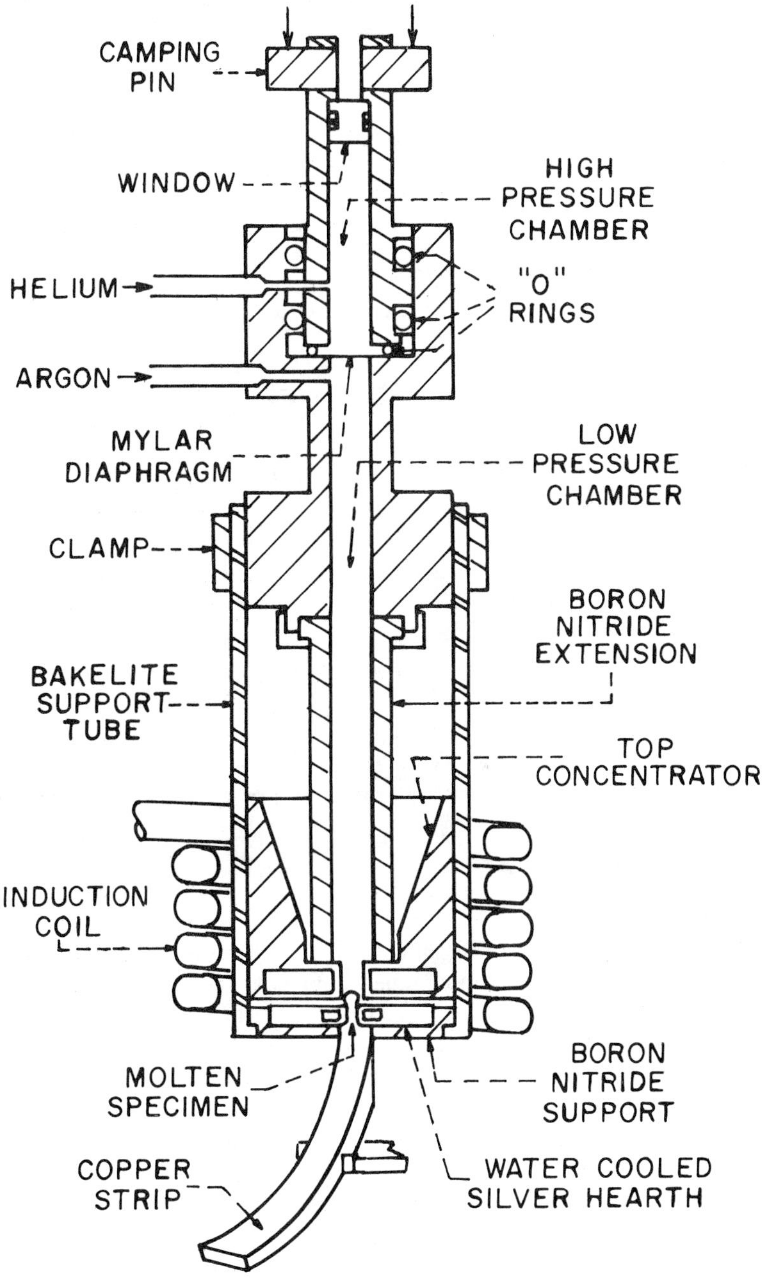

Fig. 3.13 Schematic of the rapid-quenching apparatus.[23]

Table 3.1

SUPERCONDUCTING TRANSITION TEMPERATURES AND STRUCTURES FOR SOME METASTABLE COMPOUNDS[25]

Alloy	*Equilibrium Structure*	*Transition Temp* (°K)	*Rapid-Quench Structure*	*Transition Temp* (°K)
WC	Hexagonal (B_h)	N(0.3)	*fcc*(B1)	10.0[8]
MoC	Hexagonal	9.26[9]	*fcc*(B1) a = 4.2777±5Å	14.3[8]
MnC	Two phase Mn_7C_3 trigonal + graphite	N(1.8)	*hcp* a = 2.778±2Å c = 4.475±4Å c/a = 1.61	N(1.7)
$(Mo_{0.98}Mn_{0.02})C$	–	–	*fcc*(B1) a = 4.275±1Å	N(1.7)
Nb_3Al	Cubic (A15)	18.0[10]	*bcc*(A2) a = 3.274±3Å	3.1
Ta_3Al	a phase ($D8_b$)	N(1.02)[11]	*bcc*(A2) a = 3.283±1Å	1.59
Ta_3Au	Two-phase complex	0.97, 0.4	*bcc*(A2) a = 3.283±5Å	0.82

Spray Drying

This process is carried out in order to obtain a granular product either from a solution or from a suspension. It consists of a simultaneous heat transfer, from a hot gas to the liquid droplet surface, coupled with mass transfer—the solvent from the droplet surface to the gas stream. A series of high-speed photomicrographs is shown in Fig. 3.14 in which a droplet of salt solution is allowed to evaporate in a suspended position. The variation of droplet temperature and size with time is shown in Fig. 3.15.[7]

Neither the rate of evaporation nor the rate of heat flow is uniform around the droplet. In fact, it is known from boundary layer theory that the rate of evaporation will be a maximum on the area facing the incoming gas stream, it will be a minimum near to the boundary layer separation point and it will rise to a maximum that is smaller than the first one on the surface furthest away from the incoming gas stream. This has been confirmed experimentally as is shown in Fig. 3.16.[27]

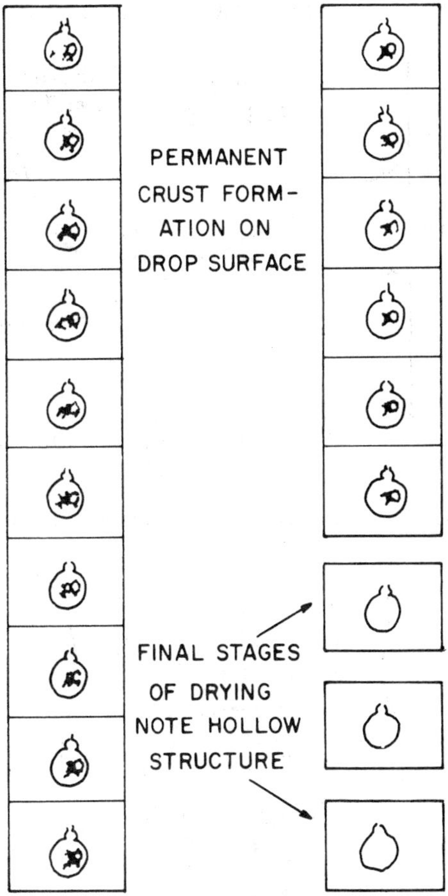

Fig. 3.14 Frames of a motion picture of an evaporating drop of sodium chloride illustrating crust formation and final structure.[7]

Drag Factor Plot

The sketches shown in Fig. 3.17 illustrate in a qualitative way the variations in the boundary layer around the drop as a function of the Reynolds Number on a so-called "drag factor plot." The flow lines of the gas stream

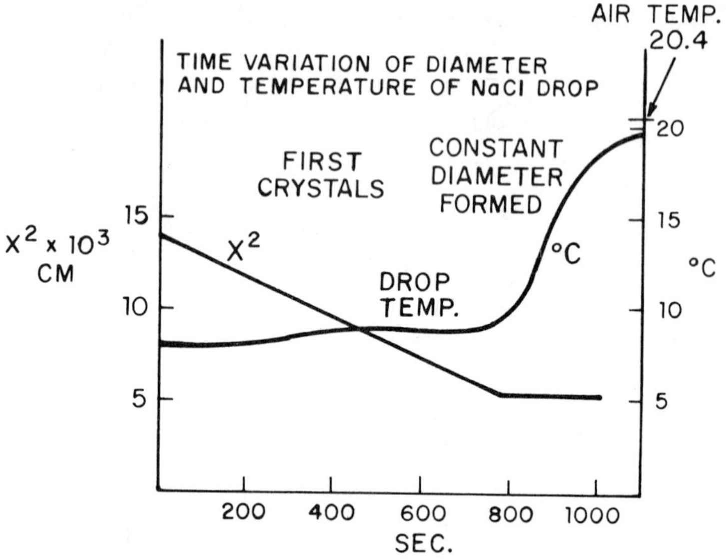

Fig. 3.15 Variation in drop diameter and temperature in the drying of a drop of sodium chloride.[7]

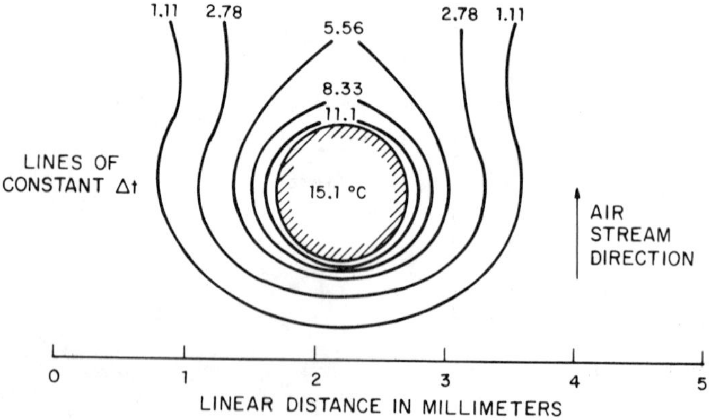

Fig. 3.16 Temperature exploration around an evaporating water drop.[27]

are drawn in, as are the shadings which are intended to show the nonuniformity of heat and mass transfer around the drop, as a function of the Reynolds Number.

A more analytical insight is given by the relationships between the Nus-

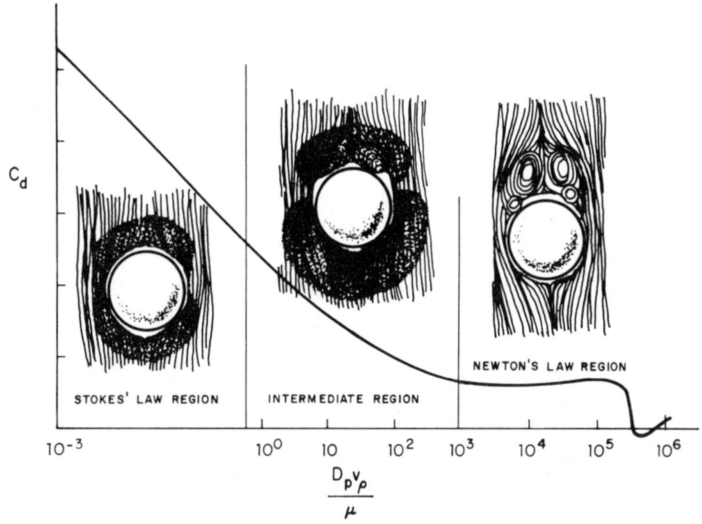

Fig. 3.17 Nonuniformity of heat and mass transfer around an evaporating drop suggested on a drag-factor plot. The flow patterns set up around the drop in each region create variations in the mass and heat transfer rates over the drop surface.[7]

selt, Prandtl, Schmidt, and Reynolds Numbers

$$N_{\mathrm{Nu}} = 2 + 0.6 N_{\mathrm{Sc}}^{1/3} \cdot N_{\mathrm{Re}}^{1/2} \tag{3.7}$$

and

$$N_{\mathrm{Nu}}' = 2 + 0.6 N_{\mathrm{Pr}}^{1/3} \cdot N_{\mathrm{Re}}^{1/2} \; . \tag{3.8}$$

A more detailed version of the two equations above is given below. For more information the reader should consult reference 7 et seq.

$$N_{\mathrm{Nu}} = \frac{h_{cx}}{k_f} = 2 + 0.6\left(\frac{c_p\mu}{k_f}\right)^{1/3}\left(\frac{v_a\rho x}{\mu}\right)^{1/2} \tag{3.9}$$

Eq. (3.9) deals with heat transfer;

$$N_{\mathrm{Nu}}' = \frac{k_g M_m x P_f}{D_v\rho} = 2 + 0.6\left(\frac{D_v\rho}{\mu}\right)^{1/3}\left(\frac{V_a\rho x}{\mu}\right)^{1/2} \tag{3.10}$$

Eq. (3.10) deals with mass transfer. Both equations are applicable over a range of conditions.[7]

In these equations

h_c = convection heat transfer coefficient,
x = droplet diameter,
k_f = average thermal conductivity of gas film,
k_g = mass transfer coefficient,
M_m = molecular weight,
P_f = average vapor pressure of a nondiffusant (dimensionless),
D_v = diffusion coefficient,
c_p = heat capacity at constant pressure,
μ = viscosity,
ρ = density, and
V_a = air–liquid relative velocity.

Spray Dryer Flow Patterns

In an actual spray dryer, the flow may be either horizontal or vertical. In general there are three major classes of spray dryer designs:

1. concurrent spray dryers (Fig. 3.18, not to scale),[7]

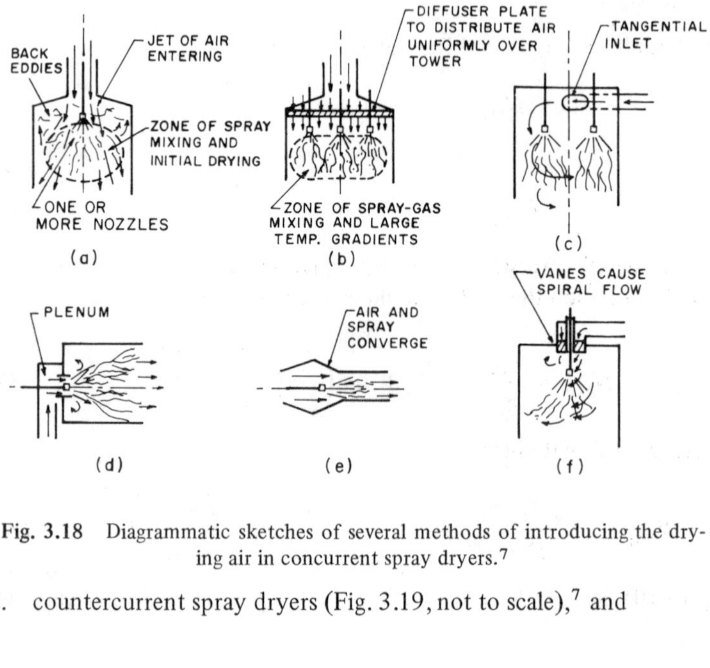

Fig. 3.18 Diagrammatic sketches of several methods of introducing the drying air in concurrent spray dryers.[7]

2. countercurrent spray dryers (Fig. 3.19, not to scale),[7] and

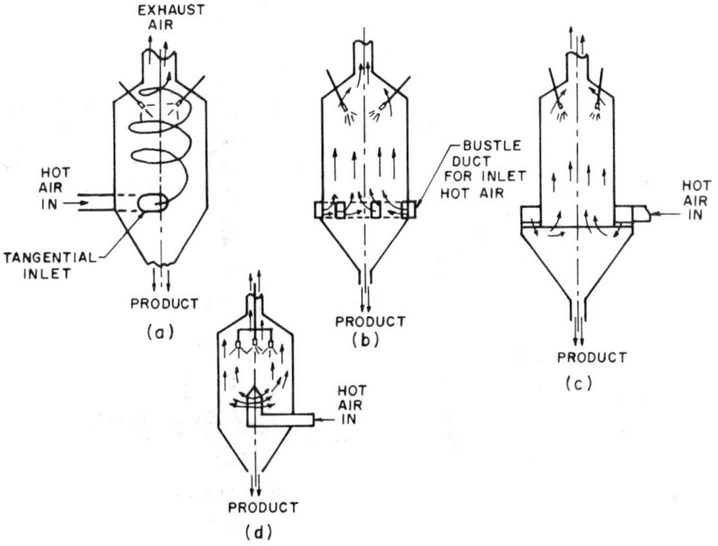

Fig. 3.19 Sketches of various inlet arrangements for countercurrent air flow in spray dryers.[7]

3. mixed current spray dryers (Fig. 3.20, not to scale).[7]

Air-Product Removal

Sketches showing possible ways of effecting air-product removal from the spray dryer are shown in Fig. 3.21.[7]

But once removed from the spray dryer, the particles and the gas phase have to be separated so that the product can be recovered. Various ways for achieving this separation are suggested in Fig. 3.22.[7]

Because there are so many ways of doing spray-drying, there is some divergence in results when the effects of processing variables upon product properties are ascertained. For example, a most useful and often quoted property is that of bulk density. It is generally agreed that increasing the air temperature decreases the bulk density of the product. However, it is not at all clear what the effects of either feed concentration or solid temperature upon product bulk density are.[7]

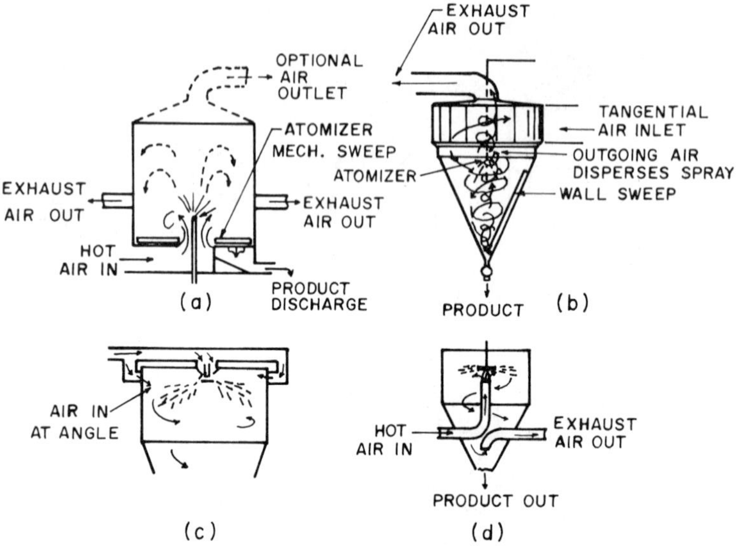

Fig. 3.20 Various air-inlet arrangements for mixed-flow spray dryers.[7]

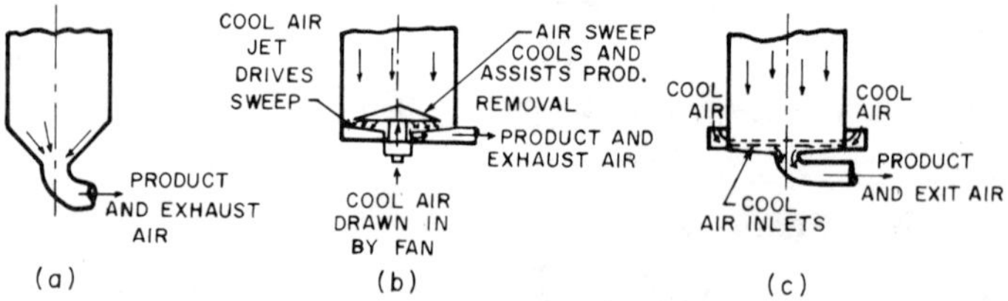

Fig. 3.21 Diagrammatic sketches of air-product removal from a common outlet. Sketches (b) and (c) suggest methods of cooling products with secondary air.[7]

3.4 COMMINUTION

The word comminution is a rather general term for size reduction in so far as it does not signify the mode of size reduction actually occurring. For example, two subterms might be stated as "crushing" and "grinding." These two words are intended to convey the mechanisms of fracture by compression and fracture by shear, respectively. Analysis of modes of size reduction indicates that any single or multiple of the following modes might be operative: tension, compression, torsion, shear, flexure, and attrition. These may or may not incorporate interactive effects with the environment (e.g., the presence of liquid wetting agents).

A general scheme of classification for size reduction or comminution equipment is shown in Table 3.2.

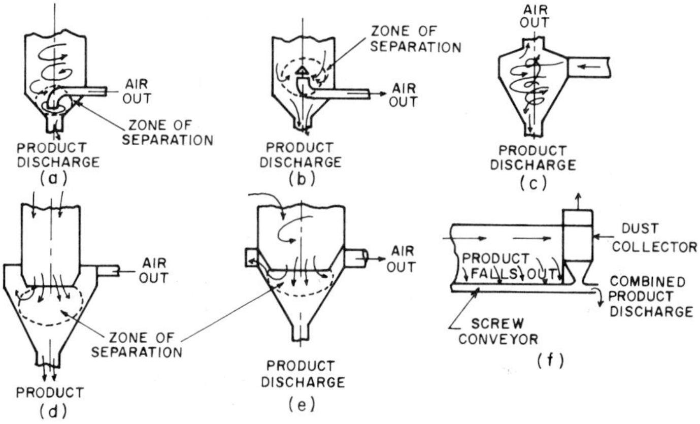

Fig. 3.22 Various schemes for effecting a product separation from the exhaust air in spray dryers.[7]

Table 3.2

CLASSIFICATION OF COMMINUTION EQUIPMENT[28]

Compression or nipping	*Blow or impact*	*Tumbling*	*Cutting or shredding*	*Attrition*
Jaw	Hammer	Rod-loaded	Shears	Buhrstone
Cyratory	Pin	Ball-loaded	Dicers	Colloid
Roll	Stamp	Autogenous		Fluid-energy
Disc	Vibration			Sand
Pan				
Ring roll				
Ring ball				

Reprinted with permission of G. C. Lowrison, *Crushing and Grinding,* CRC Press,1974.

Choice of Equipment

There are numerous factors or variables that must be considered in trying to decide what type of equipment should be chosen for any particular comminution operation. These factors are: the necessity (it is not only a costly process in terms of money, it is also costly in terms of energy), the degree of size reduction, the rate of comminution, the size of the feed stock, the material properties of the feed, moisture, temperature, chemical reactivity, and location of the equipment. These are briefly discussed:

Necessity. Quite apart from financial and energy costs, it may be more advantageous to buy already sized feed or to alter some other stage of the processing, in order to avoid comminution altogether.

Reduction. Size reduction ratios may vary from less than 5 to more than 500. It is particularly important that the size range be specified and that the mean size be actually required for sound technical and/or economic reasons.

Rate. The rate of comminution depends upon a variety of factors such as the fluctuation of the supply level of incoming feed stock, the availability of storage, and the feasibility of storage.

Feed Stock. This may be coarse or fine, it may have a broad size range, the shapes may be many and varied: there may be very wide fluctuations within the feed stock regarding all of these properties. It is possible that any moisture present may promote serious handling difficulties and even more serious storage difficulties. Really difficult problems may arise in the case of feed stock that is reactive. For example, the plant construction itself may be corroded. One convenient way of defining the material properties of the feed stock is by means of Mohrs Hardness. Last but not least, comminution generally raises the feed stock temperature and this may change the substance in some undesirable way.

The above considerations have to be considered carefully in each individual case and Table 3.3 may be useful in this regard. Following preliminary evaluation, actual trials should be conducted before a final decision is made. It is advisable to consider the extreme effects during these trials in order to be able to assess the severity of the problems that can be experienced in the full scale plant.

Nipping Machines: Jaw Crushers

There is a great variety of these jaw crushers: Blake, Dodge, and Universal. The Blake Crusher consists of two jaws, one fixed and one move-

Table 3.3[28]

THE LOWRISON CHART

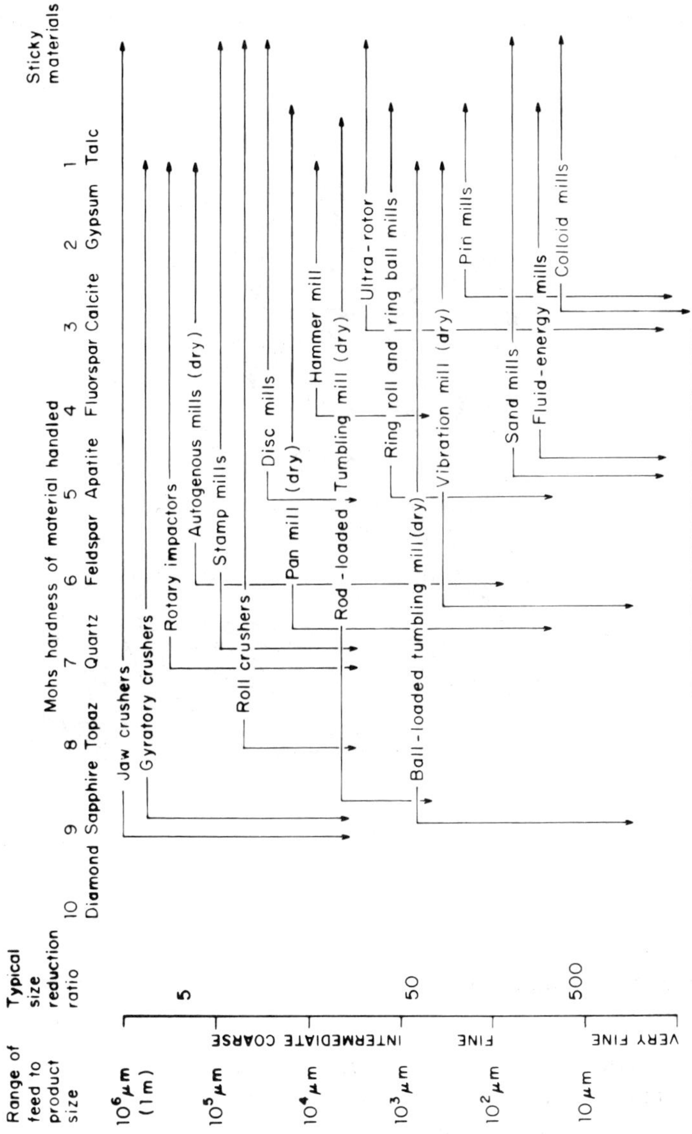

Reprinted with permission of G.C. Lowrison, *Crushing and Grinding*, CRC Press, 1974.

able. The latter is pivoted at the top so that the feed intake area is fixed. In the Dodge Crusher, the outlet area is fixed. The Universal Crusher has both areas variable. The Blake Crusher may be either a double toggle or a single toggle type. (The toggle is a tie-cum-push bar.) The single toggle Blake is shown in Fig. 3.23.

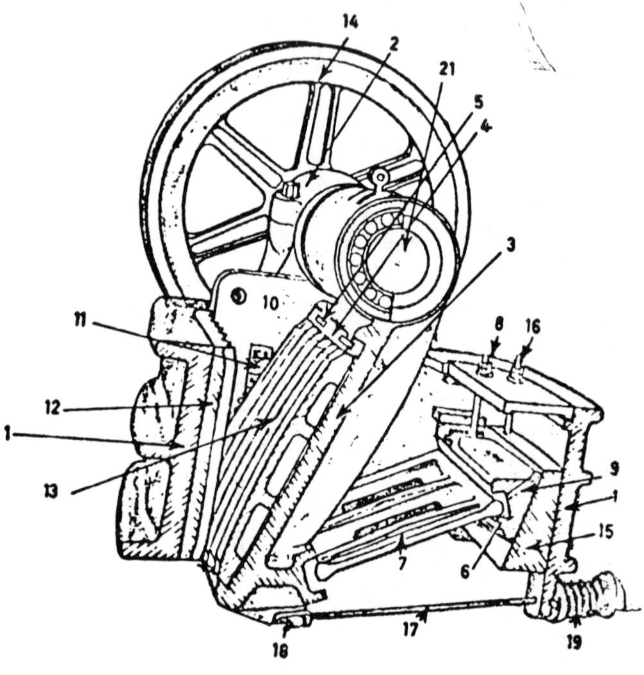

Fig. 3.23[112] The single toggle Blake (Courtesy Newell Dunford Engineering, Ltd.).

It is interesting to note that for the same reduction ratio, the Blake Crusher out-performs the Dodge and Universal. For example, for the same kW rating, its output per hour is twice that ot the other two crushers (ratio of 10, kW rating of 45, 25 tons per hour versus 50 tons per hour).

(a) Gyratory Crushers. These crushers are constructed to handle up to 5000 tons per hour. They occupy small headroom, are adjustable for product size, crush wet (but not sticky) material, and produce sharply angled particles. An example is shown in Fig. 3.24.

(b) The Bond and Wang Relationship. The power consumed per ton of

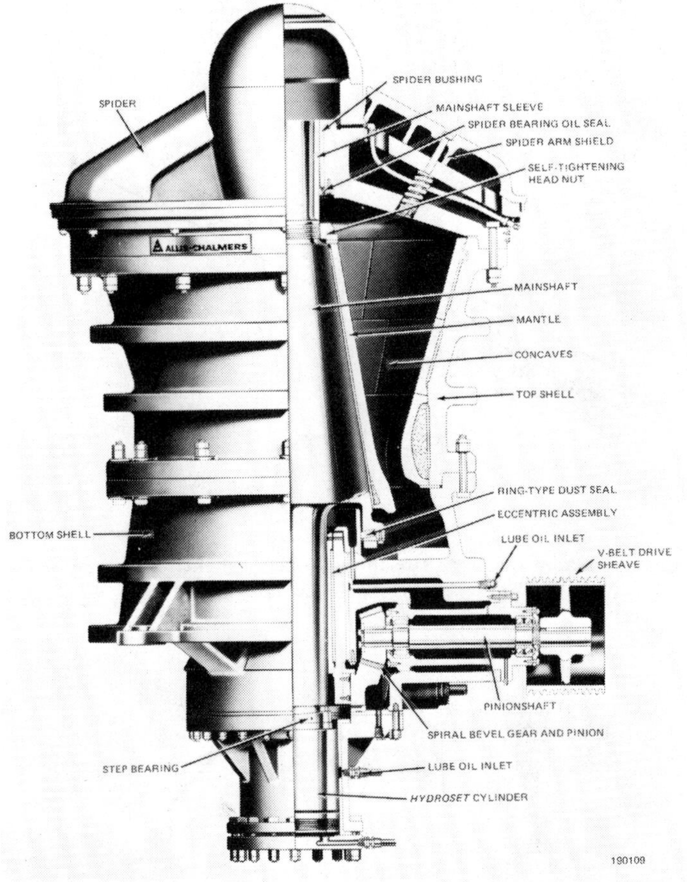

Fig. 3.24[115] Hydroset mechanism (Courtesy Allis-Chalmers Ltd).

material comminuted varies according to the ratio $(n^{1/2}P)^{1/2}$ where n is the reduction ratio and P is the product size.[29] As shown in Fig. 3.25, this relationship appears to hold over a wide range of equipment types.

(c) Roll Crushers. These consist essentially of at least one rotating cylinder. The feed stock is fed down through a gap between the surface of the rotating cylinder(s) and abutting surface(s). A diagram illustrating a general arrangement of such a device is shown in Fig. 3.26.

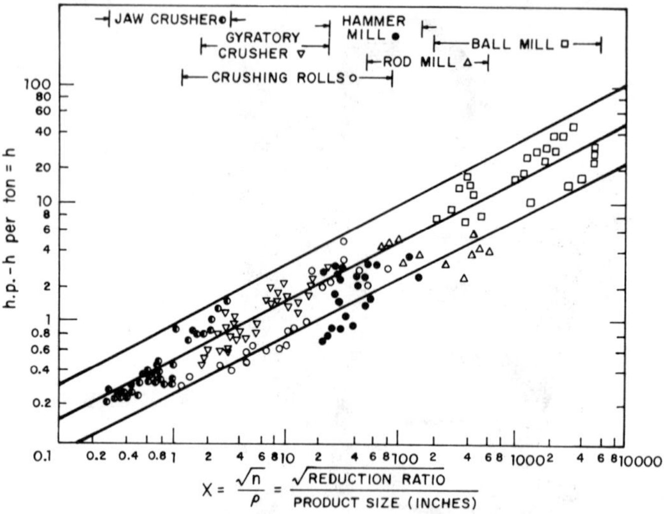

Fig. 3.25 Bond and Wang's relationship between energy consumption by equipment type and reduction ratio.[29]

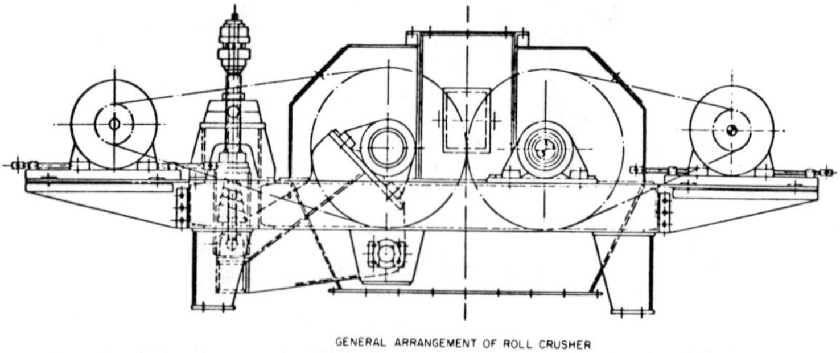

Fig. 3.26 General arrangement of roll crusher (Courtesy Aulman and Beckschulte, 5275 Bergneustadt 1, W. Germany).

One clear advantage that this machine has is that it will handle sticky material (even frozen feed stock can be handled). Another advantage is that the size distribution may be quite narrow (say 4 to 1). However, the usual degree of size reduction is only about 4, although high yields of up to 85% may be obtained and still within a narrow size range. For feed stock that might tend to slip through between the crushing surfaces (such as flakey materials, for example) the rolls may be roughened up by providing them with teeth. A comparison between the Jaw, Gyratory and Roll crushers is given in Table 3.4.[28]

Pan Mills. The pan mill is a very flexible machine which is widely used in many industries. It lends itself to small scale operation also. A sketch of a mill with a rim discharge is shown in Fig. 3.27. The process may be conducted wet or dry; the load (the crushing load) on the rolls may be varied; the comminution occurs by a combination of rolling and sliding and the depth of feed is controlled by the rate of feed coupled with the take-off of finished product. There have been many investigations into the operation of this device (see elsewhere for a comprehensive report of these,

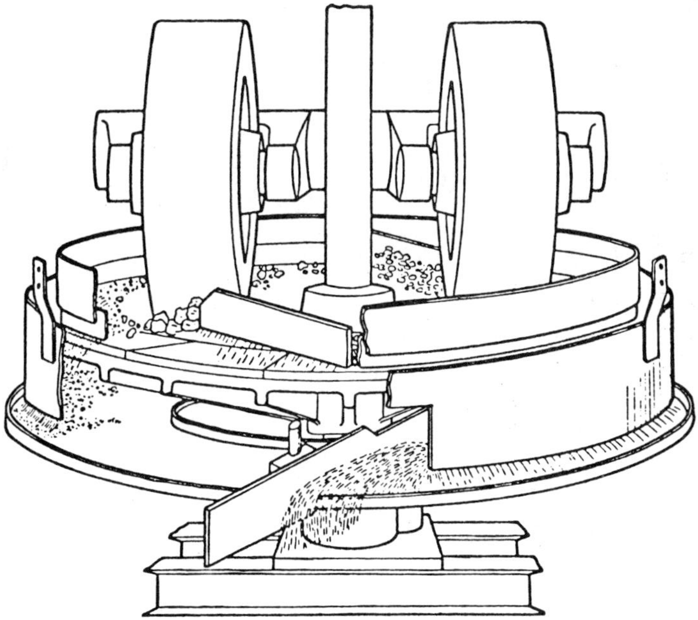

Fig. 3.27[116] Pan mill with rim discharge (Courtesy Craven Fawcett (Wakefield) Limited, formerly Bradley and Craven, Limited).

Table 3.4

COMPARISON OF JAW, GYRATORY AND ROLL CRUSHERS BASED UPON SAME FEED AND SAME SIZE OF PRODUCT[28]

Per unit quantity of product	*Jaw*	*Gyratory*	*Roll*
Capital cost including installation	–	Lowest	Highest
Weight (proportional to installation)	Heaviest	–	Lightest
Power consumption	Highest	–	Lowest: balanced mechanism, low starting peak
Idling load percent of total	45	30	30
Headroom required	–	Highest (maintenance to lift out head)	–
Lubrication	Least simple	–	Most simple
Robustness	–	–	Most robust
Type of material processed	Blocks, massive rocks. Clayey, sticky, spongey rocks can be handled. Disposition to bridge. Product tends to be tabular	Slabs and stratified stone handled. Product tends to be cubical	Any kind of material. Shape can be controlled by type of roll surface
Feeding	–	Simplest: can be buried in feed, if necessary, and always fed from two sides	Most fastidious: requires very even feed across rolls
Wear and maintenance	Retains adjustment even when worn. Wear is localized and intense. Requires crane with two-way travel	Maintenance costs highest	Easy to maintain *in situ*. Maintenance costs lowest
Safety guarding	–	Least extensive	–

Reprinted with permission of G. C. Lowrison, *Crushing and Grinding*, CRC Press, 1974.

reference 28). One result that is of interest is the relationship of the roll pressure to the finished properties:[30]

$$P = C\left(\frac{Q}{BC[D(1 - .33\mu)]^{1/2}}\right)^{\mu/\mu + 0.5} \tag{3.11}$$

in which

P is the roll pressure,
C and μ are constants whose values depend upon the pan loading conditions,
Q is the weight of the roll,
D is the diameter of the roll,
B is the width of the roll, and
$P = Cy^{\mu}$, where y is the feed stock compressibility.

The power requirements for an edge runner were not included in the data of Bond and Wang shown in Fig. 3.25, but, at a ratio of 4, they are on the high side of the data band shown.

Impacting Machines

These machines, particularly the rotary type (see Table 3.5 for the classification[28]) are very widely used. They are intermediate as a class in energy consumption and in their reduction ratio, as shown in Fig. 3.25.

Rotary Impact Mills. Rotary impact mills are manufactured with any desired capacity, from the very little to as high as 2000 tons per hour. One particular advantage of the process of high impact is that there is apparently little energy stored in the material of the feed stock, as compared with the nipping machines.[28] Most materials can be comminuted in rotary impact machines, with few exceptions. Perhaps the single most important limitation is that hard feed stock may cause too much wear on the equipment. There are some possibilities for alleviating this problem with hard facing in some portions of the equipment. Other methods, including construction material substitution for particularly difficult areas may be practiced. Caution is necessary because there is always a possibility of dust explosion if sparks created by the impacting process are allowed to escape from the equipment to a locality in which ignitable dust is present.

Rotary Hammer Machines. Two types of these are illustrated in Fig. 3.28. On the left-hand side is shown a common swing hammer mill. On the right-hand side is shown the familiar garbage disposal type of macerator. It should be noted in Fig. 3.25, that the rotary hammer mill has the highest reduction ratio next to the tumbling mills, for high tonnage mills. The size of the product may be produced as small as 100μ and by appropriate recycling even smaller than this (i.e., by sieving the product and recycling the oversize).

Table 3.5

CLASSIFICATION OF IMPACT MILLS[28]

IMPACT MILLS		
ROTARY		OTHER
Nature of rotor	*Additional breaking device*	
Impactors set round periphery of rotor disc:		Stamp, Vibration
Fixed hammers	Breaking plate	
Pivoted or swing hammers	Breaking bars or grid	
Short chains	Drum	
Rotor in form of drum:	Cage	
Rotor in form of cage:	Concentric cages	
Pieces set over surface of drum or disc		
Lugs		
Rods (pins)		

Reprinted with permission of G.C. Lowrison, *Crushing and Grinding*, CRC, 1974.

Other Impact Mills. Two other impact mills of particular interest are the pin mill and the vibratory mill. The pin mill consists of two discs on which square pins are set in concentric rows, each of which is on the other disc. Either one or both discs rotate in opposite directions and in this way the comminution action is achieved. The exact mechanism by which the particles are produced is not clearly understood.[28] The mill produces a closely graded product and can achieve fine particles sizes (below 50 microns) in large quantities at fairly high energy cost. For example, limestone 3 mm feedstock is comminuted to 96 percent–50-micron sized product at a rate of 720 kg/h at an energy cost of 21 kWh/ton.

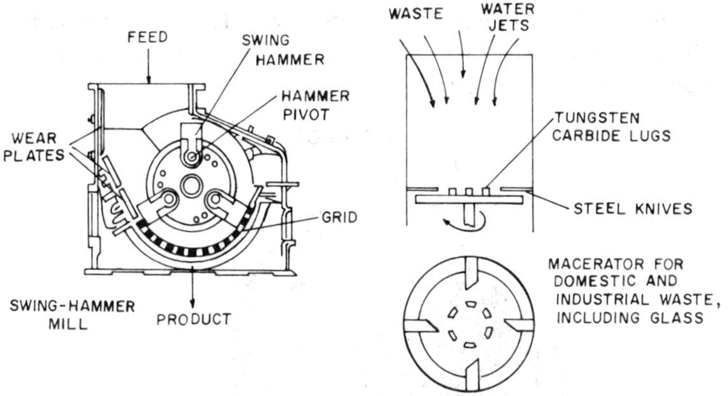

Fig. 3.28[28] Rotary hammer machines (Reprinted with permission of G. C. Lowrison, *Crushing and Grinding,* CRC, 1974).

The vibratory mills are much more energy economical to use than are the tumbling mills and, in addition, they are lighter in construction, use much less ball charge, occupy less work space, and produce a more closely graded product than do conventional tumbling mills. A vibration mill comminutes mica, which a conventional mill does not. However, the vibratory mill is sensitive to moisture content of the feedstock and the feedstock size (being unable to accept particles greater than a certain critical diameter). In addition, the vibrational action makes the mill perennially predisposed to self-destruction. It is said that the upper limit of size is 5 tons/hour.[28] Heat-sensitive, tough material is not appropriate feedstock.

Tumbling Mills

An important portion of finely ground materials is produced in tumbling mills. In general these mills consist of a long rotating chamber in which the material to be ground is mixed with grinding media such as balls, rods, etc., and is ground. A sketch of the various regions of such a machine is shown in Fig. 3.29.

The action of this mill may be described as follows. The total content of the mill is carried up the side by abrasion friction as the drum rotates. In order to minimize sliding, which may tend to promote surging of the charge, the mill may be fitted with lifters which are schematically indicated in Fig. 3.29. The comminution action occurs during the travel of the mill content up the side of the drum, during the cascading and especially during the cataracting of the contents as it leaves the mill wall and travels downwards. The foot impact region should be noted at the end of the travel of the cataracting material.

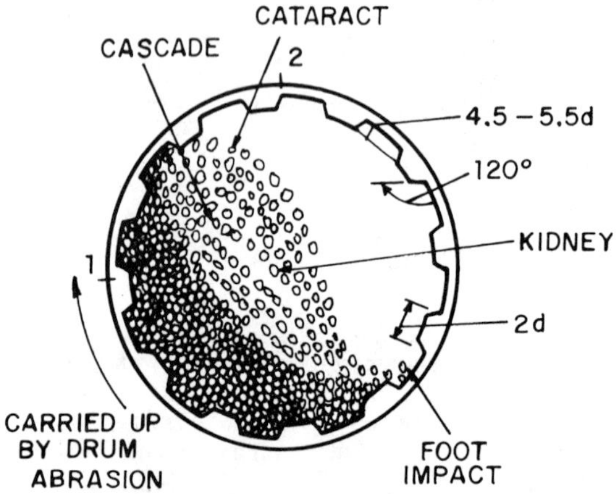

Fig. 3.29[28] Motion of loose units in tumbling mill. Dimensions of lifters in lining, d = maximum diameter of loose unit (Reprinted with permission of G. C. Lowrison, *Crushing and Grinding,* CRC, 1974).

(a) Critical Speed of Rotation. The speed of rotation of a tumbling mill is usually considered in relation to the critical speed, which is that speed of rotation at which the charge of the mill centrifuges as the latter rotates.[31] The critical speed may be calculated by equating the centrifugal force with the weight of the media as follows

$$\frac{Mv^2}{R} = Mg$$

and

$$v = N2\pi R$$

Consequently,

$$N_c = \frac{1}{2\pi}\left(\frac{g}{R}\right)^{\frac{1}{2}} \qquad (3.12)$$

in which

N is the number of revolutions of the mill and N_c is the critical

speed,
v is the velocity of the mill rotation,
R is the radius of the mill in centimeters, and
g is the acceleration due to gravity and is 981 cm/s.

If the mill speed is calculated in revolutions per minute, the critical speed N_c is $300/R^{1/2}$. A more conservative approach takes into account the friction μ between the rotating mill content and the mill wall.[32] In this case it has been argued that a critical point is found at the location marked 1 in Fig. 3.29 [rather than at point 2 as in the case in Eq. (3.12)]. By equating the frictional force with the gravitational force it is found that

$$N/N_c = 1/\mu^{1/2} \tag{3.13}$$

Ball Mills

As is shown in Fig. 3.25, rod mills and ball mills are capable of high reduction ratios and they also consume a great deal of energy in achieving their ends. Some common types of ball mills are shown in Fig. 3.30.[31]

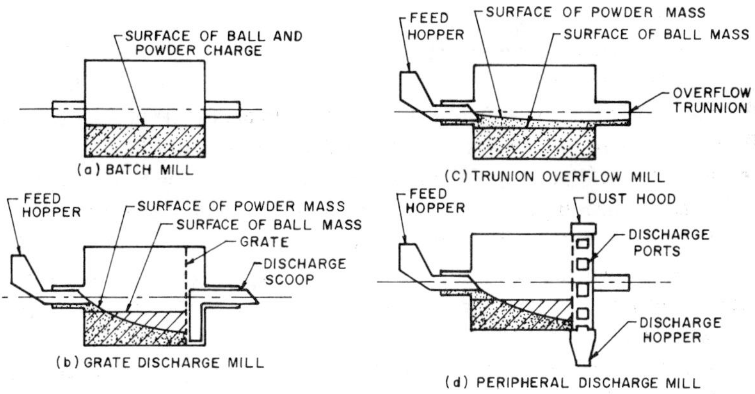

Fig. 3.30 Some common types of ball mill.[31]

A batch mill may be used in processes with fairly small throughput tonnages or it may be used in the case of special atmosphere, or in volatile additive situations. In the grate discharge mill, the grinding balls are confined as shown. The trunnion overflow design is also used in mills in which product removal is achieved by an air lift. Although much of ball milling is carried out dry, wet milling is also practiced in which case air lifting is not possible. The product may or may not be classified and the coarse fraction reprocessed. A once-through process is termed open circuit and a recycling system is called closed circuit.

Of course, the ball mill structure itself has to be protected from the comminution/wear action of the mill content by the incorporation of a lining of suitable construction. Materials suitable for this purpose include Hadfield-type manganese steel, Ni-Hard, silica, silex, etc.

(a) Rate of Grinding. The rate of grinding can be related to the ball diameter and the mill diameter in a number of ways. For example, the rate of grinding in $(\text{meter})^2$ per gram per hour is[34]

$$\text{Rate} = 0.045/d + 0.055D^{0.65} \tag{3.14}$$

Alternative equations are available such as[35]

$$\text{rate (I)} \quad c\rho D^{1/2} b d^{-2} \tag{3.15}$$

and

$$\text{rate (II)} \quad c\rho D^{1/2} b d^{-1} \tag{3.16}$$

in which

d is the ball diameter,
D is the mill diameter,
ρ is the ball density,
b is the particle diameter, and
c is the constant.

Eq. (3.14)-(3.16) are not exhaustive; there are others in the literature. However, there is a general agreement that the rate of grinding varies directly with the diameter of the mill and inversely with the ball diameter.

This conclusion permits us to make some interesting observations. Thus, in a large mill for grinding cement clinker, D = a constant and therefore the mill is divided into compartments. As the charge passes along it becomes finer and to keep up the grinding rate, d has to be reduced. This is shown in Fig. 3.31.

This idea has been further developed in the Hardinge Conical Ball Mill. In this mill, the body gradually tapers toward the product end and the ball is medium classified with the largest balls at the input end. Therefore, as the product becomes finer and moves along the mill the grinding balls become finer also. A sketch of a Hardinge Ball Mill is shown in Fig. 3.32. A diagrammatic layout of a typical closed-circuit plant is given in Fig. 3.33. It is self explanatory.

(b) Wet versus Dry Grinding. The importance of the flow characteristics of the feed are underlined by comparing dry and wet milling. As the

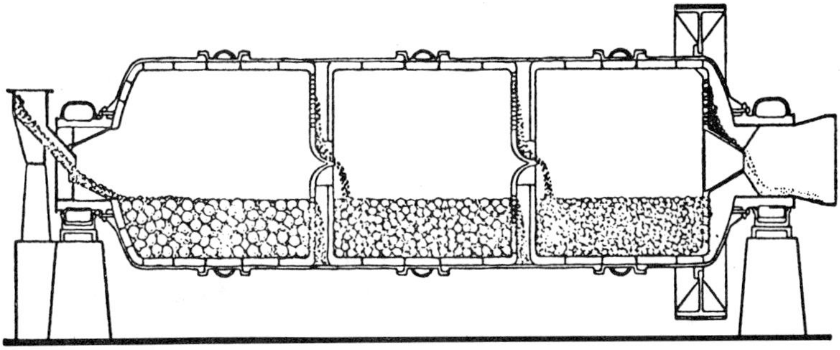

Fig. 3.31 Three-compartment tube mill (Courtesy Koppers Co., York, Pa.).[114]

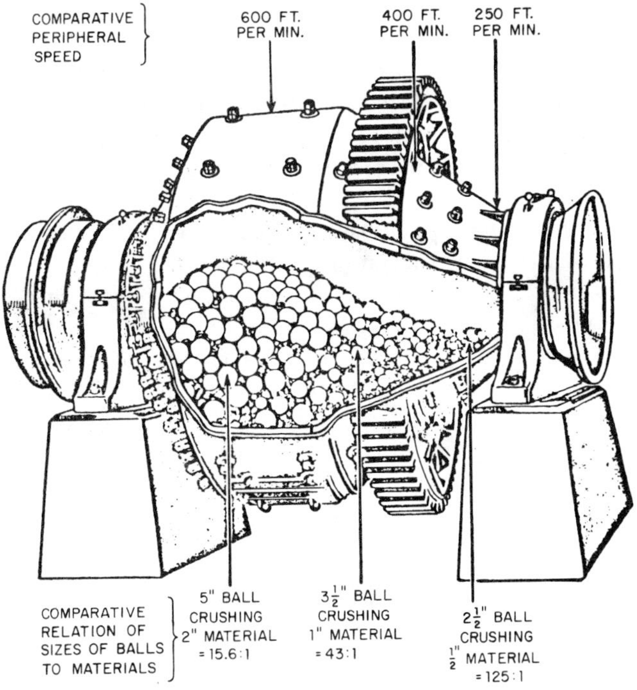

Fig. 3.32 The Hardinge mill: cutaway showing how various sizes of balls tend to be located[114] (Courtesy Koppers Co., York, Pa.).

moisture content is increased flow of the charge almost ceases but when the moisture content increases such that free flow of the charge occurs, the power requirement may be as little as 2/3 of the dry grinding requirement. Other advantages include the already wetted product (where this is applicable as an advantage), potential use of surface action grinding agents, and reduction of dust in the workplace.

But there are disadvantages, for example, the wear and tear of the mill and balls, etc. is much increased, possibly by as much as eight-fold.[28] Additionally, the fact that the product is wet may be a disadvantage as it may have to be dried for further processing. Another problem is that shutdown time is much longer than in the case of dry grinding because of the necessity of cleaning.[28] However, this may be offset by the higher production rate that can be achieved by wet grinding.

(c) Cascading, Cataracting, and Surging. The dynamics of ball milling include two important operations:

1. Cascading and cataracting, and
2. Surging.

During ball milling at low speeds the surface of rotation of the charge becomes unstable at c.a. 30° and *cascades* or collapses. At higher speeds of rotation the charge is projected into space and *cataracting* occurs. The relationship between mill filling (J) and speed of rotation (N) relative to the critical speed (N_c) is given in Fig. 3.34. Here the mill filling J is the fractional ball filling, (volume of balls – volume of voids)/volume of mill, and λ is the coefficient of friction of the mill and the charge. The practical importance of cataracting is that if the balls hit the mill liner, excessive wear occurs.

(d) Surging. This is a phenomenon in which the total charge slips against the side or shell of the mill. As rotation proceeds this produces a pendulum effect. In practice surging rarely occurs in the milling of materials which exhibit high friction on the steel of the lining. Also mills fitted with lifters rarely show surging.

The amplitude $\underline{\underline{\alpha}}$ of the surging phenomenon would be expected to depend on a number of factors. The mill filling (J), the ball (d), the mill (D) diameters; the speed of rotation (N), the coefficient of friction (μ), and the period of oscillation of the mill charge (t). It is assumed that the variation of friction with the velocity of relative motion of the charge and shell is negligible. It can be said that[36]

$$\underline{\alpha} = f(D, d, t, N, \mu, J) \tag{3.17}$$

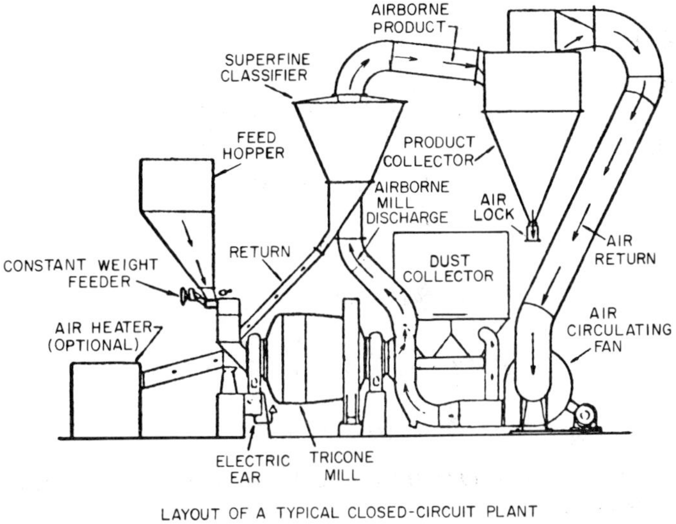

Fig. 3.33 Conical Mill®, Gyrotor® Air Classifier, Product Collector and Dust Filter (Courtesy Koppers Co., York, Pa.).[114]

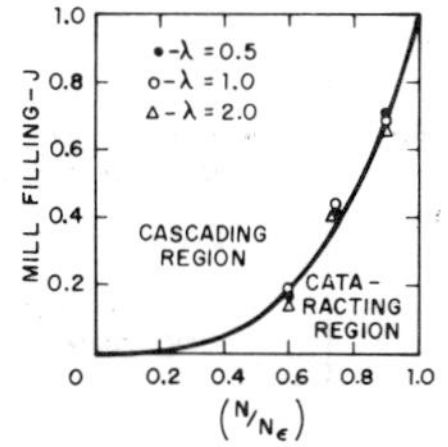

Fig. 3.34 Illustrating cascading and cataracting[36] (Reprinted by permission of the Council of the Institution of Mechanical Engineers from Proceedings of the Institute of Mechanical Engineers, 1956).

and

$$D \overset{d}{=} L$$

$$d \overset{d}{=} L$$

$$t \overset{d}{=} T$$

$$N \overset{d}{=} T^{-1}$$

$$\mu \overset{d}{=} 0$$

and

$$J \stackrel{d}{=} 0$$

Therefore,

$$M^{\circ}L^{\circ}T^{\circ} \stackrel{d}{=} L^{a_1} L^{a_2} T^{a_3} (T^{-1})^{a_4}$$

in which

$$L \quad a_1 + a_2 = 0$$

$$T \quad a_3 - a_4 = 0$$

Therefore

$$a_2 = -a_1 \quad \text{and} \quad a_3 = a_4 .$$

and

$$M^{\circ}L^{\circ}T^{\circ} = L^{a_1} L^{-a_1} T^{a_3} (T^{-1})^{a_3}$$

so

$$\underline{\alpha} = f(D/d, Nt, \mu, J)$$

Now $N = KN_c$ when $N_c \alpha (1/D)^{1/2}$ and $t \alpha D^{1/2}$; therefore Nt is independent of D and therefore

$$\underline{\alpha} = f(D/d, \mu, J) . \tag{3.18}$$

The results of an experimental investigation of these relationships is summarized in Fig. 3.35.[36]

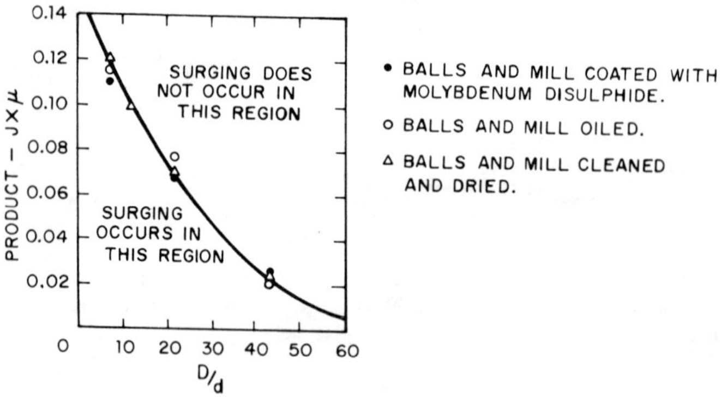

Fig. 3.35[36] Surge Control (Reprinted by permission of the Council of the Institution of Mechanical Engineers from Proceedings of the Institute of Mechanical Engineers, 1956).

(e) Power Requirement. Over the years there has been much interest shown in relating this requirement to the practical variables of milling, particularly in the case of ball milling. Various investigators have stated that the power required to drive a ball mill is essentially related to the torque

required to raise the charge above the center of gravity of the system. For example, the power is proportional to[37]

$$d^3 \rho D \tag{3.19}$$

the power is proportional to[35]

$$\rho D^{2.5} \tag{3.20}$$

the power is proportional to[28]

$$\rho L D^{2.5} \tag{3.21}$$

in which

d is the ball diameter,
D is the mill diameter, and
ρ is the ball density.

(f) The First Relationship. The power required may be calculated according to two relationships. The first one simply assumes that much of the power requirement is that which is needed to lift the contents above the centroid, thus[38]

$$P = 0.32 WNy \tag{3.22}$$

in which

P is the power in kilowatts,
W is the total weight of contents,
N is revolutions per minute, and
y is the displacement (in feet).

(g) The Second Relationship. This is much more complicated and it is introduced later in more detail. The power required to drive a ball mill loaded with balls and powder excluding power losses in the bearings, gears, etc., is of interest both to the designer and operator. Clearly the number of variables involved is very large and if only for this reason a theoretical deduction of the problem has not been undertaken. However, an experimental determination of the relationship between the variables and the power requirement has been carried out using the principles of dimensional analysis.[39,31] The gist of this investigation is introduced here. The variables include mill length (L), diameter (D), ball diameter (d), density (l), fractional charge volume (J), speed of rotation (N), (g); coefficient of restitution (e), the coefficient of friction (f), powder particle diameter (b), the energy necessary to achieve a unit increase of specific surfact (E), and fraction of ball voids occupied by powder (V). In the case of wet milling, P the power would also depend upon the mix kinematic viscosity (ν), the mix density (σ), the solid–liquid volume ratio (U), and the number of

lifters (n) and their height (h). Thus

$$P = \phi(L, D, d, \ell, N, g, b, E, \nu, \sigma, h, \underbrace{J, f, e, V, U, n}_{\text{no dimensions}}) \tag{3.23}$$

in which the dimensions are as shown below:

$$\begin{array}{ll} P \overset{d}{=} ML^2T^{-3}, & g \overset{d}{=} LT^{-2}, \\ L \overset{d}{=} L, & b \overset{d}{=} L, \\ D \overset{d}{=} L, & E \overset{d}{=} ML^2T^{-2}, \\ d \overset{d}{=} L, & \nu \overset{d}{=} L^2T^{-1}, \\ \ell \overset{d}{=} ML^{-3}, & \sigma \overset{d}{=} ML^{-3}, \\ N \overset{d}{=} T^{-1}, & h \overset{d}{=} L \end{array}$$

In order to develop the dimensionless ratios that are sought, the dimensional form of Eq. (3.23) is stated

$$M^\circ L^\circ T^\circ = \underset{P}{(ML^2T^{-3})^{a_1}} \underset{L}{L^{a_2}} \underset{D}{L^{a_3}} \underset{d}{L^{a_4}} \underset{\ell}{(ML^{-3})^{a_5}} \underset{N}{(T^{-1})^{a_6}} \underset{g}{(LT^{-2})^{a_7}} \underset{b}{L^{a_8}}$$

$$\underset{E}{(ML^2T^{-2})^{a_9}} \underset{\nu}{(L^2T^{-1})^{a_{10}}} \underset{\sigma}{(ML^{-3})^{a_{11}}} \underset{h}{L^{a_{12}}}$$

The solution is obtained by equating the dimensions of M, L, and T with their respective items on the right hand side. Like terms are then collected (i.e., terms incorporated by the same superscript a_i) and the dimensionless groups established (note the solution is produced by solving for a_3, a_5, and a_6).

The resultant relationship is stated in abbreviated form as

$$P/D^5N\ell^3 = (1 + 0.4\sigma/\ell) \cdot L/D \cdot \phi_1(N_c/N) \cdot \phi_3(J) \cdot \phi_4(d/D) \cdot \phi_5(n) \cdot \phi_6(h/D) \cdot \phi_9(b/D) \tag{3.24}$$

The values of the various dimensionless factors, for given sets of appropriate conditions are determined experimentally. For example, consider the case of the factor $\phi_3(J)$. Using small laboratory sized mills of varying diameter the graphical relationship shown in Fig. 3.36 is obtained.[31]

The results obtained with this formula for the power requirement to

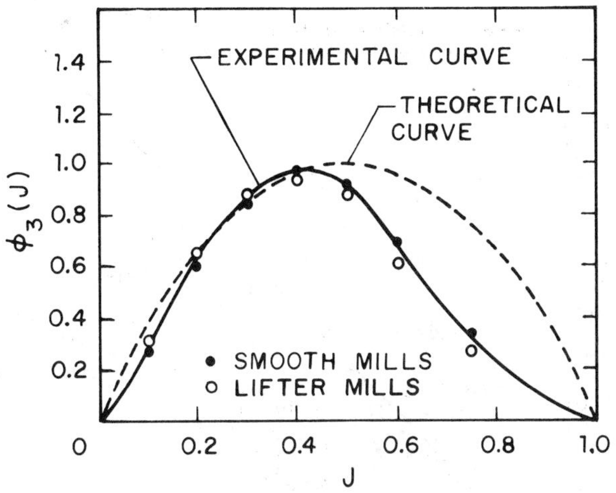

Fig. 3.36 Experimental determination of dimensionless factor $\phi_3 J$.[31]

operate a ball mill are compared with actual data for their operation for many specific mills. The comparison is shown in Fig. 3.37. The agreement is clearly impressive.

Attrition and Cutting Machines

Perhaps the most well-known attrition machine is the fluid energy mill. A sketch plus a possible plant layout is shown in Fig. 3.38. This machine has a distinct advantage in that very fine particle sizes can be produced in a narrow range and because there are no moving parts, the product is clean of wear debris from the material of construction of the plant itself. However, the device is extremely energy expensive. For example, to reduce a product from 80 to 3.5 microns takes 2500 kWh per ton (calculated from steam usage).[28] This value is clearly way off the scale of Fig. 3.25. On the other hand, it is the only figure quoted for obtaining such a small size of product. There is some debate about just exactly how the comminution occurs.[28] But interparticle interactions clearly predominate in the process(es).

For many years now, large quantities of steel granules (for shot peening) and nonferrous granules (for filter production) have been made by a rela-

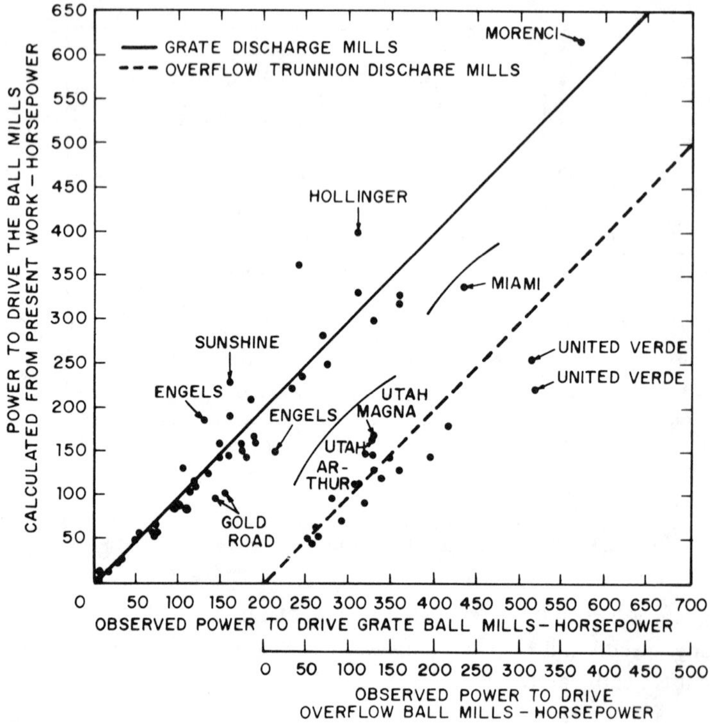

Fig. 3.37 Power to drive ball mills. Comparison between data published by Taggart and corresponding values as calculated from the present work.[31]

tively simple process of wire-cutting using a cutting mill that is illustrated in Fig. 3.38. The size range of the product is extremely narrow.

Fracture

In this generalized treatment of grinding and particle fracture, a whole variety of materials and a wide range of comminution methods are dealt with. Materials contain large numbers of defects, including surface irregularities and clefts, dislocations, grain boundaries, phase boundaries, impurity concentrations, anisotropy, porosity, gross inhemogeneity, cracks and fissures. In addition they may be capable of extensive plastic deformation before losing structural integrity; alternatively they may suffer only elastic deformation prior to fracture.

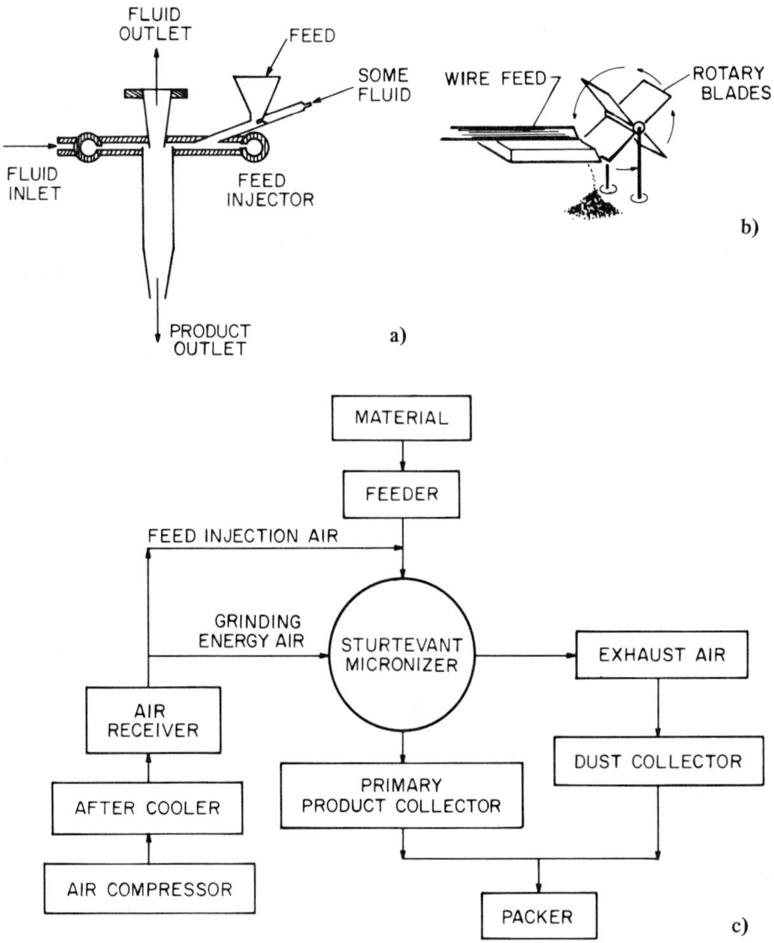

Fig. 3.38 a) Cutting mill; b) fluid energy mill[28] (Reprinted with permission of G.C. Lowrison, *Crushing and Grinding*, 1974); c) Sturtevant "Micronizer" diagram showing plan of a typical layout incorporating this fluid energy mill.

A material which is brittle under one set of conditions may become somewhat ductile if these conditions are changed and vice versa. For example, plastics are ground in a low temperature environment. The fracture process can be materially aided if the surface energy required to form the new crack surface is lowered. This is one of the benefits gained from comminution in the presence of salt solutions.

As the particles that are being ground become finer, they are more difficult to grind. One reason for this is that as the particles become smaller, the flaws become less dense and so fracture is progressively more difficult. Consider a brittle nonporous material: in its initial condition an inch cube of this material may contain a small number (< 10) of gross cracks and a larger number of minor cracks and similar flaws (arbitrarily assume ~50). The cube quickly splits into ten pieces and later into 500 pieces, for example, but thereafter the fracture mechanism depends upon other factors for its success. Clearly the process of fine grinding is intimately connected with the characteristics of the bulk material.

Energy Requirements. Numbers indicating the energy requirement for the fracturing of individual particles have been obtained as shown in Table 3.6.[40]

Table 3.6

ENERGY UNITS ASSOCIATED WITH PARTICLE SIZE PRODUCTION[40]

Material	Particle size μm			
	100	1000	10000	100000
Coal	1.43	0.53	0.28	0.13
Quartz	3.1	2.1	1.45	0.7

However, the instrinsic properties of the material of the particle and the size and morphology of the particle are not the only considerations. Breakage and selection are essentially independent events, and the process that determines which particle, out of a given set, will be selected and comminuted is essentially random.

Selection Function and Breakage Function

Let $S(u)$ be the selection function for the feed with size range u through $u + \delta u$. It is the proportion of particles of sieve size x which are broken in a unit comminuting action.*

*In general this is defined as a single, identifiable cycle of the equipment. It varies from equipment to equipment.

Let $B(x, u)$ be the proportion of that feed material selected for comminution that is broken down below size x (the breakage function).

If we symbolize the total mass of feedstock by F and the feed between u and $u + \delta u$ by $\delta F(u)$ then the weight of the material that is selected and actually broken down below x must be the product of the two functions multiplied by the quantity of feedstock $\delta F(u)$ that was originally available to be processed. This is equal to $B(x, u) \cdot S(u) \cdot \delta F(u)$ and so the total amount of the material from the original feed that has been broken down to below size x is

$$\sum_{u=x}^{u=\alpha} B(x,u) \cdot S(u) \cdot \delta F(u) + F(x) \qquad (3.25)$$

in which $F(x)$ is the amount smaller than size x in the original feedstock.

The breakage function is dependent upon the following factors:

1. *Reduction ratio.*[28] For example, in the case of coal,[41] increasing the ratio from 0.1 through 0.8 increases the breakage function from 0.02–0.5.
2. *Size of feed.*[28] For example, in the case of fertilizer granules,[42] increasing the feed from 3 mm through 8.5 mm increases the breakage function from 0.15 through 0.38 at a reduction ratio of 0.2.
3. *The rate of comminution.*[28] For example, in the case of a rotary hammer mill,[42] increasing the rpm from 1000–3000 increases the breakage function from 0.01 through 0.15 where the feed stock of fertilizer granules has a size of 2.7 mm and a reduction ratio of 0.2 was used.

The selection function also varies with certain parameters as indicated:

1. *The size of the feed.*[28] In the case of a Hardgrove machine, as the size of feed coal is increased from 100 through 1000 microns, the selection function increases from 0.008 through 0.10.[41]
2. *The rate of comminution.*[28] Increasing the rate of a swing hammer from 1000–3000 rpm increases the selection function–in the case of the 3000-micron, incoming feedstock size (fertilizer), from 0.35 through 0.77.[42]

Specificity of Equipment

The specificity of each type of comminution equipment is illustrated in

Fig. 3.39. Thus abrasion leads to a narrow size range in the finished product, compression leads to a broad size distribution in the finished product, and impact leads to an intermediate condition.[28]

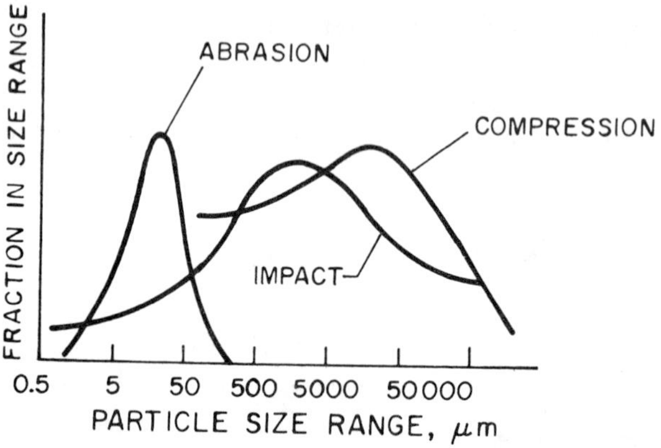

Fig. 3.39[28] Characteristic size distributions from comminuting action (Reprinted with permission of G. C. Lowrison, *Crushing and Grinding*, 1974).

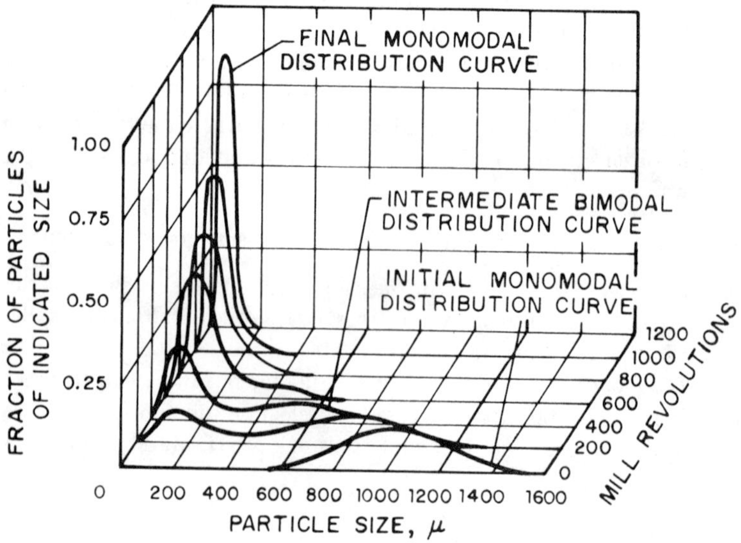

Fig. 3.40 Shifts in particle-size distribution with grinding progress.[43]

The effect of repetition of the comminution cycle is shown in Fig. 3.40.[43]

Prediction of Size of Product

In industrial practice, closed circuit grinding is carried out in ball milling, gyratory crushing, and jaw crushing. A treatment using the concepts of selection function and breakage function has been conducted to determine the final size distribution in the finished product.[44,45] Consider a closed circuit mill in which the feed is m, the mill product is p, the output is q, and the total load on the mill is k. Then, by simple reasoning

$$k = m + p - q \tag{3.26}$$

If c is designated the separation efficiency of the system, then

$$k = m + p(1 - c) \tag{3.27}$$

in which

$$q = cp\ .$$

Now if the total load on the mill is divided up into products of various sizes and if each of these respective size ranges is comminuted into other size ranges, then the matrix form of this relationship can be written as

$$P = AK \tag{3.28}$$

in which the product consists of

$$k_1(a_{11} + a_{21} + a_{31} + a_{41} + \ldots + a_{j1}) + k_2(a_{22} + a_{32} + a_{42} + \ldots + a_{j2}) + \ldots + k_j(a_{jj})$$

The subsequent derivation of this set of relationships leads to the matrix expression

$$Q = CG^{-1}AM \tag{3.29}$$

This relationship was found to give the correct size distribution of the product.[45] Thus if one knows the details of Eq. (3.28) and the separating efficiency for each fraction (all for a given total load), it is possible to control the comminution system to give a desired final product from a given feedstock size.

Energy Consumption Equations

The overall energy efficiency of comminution is absurdly low[46] as shown below:

Table 3.7

ENERGY CONSUMED IN ROCK GRINDING[46]

Grinding of rock	
Heating of rock	24 percent of total (28 kW)
Heating air circulation	18 percent of total (21 kW)
Drying	35 percent of total (41 kW)
Radiation, etc.	20 percent of total (23 kW)
Actually used to comminute	3 percent of total (4 kW)

It has been observed that most of the energy is lost because of the difficulty of transmission of the applied forces to the particles in the mill contents that most need to be comminuted.[47] The culprit is interparticle friction. This is coupled with the fact that most materials processed by comminution are substantially weaker in tension than in compression and, for the most part, the forces are not tensile in comminution. Despite these difficulties, considerable effort has been made to relate energy consumption to the product characteristics. There are three relationships:

1. *Rittingers Law:*[48] Rittinger assumed that the energy required to reduce a particle to a number of smaller particles is directly proportional to the new surface produced. This theory ignores elastic and plastic deformation.
2. *Kick's Law:*[49] Kick assumed that the energy required to achieve a given subdivision is constant. That is, if one unit of energy is required to reduce one particle to two particles, exactly one more unit of energy will be required to reduce the 2 particles to 4 particles. This theory has the underlying assumption that the material defects are uniformly distributed and this is not correct.
3. *Bond's theory:*[50] Bond assumed that the work is used to produce crack tips and that it is therefore directly proportional to the total crack length formed. According to Bond the crack length is proportional to (½ new surface area formed)$^{1/2}$ or total work $\alpha\, 1/(D$

product)$^{1/2}$. (Note: for a cube cracked in half, area = $2 \cdot 1^2$; therefore crack length $= \sqrt{2 \cdot 1^2/2} = 1$).

In equation form these three laws may be written as

Rittinger $\quad E = K(1/d_2 - 1/d_1) \quad$ (3.30)

Kick $\quad E = K(\log d_1/d_2) \quad$ (3.31)

Bond $\quad E = 10W_i(1/\sqrt{P} - 1/\sqrt{F})$

or $\quad E = W_i \dfrac{\sqrt{F} - \sqrt{P}}{\sqrt{F}} \sqrt{100/P} \quad$ (3.32)

in which W_i = work index; 80 percent of the feed passes F microns; 80 percent of the product passes P microns (80 percent is quite arbitrary).

In general the work required to diminish a particle in size is a function of the diameter. This may be written as[51]

$$dW/dD = -CD^n \tag{3.33}$$

in which

D = diameter,
C = constant, and
n = integer.

Let $n = -2$ and integrate between D_0 and D_1

$$\int_0^W dW = -C \int_{D_0}^{D_1} D^{-2} dD \quad \therefore \quad W = -C(-1/D)_{D_0} \, .$$

Therefore,

$$W = C(1/D_1 = 1/D_0)$$

This is Rittinger's Law.

Let $n = -1$ and integrate

$$\int_0^W dW = -C \int_{D_0}^{D_1} D^{-1} dD \, .$$

Therefore

$$W = -C \ln D \Big|_{D_0}^{D_1}$$

$$W = -C(\ln D_1 - \ln D_0)$$

and

$$W = C \ln D_0/D_1 \ .$$

This is Kick's Law.

Let $n = -3/2$ and integrate

$$\int_0^W dW = -C \int_{D_0}^{D_1} D^{-3/2} dD$$

Therefore,

$$W = -C \left[\frac{1}{-\frac{1}{2}} D^{-1/2} \right]_{D_0}^{D_1}$$

and

$$W = 2C \left[\frac{1}{D_1^{\ 1/2}} \ \frac{-1}{D_0^{\ 1/2}} \right] .$$

This is an alternative form of Bond's law.

This means that the only difference numerically between the three laws of comminution is in the value of the exponent n. There are other views, however.[28] The use of Bond's work index has been popularized and tables of "average" values of W_i are available.[28,51]

Range of Applicability of Equations

Each of the three "laws" has a range of particle size for which it is the most applicable.[52] The correlation of this statement is shown in Fig. 3.41 and a number of minerals follow this composite curve closely.[52]

Theoretical Size Distribution of Single Particles[53]

In recent times several attempts have been made to predict the size distribution that would result from the comminution of a single morsel of

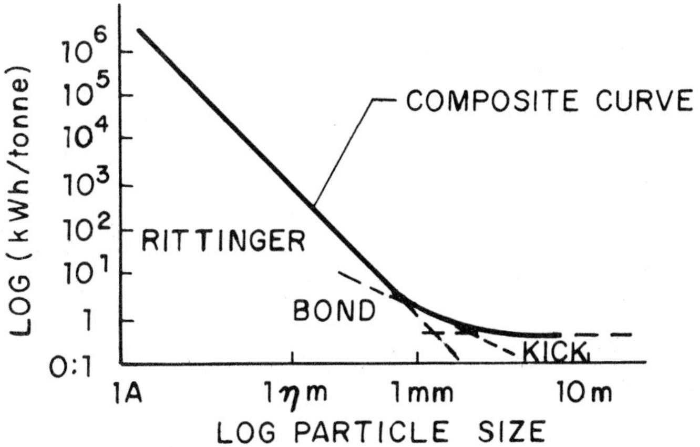

Fig. 3.41 Composite Rittinger, Bond, Kick energy-particle size curve.[52]

solid, brittle matter. These take the form of mathematical models; four of these are discussed:

The Flaw Model[54] assumes that line, surface, and volume defects can participate in the particle forming processes during comminution. The theory is that the resulting size distribution creates the activation and interaction of these various flaws. It has the form

$$B(x) = 1 - \exp\left[-\left(\frac{x}{i}\right) - \left(\frac{x}{j}\right)^2 - \left(\frac{x}{k}\right)^3\right] \tag{3.34}$$

This equation means that the cumulative mass [signified by $B(x)$] which is smaller than x is a function of the respective flaw densities (i signifies linear, j signifies area, and k signifies volume).

The Crack Density Model[55] makes the assumption that the crack density at any location is proportional to the peak energy in the solid at the time of comminution. The breakage function has the form

$$B(x) = 1 - (1 - x)^r \tag{3.35}$$

in which r is the average number of cuts on an imaginary line passed through the original specimen.

The Blocky Particles Flaw Model[56] was derived using a spherical (blocky) approximation for the product particles instead of a cubical one and the model has an added term as follows:

$$B(x) = 1 - (1-x)^r - \frac{rx}{2}(1-x)^{r-1} \quad (3.36)$$

The Extrusion Flaw Model[57] develops an analogous form of equation for the breakage function. It has been observed that the further development of these models awaits our improved ability to precisely characterize particle morphology.[53] Finally, the reader may consult reference 58 for a more recent discourse on fracture. Other important sources include circuit grinding control (Herbst, Utah); physics of fracture (Schönert, Karlsruhe).

3.5 CRYSTALLIZATION

Crystallization is defined as the formation of solid particles within a homogeneous phase. One well-known and excellent example of this is the production of the carbonyl iron. In this process, Fe is heated in a stream of CO to form gaseous $Fe(CO)_5$ which is decomposed in another chamber to form little nuclei. These nuclei grow to form small particles of a typical microstructure termed "onion structure." This title indicates that the microstructure consists of enveloping and distinctive layers of metallic iron. Particle sizes can be produced from ten or more microns down to the submicron range. The shape is spherical.

However, the bulk of the production of granular matter by crystallization is obtained by crystallization from liquid solutions. One particular advantage that this method affords is that despite the impurity of the mother solution, the crystals consist of pure substance. The entrained mother liquor is then removed by subsequent washing procedures of a standard type. The size and shape of the crystals can be controlled to serve the needs of different markets.

In many operations, the solidifying crystals and the parent solution are in contact long enough to attain equilibrium, which is achieved when the liquor is saturated. In most cases, the solubility increases with the rise in temperature and the yield can be directly calculated from a knowlege of the original concentration and the solubility at the original temperature. Chemical engineering practice also requires the calculation of the heat balances, etc.

Nucleation and Growth

The two main stages involved in crystallization are called nucleation and growth. The nucleation is divided into clustering, embryo formation, and nucleation proper. If the solution is saturated, the mechanism of nucleation is called heterogeneous nucleation in which foreign nuclei, equipment surfaces, and other influences play a role in the process. If the solution is not at all saturated and if crystals are nucleating rapidly, the mechanism is called homogeneous nucleation (which is essentially spontaneous). In both cases, growth steps on the surface of the crystal play an important role in the process. These originate in the crystal as a result of the incorporation of screw dislocations in the new phase.

The nucleation step should lower the free energy (G) of the system. However, the formation of the embryo-solution interface results in an overall increase in the free energy of the system (see Fig. 3.42).

At a certain critical size r_n, the increase in G due to the increase of surface area, is just balanced by the decrease in G due to the energy difference between the crystal and its liquor. This stage represents the end of nucleation and the beginning of growth. It is located at the maximum point on the total free energy change curve shown in Fig. 3.42.

This maximum represents an energy barrier, which has to be overcome for growth to occur.

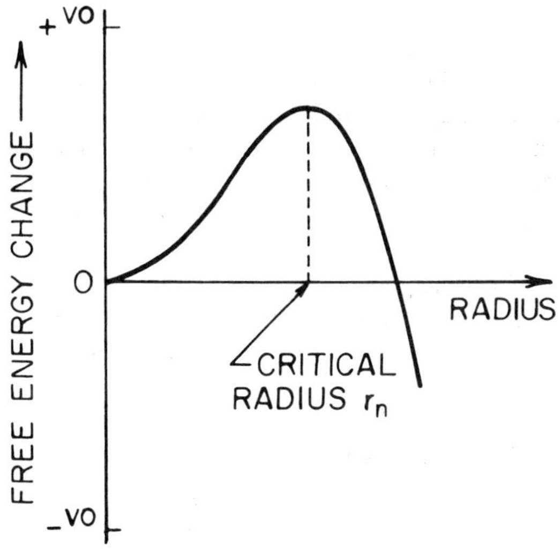

Fig. 3.42 Change in overall free energy as a nucleus grows.[59]

Degree of Saturation

The rate of nucleation is particularly sensitive to the degree of saturation as is demonstrated in Fig. 3.43.

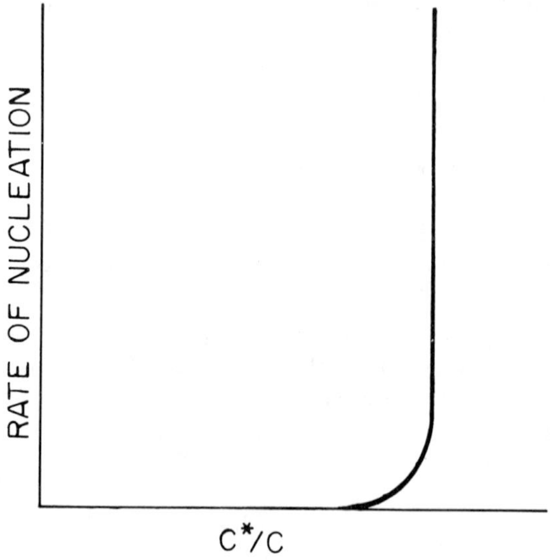

Fig. 3.43 Change in the rate of nucleation with increasing supersaturation.[59]

The origin of this supersaturation effect has been explained simply. The rate of nucleation is defined as a product of two independent factors. One of these, as already outlined, deals with the probability associated with the formation of a stable nucleus and this is proportional to exp $(-\Delta G_{rn}/kT)$. The second factor is concerned with the supply of incoming species from which the growing nucleus is made. This is proportional to exp $(-Q/kT)$, in which Q is the activation energy for the transportation process involved. The overall nucleation rate, therefore, is

$$J = \text{constant} \cdot \exp(-\Delta G_{rn}/kT) \cdot \exp(-Q/kT) \qquad (3.37)$$

In the above relationship, the free energy change (the activation energy for the process) is related to the degree of supersaturation in the following way:

$$\Delta G_{rn} = \frac{16\pi\sigma^3 \nu'}{3(\mu-\mu^*)^2} \qquad (3.38)$$

in which

σ is the surface energy of nucleus/solution interface,

ν' is the molecular volume, and

μ are the chemical potentials of the solute in the saturated and the supersaturated (*) solutions, respectively.

The chemical potentials are related by the well-known expression

$$\mu - \mu^* = kT \ln c/c^* \tag{3.39}$$

Therefore, as the ratio of the respective chemical potentials increases beyond a critical value, the rate, J, shoots up.[59]

Batch and Continuous Crystallizers

Batch crystallization suffers from the disadvantage that when the required degree of supersaturation for crystallization is attained (it is brought about suddenly), many crystals form and the degree of supersaturation falls to zero. The conditions are therefore continuously varying in batch crystallization. This makes theoretical predictions concerning particle size, size distribution, etc., difficult. Modern methods of mass production, especially chemical engineering style, dictate continuous processing, whereever possible. Two typical examples of continuous crystallization units are shown in Fig. 3.44.

The particle size and the particle size distribution in continuous crystallization are related to the fundamental parameters of the process.[60] It is assumed that the tank operates under steady state conditions. Three equations are important. The first expresses the fraction of the crystals of a given size. The second converts the first to a weight basis. The third equation expresses the growth rate for the overall process. These are quoted below as equations (3.40, 3.41, and 3.42) respectively.

$$n(r) = \frac{J\nu}{g(S)} \text{esp} \left[-\nu r/Vg(S)\right], \tag{3.40}$$

$$w(r) = \frac{J\nu}{g(S)d} \exp \left[-\nu r/Vg(S)\right] \cdot \gamma r^3, \tag{3.41}$$

and

$$\log \frac{S_o - S}{[g(S)]^3} = \log \frac{8\pi V^4}{dc^* \nu^3} + \log(\text{constant}) - \frac{Q}{kT} - \frac{16\pi\nu'^2\sigma^3}{3k^3T^3\log^2 S} \tag{3.42}$$

in which

$n(r)$ is the number of crystals between size r and $r\,dr$ leaving the crystallizer per second,
J, v', Q, constant, k, T, c^* are as before,
$g(S)$ is the rate of growth of the crystal (this is constant for steady state, as is J),
$w(r)$ is the weight of crystals between r and $r\,dr$,
d is the crystal density,
V is the vessel volume,
v is the rate of efflux of the solution,
S is the actual supersaturation,
S_o is termed the hypothetical supersaturation, without precipitation, and
γ is the shape factor for the crystals.

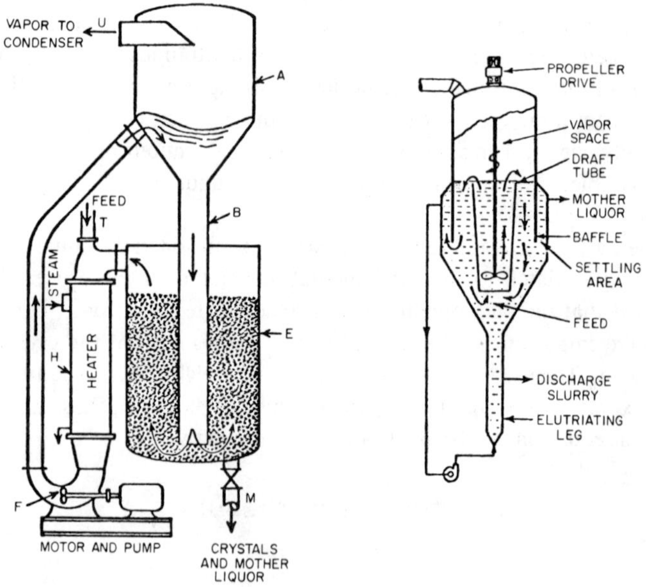

Fig. 3.44[1] (From *Unit Operations of Chemical Engineering* by McCabe and Smith, © 1967, second edition, used with permission of McGraw Hill and Co., NY).

It is stated that although this set of equations gives similar results to experimental data [if S is known, J and $g(S)$ can be obtained and then plugged into Eqs. (3.40) and (3.41)], in practice very simple deviations from the ideal can render the set of equations inapplicable. However, even if the system is *not* crystallizing by homogeneous nucleation, this can be taken into account by simply introducing a factor f_o/V in place of J, when f_o is the number of foreign nuclei fed in per second.[59] Note also that the values of S for homogeneous and heterogeneous nucleation are not the same.[59]

During the last decade there has been considerable research activity in the area of analysis and techniques of continuous crystallization.[61] Specific concerns include: population balance, crystal size distribution, dynamics and stability of crystallizers, nucleation, and growth rate kinetics. The basis of this fruitful approach is population balance (i.e., the numbers balance) which is used to analyze size distributions of crystals resulting from a crystallization process in which both nucleation and growth are coterminous. A number of texts on crystallization are given in reference 62. The reader interested in further study should consult these and reference 61.

3.6 PRODUCTION OF FINE POWDERS

Methods for the production of fine powders have been divided into those that comprise the breaking down processes and those that include the building up processes.[63] Breaking down procedures include atomization, comminution, and micronization which have already been discussed in some detail. The reader is referred to the Fig. 3.13 and Table 3.1, for example. Much work has been reported on the fine grinding of feedstock such as calcite, kaoline, marble, MoS_2, etc.[64-68]

For example, overgrinding can occur so that after very long grinding times (in excess of 400-500 hours) the particle size of the product starts to increase[64,68] due to agglomeration. The effect of the addition of surface active chemical reagents in increasing the grinding efficiency is usually explained in terms of Griffiths Crack Theory (see Eq. 2.22 et seq for further detail). It has been observed that the subsurface structure of particles produced by fine grinding tends to become amorphous. For example, quartz has this subsurface structure down to a depth of 100 atoms.[69] In general it appears that increasing the degree of fineness by grinding gradually increases the chemical reactivity of the material.[63]

Other breaking down processes that have been used include: production by electrical discharge using the methods of electric spark machining[70] and atomizing blasts of fused wire.[71] It appears that both of these methods produce a rather coarse product. A much finer product has been obtained by the controlled reduction of Cr_3O_3 with Mg or Li vapors in a vacuum furnace. The fine fraction reported was approximately 0.0068 microns.[72]

The mechanisms that are utilized in a variety of processes for the production of fine powders have been reviewed elsewhere.[63] A brief resume of the information from that reference is given in Table 3.8.

It has been observed that the potential production rate of fine powder via aerosol technology is so small that other methods must be used.[63] On an industrial scale, there are three types of processes involving vapor phase reactions for the production of fine powders:[63]

1. the production of TiO_2 and SiO_2 by the oxidation or the hydrolyzing volatile chlorides,
2. the production of oxide smoke (ZnO, Sb_2O_3), and
3. the production of carbon black by the thermal decomposition of hydrocarbons.

Other methods involving the use of vapor phase reactions have also been listed.[63] In the case of precipitation methods, it is necessary to keep the growth rate down in order to obtain a large crop of extremely fine particles. The methods for achieving this end vary with the substances being produced.[63] Some complicating factors include: the higher solubility of the finer particles, cementing together of interparticle necks due to recrystallization, and the flocculation of small particles. All of these factors tend to result in a coarser sized product than otherwise would be obtained.[63] The reader is referred to the patent literature for additional details of these processes.[63]

3.7 GRANULATION

The subject matter of this chapter is the formation and production of particles. Particles range in size from the very small (e.g., nuclei discussed in this chapter) to the large. Many particles are not individual morsels of matter but consist of accumulations of many individual morsels, all held together by various forces. These forces can be considered in four categories:[80]

1. Van der Waal's forces of attraction, which are of the same order of

Table 3.8

PROCESSES USED FOR FINE POWDER PRODUCTION[63]

Mechanism of Particle Formation	*Description*	*Characteristics*	*Process*	*References*
vapor→liquid→solid	two step condensation	spherical, 0.005–0.1 micron	high intensity arc oxide (in air)	mixed oxides[73] mixed carbides[74]
vapor→solid	one step condensation	angular, see above	high intensity arc, see above	see reference 63
vapor + vapor → solid (+ vapor)	reaction condensation	varied	condensation in special atmosphere	see references 63, 75, and 76 (see Table 5)
Liquid→solid	vacuum condensation	spherical	carbonyl Fe, Co, Al	see reference 77
	precipitation	varied	very varied conditions depending upon product	see reference 63
Solid→solid	thermal decomposition	oxide	atomizing solution into flame	see reference 78
		monohydrate	dehydrate in vacuum	see reference 79
		various	various	(see Table 10)[63]

Table 3.9

AN OUTLINE GUIDE TO AGGLOMERATION PROCESSES AND EQUIPMENT

Operation	*Equipment*	*Limitations, Comments*	*Product Form*	*Representative Products*
COMPACTION				
Tableting–two opposing plungers operating within a cavity compress loose material into a product of uniform density. More complex plunger systems can also be used.	Single-punch press Rotary press Layer press	Material should have good flow characteristics and cohesive properties. Lubricants and binders can be used to impart these characteristics.	Cylindrical	Pharmaceuticals, catalysts, industrial chemicals, ceramics, metal powders
	Molding press	Requires close control of quality and quantity of ingredients. Die tooling costs are higher and some designs require sophisticated systems for multi-directional compression.	Determined by shape of die.	Plastic preforms, small machine parts from metal powders (cams, gears, gaskets)
Roll Compaction–two rolls rotating counter to each other compact material into preferred shapes.	Smooth roll-type presses: Breaker	Material needs to become semi-plastic under pressure and breakable after compaction.	Flakes	Clay-type minerals
	Granulator	Material needs to become bonded under pressure and be crushable afterward.	Granules Pellets Powders	Potassium chloride, sodium chloride, urea, diammonium phosphate.

Table 3.9 *(cont'd.)*

Operation	*Equipment*	*Limitations, Comments*	*Product Form*	*Representative Products*
	Smooth or corrugated roll-type press, special feeder.	Material needs to become plastic and well bonded under pressure.	Sheets, bars	Plastics, metals
	Briquetting (pocketed roll-type press)	Material should be formable at low unit pressures.	Thick pillows Spheroids	Clay-type minerals, organic compounds, soft metals
	Special feeder	Material should be bondable under pressure.	Thin pillows Almonds	Lime, magnesia, titanium, sponge, metal powders
	Premixer	Almost any particulate matter can be handled with a suitable binder.	Thick pillows Spheroids	Ores, charcoal, metal powders
	Preheater	For material that needs to be softened prior to briquetting.	Pillows Sticks	Iron ores, metals, phosphate rock
Extrusion–forcing of material through a die by means of a roller, screw or plunger.	Pellet mill	Material for processing should be free flowing.	Spahetti-like product. Peripheral shape determined by shape of the die. Knives cut pellets into desired length.	Pharmaceuticals, plastics, clays, carbon, charcoal, industrial chemicals, fertilizer, rubber products.
	Plasticizing pan	Material should be plasticizable	As above	Phosphate, carbon black, clays
	Screw extruder or	Material becomes plastic under heat conditions.	As above	Bauxite, plastics, rare earth fluorides, clays, electrodes

Table 3.9 *(cont'd.)*

Operation	*Equipment*	*Limitations, Comments*	*Product Form*	*Representative Products*
AGITATION				
Balling–the formation of tightly bound aggregates by the collision and coalescence of particles without a chemical reaction and usually in the presence of a wetting agent.	Rotary drum (cylinders)	Material tends to slide instead of roll. Amount of recycle is higher than that of disk or cone. Wets particles indiscriminate of size.	Spheroidal particles. Particles formed from pastes or slurries are stronger and harder than those formed with little or no moisture.	Low-grade iron ores, nonferrous ore, fertilizers, mineral and clay products, carbon black.
	Rotary pan (Disk)	Not well suited for processes involving chemical reactions.	As above	As above
	Cone	Recycle rate is between that of drum and pan.	As above	Magnetite ores
Granulation–spheroidal particles formed through interaction in a liquid phase to form seed nuclei particle growth takes place on these nuclei either due to cohesion (promoted by chemical reaction and/or temperature increase) or due to layering. Granulation is used extensively in the fertilizer industry.	Rotary drum	For general purpose granulation; not well suited for highly soluble salts.	Hard spheroid	Diammonium phosphate, nitric phosphate, granular super-phosphate, compounded fertilizers.
	Rotary pan	For granulation of highly soluble salts. Not well suited for ammoniation.	Hard spheroid	Ammonium nitrate, urea, ammonium phosphate nitrate, granular superphosphate.
	Pugmill	Not well suited for ammoniation; high recycle requirements.	Hard spheroid	Granular superphosphate.

Table 3.9 *(cont'd.)*

Operation	*Equipment*	*Limitations, Comments*	*Product Form*	*Representative Products*
Clustering–forming of loosely bound aggregates by surface wetting and by inducing collision and cohesion in a turbulent stream.	Conical blender	Suited for dry powders to give a spongelike structure that will dissolve readily to form a colloidal dispersion in a suitable solvent.	Irregularly shaped spheroid	Coffee, dextrin products, nonfat dry milk, starches.
HEAT REACTION AND BONDING Indurating the heat hardening of balls of ore or minerals by removal of moisture and volatile impurities. The forming and hardening of balls is sometimes called pelletizing.	Horizontal grate Shaft furnace Grate kiln Circular grate	Suitable for materials where moisture content is undesirable due to economics of shipping, or where volatile impurities are present that would interfere with some subsequent process.	Spheroids	Ferrous and nonferrous ores, minerals
Sintering–the formation of agglomerates through partial fusion of the material itself or an added binder.	Stationary hearth Rotary kiln Moving bed	Sinter is highly abrasive and breaks more easily than pellets during shipment.	Irregularly shaped clusters of particles	Ferrous and nonferrous ores, minerals
Nodulizing–the formation of hard lumps by partial fusion of individual particles of like or unlike materials. May be accompanied by chemical reaction.	Rotary kiln	Highly abrasive. Offers poor economics for shipment due to large void spaces between lumps.	Hard, irregularly shaped clusters	Cement clinker

magnitude as the weight of the particles at a 60-micron diameter;
2. electrostatic forces of repulsion or of attraction—between particles of like charge, these are repulsive;
3. friction-cum-mechanical forces due to a combination of interlocking and welding bridge formation; and
4. surface tension forces due to absorbed films or due to free liquid entrained within the powder mass.

It should be borne in mind that all of these forces are operative in a gravitational field.

There are three broad sets of methods for the agglomeration of particulate matter:

1. processes in which agglomeration occurs when individual particles are encouraged to touch, usually in the presence of a liquid;
2. processes in which agglomeration is achieved by the application of mechanical forces in tabletting and briquetting production; and
3. processes in which agglomeration is achieved through the application of heat.

An outline guide to agglomeration processes and equipment is given in Table 3.9.[81]

Mechanisms of Granulation

The mechanisms that occur during granulation are:

1. crushing and layering,[82,83,84]
2. agglomeration or coalescence,[82,85] and
3. layering or snowballing.[82,86]

It has been observed that the actual kinetics of the granulation process can occur by any one of these three mechanisms either alone or in some combination (either simultaneous or sequential).

The crushing and layering mechanism occurs if the granules are fairly weak and susceptible to fracture in the rotating granulator. Conditions that appear to favor this include a coarse feedstock powder which is closely graded and a low or deficient binding liquid. The equation relating the number of granules of a certain size occurring at time t contains a breakage function and a growth function defined as

$$\frac{\partial n, (x, t)}{\partial t} = -B(x, t)n(x, t) - \frac{\partial}{\partial x}[G(x, F)n(x, t)] \,, \qquad (3.43)$$

in which

$n(x, t)$ is the number of granules of size x at time t;

$B(x, t)$ is the breakage function, the fraction of the granules of size x broken per unit time at any time t; and

$G(x, F)$ is the growth function, the rate at which granules are growing to be greater than size x by picking up the crushed material whose volume is $F(t)$.

This means that the crushed material volume balance is given as

$$\frac{dF}{dt} = \int_0^\infty B(x, t)n(x, t)dx - \int_0^\infty G[x, F(t)]\,n(x, t)dx \qquad (3.44)$$

In the case of wet particulate systems, agglomeration is likely to occur by the mechanism of coalescence, especially in the nuclei growth region (see Fig. 3.45).[88]

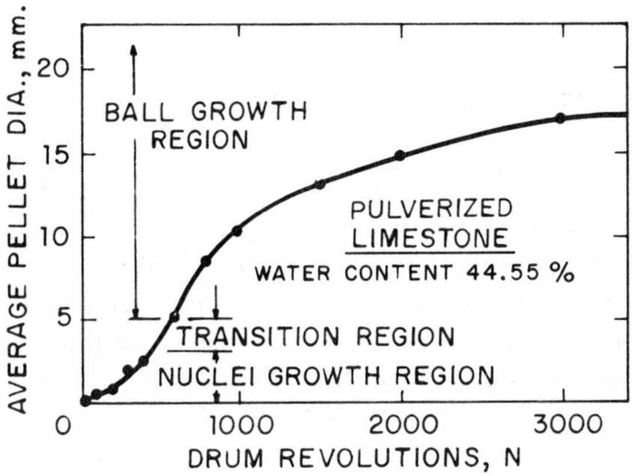

Fig. 3.45 Average pellet diameter as a function of the number of drum revolutions for pulverized limestone.[88]

Three Stages

The process of agglomeration is divided into three stages:[88]

1. the nuclei growth region is formed by simple coalescence of the rotating agglomerates;
2. the clump deforms into a rounded shape, and water is expelled because of reduced porosity; and

3. in the ball growing region, two or more balls clump together. This is coupled with abrasion and interball material transfer.

The equation for agglomeration by pure coalescence is:

$$\frac{dn(x,t)}{dt} = -\int_0^\infty \frac{\lambda(t)}{M_o(t)} n(x,t)n(\tilde{x},t)d\tilde{x} + \frac{1}{2}\int_0^x \frac{\lambda(t)}{M_o(t)} n(x-\tilde{x},t) \cdot n(\tilde{x},t)d\tilde{x} \tag{3.45}$$

in which $M_o(t)$ is the total concentration of particles at time t and is expressed as

$$M_o(t) = \int_0^\infty n(x,t)dx\ . \tag{3.46}$$

where

$\lambda(t)$ is the specific coalescence rate function.

Self-Preserving Distribution

It has been observed that during batch granulation, a constant ratio between the diameter of the largest particle and the diameter of the smallest particle is developed. This in effect means that a self-preserving granule size distribution is established.[89] However, in those processes in which there is a continuous addition of fines, the ratio between the largest and smallest particle diameter is not a constant (see Fig. 3.46). This mechanism of growth is common in large scale balling operations and is termed "snow balling."[86] As shown in Fig. 3.46, this mechanism leads to a narrow size distribution which is very desirable in many applications (e.g., iron ore pelletizing).

General Theory

A general theory of granulation in which the general population balance equation for the kinetics of granulation by the simultaneous operation of two of the mechanisms, crushing and layering, and coalescence, has been developed. Self-preserving solutions are obtained and a self-preserving size distribution can be developed with the simultaneous operation of the two mechanisms.[87]

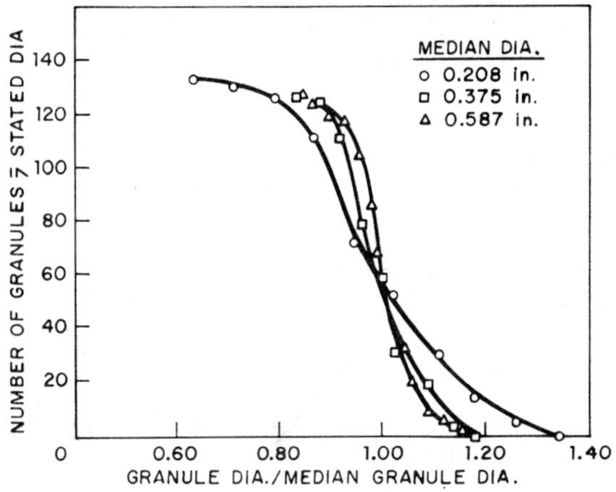

Fig. 3.46 Showing the size distribution changing in a continuously fed agglomeration process. Iron Ore feedstock. Illustrating the snowballing mechanism.[78,79]

It has been demonstrated that this general approach can be applied to aerosols undergoing simultaneous coagulation, turbulent diffusion, and growth by gas-to-particle condensation. Other systems and processes may also have appropriate applications in this approach. These include: grinding of solids, crystallization, aging of catalysts, etc.[89]

Industrial Scale Study

On the industrial front, a great deal of attention has been paid to the chemistry of various granulation processes but little concerted effort has been applied to modeling such systems.[90] A successful study of this has been reported in the case of the manufacture of diammonium phosphate.[90] Industrial practice is characterized by the incorporation of a crusher in the process circuit. This was assumed to be the source of seed particles. Another assumption was that the relative length of the residence time in the granulator was essentially random. The authors[90] were able to relate the size distribution data for the input material with that of the recycle material and then calculate the recycle ratio.

Strength of Agglomerates

As shown in Fig. 3.47[91] the level of strength exhibited by an agglomerate depends upon the type of bonding established or predominating in the agglomerate mass. It also depends upon the size of the individual particles of which the agglomerate is constructed.

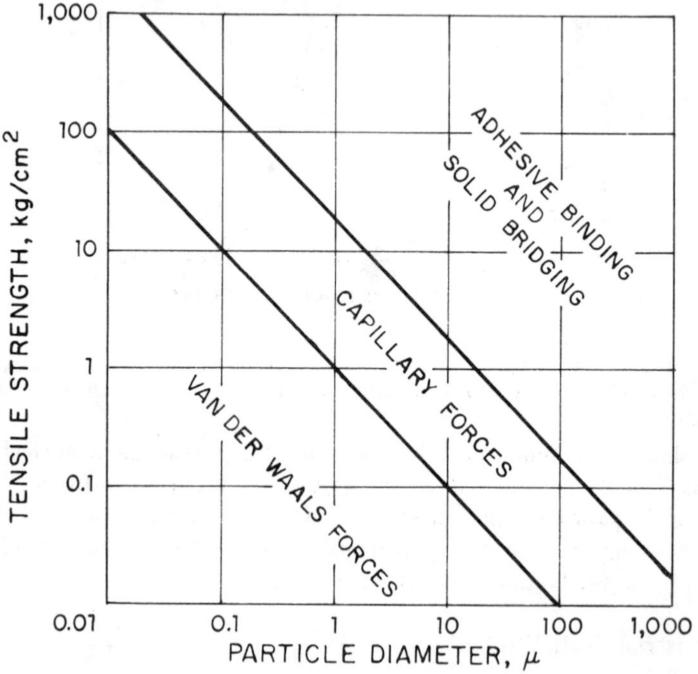

Fig. 3.47 Particle-size and agglomerate-strength regions in which various binding mechanisms predominate.[91]

Some interesting work has been reported in which it is demonstrated that simple modeling of the different bonding mechanisms conceived as being relevant in agglomeration can lead to a deeper understanding of the stress–strain behavior of agglomerated masses of particulates.[92]

Wet and Dry Agglomeration

There are two general types of agglomeration: dry agglomerates and wet aggomerates. Four distinct subtypes of wet agglomerates are shown in Fig. 3.48.[92]

The condition referred to as the pendular state has been successfully modeled and the following equation has been proposed and tested:[91,92]

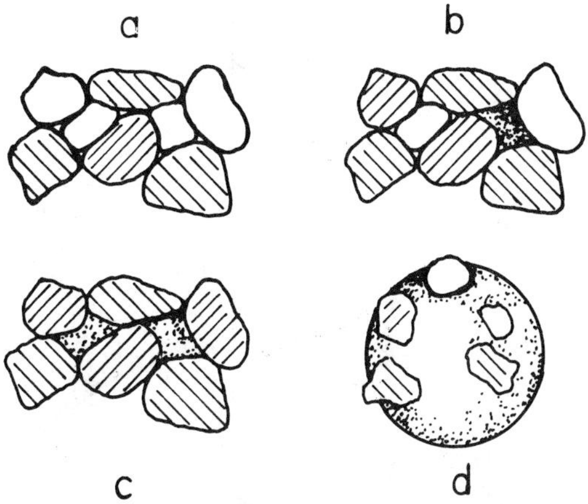

Fig. 3.48 Distribution of liquid in agglomerates: a) pendular state, b) funicular state, c) capillary state, d) liquid drop with particles inside or at its surface.[92]

$$\sigma_v = \frac{9}{8} \frac{(1-e)}{\pi d^2} \text{ K. H.} \tag{3.47}$$

in which

σ_v = agglomerate tensile strength,
e = void volume fraction,
K = a coordination number,
d = particle diameter, and
H = binding force of a point contact.

The coordination number K depends on porosity and on the product Ke which may take a value approximately equal to 3.1 or π. Values of d and e can be obtained from size distribution analysis and from the density of packing whereas in an idealized model of a liquid bridge structure (Fig. 3.49), H can be expressed simply as

$$H = \alpha df \cdot (\delta\ \beta)\,, \tag{3.48}$$

in which

α = surface tension,
δ = contact angle, and
β = sector angle of the liquid ring.

The model for the strain behavior of a moist agglomerate in the pendular

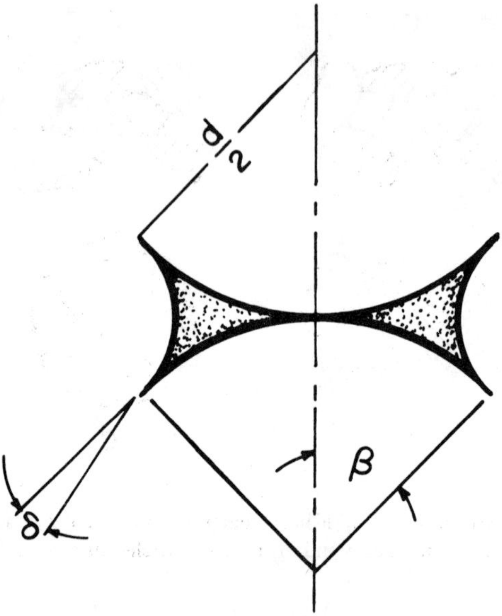

Fig. 3.49 Model of the liquid bridge.[91]

state and the model of fracture are both shown in Fig. 3.50.[92]

The stress–strain behavior can be predicted according to this model and experimental agreement can be observed.

The funicular and capillary states of agglomeration [see Fig. 3.48 b) and c)] deform to an extent that depends upon the degree of liquid saturation S so that the capillary pressure function P_c is

$$P_c = f(S) \tag{3.49}$$

It is interesting that the exact form of the relationship between agglomerate fracture strain and degree of saturation S depends upon whether the liquid is coming or going. Specifically, if drainage is occurring, the deformation is much less than if imbibition is occurring. This is illustrated in Fig. 3.51.[92]

Other work on the dry agglomerates (which were prepared by the drying of moist agglomerates) showed that they tend to brittle fracture because of their low strain at fracture.[92]

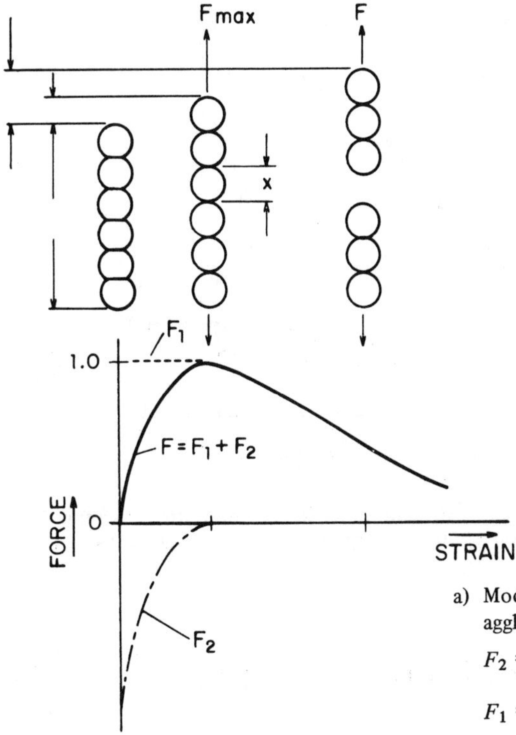

a) Model for strain behavior of moist agglomerates in the pendular state.

F_2 = repulsion due to sphere deformation

F_1 = adhesive force due to liquid bridge

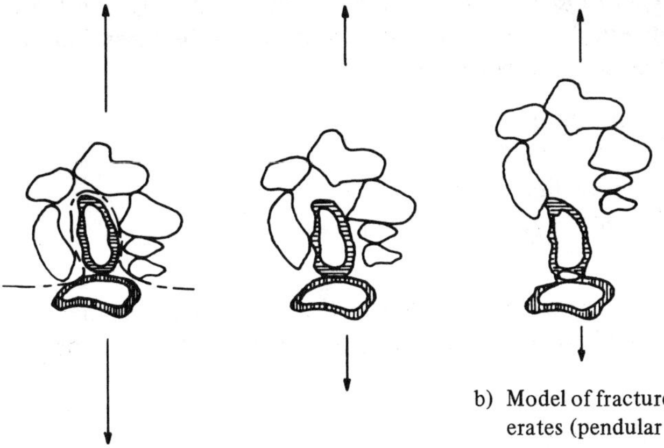

b) Model of fracture of moist agglomerates (pendular state).

Fig. 3.50[92]

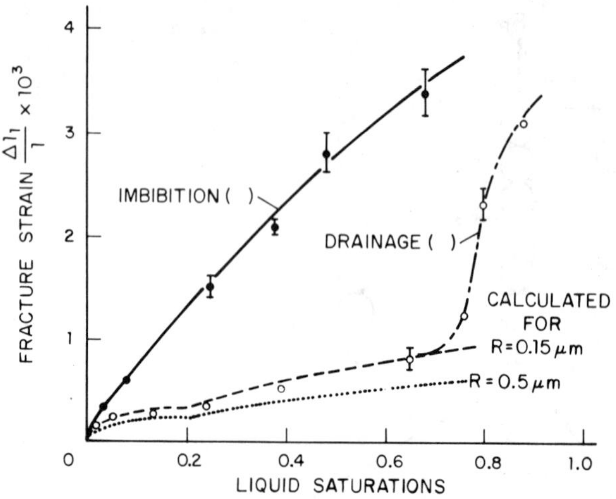

Fig. 3.51 Fracture strain $\Delta \int_1 / \int$ (= strain at maximum tensile stress) as a function of the liquid saturation S, for moist agglomerates in the cases of drainage and imbibition.[92]

3.8 AEROSOL PARTICLE GENERATION

There are numerous types of generating devices for the production of very fine particles of dioctyl phthalate (DOP) (this is a popular substance that produces spherical particles) and numerous other substances that can be obtained in solution. Among these are:

1. the Collison Atomizer,[93]
2. the DeVilbis Atomizer,[94]
3. the Lovelace Nebulizers,[95]
4. the vibrating orifice generator,[96] and
5. the atomizer generator.[97]

Because aerosol particle size is a function of the concentration of the solution, items 1-3 tend to offer a changing sized dispersion during operation.[98]

Vibrating Orifice Generator

A sketch of the vibrating orifice monodisperse aerosol generator is shown in Fig. 3.52.[96] This device produces particles over 0.5 micron diameter. The ratio of measured to theoretical concentration is about 80 percent up to a diameter of approximately 6 microns, then it drops off thereafter as shown in Fig. 3.53.[96] Details of the equipment characteristics are given in Table 3.10.[96]

Table 3.10

CHARACTERISTICS OF THE VIBRATING ORIFICE MONODISPERSE AEROSOL GENERATOR.[96]

Diameter of liquid orifice (μm)	*Nominal frequency* (kHz)	*Droplet diameter* (μm)[a]	*Particle diameter range* (μm)[b]	*Nominal concentration* (particles/cc)
5	450	15	0.6–15	273
10	225	25	1.0–25	137
20	60	40	1.8–40	36

[a] Continuously adjustable over an approximate 25 percent range by varying the frequency.
[b] Obtainable by the solvent evaporation technique.
[c] Theoretical concentration based on the nominal aerosol output 100 lpm.

Atomizing Generators

The aerosol generating equipment of the atomizing type is shown in Fig. 3.54.[97] It consists of an atomizer, an evaporator-condenser conditioner, an electrostatic classifier with an electric aerosol detector and a neutralizer.[98]

The atomizer can generate a moderately monodisperse aerosol from nonvolatile liquids (such as DOP) dissolved in ethyl alcohol and from aqueous solutions such as methylene blue. Suspensions of presized latex spheres and carbon (nonspherical) and iron oxide (spherical) particles with a mean particle size in the range of 0.01–0.3 μm may also be produced by mixing these particles in pure demineralized water, or in alcohol.

The evaporator-condenser is included to convert the moderately monodisperse aerosols produced by the atomizer from liquids (like DOP) into a more monodisperse suspension by an evaporation process followed by a condensation process.

The electrostatic classifier serves as the final stage in selecting particles,

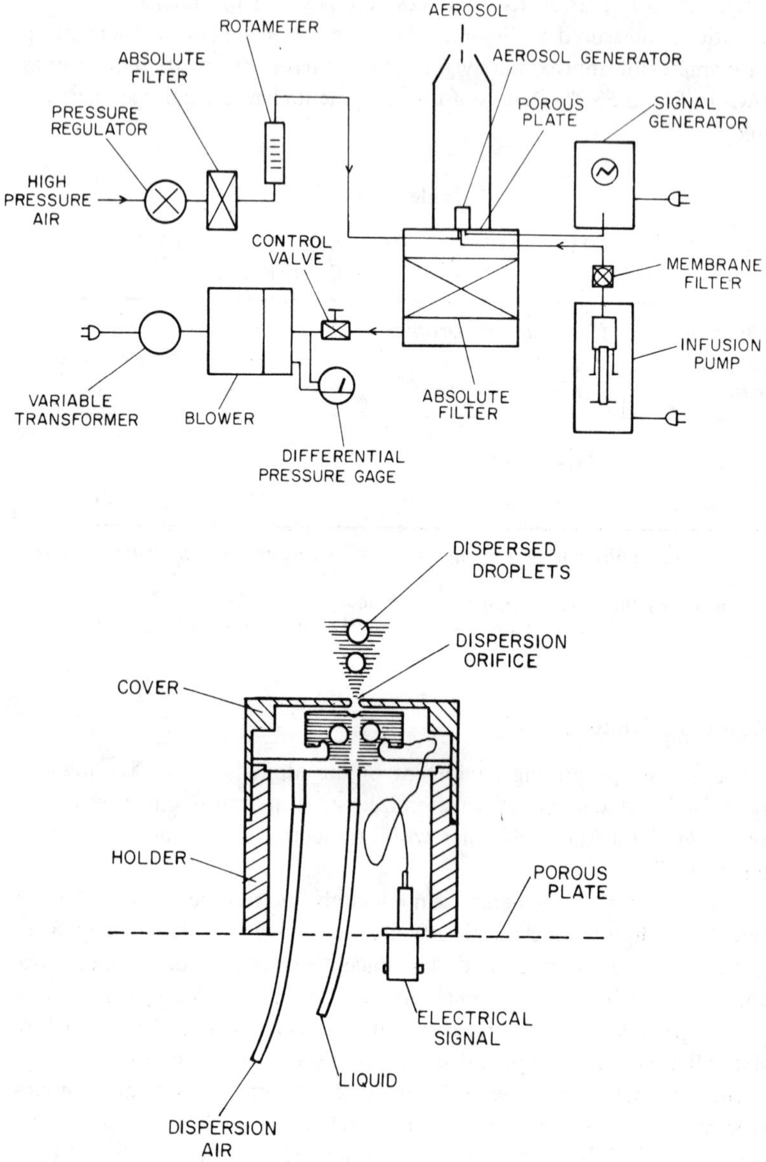

Fig. 3.52 Vibrating orifice monodisperse aerosol generator (Top: schematic of system. Bottom: generator head).[96]

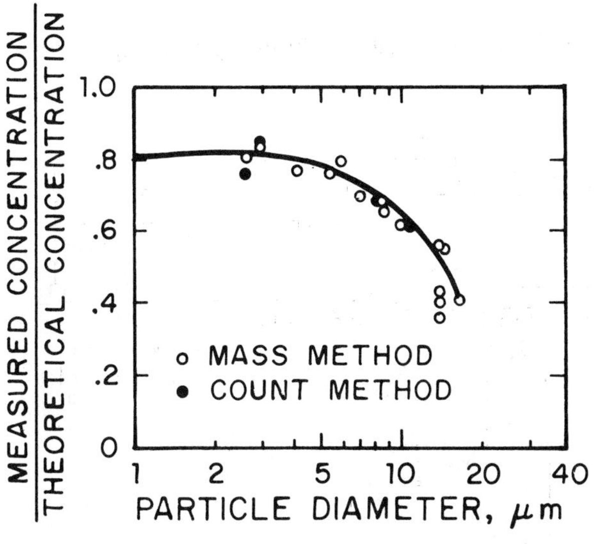

Fig. 3.53 Output aerosol condensation of the vibrating-orifice aerosol generator as a function of particle diameter.[96]

on the basis of their electric mobility, in order to make up a truly monodisperse aerosol. The classifier is a Kr-85 bipolar charger. This establishes an equilibrium bipolar charge distribution among the particles; part of the classifier, also, is a differential mobility analyzer which extracts only those particles with mobilities that lie within a narrow range. Since particles below approximately 0.3 μm are singly charged, the electric mobility of the charged particles is related to the particle size. Hence, the particles extracted with a narrow mobility range are highly monodisperse (and are charged). (The particle diameter is determined from the known voltage applied in the mobility analyzer section.) An electric aerosol detector is included so that the particle concentration of the generated monodisperse aerosol can be determined when desired.

A Kr-85 neutralizer can be used when a neutral aerosol is desired for the coagulation tests. This system is capable of producing uniform particles from 0.01 to 0.5 μm at concentrations up to 10^6 cm^{-3}. Concentrations up to 10^7 are possible with DOP aerosols.

The particle diameter of the aerosols generated by this system can be determined to within an accuracy of about 2 percent and the concentration to within about 5 percent. The relative standard deviation of the size distribution for aerosols produced by this system ranges from about 0.04 to 0.08.[99]

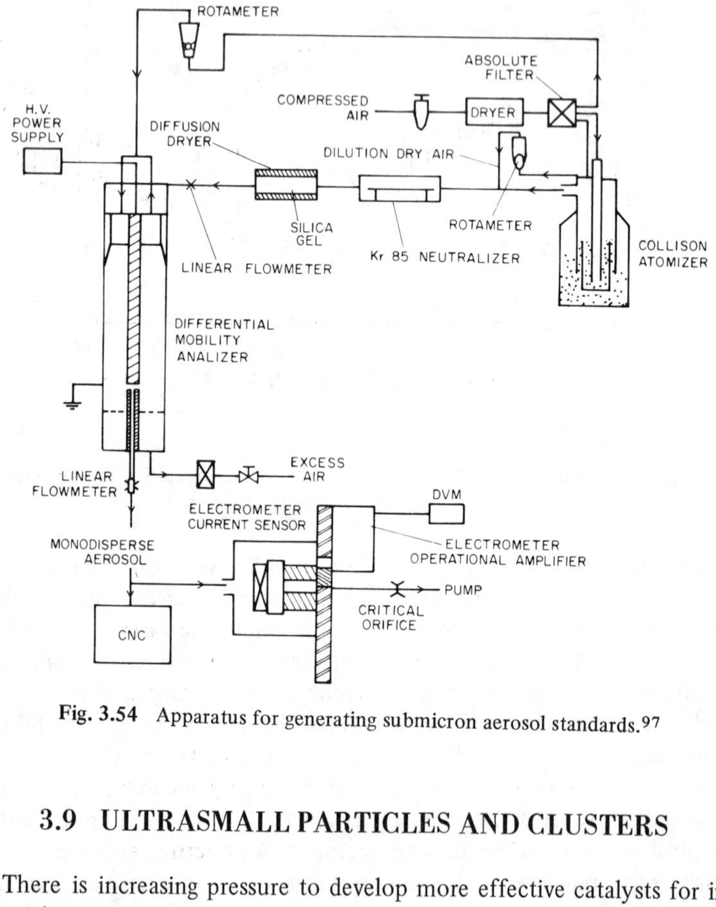

Fig. 3.54 Apparatus for generating submicron aerosol standards.[97]

3.9 ULTRASMALL PARTICLES AND CLUSTERS

There is increasing pressure to develop more effective catalysts for industrial reactions due to increased interest in reducing waste; and, at the same time, increasing the efficiency of energy utilization is also needed.[100] Catalysts are divided into two groups:

1. homogeneous catalysts, and
2. heterogeneous catalysts.

Heterogeneous catalysis usually involves the exposure of a metal or metal oxide with high surface area per unit quantity of catalyst to gaseous or liquid reactants. The latter becomes chemisorbed onto the catalysts surface where the reaction(s) occur(s). These catalysts are relatively stable and can be separated from the reaction products (or vice versa) to be reused again and again.

On the other hand, homogeneous catalysts are discrete entities which dissolve in the reaction medium as do the reactants themselves. Their recovery is therefore much more difficult than is the recovery of heterogeneous catalyst. The latter is the more popular of the two groups of catalysts. However, the homogeneous catalysts have a very strong advantage in that they are very specific in the reactions they promote.

Ultrasmall particles have been prepared by a number of methods including the following.

Codeposition of Metal Vapors

Codeposition of metal vapors with vapors of organic or inorganic compounds has been carried out in a reaction vessel as shown in Fig. 3.55.[101]

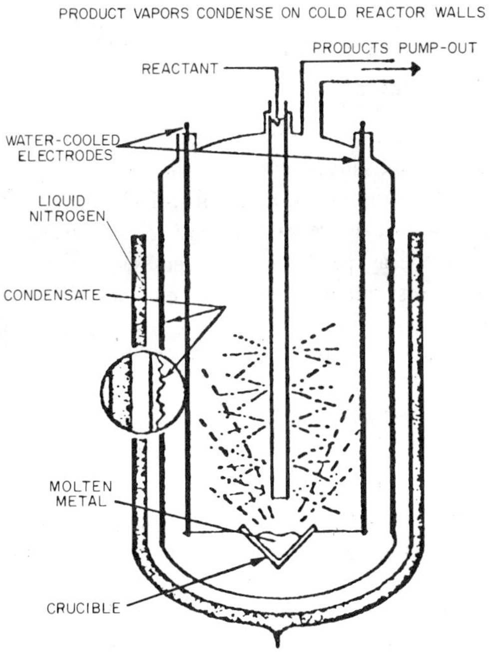

Fig. 3.55 Reaction vessel for ultrasmall particle production.[101]

A metal is vaporized in an electrically heated crucible in a vacuum flask the walls of which are cooled to -196°C. The metal atoms vaporize and migrate to the vessel wall. Reactant molecules enter the vessel and also condense on the walls. There is sufficient activity for reactions to occur to produce a solid reactant product. At least five different reaction products are possible. These are:

1. oxidation addition (insertion),
2. abstraction,
3. electron transfer,
4. simple orbital mixing, and
5. cluster formation.

Cluster Formation

Metal atoms can be dispersed in organic media such as pentane, hexane, toluene, etc. Such media are only weakly complexing and are inert to addition, abstraction, and electron transfer processes. At low temperatures, these media insulate the metal atom but as the temperature is raised there is an increasing tendency for metal atom clusters to form. It is possible to produce active metal slurries in this way.[101]

Gedenken Experiments

There has been much interest in the properties of metal clusters.[102] Theoretical calculations involve the structural and energy characteristics of clusters of metal atoms in the range of 2–100 or more.[103-108] Some of these properties are of interest in processing. In the case of Li clusters[104] the atomization energy per Li atom increases as the number of atoms in the cluster increases; but even when the cluster reaches the size of more than 100 atoms, the atomization energy is substantially less than that which corresponds to the bulk metal. See Fig. 3.56.[104]

The corresponding value of the lattice parameter is calculated and graphed in Fig. 3.57.[104]

It has been shown that ionization potential, electron affinity, and energy band spread depends upon the size of the aggregate or cluster.[103] This effect in the case of the energy bands is shown in Fig. 3.58 for Ag clusters.[103]

Furthermore it has been demonstrated that the bond energy of silver particles increases as their size increases, but at a 20-atoms size, it is only one-third of the cohesive energy of bulk Ag. This means that such clusters have the propensity to aggregate to form larger particles depending upon the circumstances.[103] Also, the calculations show that at small cluster

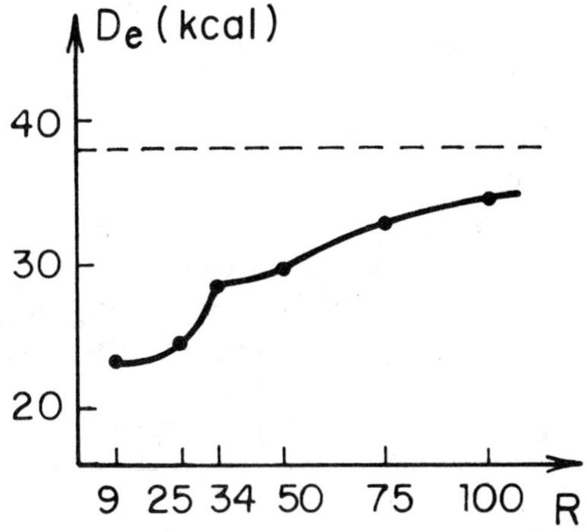

Fig. 3.56 The atomization energy per Li atom versus cluster size (the dotted line corresponds to the D_e value for the crystal).[104]

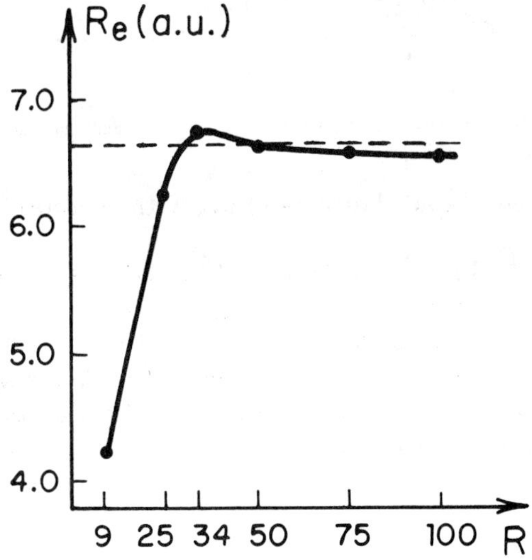

Fig. 3.57 Li BCC cluster parameter versus the size of the cluster (the dotted line corresponds to the R_e value for the crystals).[104]

sizes, linear Ag aggregates are more stable and as the cluster size increases, above c.a. 30 atoms, a transition from linear to fcc geometry is expected.[103] From this work, it is clear that the properties of clusters may be very different from those of the bulk of the same substance. This complicates the ways of measuring the properties of larger aggregate particles.

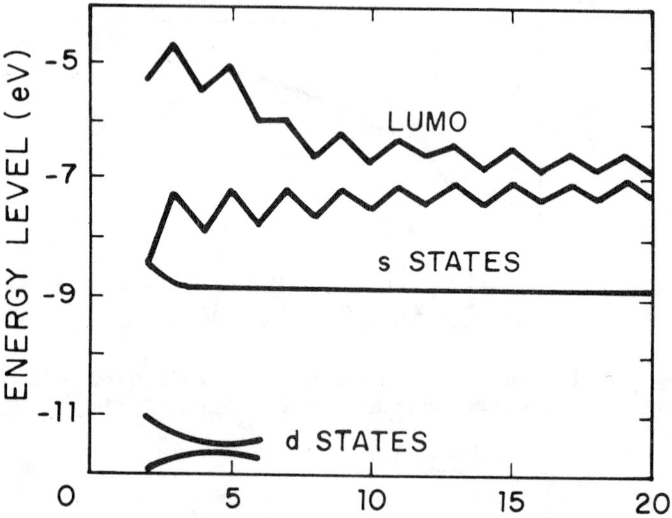

Fig. 3.58 Dependence of energy bands in Ag on aggregate size.[103]

Laser-Promoted, Rapid, Localized Energy Deposition[109]

Lasers provide the possibility for rapid localized energy deposition in the gas phase (e.g., temperatures in excess of 1000°K coupled with ignition delay and rise times of 10^{μ}s through 100^{μ}s).[101] Fig. 3.59 shows samples of some metal and metal oxide powders prepared in this way. Details are written under the figure. It should also be noted that because of the large rise and quench times available it should be possible to obtain species which are normally difficult to prepare via routes that are subject to thermodynamic restrictions.

Catalysis

Metal particles ranging in size from 5 through 100 Angstroms are prepared by simply impregnating a support with a solution of a reducible salt, followed by drying and hydrogen reduction.[110]

(a) Fe_2O_3 from $Fe(CO)_5 + N_2O$

(b) Fe from Fe(CO)

(c) Fe/Bi from $Fe(CO)_5 + Bi(CH_3)_3$

(d) Fe/Bi from $Fe(CO)_5 + Bi(CH_3)_3$

MAGNIFICATION ~ 100,000

Fig. 3.59 Electron micrographs of powders prepared by irradiating (10.6 μ, CO_2 laser) mixtures of SF_6 with various precursors.[109]

TLK Model of a Crystal Surface

The Terrace-Ledge-Kink model of a crystal surface has been utilized in a very interesting study of the reactions occurring on single crystal surfaces.[111] For example, the effects of A step density and B kink density (at a constant step density) upon the initial rate of dehydrogenation of cyclohexane to benzene on a platinum single crystal surface is shown in Fig. 3.60.[111]

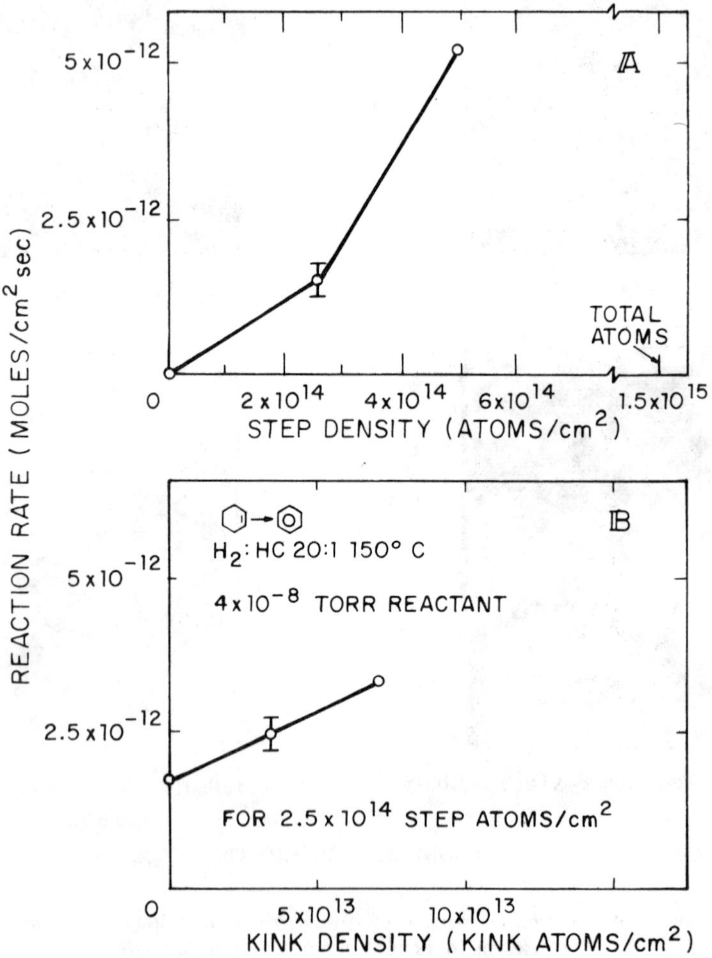

Fig. 3.60 Initial rate of cyclohexene dehydrogenation to benzene on platinum single-crystal catalysts as a function of (A) increasing step density and (B) increasing kink density at constant step density. The reaction conditions are 5×10^{-8} Torr of cyclohexene, 1.0×10^{-6} Torr of hydrogen and 423 K catalyst temperature.[111]

REFERENCES

1. McCabe, W. L., and Smith, J. C. *Unit Operations of Chemical Engineering.* second edition, New York: McGraw-Hill, 1967.
2. Beddow, J. K. *The Production of Metal Powders by Atomization.* London: Heyden and Sons, 1977.
3. Lubanska, H. J. *J. of Metals,* 22 (1970) pp. 45–49, ©American Institute of Mining, Metallurgical and Petroleum Engineers, 345 E. 47th St., NY, NY 10017.
4. Rao, P. "Shape and Other Properties of Gas Atomized Metal Powders." Ph.D. thesis, Chemical Engineering Department, Drexel University, Philadelphia 1973.
5. Gummeson, P. U. "High-Pressure Water Atomization." *Powder Metallurgy for High Performance Applications,* 18th Sagamore Army Materials Research Conference, ed. by John J. Burke and Volker Weiss (Syracuse: Syracuse University Press, 1972), p. 35.
6. Grandzol, R. J. "Water Atomization of 4620 Steel and other Metals." Ph.D. thesis, Drexel University, 1973. See also, Grandzol, R. J. and Tallmage, J. A. *Int. J. P. Met.* and *P. Tech.* 11, 1 (1975): 103–114.
7. Marshall, W. R., Jr. "Atomization and Spray Drying" *Chem. Eng. Prog. Monograph Series* L, 2, New York: AIChE, 1954.
8. Lord Rayleigh. *Proc. London Math. Soc.* 10 (1878): 4–13.
9. Plateau, "Statistiques Des Liquides." 1873.
10. Tyler, E. *Phil Mag.* 16 (1933): 504.
11. Work of Gerrard reported in *Chem. and Eng. News*, p. 114, Nov. 26, 1962.
12. Weber, C. *Zfur Angewand Math and Mech.* 11, 2 (1931): 136–154.
13. Ohnesorge, W. *Z Angew Math and Mech.* 16 (1936): 355.
14. Monk, J. *Applied Physics* 23 (1952): 288.
15. Lawley, A. Education Luncheon Address, Powder Metallurgy International Meeting, Chicago, 1976.
16. Hinze and Milburn. *J. App. Mech.* 17 (1950): 145.
17. Adler, C. R., and Marshall, W. R. *Ch. Eng. Prog.* 47, 515, 601 (1951).
18. Lane, W. R. National Adv. Comm. Aeronautics Report 425, 1932.
19. Schuebel, F. N. Nat. Adv. Comm. Aeronautics Technical Memo 644, 1931.

20. Lane, W. R. *Ind. and Eng. Chem.* 43 (1951): 1312.
21. Castleman, R. A. Res. National Bureau of Standard 6, RP. 281, 369, 1931.
22. Grant, N. J. *Fizika* 2, 16 (1970). See also Doherty, R. D., Feest, A., and Holm, K. *Met. Trans.* 4, 115 (1973).
23. Duwez, P., and Willens, R. H. *Trans. Metallurgical Society of AIME* 222 (1963): 362–365 © American Institute of Mining, Metallurgical and Petroleum Engineers, 345 E. 47th St., NY, NY 10017.
24. Moss, M. *Acta. Metallurgica* 16 (1968): 321–326.
25. Willens, R. H. and Buehler. *Trans. Metallurgical Society of AIME* 236 (1966): 171–174. © American Institute of Mining, Metallurgical and Petroleum Engineers, 345 E. 47th St., NY, NY 10017.
26. Martin, J. W., and Doherty, R. D. *Stability of Microstructure in Metallic Systems,* Cambridge: Cambridge University Press, 1976.
27. Ranze, W. E., and Marshall, W. R. *Chem Eng. Progr.* 48, 141, 173 (1952). See also Frossling, N. Gerlands Beitr, *Geophys.,* 52, 170 (1938).
28. Lowrison, G. C. *Crushing and Grinding.* Cleveland: CRC Press, 1974. Address of author: 4 Cranbourne Rd., Bradford BD9 6BH, Yorkshire, England. See also Snow, R. H. and Luckie, P. T. *Powder Technology.* 13 (1976): 33–48.
29. Bond, F. C. and Wang, J. T. *Trans. AIME* 187 (1950): 871. © American Institute of Mining, Metallurgical and Petroleum Engineers, 345 E. 47th St., NY, NY 10017.
30. Makarov, V. I., and Larin, A. P. *Ogneupory* 9, 18 (1968).
31. Rose, H. E., and Sullivan, R. M. E. *Ball, Tube and Rod Mills.* London: Constable, 1958.
32. Hukki, R. T. *Trans. AIME.* May 1958, 581.
33. Pit and Quarry Handbook Chicago, 1958.
34. Berry and Kamack, Proc. Int. Cong. Surface Activity 2nd (IV), pp. 196–202, London: Butterworth, 1951.
35. Nigman, J. Dechema Zerkleinern Symposium 249, 1962.
36. Rose, H. E., and Blunt. *Proc. Inst. Mech. Eng.* 170, 23 (1956): 793–800.
37. Joisel, A. Dechema Zerkleinern Symposium, 49, 1962.
38. Carey, W. F., and Stairmand, C. J., Inst. Min. and Met. Symposium on Mineral Dressing, 1952.
39. Rose, H. E., and Evans. *Proc. Inst. Mech. Eng.* 170, 23. (1956): 775.
40. Carey, W. F., and Stairmand, C. J. "Recent Advances in Min-

eral Dressing." Symposium Inst. Min. and Met. on Mineral Dressing, 1952, 10.
41. Gardner, R. P., and Austin, L. G. Dechema Zerklienern Symposium, 1962, 218
42. Garside, J., and Wilsmith, J. A. *Trans. Inst. Chem. Eng.* 47 (1969): 285.
43. Kelleher, J. "Comminution Theory and Practice." *Brit. Chem. Eng.* 4,472, (1959).
44. Broadbent, S. R., and Callcott, T. G. *J. Inst. Fuel* 29 (1956): 191.
45. Brown, R. L. *Br. Chem. Engr.* 4 (1959): 463.
46. Carey, W. F., and Halton, E. M. "Energy Flowsheet in Milling Practice." *Trans. Inst. Chem. Engr.,* 1947. Quoted by Stairmand in *Powders in Industry.* Society of Chemical Industry Monograph 14, 1961, London p. 23.
47. Stairmand, C. J. discussion contribution, *Powders in Industry.* Society of Chemical Industry Monograph 14, London (1961): 23.
48. Von Rittinger Lerhb d. Aufbereitung Kunde, Berlin, 1867.
49. Kick Das Gesctz d. Proportionalen Widerstande u Sein Ardwerdung Leipzig, 1876.
50. *F. Bond Chem. Eng.* 59 (1952): 242.
51. Orr, C. *Particulate Technology.* New York: Macmillan, 1966.
52. Hukki, R. T. *Trans. AIME* 223 (1962): 402. © American Institute of Mining, Metallurgical and Petroleum Engineers, 345 E. 47th St., NY, NY 10017.
53. Meloy, T. P. "Fine Grinding–Size Distribution, Particle Characterization and Mechanical Methods" in *Ultrafine Grain Ceramics,* ed. J. J. Burke et al, 17–37, New York: Syracuse University Press, 1970.
54. Gilvarry, J. J. "Fracture of Brittle Solids. I. Distribution Function for Fragment Size in Single Fracture (Theoretical)." *J. Appl. Phys.* 32 (1961): 391–397.
55. Meloy, T. P. "A Three Dimensional Derivation of the Gaudin-Meloy Size Distribution Equation." *Trans. Am. Inst. Mining Engrs.* 226 (1963): 447–48.
56. Meloy, T. P. AICHE, Discussions of Comminution Size Distributions and Particle Characterization, (1963). See reference 53.
57. Austin, L. G., and Klimpel, R. R. "Statistical Theory of Primary Breakage Distributions for Brittle Materials." *Trans. Am. Inst. Mining Engrs.* 232 (1965): 88–94.
58. Dunning, W. J. *Powders in Industry.* London: Society of Chemical Industry Monograph 14, pp. 29–43, 1961.

59. Dunning, W. J. *Powders in Industry.* Society Chem. Indust. London, SCI Monograph 14, pp. 29–43, 1961.
60. Branson, S. H., Dunning, W. J., and Millard, B. *Disc Faraday Society* 5 (1949): 83.
61. Randolph, A. D., and Larson, M. A. *Theory of Particulate Processes, Analysis and Techniques of Continuous Crystallization,* NY: Academic Press, 1971.
62. Buckley, H. E. *Crystal Growth.* NY: Wiley, 1951.
Larson, M. A. ed., "Crystallization from Solution: Factors Influencing Size Distribution." *Chem. Eng. Progr. Symp. Ser.* 110, 67 (1971).
Nielsen, A. E. *Kinetics of Precipitation.* NY: Macmillan, 1964.
Mullin, J. W. *Crystallization.* London: Butterworth, 1961.
Palermo, J. A., and Larson, M. A. ed., "Crystallization from Solutions and Melts." *Chem. Eng. Progr. Symp. Ser.* 95, 65 (1969).
Strickland-Constable, R. F. *Kinetics and Mechanisms of Crystallization.* NY: Academic Press. 1968.
63. Veale, C. R. *Fine Powders.* New York: Wiley (with permission of Her Majesty's Stationery Office), 1972.
64. Gregg, S. J. *Chem and Ind.* (1968): 611.
65. Bickle, W. H. "Powder Production by Fine Milling" in *Powders in Industry.* London: Society of Chemical Industry Monograph 14, pp. 3–22, 1961.
66. Meloy, T. P. in *Ultrafine Grain Ceramics.* edited by Burke, J. J., et al., Syracuse University 17, 1970.
67. Quatinez, M., Schafer, R. J., Smeal, C. R., in *Ultrafine Particles,* edited by Kuhn, W. E., 271. New York: Wiley, 1963.
68. Rose, H. E. Dechema Zerkleinern, Symposium, 439, 1962.
69. Talbot, J. H., and Kempis, E. R. *Nature.* (1966): 197, 66. See also Hofmann, U., and Rothe, A. *Zeith An Chem.* (1968) 357, 196.
70. Grieshover, "Unique Methods for Producing Pr–20% Ir Prealloyed Powders." Battelle Columbus Labs., 1969.
71. Panlovaskaya, Lev, and Barkan. *Poroshkoyamet* 6, 8 (1963): 99–105.
72. Arias, *Powder Metallurgy* 11, 22 (1968): 411–429.
73. Holmgren, J. D., Gibson, J. O., Sheer, C., in *Ultrafine Particles.* edited by Kuhn, W. E., p. 129, New York: Wiley, 1964.
74. Everest, D. A., Sayce, I. G., Selton, B. *J. Mat. Sci.* 6 (1971): 218.
75. Powell, C. F., Oxley, J. H., and Blocher, J. M., eds. *Vapor Deposition.* New York: Wiley, 1966.

76. Shaffhauser, A. C. ed., *Chemical Vapor Deposition of Refractory Metal Alloys and Compounds.* American Nuclear Society, 1967.
77. Clough, P. J. "New Types of Metal Powders." edited by Hausner, H. H. *Met. Soc. Conf.* 23, pp. 9–19, Gordon & Breach, New York, 1964.
78. Nielson, M. L., Hamilton, P. M., and Walsh, R. J., in *Ultrafine Particles* edited by Kuhn, W. E., p. 181, New York: Wiley, 1963.
79. Hammel, F. *Ann Chem.* 11, (1939): 247.
80. Pilpel, N. Endeavor, 28, 104 (1969): 73–76.
81. Browning, J.E. *Chemical Engineering.* (Dec. 4, 1965) 147–170.
82. Capes, C. E., Danckwerts, P. V., *Trans. Inst. Chem. Engrs.* 43T (1965): 125.
83. Kapur, P. C. *Chem. Eng. Sci.* 26 (1971): 1093.
84. Capes, C. E., Danckwerts, P. V. *Trans. Inst. Chem. Eng.* 43T (1965): 116.
85. Kapur, P. C. *Chem. Eng. Sci.* 27 (1972): 1863.
86. Capes, C. E. *The Chem. Engr.* CE78, April, 1967.
87. Ramabhadran, T. E. *Chem Eng. Sci.* 30 (1975): 1027–1033.
88. Kapur, P. C., and Fuerstenau, D. W. *I & EC Process Design and Development.* 5, 1 (Jan 1966): 5–14.
89. Ramabhadran, T. E., and Shenfield, J. H. *Chem. Eng. Sci.* 30 (1975): 1019–1025.
90. Han, C. D., and Wilenitz, I. *Ind. Eng. Chem. Fundam.* 9, 3 (1970): 401–411.
91. Rumpf, H. in *Aggomeration.* edited by Knepper, W. A., p. 399, New York: Wiley, 1962.
92. Schubert, H., Herrmann, W., and Rumpf, H. *Powder Technology.* 11 (1975): 121–131.
93. Collison Atomizer, British Standards Institute, British Standard 2577, 2 Park Street, London, W 1, 1955.
94. DeVilbis Atomizer, the DeVilbis Co., Somerset, Penn., USA.
95. Lovelance Nebulizers, Retec. Development Lab., 9730 S. W. Schools Ferry Road, Portland, Oregon, 97223, USA.
96. Liu, B. Y. H. *APCA Journal.* 24, 12 (Dec. 1974): 1170–1172.
97. McCabe, W. L., Smith, J. C. *Unit Operations of Chemical Engineering.* New York: McGraw-Hill, 2nd ed., 1967. (Liu, B.Y.H., and Pui, D.Y.H. *J. Coll. and Interface Sci.* 47, 1 (1974): 155).
98. Lui, B. H., and Lee, K. W. *Am. Ind. Hyg. Ass. J.* (Dec. 1975): 861–865.
99. Scholz, P. D. private communication. Feb. 1977.

100. Skinner, K. J. *Chemical and Engineering News* (Feb. 1977): 18–19.
101. Murray, C. *Chemical and Engineering News* (Jan. 24, 1977): 23–24.
102. Stwalley, W., personal communication, Jan. 1977.
103. Baetzold, R. C. *J. Chem. Phys.* 55, 9 (Nov. 1971): Part I, 4355–4363. Part II 4363–4370.
104. Borisov, Yu A. *Chem Phys. Letters.* 44, 1, 15 (Nov. 1976): 16–19.
105. Whitehead, J. C., and Grice, R. *Mol. Phys.* 26, 2 (1973): 267–280.
106. Gelb, A., Jordan, K. D., and Silbey, R. *Chem. Phys.* 9 (1975): 175–182.
107. Ellison, F. O., and Delle-Donne, M. J. *J. Chem. Phys.* 59, 11, 1 (Dec. 1973): 6179–6180.
108. Raffenetti, R. C., and Ruedenbuerg, K. *J. Chem. Phys.* 59, 11, 1 (Dec. 1973): 5978–5991.
109. "The Laser Revolution in Energy-Related Chemistry." ed. J. Steinfeld, National Science Foundation, Washington, D.C., May 9–11, 1976.
110. Bond, G. C. *The Physical Basis for Heterogeneous Catalysis.* pp. 53–71, New York: Plenum Press, 1975.
111. Samorjai, G. A. *Accounts of Chemical Research.* 9 (1976): 248–256.
112. Newell & Dunford Engr., Ltd., Misterton via Doncaster DN 10 4DN, South Yorkshire, England.
113. Klar and Schaeffer. *Powder Metallurgy for High Performance Applications.* Syracuse: Syracuse U. Press, 1971, pp. 57–68.
114. Koppers Co., Inc., 1–83 Industrial Park, York, Pennsylvania 17405.
115. Allis-Chalmers, Box 2219, Appleton, Wisc. 54913.
116. Craven Fawcett (Wakefield) Limited, PO Box No. 21, Dewsbury Rd., Wakefield WF2 9BD, England.

Chapter 4

THE PROCESSING AND HANDLING OF PARTICULATE MATTER

4.1 CURRENT STATE OF THE ART

This chapter presents a brief review of some of the industrially important processes in powder technology so that the current state of the art can be appraised. With this knowledge, a judgement between the fundamental-scientific approach and the empirical-macroscopic approach in these variously important branches of technology can be made along with developing a whole-field view of fine particle science and technology.

In point of fact, the empirical-macroscopic approach has been quite successfully applied until recent time. Now there are growing pressures for the development of new and better technology in response to the demands for improvements in energy utilization, raw materials utilization, and environmental quality control, to name but three, and to push the technology forward requires a much more forceful application of science. Great advances have already taken place in the fields of solids, liquids and gases. Moreover, our knowledge, coupled with our other allied resources, is cumulative and thereby (one would expect) even more potent today when applied to current problems. It is important to realize that the problems of powder technology are more fully manifest in highly industrialized societies, because enormous quantities of matter in finely divided form are handled and processed in our modern plants.

In addition to the pressure for the development of new technology mentioned above, already established industries (chemicals, plastics, food) are reappraising their own situations because there is a growing realization that plugging numbers into some handy formula is just simply not good enough when one is faced with the job of constructing a new plant, improving the design of an old one, or making pilot plant studies.

A whole-field view of particle science and technology must be developed in order to advance our processing. There is not enough understanding of what is actually going on in most, if not all, of our processes. To gain a better understanding, new applications of science in crucial areas must be developed, our imaginations must be tapped in order to develop new models of our processes which can lead to improved technology. One of the most crucial scientific areas is that of particle morphological analysis. And our new models might well be based upon the concept that a particulate material has a structure which is changed during processing, whether it be flow, mixing, pneumatic, or hydraulic transport. It is more than a coincidence that both of these extremely important areas are concerned with pattern recognition.

Every one who has either operated a plant or worked in one is aware of disastrous structures of storage hoppers, silos, and bunkers that unwitting designers have foisted on the unwary customers. Indeed, the world is littered with nonfree flowing hoppers. The drive to try to change this situation was implemented in the 1950's[2] but the difficulties are still with us. This illustrates the very slow pace of improvements in powder technology during the last 20 years. Research in mixing has been going on for many years and the general approach has been to utilize statistical methods and/or analogs of the behavior of matter in other forms. Only recently is it being suggested that mixing might be approached more directly by measuring the properties and structure of the particulate matter in the mixing process in relationship to the imposed changes in the environmental conditions.[53,59] This concept of the particulate matter developing a structure in response to the imposed changes of the environment within which the matter is contained might find use in model development in many other dynamic systems. However, at the present time the dominant theme in powder technology is empirical-macroscopy which is discussed in this chapter.

The recent achievements should not be underestimated. The ingenuity of humankind in solving the many, varied, and often frustrating problems in the processing and handling of bulk solids is outmatched only by the subtlety of nature herself. Bulk particulate materials must be conveyed from place to place, both inside and outside of plant buildings. They often have to be loaded (or unloaded) into a variety of conveyances for transportation. In all of these operations, spillage must be minimized, and cleaned up afterwards. Related health and safety problems increase in their importance as society becomes increasingly aware of hazards. Once delivered, bulk finely divided solids may have to be stored in silos, bins, hoppers, etc. The storage of bulk solids can be a particularly aggravating problem because of several factors: inadequate design criteria in the initial plant construction; inability to see exactly what is going on inside the storage facility; subjection of the particulate mass when stationary to a variety of external influences including the pressure of the mass of material above; and the effects of air, moisture, and chemical agents in general as they permeate the particulate mass. A perennial problem is the tendency for feed materials to segregate. The resultant nonuniformity of the product may aggravate later processing steps to the extent that mixing to restore the required uniformity is necessitated. Mixing is an important part of powder technology.

The angler usually tells the famous story of the "big one that got away." In the handling of particulates, the little ones are most likely to escape, with potentially very serious consequences. Separation science and technology are complex areas of science and engineering that seek ever more efficient and effective ways to remove offending particulates from effluence. Some interesting problems are illustrated in Fig. 4.1.[1]

The topics discussed in this chapter include: storage, conveying and flow, beds, mixing, separation, compaction and sintering. They present an interesting comparison as to their composition regarding the relative proportions of the fundamental-scientific approach and the empirical-macroscopic approach. For example, as already mentioned, the storage of finely divided matter is currently dealt with at a macroscopic level, the mixing of particulate solids is moving in the fundamental-scientific direction (as is the compaction of particulate materials); while the sintering of fine particles has borrowed heavily from the advances made in solids during the past 25 years and is therefore much more heavily oriented towards the basic science side. These topics are widely representative of processing carried out in many industries. They are discussed in terms of the present state of the art. Thus, storage is essentially macroscopic and it might therefore tend to present an image that could be popularly labeled "design." However, in the sense that design is an attempt to impose a pattern of convenience upon an already existing natural order, virtually all of the topics of this chapter are dealing with design at one stage or another.

4.2 FLOW AND STORAGE OF PARTICULATE SOLIDS

In designing storage bins, two major problems are presented ensuring a satisfactory flow of the material through the bin and removing the material successfully from the bottom of the bin. There are two excellent reference

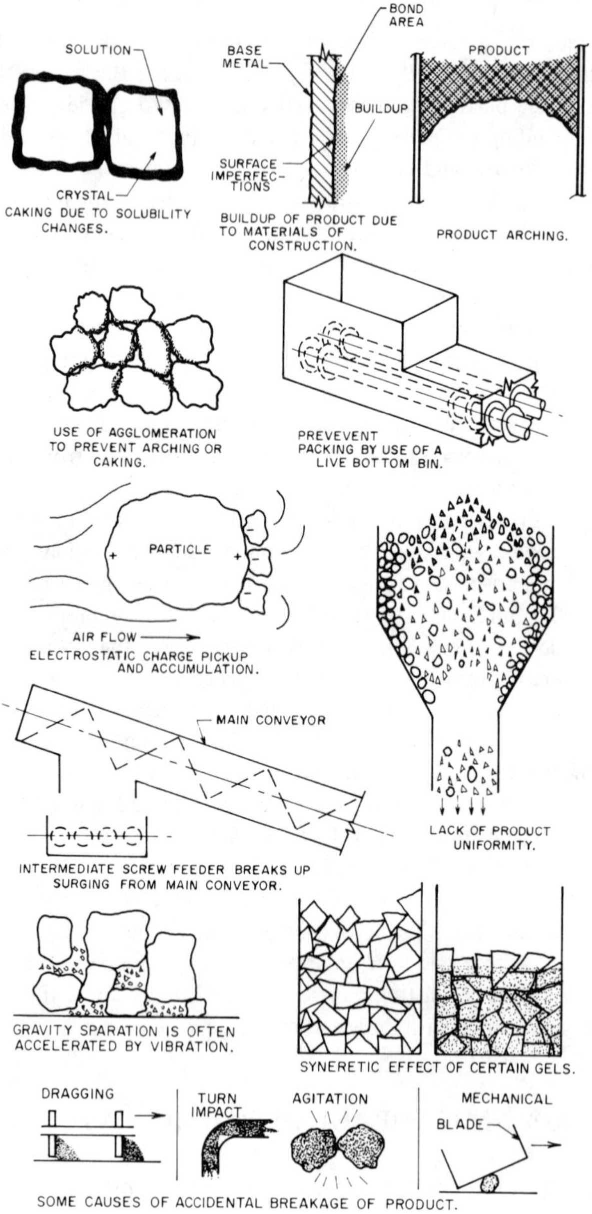

Fig. 4.1[1] Illustrating some phenomena in particulate technology.

works available on this subject.[2,3] The latter contains 333 references and a list of 61 manufacturers, world wide. Major difficulties related to bin design include:

1. No flow–this is clearly a very severe problem.
2. Erratic flow–this can vary all the way from no flow to a completely uncontrolled flow. It can be associated with channeling.
3. Flushing–otherwise known as fluidizing or flooding. This is associated with finer powders and it tends to occur if the design permits aeration.
4. Segregation–is usually associated with particle size differences and may be accentuated by core removal of the bin contents.
5. Insufficient design capacity–this may be due to an error but, of course, more often it is due to an increased rate of throughput.
6. Inadequate level control devices–these are many and varied. The old adage that the simpler the design the better off one is may not always be true. For example, a plumb bob is simple but it may be ineffective as a level indicator in a funnel flow bin.

Three types of bin have been identified: the funnel flow type, the mass flow type and the expanded flow type. These three are illustrated in Fig. 4.2.

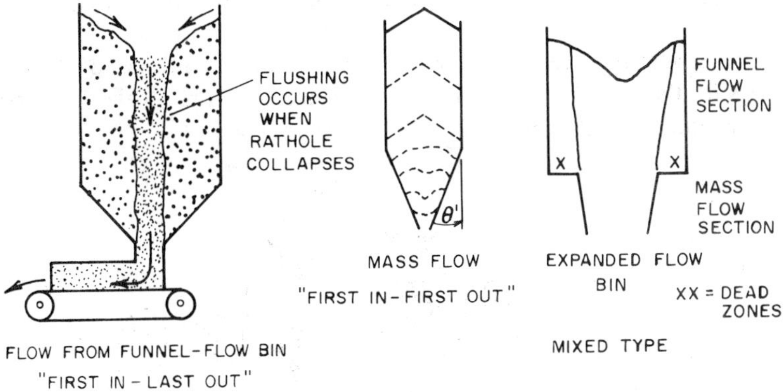

Fig. 4.2 Types of storage bins.

The funnel flow bin in which powder is stored is prone to the formation of a pipe or rat hole which renders it unstable. When this collapses, the outflow may be fluidized. Noise and vibration may result to an extent that depends upon the size of the plant and the loading. Whereas the funnel flow bin tends to promote segregation, the mass flow bin actually causes the contents to remix before exiting. In addition, the mass flow contents deaerates to an extent that depends upon the dwell time. Whether the bin is a mass flow type or a funnel type depends upon the angle of the hopper slope. This is denoted by θ' in Fig. 4.2. The relationship between this angle and the angle of friction of the powder is shown in Fig. 4.3.

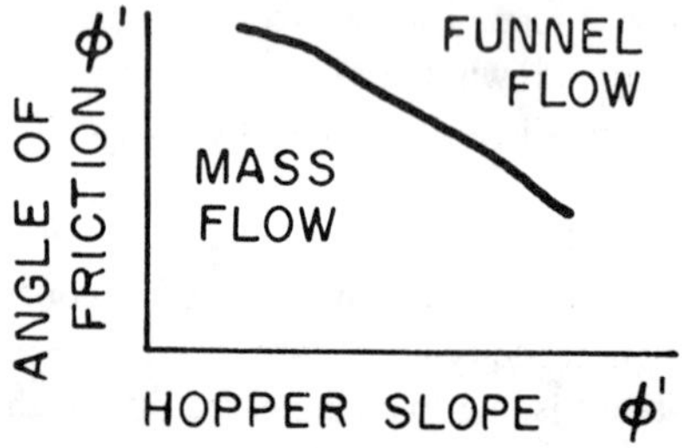

Fig. 4.3 The effect of hopper slope on bin flow. Note: The chart varies with the assymmetry of the flow.

In general, if the slope of the hopper is of the order of 45°, the bin will be a funnel flow type in 90 percent of the cases. If the hopper slope is 20°, the bin will be a mass flow type in 90 percent of the cases.[4]

It appears that the primary factor that is responsible for the variation of the flow properties is the particle morphology.[3] This is indicated in Table 4.1.[3]

From what has been said above, one can see that as the cone of the bin draws in, there is a tendency to arch mechanically. A photoelastic model of this situation is shown in Fig. 4.4.[5]

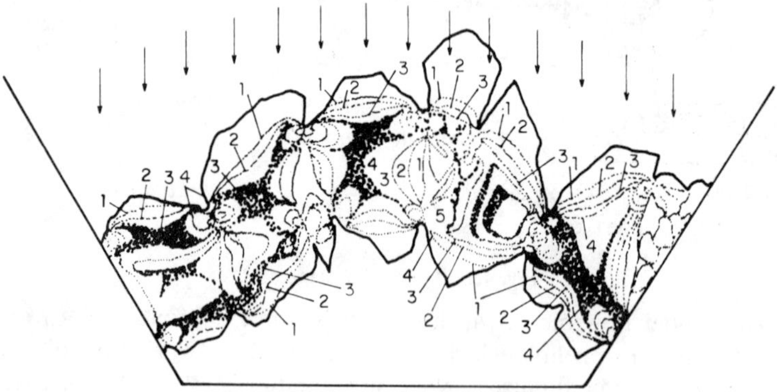

Fig. 4.4 A photoelastic demonstration of mechanical arch, force transmission.[5]

Table 4.1

CLASSIFICATION SYSTEM FOR MASSES OF PARTICLES BASED ON AVERAGE RELATIVE FLOW PROPERTIES.[3]

Description of class	*uniformity coefficient*	*powder fraction (%)*	*flowability score*	*floodability score*
I. Corpuscular				
A. Granules				
1. uniform	1–10	<5	70–100	0
2. nonuniform	15–30	<5	60–75	0
3. very nonuniform	30 +	<5	50–70	0
4. damp	30 +	<5	30–65	0
5. soft (or sticky)	–	<5	40–75	0
B. Powdered granules				
1. uniform	1–12	<30	70–80	0
2. less uniform	8–18	<30	55–70	0
3. nonuniform	15–30	<30	50–65	0
4. very nonuniform	30 +	<30	30–60	0
5. damp	30 +	<30	40–65	0
6. soft (or sticky)	–	<30	50–70	0
C. Fluid powders				
1. very				
a) granular (some)	–	60–90	55–70	70–90
b) gran. and powder	–	30–60	50–65	30–70
c) powder	–	95–100	45–65	80–95
2. fluid				
a) granular (some)	–	60–90	45–65	50–75
b) gran. and powder	–	30–60	40–60	20–60
c) powder	–	95–100	35–50	45–75
3. fluid cohesive				
a) granular (some)	–	60–90	20–45	25–60
b) gran. and powder	–	29–60	20–40	10–35
c) powder	–	95–100	10–40	15–45
D. Cohesive powders				
1. powder	–	95–100	5–25	0–20
2. gran. and powder	–	30–60	5–40	0–20

	average size range		
II. Laminar			
A. Micaceous			
1. thin	1/2 to 3 in.	1–15	0
2. thick	1/4 to 1 in.	10–20	0
3. powdered	-200 mesh	30–50	35–60

Table 4.1 *(cont'd.)*

Description of class	*average size range*	*flowability score*	*floodability score*
B. Film			
1. very thin	1/16 to 1/2 in.	1-10	0
C. Chips			
1. fine, uniform	+ 100 to + 10 mesh	70–80	0
2. fine, nonuniform	-60 to + 10 mesh	60–70	0
3. large, nonuniform	-10 to 3 in.	30-40	0
D. Flakes			
1. thin	- 100 to 1 in.	1–20	0
2. fine, uniform	- 20 to 3/8 in.	50-80	0
3. fine, nonuniform	- 40 to 3/4 in.	45-50	0
4. powdered	- 200 to 1/4 in.	50–70	0
III. Fibrillar			
A. Stems			
1. very short	1/4 to 3/8 in.	50-60	0
2. short	1/4 to 1 in.	10–20	0
3. long	2 to 3 in.	1-10	0
B. Fibrous			
1. bunches	1/2 to 2 in.	5-20	0
2. fine bunches	- 100 mesh	20-30	0
3. powder, coarse	- 200 to + 10 mesh	40-55	0
C. Acicular			
1. fine	+ 100 to + 40 mesh	65-75	0
2. fine, nonuniform	- 100 to + 10 mesh	55-70	0
3. medium	+ 20 to 7/8 in.	30-40	0
4. powder, fluid	- 200 mesh	35-50	40-60

Note: Uniformity coefficient is ratio of sieve aperture passing 60% divided by that passing 10%.

The most widely used method for hopper design is based upon a continuum-mechanical approach[2] to the behavior of the particulate matter. This is the basis upon which the work was done that is reported here. However, when the material flows, its continuity seems to break down as evidenced by the abrupt appearance of shear planes accompanied by the immediate surge of the particulate mass.*

It is important to note that arching occurs only when the green strength

*See for example some of the work by Cowin, Lee Templeton reported in American Society of Rheology and entitled Kinematics of Hopper Flow.

of the material (under extant conditions) equals or is greater than the stress at the arch abutment. There are basically three important aspects of the problem which must be clearly understood so that the steps necessary to determine the hopper dimensions can be understood. These are:

σ_1 the normal pressure or the consolidation stress
$\bar{\sigma}_1$ the stress at the arch abutment acting on the arch
f_c the green strength of the particulate mass under the extant conditions.

The trend of these three parameters in the case of a conical bin is shown in Fig. 4.5.[3]

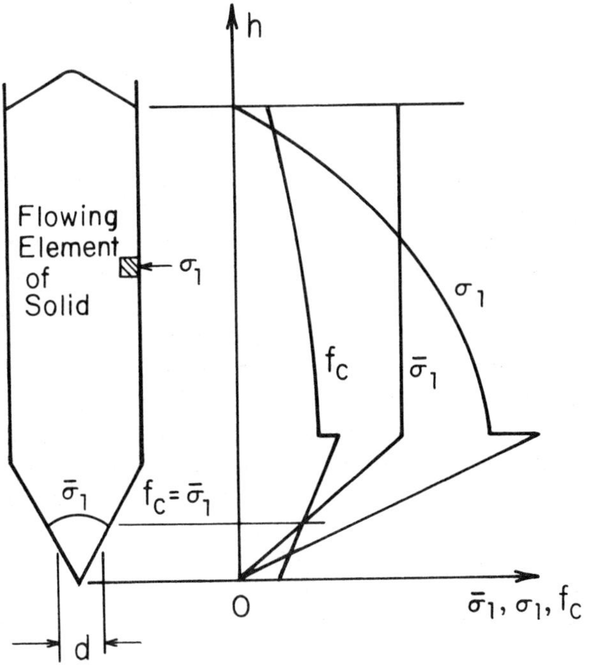

Fig. 4.5 A schematic outline of the trends of stress and strength in a bin filled with granular material to illustrate mechanical arching.[3]

Strength of a Granular Material

If it is necessary to know the strength of the granular solid under the extant conditions, how can this be determined? The procedure is called a shear cell test. One of these testers made by Jenike is shown in Fig. 4.6.

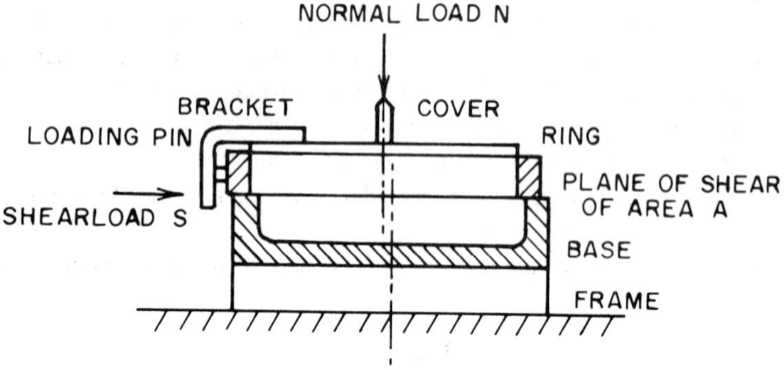

Fig. 4.6 Illustrating a Jenike Shear Cell Tester.[2]

The particulate material is placed in the body of the tester, leveled, and covered. The cell contents are then subjected to a normal load N in order to achieve a consolidated condition. The normal load is removed and the sample is then sheared off under a reduced normal load. The shear load is converted to shear stress and the normal load is converted to normal stress. The experiment is repeated with another sample of particulate consolidated under the same normal load N, but this time using a different reduced normal load. In this way, by repetition under different reduced normal loads each time, the yield locus is built up as shown in Fig. 4.7. The end of the line is located when the shear cell test is conducted under τ/σ conditions in which there is no volume change (as indicated by a dial indicator). By using different consolidating loads N, and by repeating the procedure in a similar manner, a whole family of curves can be built up as shown in Fig. 4.8.

The slope of the line δ' characterizes the failure conditions in the case of each of the conditions of preconsolidation.

Redrawing Fig. 4.7 for a moment and considering the similar triangles shown leads to Eq. (4.1) (see Fig. 4.9). Take the ratio of the vertical sides of the two triangles and because the two triangles are proportionate, make the ratio of the sides equal with the ratio of the bases, thus

$$\frac{\tau}{c} = \frac{\sigma + T}{T} \tag{4.1}$$

Of course, the locus is not a straight line and the curvature must be corrected for. This is done by the insertion of a factor of the power n as shown in Eq. (4.2).

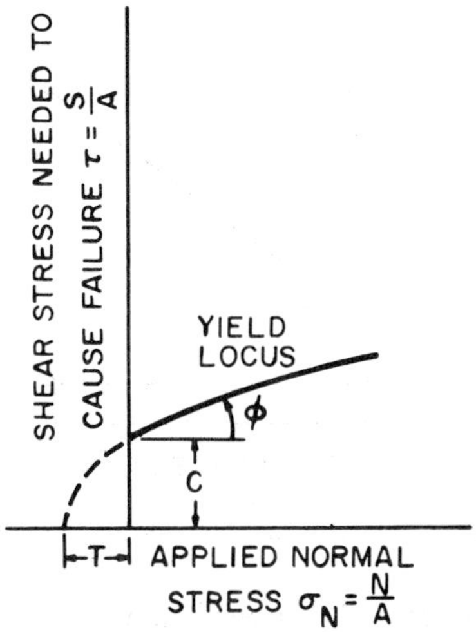

Fig. 4.7 A yield locus.

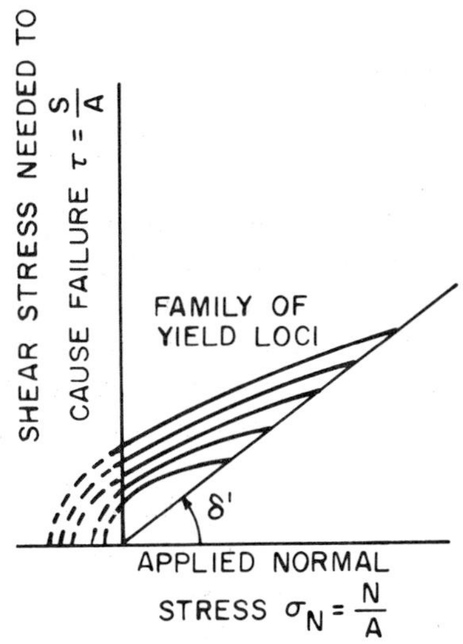

Fig. 4.8 A family of yield loci for one material.

$$\left(\frac{\tau}{c}\right)^n = \frac{\sigma + T}{T} \tag{4.2}$$

This is known as the Warren Spring Equation.[6] Its experimental deter-

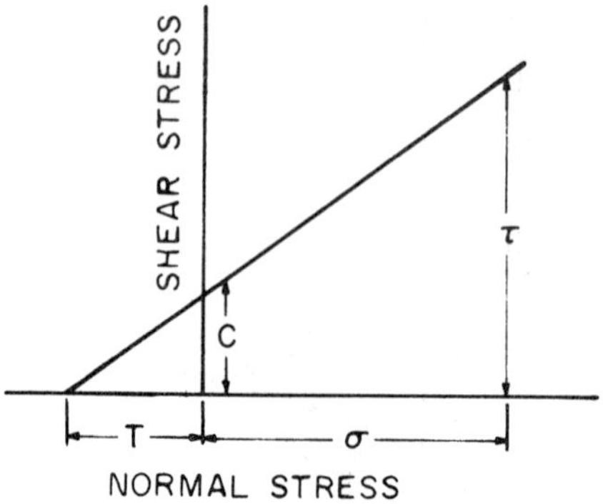

Fig. 4.9

mination necessitated the invention of a device for measuring the tensile strength of a powder.[7]

Determination of Critical Hopper Dimensions

To carry out this design task, steps involving testing and calculations are required as follows:

Step 1. Carry out the shear test and determine the yield locus and Particulate Flow Function PFF.

Step 2. Determine the effective angle of internal friction δ and, the angle of wall friction ϕ'.

Step 3. Select the angle of the hopper from the chart of ϕ' versus θ'.

Step 4. Step 3 has effectively fixed the value of the Hopper Flow Factor (HFF) which is now selected.

Step 5. Using the method of construction of the Mohr Circle of Stress, and plotting the Particulate Flow Function PFF over the Hopper Flow Factor HFF, determine critical f_c and σ_1, as previously defined.

Step 6. Calculation of the minimum size of outlet diameter with f_c critical and $H(\theta')$, where $H(\phi')$ is one for a wedge shaped hopper and two for a circular one.

The details of these steps are now discussed:

Step 1 (PFF). Step 1 has been partially outlined earlier. To determine the value of f_c and σ_1 use Mohr's Circle. Thus at the free surface both the normal stress and the shear stress are zero. What this means is that Mohr's Circle drawn at the end of the yield locus (i.e., tangent to the end point) gives the value of σ_1 at the x axis, where the shear stress is zero. Another Mohr's Circle drawn through the origin and tangent to the yield locus gives the value of f_c, also on the x axis, where the shear stress is zero. The idea behind putting the circle through the origin is that, at zero consolidation, there is zero strength, which is reasonable. This construction is shown in Fig. 4.10.

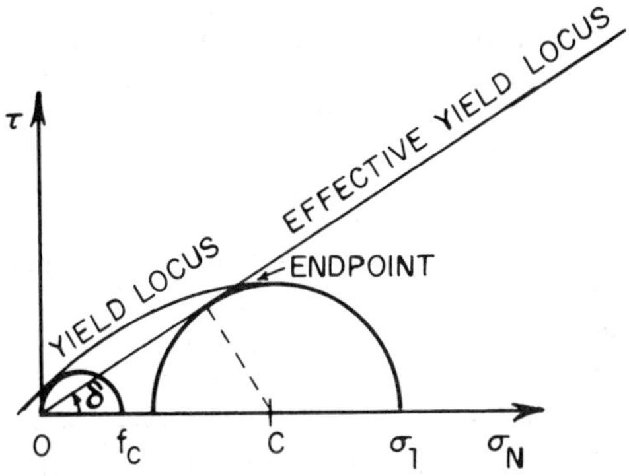

Fig. 4.10 Showing the use of Mohr's Circles to determine f_c and σ_1.

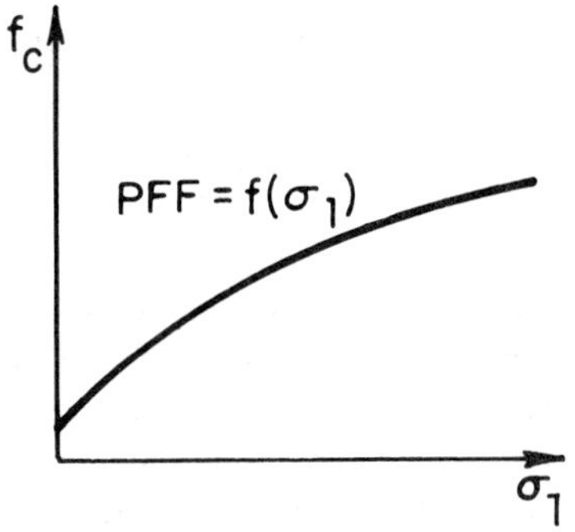

Fig. 4.11 Illustrating the form of the Particulate Flow Function for a particulate material.

The ratio of the values of σ_1 and the corresponding values of f_c defines the Particulate Flow Function relationship

$$\text{PFF} = \frac{\sigma_1}{f_c} \tag{4.3}$$

The Particulate Flow Function is clearly a property of the particulate material and its degree of consolidation. The form of PFF for a particulate material is shown in Fig. 4.11. The Particulate Flow Function has been classified on a scale of from less than two to more than ten in the following way:[2]

Table 4.2

FUZZY CLASSIFICATION OF FLOW PROPERTIES OF PARTICULATES BY MEANS OF PARTICULATE FLOW FUNCTION VALUES[2]

Value of PFF of Particulate	*Description of Particulate*
<2	nonflowing; very cohesive
2–4	cohesive
4–10	easy flowing
>10	free flowing

Step 2 (δ). Referring to Fig. 4.10 once more, the effective yield locus is the tangent to the largest Mohr Circle, which is itself tangent to the yield locus at the end point. The idea behind this is that the end points of the yield loci are the extant conditions in the free flowing particulate mass. Furthermore, the angle δ gives a simple relationship between the two principal stresses:

$$\frac{\sigma_1}{\sigma_2} = \frac{1 + \sin \delta}{1 - \sin \delta} \tag{4.4}$$

This is derived by simple trigonometry in the following way: One end of the large Mohr's Circle in Fig. 4.10 is σ_1 and the other end is σ_2. The tangent to the circle from the origin, the effective yield locus, is connected to the center of the circle by the dashed line shown. This dashed line is perpendicular to the tangent and is numerically equal to the radius of the circle. The value of the sin δ is equal to the radius divided by the hypotenuse of the triangle (OC) and is equal to r/h. Now r is equal to $(\sigma_1 - \sigma_2)/2$ and h is equal to r plus σ_2. Substituting these values gives Eq. (4.4), as before.

Step 2 (φ'). The angle of wall friction, ϕ' is determined in a manner that is exactly analogous to that used for the determination of the effective angle of internal friction δ. The only difference is that the base upon which the powder mass rests is made of the material of construction of the wall of the hopper and the shear test is conducted in order to find the shear stress necessary to move the powder mass across the base under a known consolidating load. The curve of shear stress versus normal stress is usually a straight line, so the value ϕ' can be read off.

Step 3 (θ'). The angle of the hopper can be read off from a chart of ϕ' versus θ' (see Fig. 4.3). In order to ensure that the type of flow required is surely obtained, a value of θ' well within the required region on the chart should be selected.

Step 4 (HFF). Whereas the Particulate Flow Function PFF is a property of the particulate material, the Hopper Flow Factor HFF is a system characteristic. HFF is equal to $\sigma_1/\bar{\sigma}_1$, in which $\bar{\sigma}_1$ is the stress which acts on an obstruction. The Hopper Flow Factor HFF is defined as

$$\text{HFF} = 1 = m \frac{S(\theta')(1 + \sin \delta)}{2 \sin \theta'} = \frac{\sigma_1}{\bar{\sigma}_1}, \quad (4.5)$$

in which

HFF is the Hopper Flow Factor,

m is either one for axialsymmetric flow (a cone) or zero for plane flow (a slot outlet), and

$S(\theta')$ is a function of θ'.

Values of HFF for a wide range of conditions have been calculated and charted.[2] Two of these charts are given in Fig. 4.12 and 4.13 respectively. Comparison of these two figures shows the rather large differences between conical and plane flow channels.

Step 5 (f_c critical). Determining critical f_c is done easily by plotting the Hopper Flow Factor HFF and the Particulate Flow Function PFF on the same graph. This is illustrated in Fig. 4.14. The logic behind this step is that f_c critical is the value of the unconfined yield strength in that particular hopper under boundary conditions specified.

Step 6 (d). The minimum size of the hopper outlet is d. It is determined by substitution in the following equation:[2]

$$d = \frac{H(\theta') f_c \text{crit.}}{\gamma} \quad (4.6)$$

in which

$H(\theta')$ is a function of the slope of the hopper. It is 1 for a wedge shaped hopper and zero for a circular one; and

γ is the bulk density of the material.

This is an elementary treatment of the problem. It ignores the effects of time of storage and moisture.

Many studies have been reported in the literature concerning the applicability of the above concepts and the relative merits of one design of shear cell over another. For further details, the reader should consult the appro-

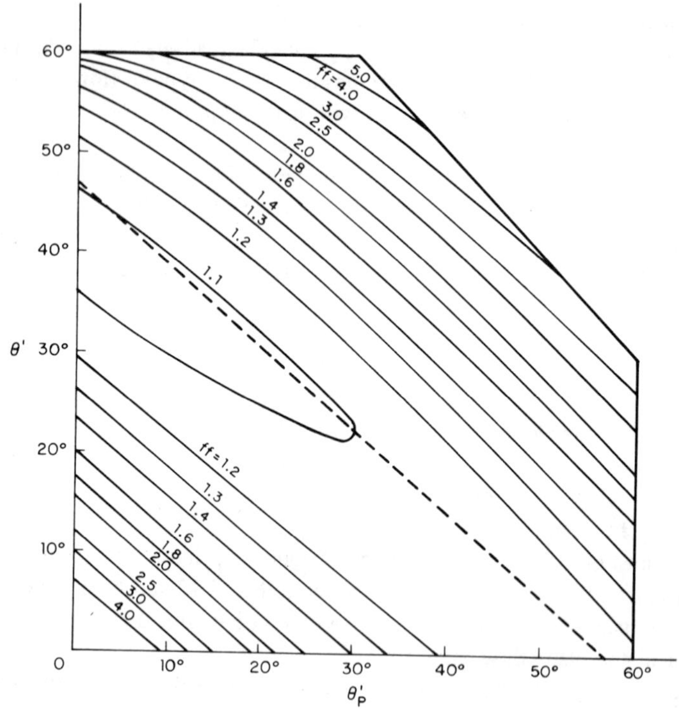

Fig. 4.12 Hopper Flow Factor (HFF) contours for symmetric Plane-flow channels, $\delta = 60°$.[2]

priate sources, reported elsewhere.[3]

One interesting design of shear cell uses a ring type of particulate bed with a motorized drive.[8] This device called a Portishead Shear Cell, has been used to study the effects of moisture upon the critical blocking diameters in the case of IDF coal. A comparison of experimental results with the prediction from theory was said to show an improvement over the standard Jenike approach. Fig. 4.15 illustrates the data in graphical form.[9]

Other workers have developed special testers in order to determine the

Real Storage Bins

The theoretical and experimental developments which have just been very briefly reviewed now make it more likely that a design engineer can develop a proper design of storage bin for a required purpose. This can

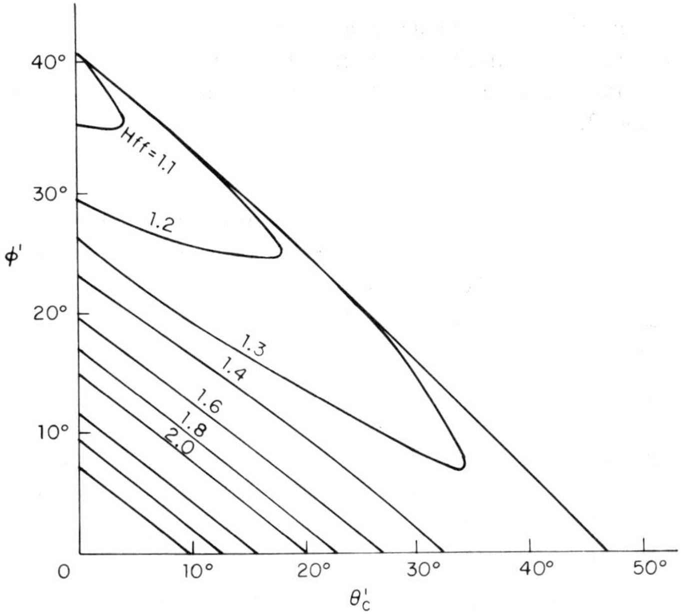

Fig. 4.13 HFF contours for conical channels, $\delta = 60°$.[2]

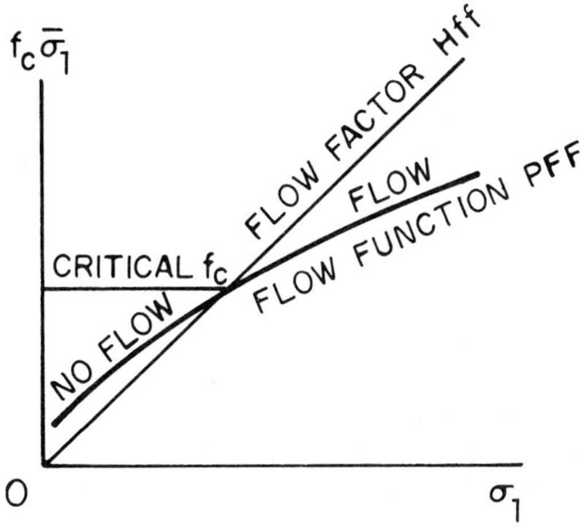

Fig. 4.14 Showing plot of Hopper Flow Factor HFF and Particulate Flow Function PFF on same graph to determine the value of f_c critical.[3]

Particulate Flow Function in such a way that the unconfined yield strength is determined directly.[10] It was reported that the results agreed closely with those determined from a Jenike type shear cell in the case of TiO_2 type RSM-2.

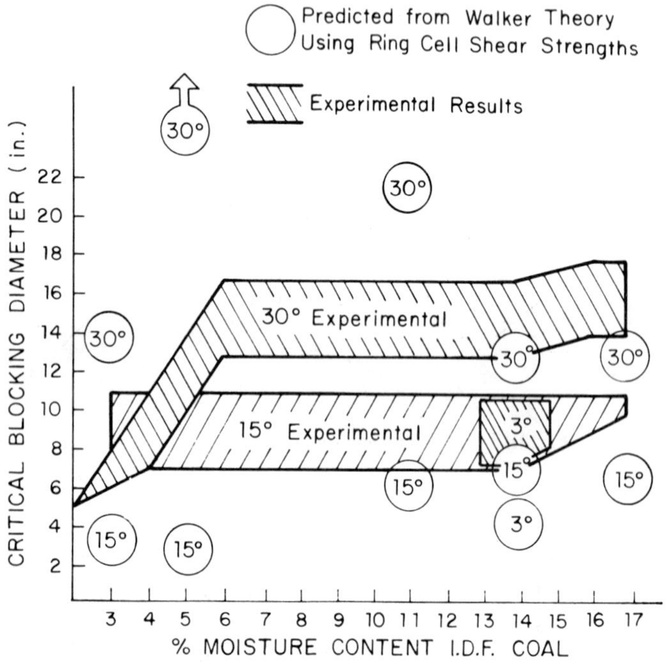

Fig. 4.15 Critical blocking diameters.[9]

only come about, however, if the designer is fully aware of these developments in this field. Unfortunately this is not always the case. In addition, there are many, many storage hoppers, silos and bins which were built with little real understanding of the factors involved in bin operation. Consequently, many industrial installations are plagued with erratic flow conditions, sometimes accompanied by very noisy collapse of unstable rat holes, which can be alarming to an operator. Sometimes this instability is accompanied by flooding out of the particulate from the base of the bin.

Solutions that have been developed to some of these problems include:

1. The well-known attachment of vibrators of ever increasing strength on the sides of the offending hopper. Often this is accompanied by the use of encouraging and, usually, violent hammering by the bin's attendant. In desperate situations, the latter have been known to remove the base plate and poke big sticks around to try to dislodge the arched particulate. This should be discouraged. Acts of paralleled desperation include the actual physical entry into the bin atop of the bin contents. As before, this procedure should be discouraged because it has been known to cause the death of the operator due to smothering. (The operator causes the collapse of the bin contents and is buried alive therein.) Large compressed air devices are used to impart a sharp mechanical shock to the bin hopper.
2. The offtake from the bin may be carried out using a moving belt. What is not so obvious is that the belt should be sloped gently away from the feed input end. Another solution to the problem of controlling outflow from a bin is to design and install so-called live bottom feeders. Many designs may be utilized including the two shown in Fig. 4.16 a and b.

Gravity Flow of Powders

The flow pattern in a particulate flowing through a hopper under gravity, and discharging through an orifice is shown in Fig. 4.17.[11]

In general the particles in region A slide in turn over B which in turn slide over the stationary ones. In zone C the particles are moving in arcs and accelerating. In zone D, which varies in size, the particles are moving at about 1 percent of their terminal velocity and pass through the orifice center. The central region through which the particles move is surrounded by an annular ring which is essentially devoid of particles. The width of this annulus, $k/2$ has been related to the size of the aperture when blocked thus:[13]

$$4k = D \tag{4.7}$$

The mass rate of powder flow ($\dot{M}$) is related to particulate and container properties as follows:[14]

$$\dot{M} = (\pi/4)C\rho p[(gD^5)/2 \tan \alpha]^{1/2} \tag{4.8}$$

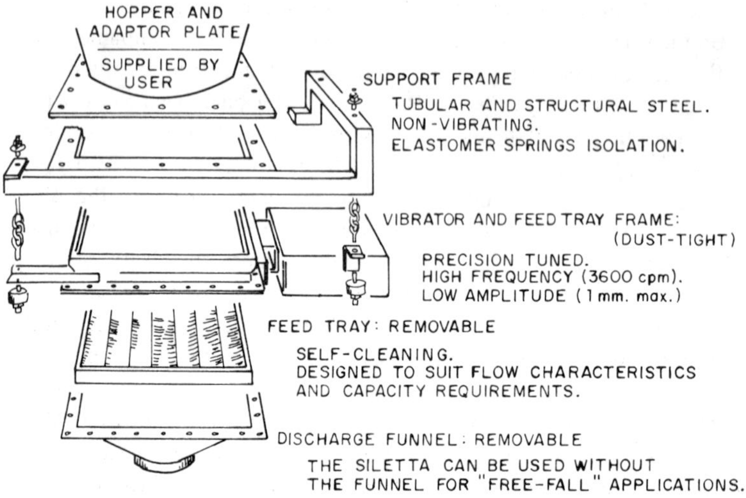

Fig. 4.16(a) Shows a live bottom feeder (Solids Flow Control Corp).

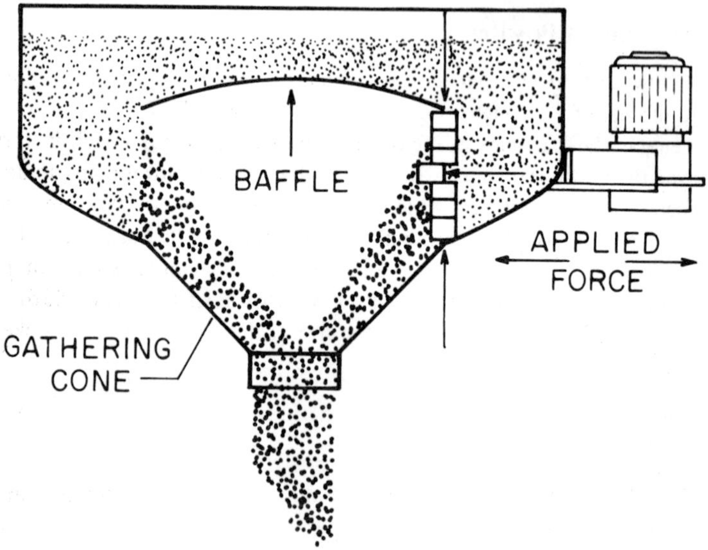

Fig. 4.16(b) Shows a vibrating baffles internal to the hopper (Vibra Screw Corporation).

in which

c = the discharge coefficient,
ρ = bulk density of the powder,
g = acceleration due to gravity,
D = orifice diameter, and
α = angle of repose.

The flow stream contracts just below the orifice to about $0.5D$ – cf, the Vena Contracta for liquids.

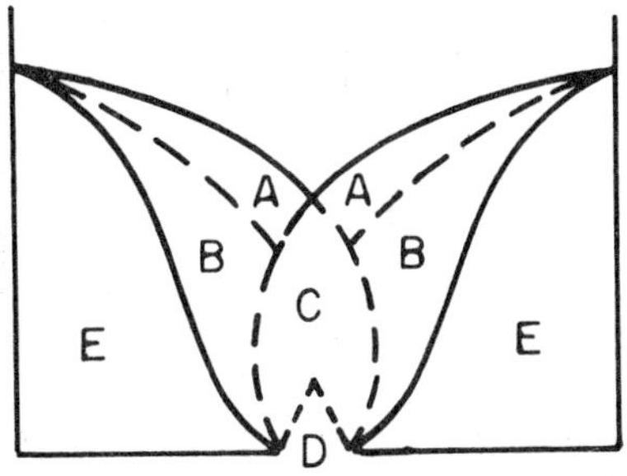

Fig. 4.17 Pattern of gravity flow in a hopper.[11]

The consequences of this flow pattern within the bin are most interesting. The ideas are simple but they demonstrate the great insight of their originators.[12]

The solids pressure within the particulate in the bin increases with the size of the head, as shown in Fig. 4.18. It becomes constant (showing the influence of the wall) and eventually drops off towards zero at the bin orifice. The bulk density of the particulate follows a similar trend so that the volume of the interstitial voidage first decreases and then finally increases as the discharge point of the hopper is reached. It is in this region, where the individual particles slide over each other, that the interstitial voidage increases. This increase of voidage causes the pressure of the gas (air) entrained in the voidage to drop below atmospheric and the negative gradient $-dp/dz$ tends to hold up the flow of the particulate out of the bin orifice. In simple terms, the negative pressure has to be equalized by air bleeding into the interstitial voidage in order that the powder mass can flow out of the orifice. A similar effect can be observed in the case of a liquid flowing out of a bottle. The familiar "glug, glug" as the air enters the bottle to replace the water that has flowed out is quite analogous, except that, in the case of a bin filled with a particulate, the air can not move up through the powder mass to supply the required gaseous volume to equalize the pressures. This air has to enter the bin through the orifice counter to the exiting powder. Although this effect is quite small in the case of coarse granular materials, it is very marked in the case of fine particulates (such as fine anthracite[12]).

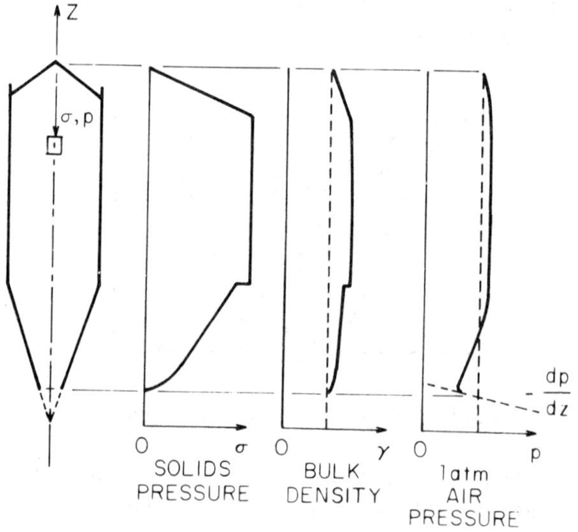

Fig. 4.18 Pressures and density during flow.[12]

The solution to this problem is to use aeration techniques by flowing air into the hopper section of the bin. In this way, the interparticle contacts are reduced and a free flowing particulate may be obtained. A commercial design is shown in Fig. 4.19.

4.3 CONVEYANCE AND FLOW OF PARTICULATE SOLIDS

Probably few industries have no powder handling stage during the processing of their product. The pneumatic handling of grain at the docks is a familiar sight; also powdered food components such as milk and sugar are often "untouched by hand" because they are conveyed from process to truck and from truck to the consumers process by pneumatic means. Powdered coal, inorganic fertilizer, and many other particulate solids are trans-

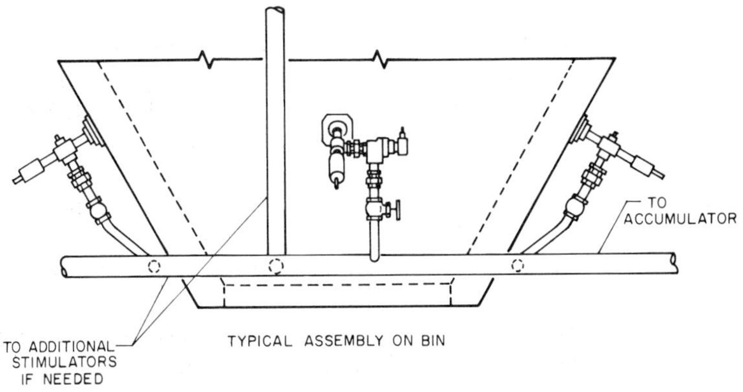

Fig. 4.19 An aeration device installed (Dynapac Mfg., Inc., Stanhope, New Jersey).[200]

ported hydraulically. In the future, more particulates will be moved either hydraulically or pneumatically.

In this section both hydraulic and penumatic transport are discussed. In the former, it is important to be able to differentiate between different types of slurry behavior as evidenced by their different shear stress/rate of shear relationships. The appropriate equations and conditions for their application in relationship to the flow of slurries along pipes are discussed along with the sound economic reasons for the consideration of this process. Similar considerations apply in the case of pneumatic conveying of finely divided matter which is discussed later in this chapter.

Hydraulic Transport

There is a pattern to a mixture of fine particles within a connecting fluid. Which is to say that such mixtures possess a structure. This structure is a function of the characteristics of the particles as well as the fluid. When the slurry is mechanically disturbed the structure may be changed and this alteration of structure may be gradual, instantaneous, or delayed. Because the behavior of slurries in motion has many analogies to fluid flow, much has been borrowed from the field of hydraulics in order to explain the hydraulic transport of slurries. As in the case of flow and storage of particulate matter, the approach has been generally one of macroscopic scale analysis. It remains to be seen if a more fine scale level of investigation will be undertaken and what the results would be.

A slurry is a mixture of finely divided particles and liquid such that it retains its characteristic of behaving like a fluid after considerable settling time. It is moved under both laminar and turbulent flow conditions (the dividing line between laminar and turbulent flow condition is defined by the value of Reynolds Number Re).*

Alternatively a suspension is unstable in laminar flow conditions and the particles settle. It has different flow properties in the vertical and horizontal directions. A main difference between a slurry and suspension is in particle size—a slurry is often $<30\ \mu$ and a suspension often has particles $>30\ \mu$.

In fluid flow there are two main origins of friction—that caused by intermolecular action and that caused by interaction of large molecular groups or "particles." Because of attraction between the fluid and the container, the fluid at the container walls has the same velocity as the wall.

In the case of a solid material the shear stress $= G$ (the shear modulus) $x\gamma$ (the strain). In the case of a liquid: the shear stress = viscosity coefficient multiplied by the rate of deformation (dv/dy).

$$\tau = \mu \frac{\Delta v}{\Delta y} \equiv \mu \frac{dv}{dy} \tag{4.10}$$

A plot of shear stress versus the rate of deformation (where the slope of each line is the viscosity, μ) is a convenient way of indicating graphically the properties of slurries. Fig. 4.20 shows an illustration of some quite different types of slurry behavior:

1. Newtonian,
2. Ideal plastic,
3. Pseudoplastic,
4. Dilatent,
5. Thixotropic, and
6. General non-Newtonian.

Newtonian. In this fluid the relationship between shear stress and strain rate is constant because the structure of the fluid is not affected by the displacement: e.g., water, oil, and air.

*An important or critical value of this quantity, Re, was found to be ≈ 2000 at which turbulent flow occurred; below 2000 laminar flow occurred. The

$$\text{shear stress in turbulent flow} = \tau_t = \mu(dv/dy) + \rho l^2 [dv^2/dy] \,, \tag{4.9}$$

in which l = the Prandtl mixing length, the distance of travel sufficient to get mixing between the molecules of two fluid particles.

Ideal plastic or binghamite plastic. These fluids consist of interlocking particles intermixed with liquid. The interlocking effectively prevents any shear process from occurring until a certain yield stress t_s is reached. From then on the behavior is essentially Newtonian. Examples of this fluid are suspensions of sand, coal, ores, and rock in water, and many manufactured foods.

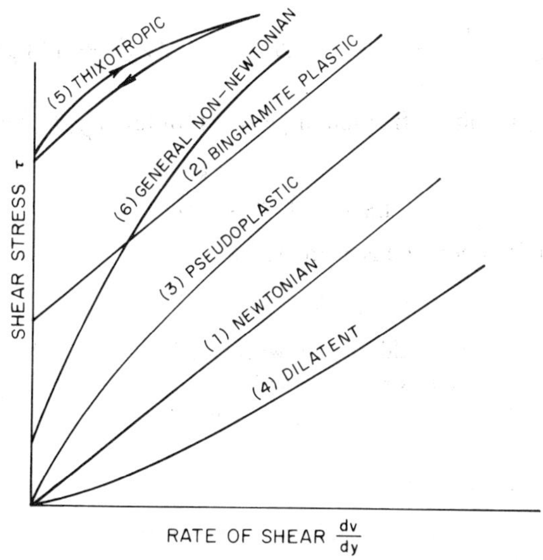

Fig. 4.20 Illustrating six different types of slurry behavior due to differences in the structures of different slurries.

Pseudoplastic fluids. These include systems such as graphite-H_2O, paper pulps, and some paints. In general as the shear stress increases, the rate of shear progressively decreases. These materials eventually reach a limiting value of viscosity.

Dilatent fluids. These are such that, under the action of the stress, the particles move from their most stable position into a less close packed configuration. This rearrangement process apparently starts as soon as the stress is applied. The fluid eventually reaches a condition of dynamic equilibrium as shown by the linear portion of the curve. Dilatent fluids include starch, mica-water mixes, and quicksand.

Thixotropic fluids. These have a micellular type of structure which is partially destroyed by shear but which reforms upon standing. Thixotropic paints are well known as "Nondrip" paints. A certain level of τ is required after which the fluid shows a progressively increased rate of shear as the stress rises.

Viscosity and Fluid Composition

There are several empirical relationships relating viscosity to particle concentration. One of these is[16]

$$\mu_{\text{relative}} = \frac{\mu_{\text{solid}}}{\mu_{\text{liquid}}} = 1 + 2.5V_p + 7.2V_p^2 + 16V_p^3 \qquad (4.11)$$

in which V_p = volume fraction of particles for low V_p; V_p^2 and V_p^3 drop out so that

$$\mu_{\text{relative}} = 1 + 2.5V_p \qquad (4.12)$$

The effect of temperature is of the form:[1]

$$\ln \mu_{\text{relative}} = a + b\theta \qquad (4.13)$$

These relationships should be used with care because they are based on the assumption that there is no interparticle interaction and that only spheres are present. Complete wetting is also assumed.

The Flow of Slurry in Pipes

In the flow of slurry through pipes, it is clearly important to know just how much friction loss in the pipe there will be and how this friction loss is related to the properties of the slurry and the velocity of flow of the slurry in the pipeline. Mainly empirical methods of calculation are available and the outline of the procedure is described here. But first it should be noted that most pipelines are operated under conditions such that the friction loss is minimized. This condition is obtained at a velocity of from four through ten feet per second. The general situation is illustrated in Fig. 4.23.[17] In the case of heterogeneous slurry flowing in a pipe, if the velocity is too low, the solids tend to separate out on the bottom. At the lowest velocity, moving bed flow may tend to occur. As the velocity is increased, saltation flow occurs. At even higher velocities (see Fig. 4.23) the system is still heterogeneous and the friction loss starts to increase again as the velocity of flow is raised. To maintain a homogeneous suspension requires an even greater amount of energy expenditure.

Inspection of Table 4.3 reveals that there are many pipelines in operation for the transportations of coal, iron ore, tailings, concentrates of vari-

Table 4.3

SLURRY PIPELINES[17]

	Location	*Material*	*Length* mi.	*Diameter* in.	Million Tons/ Year
		Constructed			
1	Ohio, U.S.	Coal	108	10	1.3
2	Utah, U.S.	Gilsonite	72	6	0.4
3	Arizona, U.S.	Coal	273	18	4.8
4	USSR	Coal	38	12	1.8
5	France	Coal	5.5	15	1.5
6	Poland	Coal	126	10	–
7	Colombia, S.A.	Limestone	9.2	5–6	0.4
8	Trinidad	Limestone	6	8	0.6
9	England	Limestone	57	10	1.7
10	Colombia, S.A.	Limestone	5.8	7	0.7
11	Colombia, S.A.	Limestone	2.4	8	1.2
12	California, U.S.	Limestone	17	7	2.0
13	Georgia, U.S.	Kaolin	16	8	0.6
14	Tasmania, Australia	Iron Concentrate	53	9	2.3
15	Waipipi, New Zealand	Magnetite	4	8	1.0
16	Waipipi, New Zealand	Magnetite	1.8	12	1.0
17	Bougainville	Copper Concentrate	17	6	1.0
18	West Iran	Copper Concentrate	70	4.5	0.3
19	Turkey	Pyrite-Chalcopyrite	38	5	1.0
20	South Africa	Gold Tailings	22	6&9	1.1
21	Japan	Copper Tailings	40	8	1.0
		Under Construction			
22	Las Truchas, Mexico	Iron Concentrate	25	10	1.5
23	Peña, Colorado, Mexico	Iron Concentrate	30	8	2.5
		Planned			
24	U.S.	Coal	1040	38	25.0
25	U.S.	Coal	1350	–	–
26	U.S.	Coal	600	28	12.5
27	Canada	Coal	500	24	12.0
28	Canada	Sulphur/Hydrocarbon	800	12–16	–
29	Africa	Iron Concentrate	165	16	4.0
30	Australia	Phosphate	180	16–22	–
31	Argentina, S.A.	Iron Concentrate	20	8	2.0
32	India	Iron Concentrate	36	20–22	10.0
33	Brazil	Iron Concentrate	300		–
34	Brazil	Bauxite	200		6.0
35	Africa	Bauxite	90		1.0

ous sorts and limestone. The lengths of some of these pipelines are quite remarkable; for example, some of the planned lines are more than 1000 miles long.[17]

It needs no great ingenuity to see the main reason for the increasing usage of slurry transportation by pipeline: in one word, cost. As shown in Fig. 4.21, the longer the distance that the slurry is to be piped over, the less is the cost per unit of material moved as compared with other methods. These considerations apply to overland transportation.

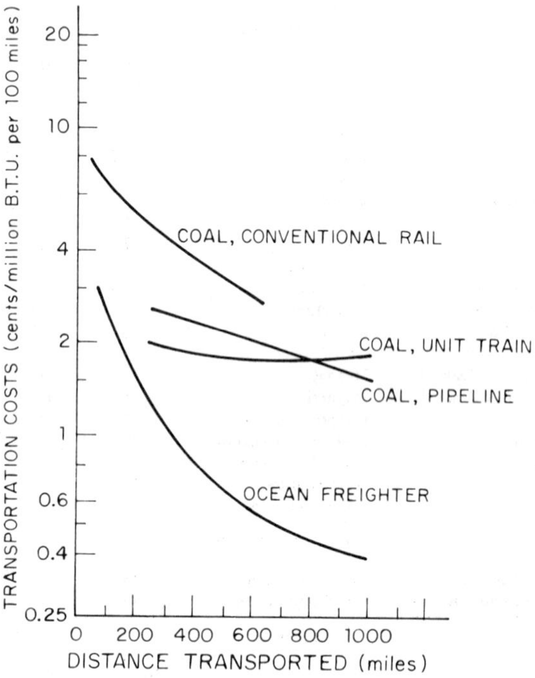

Fig. 4.21 Relative coal transportation costs.[17]

Where particle sizes are small (less than 100 microns or so), where specific gravity (of the solid) is small and concentrations are high, the distributions of the particulate through the fluid-carrying medium of the slurry is usually homogeneous. Coarser particulates tend to settle out and this separation leads to a condition of heterogeneous slurry formation. For example, clean sand in this case would be larger than 270 microns.[18] The existence of these different flow regimes, as a function of solid specific gravity and particle size, is illustrated in Fig. 4.22. Mesh size is expressed in inches and metric also.[19]

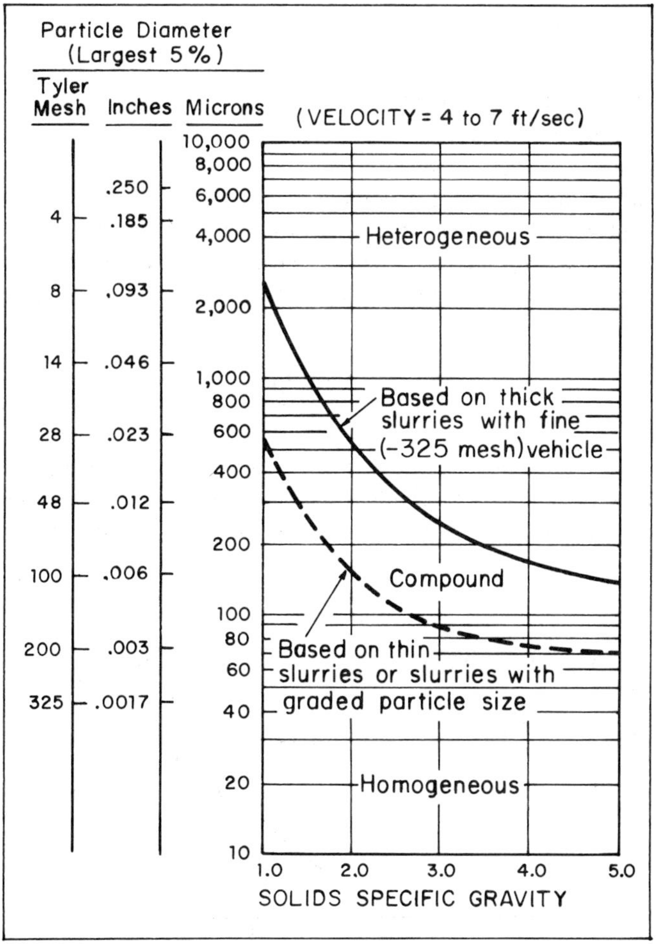

Fig. 4.22 Flow regime as a function of solids specific gravity, and particle size (After Wasp).[19]

In the case of homogeneous slurries, a Reynolds Number in the range 2100 through 4000 is the governing factor.[17]

A general model for non-Newtonian slurries is given by the following equation[20]

$$\tau - \tau_y = K\dot{\gamma}^n \tag{4.14}$$

in which

τ is the slurry shear stress,
τ_y is the slurry yield stress,
$\dot{\gamma}$ is the slurry shear rate,
K is the consistency index for the slurry, and
n is the slurry flow behavior index.

This model may be compared with those shown in Fig. 4.20. The evaluation of the above variables is carried out in laboratory rheology experiments. From this information, the modified slurry Reynolds number is determined:

$$\text{Re}_{sl} = \frac{D^n V^{2-n} \rho}{\gamma} \tag{4.15}$$

This reduces to the ordinary Reynolds Number for n equal to unity.[21]

Using this value of the Reynolds Number, a Moody Chart is consulted (see Fig. 4.23) and the value of the friction factor is read off. The friction

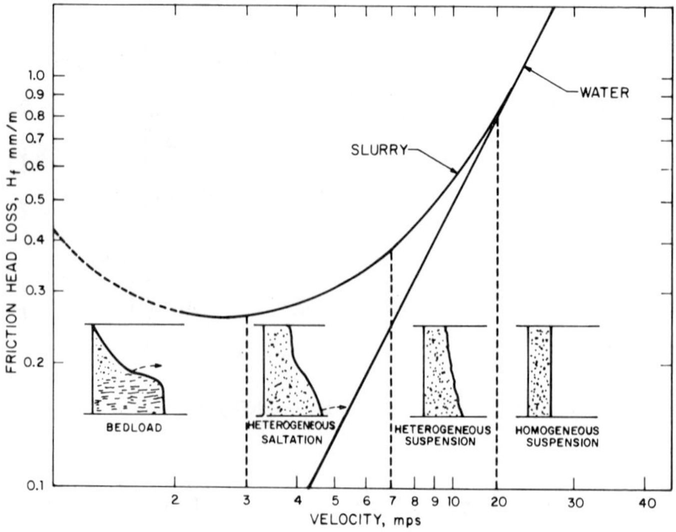

Fig. 4.23 Friction head loss vs. velocity curves for water and slurry.[17]

factor relates the head loss or pressure gradient to the fluid velocity in the pipeline as follows:[20]

$$\frac{\Delta p}{L} = \frac{2f}{D} \frac{\rho V^2}{g_c} \tag{4.16}$$

in which

Δp is the pressure drop,
L is pipe length,
f is the friction factor,
ρ is the fluid density,
V is the flow velocity,
g_c is the gravitational conversion factor, and
D is the pipe diameter.

Eq. (4.16) is the Fanning Equation.

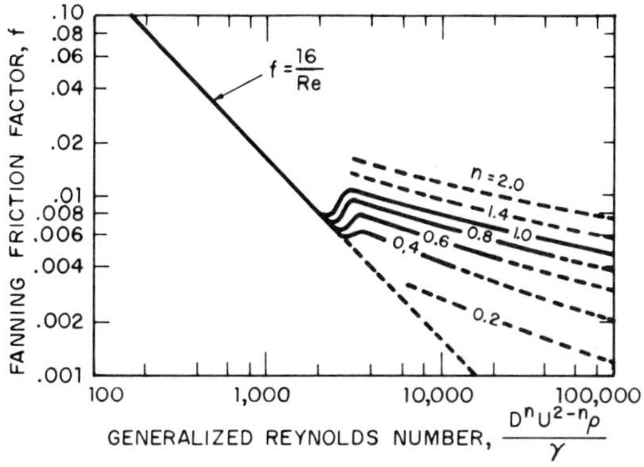

Fig. 4.24 Friction factor-Reynolds number correlation for non-Newtonian fluids. When n is 1 and the abscissa is replaced by $Du\rho/\mu$, the plot applies to Newtonian fluids.[199]

The above procedure is for calculation of the pressure drop in the case of homogeneous slurries. In the case of heterogeneous slurries the drop is calculated by summing two components of friction loss: one due to the homogeneous portion of the slurry and the other due to the remaining solids that are in the heterogeneous flow regime. The exact details of the calculation are given elsewhere (reference 20, p. 676 et seq). The calculations depend upon the method of Wasp.[19,22] (Reference 22 is used for the determination of the average volume fraction of solids in symmetric suspension.) The procedure is essentially iterative.

Three equations are quoted in the literature for the following system conditions:[17]

1. To determine the friction loss when slurries contain coarse solids.[18] This is called Durand's Equation.
2. To determine the limiting velocity in Durand's Equation as it is affected by compound slurries.[23] This is called Faddick's modification to Durand's Equation.
3. To predict the onset of bedload deposition with the Zandi-Govatos Equation.[24]

These three equations are given below:

Durand's Equation

$$i_m = i \left\{ 1 + 81C_v \left[\frac{gD(S_s - S_1)}{V_m{}^2 C_D{}^{1/2} S_1} \right]^{3/2} \right\} \tag{4.17}$$

in which

i_m is the total friction loss of the mixture,
i is the friction loss due to the vehicle,
C_v is the volumetric concentration of solids,
D is the pipe diameter,
S_s is the specific gravity of the solids,
S_1 is the specific gravity of the liquid,
V_m is the velocity of the slurry, and
C_D is the particle drag coefficient.

The Faddick modification for V_m is given by:

$$V_m/[2gD(S_s - 1)]^{1/2} = 0.2435 + 1.8786d \tag{4.18}$$

in which d is the weighted mean particle size in millimeters. In this study, C_v varied in the range from 7.2 through 44.4 percent.

The Zandi and Govatos equation is given by:

$$\frac{V_m{}^2 (Cd)^{1/2}}{C_v g D(S_s - 1)} = 11 \tag{4.19}$$

There are thus a number of ways and means available for the calculation of design criteria in the case of slurry pipelining.

An example of slurry preparation of coal is shown in Fig. 4.25. The pipeline being used is 273 miles long, the pipe is 18 inches diameter and the tonnage pumped is a little under 5 million tons.[17]

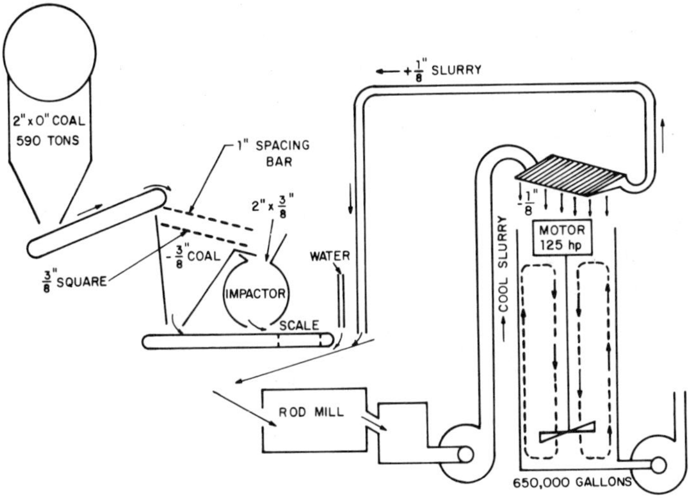

Fig. 4.25 Coal slurry preparation for black mesa pipeline.[17]

For an ideal plastic (Bingham) the relationship used is the Blasius equation:

$$f = B(\mathrm{Re})^{-b} \tag{4.20}$$

B and b are previously determined and tabulated[25] and Re is evaluated. By substitution f is found. The problem with this procedure is that one must assume that the slurry under investigation behaves as did the slurry for which the values of B and b have been determined.[26]

This section only briefly touches on the problems associated with the pipeline transportation of particulate matter. Different considerations arise when the pipe is vertical as opposed to horizontal.[20] Other particulate systems, such as suspensions, dense pastes, emulsions, and fiber pulps, present special problems of their own.

A suspension flowing along a horizontal pipe has characteristics which are strongly dependent upon velocity of flow. There are two velocities used to define the condition $V_{\mathrm{min.}}$: below this, the particles freely settle and $V_{\mathrm{stand.}}$: below this, the particles are in homogeneous distribution in the pipe.

The even distribution of particles is due to components of velocity in the y and z directions and therefore it is necessary to study the relationship between $V_{min.}$, $V_{stand.}$, and the onset of turbulence in order to successfully design a system to transport suspensions.

Dense pastes have interparticle contact and they flow neither like Newtonian or non-Newtonian fluids. Experimental work has studied the flow of total particulate material relative to the flow of solid and liquid.[26]

$$(Q_T - Q_L)V_p = Q_p \tag{4.21}$$

in which

V = volume fraction, and

Q = flow rate.

Emulsions exhibit a peculiarity in that the particles actually change their shape–size characteristics. For dilute emulsions with small particle size the viscosity has been determined by the relationship[27]

$$\frac{\mu_e}{\mu_L} = 1 + 2.5\,Vp\left[\frac{\mu_p + 0.4\mu_L}{\mu_p + \mu_L}\right] \tag{4.22}$$

in which

μ is the viscosity,

e is the emulsion,

p is the particulate,

L is the liquid medium, and

V_p is the volume fraction of particulate in emulsion.

Fiber pulps have an unusual property in that they segregate at low velocity to form a central pulp plug. At higher velocities turbulence breaks up the plug and the pulp fibers disperse to form a pseudo homogeneous mixture.[20]

Pneumatic Transport

Pneumatic transport may be either along vertical pipes (upwards) or along horizontal pipes. Because the powder is introduced into the pipe from a container it tends to form a pile underneath the injection port. A knowledge of the angle of repose of each powder allows one to specify the nature of this piling up process.

For vertical transport the fluid velocity must exceed the settling velocity of the particles. In the case of horizontal transport, a uniform velocity exerts little or no lift on the particles and a velocity gradient in the pipe is necessary. In the extreme case the velocity of the powder at the pipe wall is negligible and it will be a maximum at the center. Therefore the loading is highest at the center. In the case of smaller velocity gradients combined with electrostatic charge between suspension and pipe wall and with interparticle collisions, it is possible for the loading to be higher at the periphery than in the central core.

An experimentally determined formula that describes the velocity of a powder being transported is[26]

$$\frac{\Delta P_T}{\Delta P_A} = 1 + R/K \tag{4.23}$$

in which

ΔP_T is total pressure drop lb/ft^2,
ΔP_A is air only pressure drop lb/ft^2,
R is mass ratio particles/air,
K is a constant whose value depends on air velocity $K = 1.15$, 2.14, 3.11, and 3.5, when
$V_A = 2000, 3000, 4000, 5000$, and 6000 ft/min.

With more precise information the above equation may be modified to

$$\frac{\Delta P_T}{\Delta P_A} = 1 + R(f_A V_A / f_p V_p) \tag{4.24}$$

in which f is the friction factor.

This equation may be used in such a manner that the particle shape factor ψ can be incorporated into the relationship.[20,26]

Other Methods of Moving Particulate Material

In the case of the dense phase pneumatic conveying, the particulate mass may be fluidized and then pumped along pipelines at small energy costs. This technology is applied to bulk transport unloading, conveying to process, formula batching and product blending, and many other operations. A typical system may suffer from dune build-up further along the pipe because the fluidization is not maintained. This build-up can lead to saltation and dune motion along the pipeline. Fig. 4.26 gives an interesting example of a pump used for this set of applications.[28]

The details of the operating cycle of the process of pneumatic conveying are:

1. *Fill*—The pump is charged with material through a fill valve by di-

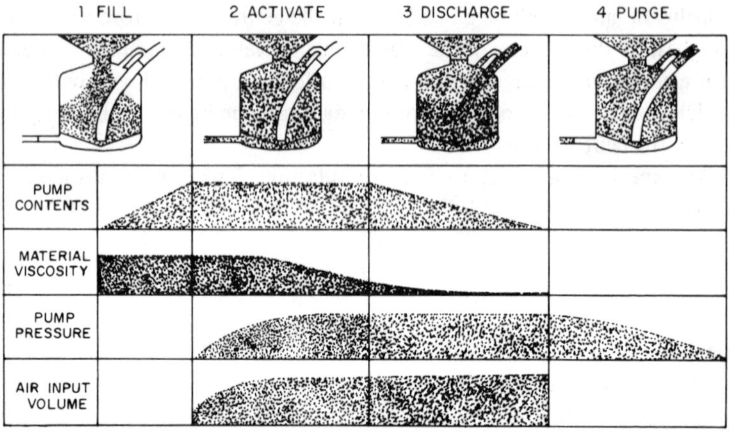

Fig. 4.26 Dense phase pneumatic conveying, Fluidflo® system[28] (reprinted with permission of Consolidated Engineering Co., Houston, Texas).

rect discharge from a storage bin or from a remote point by means of an appropriate conveyor. During the fill cycle, the pump is maintained substantially at atmospheric pressure. Air displaced by incoming material is vented through the discharge conduit or released to atmosphere through a vent valve.

Material viscosity remains at its usual flow value (slightly less than when in static storage) but still so dense that movement through even a short transport line would be possible only by extrusion, requiring very high pressure.

2. *Activate*—After filling, the fill valve is closed and air is allowed to flow into the lower pump chamber under controlled pressure. This chamber and the upper material chamber are separated by a gas permeable membrane. Due to the air pressure now existing in the lower chamber, air passes through this membrane into the material at a controlled rate. As this air is slowly diffused in minute quantities around the material particles, pressure in the pump increases to a preset value which is dependent upon the pressure differential necessary to move the material that is now fluid through the transport line.
3. *Discharge*—When the mass of material has acquired the maximum hydraulic characteristic and the pressure is sufficiently high to move the material, the discharge valve is opened and the fluid ma-

terial flows through the transport conduit under pressure. The discharge pressure is maintained in the pump automatically until all the material is discharged.

4. *Purge*—Once the pump is empty, the mass of material remaining in the line is expelled by the sustaining back pressure of air contained in the now empty pump. This may also be accomplished by diverting air around the pump and into the conduit ahead of the discharge valve. Once this line has been cleared, the pump is returned to atmospheric pressure and may be refilled.

There are many other ways to convey particulate materials, including moving belts and pitch conveyors. Typical examples of these are given in Fig. 4.27.[29,30]

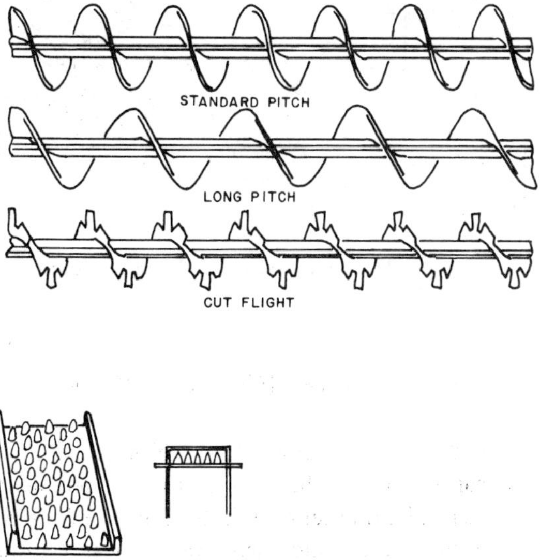

Fig. 4.27 Examples of pitch conveyors[29] and rubber conveyor belt with nubs[30] (Cambelt International Corporation).

In the case of conveying particulate matter along pipelines, rotary valves may be used for a combination of purposes including: metering, airlock provision, and feeders. Another type of valve that is very important in the operation of the plant is the diverter valve. Examples of these two devices are shown in Fig. 4.28.[31,32]

In those cases where the blades of the rotary valve may be unduly worn, replacement tips may be supplied. Special designs are available for "difficult" materials.[31]

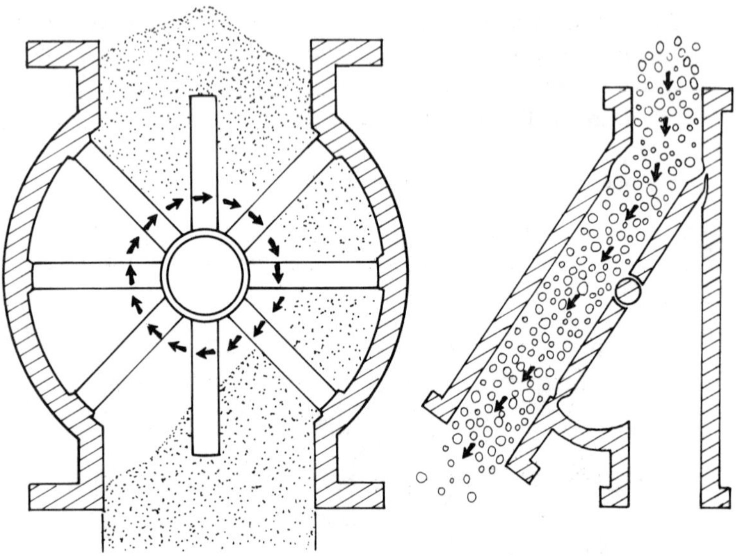

Fig. 4.28 Rotary valve[31] and diverter valve.[32]

4.4 PARTICULATE BEDS

Particulate beds offer a traditional way of obtaining thermal, chemical, or other physical interactions between matter in finely divided form (constituting the bed) and a transient fluid. Examples include beds of charcoal and sand to purify water, and many varied heat transfer applications throughout industry. In most of these applications, the particulate bed is static.

The blast furnace is an interesting type of bed. It is in reality a large particulate bed with two special characteristics:

1. The charge is top loaded while the gases (from the Tuyeres originally) are moving upwards.
2. The charge eventually melts to form slag and metal layers which are tapped off. This means that the blast furnace is a moving bed.

> Modern views of its operation tend to consider the blast furnace as a heat exchanger. The flow of the particulate charge has been studied extensively.

There are three main types of particulate beds:

Fixed Bed. The particulate is stationary and the fluid moves past the surfaces.

Spouted Bed. In this bed the particulate is disturbed to a limited extent. It is intermediate in character between the fixed and fluidized bed.

Fluidized Bed. Each particle is in motion and therefore the fluid comes into contact with all of the solid surface at some time.

The fluidized bed has found very wide application because it maximizes the contact between the particles and the fluid. Not only are these processes used in chemical reactions (including catalytic processes where the particles are catalysts), but they are also used in a wide variety of physical processes including coating processes, heat treatment, and many others.

Fixed Beds

There are two main problems with these beds. It is possible that due to unevenness of initial packing or unevenness which develops as a result of clogging during use, "channelling" develops and causes an uneven flow pattern.

The other problem is only important in beds where the ratio of bed diameter to particle diameter is small. In such cases there is an uneven packing of the particulate owing to a wall effect and this results in an annulus and central core of high velocity developing. This is shown in Fig. 4.29[33] in which it is apparent that this effect becomes quite small when the ratio D/b reaches about 100.

Fixed beds have been in use for many years and the equations relating bed variables in order to optimize conditions are known:[97]

$$\frac{\Delta p}{L} = a^1 \mu u + \rho u^2 \qquad (4.25)$$

in which

p = pressure loss,
L = bed length,
$a^1 = a/\mu$,
$a + b$ = factors related to the system,
μ = fluid viscosity,

ρ = bed density, and
u = linear fluid velocity.

The effect of bed porosity in fixed beds is incorporated as a multiplication factor in the equation.

Eq. (4.25) can be rewritten as

$$\frac{\Delta p/L}{U} = a + bG \tag{4.26}$$

in which $G = \rho U$ = mass velocity.

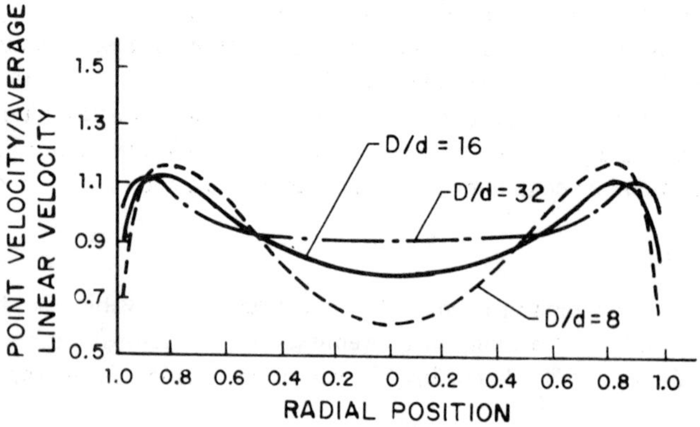

Fig. 4.29 Velocity distribution across a fixed bed as a function of tube-to-particle diameter ratio (D/d).[33]

By measuring pressure loss through a bed as a function of flow rate the values of $(\Delta p/L)/U$ may be plotted against G. The slope equals b and the intercept is equal to a. Knowing ϵ one can determine S_v. It has been shown that[98]

$$a = K_1 \mu S_v^2 \frac{(1-\epsilon)^2}{\epsilon^3} \tag{4.27}$$

and

$$b = K_2 S_v \frac{(1-\epsilon)^2}{\epsilon^3} \tag{4.28}$$

in which

$K_1 K_2$ = constants,

S_v = specific surface area,
ϵ = fractional void volume, and
μ = fluid viscosity.

If neither ϵ nor S_v is known, one can use the relationship between ϵ and the bed ρ_b and particle density ρ_p thus

$$\epsilon = \frac{\rho_p - \rho_d}{\rho_p} \tag{4.29}$$

This gives an averaging result for a specific bed.

When a fluid is moving through a fixed bed, not all of the material travels straight through and there is considerable sidetracking which causes mixing (with some resultant hold-up occurring also). The problem has been dealt with as a diffusion phenomenon and the results are presented as a series of dimensionless numbers which are called Peclet numbers. These are shown in Table 4.4 (based on reference 26).

In the top half of the table

$$P = \frac{dU}{D_r} \tag{4.30}$$

in which

d = particle diameter,
U = velocity of fluid,
D = diffusivity, and
r, A = radial and axial directions;

and in the bottom half

$$P = \frac{dU}{D_A} \tag{4.31}$$

Spouted Bed

This consists of a cone-shaped base fitted with cylindrical sides. A central axial hole admits the gas which moves up the center of the particulate and spouts. The particles fall on the outer annulus of the assembly and this completes the solid circuit. The gas flows upwards. In the central region it flows with the particulate flow and in the annulus it flows countercurrent.

Spouted beds are not easy to establish and perhaps one way of viewing

Table 4.4

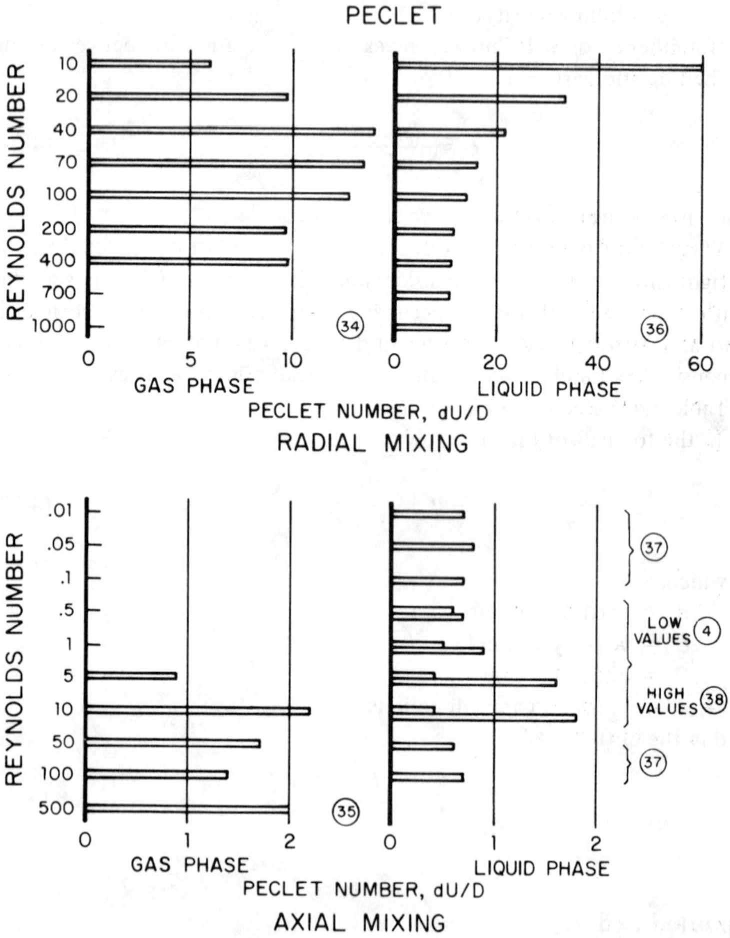

them is to consider them as essentially transient except under special conditions. The relationships between bed depth and gas velocity is shown in Fig. 4.31.

For low bed depths a fixed bed is easily converted into a spouting bed because the gas flow breaks through a thin particulate layer with little difficulty. In the case of the deep beds no spouting occurs and complete fluidization probably takes place. In the intermediate bed depths all three

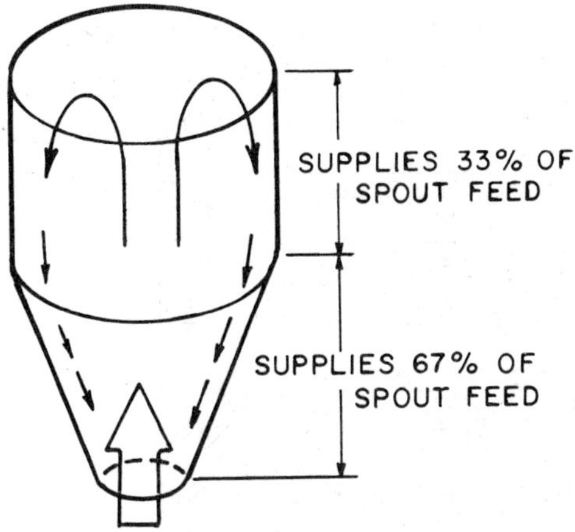

Fig. 4.30 Spouted bed.

types of behavior may occur depending upon the prevailing conditions. In the case of fairly deep beds under high gas flow conditions the particulate bumps up and down in the bed container. This is called slugging.

Fluidized Beds (Discovered c.a. 1941 A.D.)

If a particulate bed is subjected to a rising gas stream and if slugging does not tend to occur, the particles are lifted up and the increased porosity allows the gas to pass through the bed. If the gas flow is further increased the particles start to move around freely within their immediate neighborhood. This is called "incipient fluidization." If the gas flow is further increased the bed becomes fully fluidized.

A fluidized bed is essentially inhomogeneous. This is especially the case of the dispersive phase in a gas. The particles (perhaps because of electrostatic forces, etc.) tend to clump together (these are transient in identity) and therefore the gas flows via the interclump paths. This situation is referred to as aggregative fluidization. Alternatively, dispersive fluidization is more readily attained in the case of a liquid-particulate system. As the gas flow increases and full fluidization is achieved, transient streams of gas-rich material erupt through the top surface and produce a boiling appearance. At still higher gas flow rates individual particles are projected into the free space above the bed and eventually all the particles of the system may be projected in this manner. The gas velocity at which this occurs is

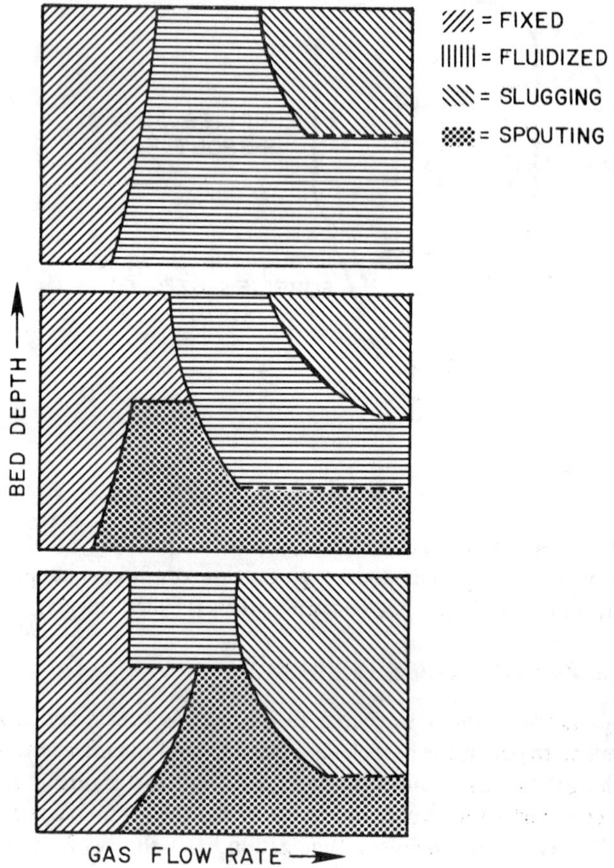

Fig. 4.31 Typical phase diagram for particle beds.[26]

termed the critical conveying point (C.C.P.). Below C.C.P. the fluidized phase is dense, above C.C.P. it is dilute. The analog of the melting of a solid is shown in Fig. 4.33.[39]

The advantages and disadvantages of dense phase and dilute fluidization are: a uniform temperature is maintained, a high rate of heat transfer is achieved, contents are easily moved, high reaction rates (large surface/volume) are achieved. But, they are not suitable for countercurrent operations because auxiliary items are expensive. Dense phase beds are more

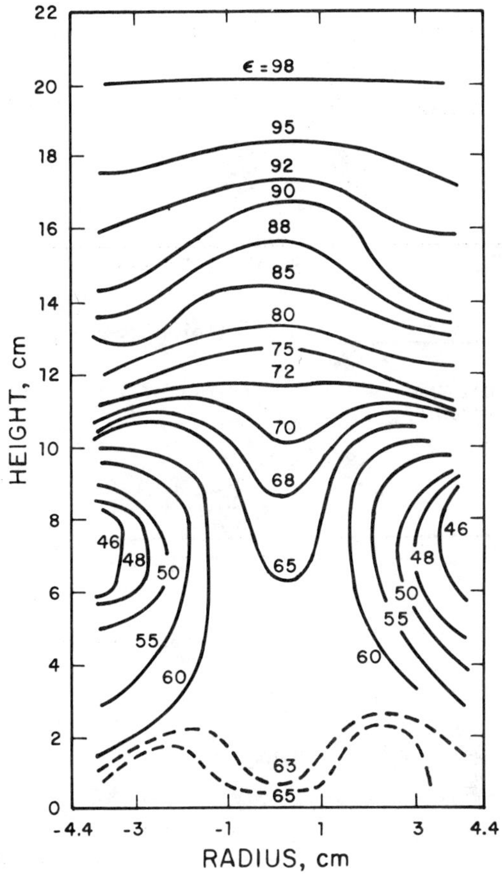

Fig. 4.32 Porosity distribution in a fluidized bed of 175–210-μ glass beads. Air velocity is 59.6 cm/s (incipient fluidization at 8.1 cm/s); bed weight is 1000 g.[43]

popular because of their smaller volume.

The relationships among pressure gradient, fluid velocity, and bed expansion are shown in Fig. 4.34.[40]

In Fig. 4.34, Li = initial bed depth, $(V - V_s)/V = \epsilon$ the void fraction, and V and V_s are the total bed and particulate volumes respectively. The regions of the curves are as follows (see Fig. 4.34):

Region A B—fixed bed, no movement of particulate;

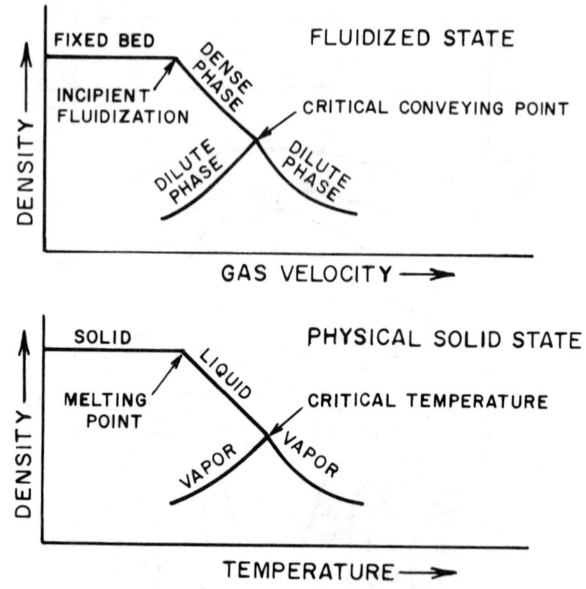

Fig. 4.33 Analogy between the fluidized and physical solid states.[39]

Table 4.5

A COMPARISON BETWEEN A FLUID SYSTEM AND
A FLUIDIZED SYSTEM – IMPORTANT VARIABLES

System	*Variables of Importance*		
Fluid system	Pressure	Volume	Temperature
Fluidized system	Pressure gradient	Bed expansion	Fluid velocity

Region B C–fixed bed, limited movements of some particulate;

Point C–the end of fixed bed conditions;

C D–some expansion of bed occurs and the pressure drops slightly. Particle clusters move around. This is incipient fluidization and it is essentially inhomogeneous at this stage;

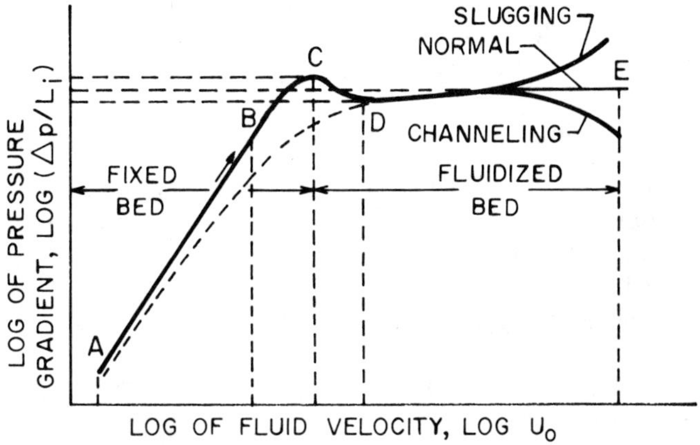

Fig. 4.34(a) Typical relationships for fluidized beds.[40]

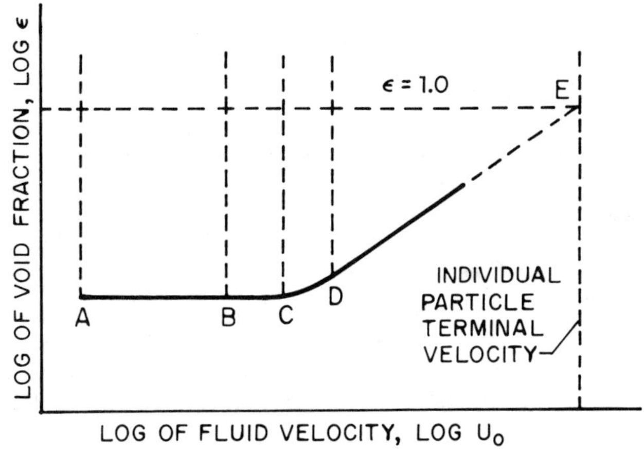

Fig. 4.34(b) Typical relationships for fluidized beds.[40]

Above D—as the gas velocity increases, the interparticle spacing increases and the attractive forces diminish in their effects. This results in only a slight increase of pressure drop with increased gas velocity. Eventually the level of pressure drop indicated by *C* will be reestablished.

However, channeling or slugging may occur.

In the case of channeling, the pressure loss is not recovered to C. If slugging occurs the pressure drop fluctuates and may well exceed C.

Design Calculations

Three characteristics are required: ϵ minimum, Δp minimum, and U minimum to define the onset of fluidization. Other factors to be considered are irregularities of bed behavior and required behavior. All of these influence bed design.

ϵ minimum is determined experimentally and approximates 0.4 for spheres to 0.8 for irregular particles.

Δp minimum is defined as the pressure drop which equals the weight of particulate per unit volume.

The equation is

$$\Delta p_{\text{min.}} = L(\rho_p - \rho_f)(1 - \epsilon)(g/g_c) \tag{4.32}$$

in which

L = bed depth with no gas flow, and

p, f = particle and fluid.

The U minimum may be determined graphically.

In general, fluidized and spouting bed design calculations are empirical and the reader is referred elsewhere for further details.[26,41]

Carry Over

If the rate of gas flow is increased, ϵ tends to approach 1 and the bed no longer exists. A limiting value of operation is $\epsilon = 0.8$ ($\not> 0.96$ for liquids). In actual operations there is always some loss of the particulate and this is augmented by attrition because the finer particles tend to flow away rapidly. This bed loss has to be replenished and the dust loss must be controlled by the use of cyclone collectors, baghouses, or similar installation. By combining dimensional analysis with experimental results, a quantitative relationship for this elutriation bed loss can be established.[42]

Bed Inhomogeneity

This is due to channeling, agglomeration, and various other defects in the bed structure. It produces variation from point to point of reactivity, heat energy, and also severe vibration can occur. Various methods have been proposed and used to study variations in bed density. The results of one of these, using a capactance measuring device, is shown in Fig. 4.34.[43]

One aspect of inhomogeneity, bubbling, which occurs as a result of a type of channeling, may or may not reduce efficiency depending upon the diffusion rates in the pocket and its vicinity. The bubble rises as the particles flow into the bottom, filling it. Because the walls are porous the gas flows outwards and the resultant drag on the particles is equivalent to surface tension in the case of bubbles of gases in liquids. The flow of bubbles has been studied by a variety of techniques.[44] Channeling occurs when a more or less permanent condition of gas leakage is set up from the bed. Redesign is usually required if this is serious. Mixing is usually no problem at a velocity above $U_{min.}$ and rates can be measured using tracer techniques.[26]

A way in which the fluidizing is achieved is illustrated in Fig. 4.35.[45]

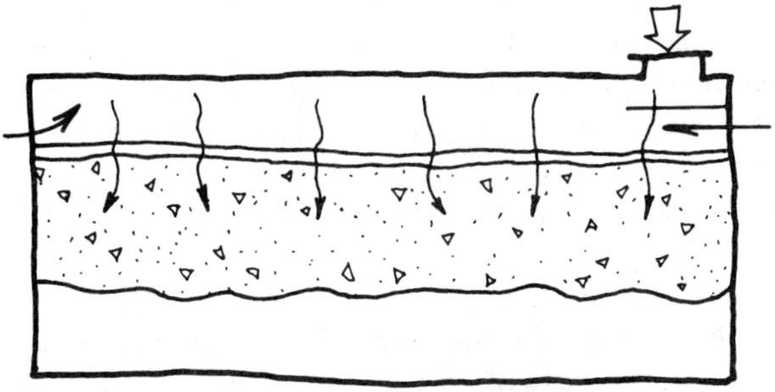

Fig. 4.35 Fluidized bed.[45] (Dynapore®) (Courtesy of Michigan Dynamics Division of Ambac Industries, Inc., Subsidiary of United Technologies Corporation).

4.5 MIXING OF PARTICULATE SOLIDS*

Introduction

There is probably no other area of study which indicates the nascency of particle science than the study of mixing particulate solids. In the first major work on powder science in 1948[46] the subject of mixing was not discussed. In the most recent text on *Particulate Technology* in 1968[47], 30 pages are devoted to this topic. This is a very interesting area of study be-

*I use the term *mixing* to cover all processes in which solid particles are intermingled.

cause from the first there has been a serious effort to approach the study of mixing from a fundamental-scientific point of view. For example, although the first published paper on the topic contained only one reference (there were no others to refer to!), this was a text book on statistics.[48] In this connection, it is worth noting that the primary thrust in mixing research and development has been of a statistical nature since that first paper was published, until recently when more concern for the inherent properties of the materials to be mixed is now being shown. This overall trend is quite clearly demonstrated in the literature which is reviewed in this section of this chapter. Thus, paradoxically, in flow, transport, and storage the initial starting point has been empirical and macroscopic with a move in the direction of a basic-scientific approach, whereas in mixing the approach has always been on a more fundamental level and the difficulty has been in relating the theory to practice. This latter problem is now being tackled.

Although mixing involves a consideration of particles of all combinations of the three stages of matter: solid, liquid, and gas, probably the major interest is in the mixing of solid particles. This is an ancient art but it is a very young science. One explanation for this slow development is that there has been little demand for change because the power requirements are not particularly large and the traditional and specialized mixer designs are well entrenched.

The most immediate saving which can be realized in the operation of particular mixes is that of mixing time. This means that the operator has to use some measurement of mixing which indicates when the optimum degree of mixing has been attained. Much of the early work was aimed at answering this very problem. Another method of attack was to develop a rate-of-mixing relationship based on a particular concept of the operative mechanism of mixing. Both of these methods of studying mixing were concerned with dry solids mixing. The mixing of liquids or gases with solids and the mixing of gases and liquids have mainly been dealt with by the use of dimensional analysis.

A bibliography compiled some years ago contained some 350 references[49] and a more recent review contains references to more than 650 works.[50] Clearly there is some considerable work being carried out on this subject at the present time and the interest is not abating as witnessed by the recent edition of *Powder Technology* devoted to the subject of mixing.[51]

However, not all are satisfied and in particular some serious criticisms of the current state of the mixing art have been leveled as follows: "Mixing indices based upon the theoretically randomized state are useless for following pharmaceutical mixing operations"[52] and further; "Mixing theory has mostly been evolved by consideration of the mixing of coarse (above 100 micron) noncohesive ideal particle such as glass beads or sand. As a result, the importance of the theory developed is questionable."[53] These comments are strong for learned journals to contain. They are quoted herein to serve as an indicator to the reader, who is probably a beginner on the subject, that although the following discussion sounds convincing, it may not always be useful in all circumstances.

In this section, the concepts of mixing versus segregation, scale of scrutiny, mixture quality, mixing kinetics, and some more recent pattern recognition techniques applied to the latter are discussed. Also considered are some reports of mixer performance studies. The last section deals with more recent ideas on mixing research which are collectively grouped under the heading of the Bridgewater Review.

Mixing and Segregation Mechanisms

Three basic mechanisms have been proposed that appear to be useful concepts.[54]

1. The transfer and exchange of groups of adjacent particles from one location to another. This is clearly related to the folding and stirring action within the body of the mixer and it has been termed *convective mixing* (e.g., this is thought to occur in a paddle or trough mixer).
2. The distribution of particles over a freshly developed surface occurs in a simple drum mixer because the particles in the free surface suffer "random" collisions from their neighbors. The situation is somewhat analogous to the mechanism of diffusion in gases and it has been termed *diffusion mixing.*
3. The setting up of slipping planes within the mass of particles is termed *shear mixing.* It is equivalent to the stirring of a fluid at low Reynolds number. Mixers that are thought to utilize this mechanism in their design of operation are sometimes known as "Geyser" types. An Archimidean spiral draws the material up the center of a hopper and it descends at the sides.

These mixing mechanisms have been considered to play a specified role in a number of batch type mixers as shown in Table 4.6.[55]

Table 4.6

ROLE OF MECHANISMS INVOLVED IN THE VARIOUS MIXERS[55]

Type of mixer	*Mechanisms*		
	Rolling or Diffusive	Shear	Convective
Rotocube			
(without impeller)	Major	Minor	–
(with impeller)	Major	Minor	Minor
V-Mixer	Major	Minor	–
Nautamix	–	Minor	Major
Ribbon blender	–	Major	Major

During mixing, segregation processes are also at work, to undo at least some of the effort that is being expended in the mixing process. A number of mechanisms have been proposed as responsible for this effect including:

1. Vibrational segregation in which larger particles tend to rise to the surface.[56]
2. Pouring segregation in which larger particles tumble further down the free surface of a growing pile.[57]
3. Percolation segregation in which the top layers of particles act as a screen through which all but the larger particles are able to pass and reach the stationary region below.[58]
4. Free flight segregation which results from particles having different trajectories.[58]

The connection between mechanisms 2. and 3. should be obvious. An alternative approach[55] is to compare the mechanisms of segregation with those suggested for mixing.

More recently, it has been noted that, although convective mixing must be an important mode of mixing, it is unrealistic to try to achieve a clear cut distinction between it and shear mixing. This difficulty arises because convective mixing occurs by the relative motion of clumps and so shearing of the interclump regions is going on. Furthermore, the term diffusive mixing is criticized on the grounds that the process can be confused with convective mixing.[59] Instead of using these approaches that are essentially modeling devices based upon analogies with other forms of matter, it is proposed that there should be a direct approach in which the actual properties of the material undergoing the mixing are studied, thus:[59]

1. Continuum Zone Properties: These include the effects of composition and also of consolidation upon the failure zone width and the failure zone spacing within the particulate mass in the mixer.
2. Microscopic Properties: These include the agglomerate strength and also the tendency for interparticle bulk percolation and free surface percolation to occur. The strength of the agglomeration is particularly important in cohesive systems.

These comments are referred to again later in this section.

The Scale of Scrutiny

This concept[60] is a formalized restatement of common sense. Thus the mixing process utilized and the time and trouble taken to ensure that adequate mixing has been achieved are directly proportional to the needs of the job. For example, in the mixing of batches of drugs which may be toxic if given in excess, a close degree of control is required. Alternatively the proportion of mixed vegetables in a canning factory may not be so critical. Examination of a random field[47] shows that the smaller the scale of scrutiny, the more likely the mixing would appear to be unsatisfactory. Similarly, the larger the sample, the nearer the result will appear to be a perfect, uniform mix. The problem of degree of mixing or mixedness and of sampling should be defined.

The perfect mixture has been defined as a given mixture of materials such that the composition of all samples taken from it, however small, are identical.[61] Thus a particulate mixture can never be perfect if the scale of scrutiny is of the order of the particle dimensions. This is why it is generally understood that "perfect" mixing of liquids is possible while that of particulate solids is not. Clearly if the scale of scrutiny is relative in both cases the "perfectness" of the mixes may well be identical.

Mixture Quality

One way of expressing the degree of mixture is by defining the variance σ^2 or S^2 in the case of a completely random mixture of black and white particles: If P is the proportion of black particles, $\sigma_r^2 = P(1 - P)/n$ in which n is the sample size and r denotes randomness.

In the completely unmixed system[54]

$$\sigma_0^2 = P(1 - P)$$

In the case of multisized particles the problem is more difficult to solve. For a particular size range (a)[54(a)]

$$\sigma_r^2 = [P(1-P)Wa + P^2(\bar{W} - Wa)]/W$$

in which

Wa = average particle weight in range a,
W = total sample weight, and
$\bar{W}$ = average particles weight in the whole = $AWa + BWb + CWc$ ..... A, B, C are fractions.

Rewriting

$$\sigma_r^2 = \frac{[P(1-P)]}{W/Wa} + \frac{[P^2(\bar{W} - Wa)]}{W}$$

Now $W/Wa = n$ and therefore this term is equivalent to σ_r^2 in a perfect mix. Because of segregation, S^2 for small particles is larger than this calculated figure.[2]

For an incompletely mixed material it is necessary to have some measure of the degree of mixedness or mixture M. One of the first ways proposed for defining M was to divide the ideal value of standard deviation by the experimental Sd:[48,62]

$$M = \sigma_r/S \tag{4.33}$$

But using this ratio, if $\sigma_r^2 = P(1-P)/n$ and if the population is completely unmixed, then $S^2 = P(1-P)$ and therefore $(\sigma_r^2/S^2)^{1/2}$. So $M = (1/n)^{1/2}$, but if M is merely a function of sample size, it cannot be used to describe a mixture.

An improved suggestion is[54]

$$M = (\sigma_o - S)/(\sigma_o - \sigma_r) \tag{4.34}$$

Now when the mixture is completely mixed $S = \sigma_r$ and $M = 1$; when it is unmixed $S = \sigma_o$ and $M = 0$.

An alternative way of expressing M is:[54]

$$M = \frac{\sigma_o^2 - S^2}{\sigma_o^2 - \sigma_r^2} \tag{4.35}$$

There are two sources of variation in the mixer. These are the variation due to the mixing process itself (σ_m^2) and that due to the sampling variation itself (σ_r^2). Therefore, the actual or experimentally determined variance $S^2 = \sigma_m^2 + \sigma_r^2$, so $\sigma_m^2 = S^2 - \sigma_r^2$, and M can be rewritten as

$$1 - M = 1 - \frac{\sigma_o^2 - S^2}{\sigma_o^2 - \sigma_r^2} = \frac{\sigma_o^2 - \sigma_r^2 - \sigma_o^2 + S^2}{\sigma_o^2 - \sigma_r^2}$$

Therefore,

$$1 - M = \frac{\sigma_m^2}{\sigma_o^2 - \sigma_r^2} \quad \text{or} \quad 1 - M = \frac{S^2 - \sigma_r^2}{\sigma_o^2 - \sigma_r^2}$$

In the case of convective mixing the population may be viewed as consisting of groups of N_g particles randomly distributed. As mixing proceeds N_g diminishes and the limit = 1. The number of groups per sample $= N/N_g$ and therefore,

$$S^2 = \frac{P(1-P)}{N/N_g} = N_g \sigma_r^2$$

It is known that $\sigma_o^2 = n\sigma_r^{2\,*}$; therefore,

$$1 - M = \frac{N_g\sigma_r^2 - \sigma_r^2}{N\sigma_r^2 - \sigma_r^2} = \frac{N_g - 1}{N - 1} = \quad \text{or} \quad = \frac{N_g}{N} \tag{4.36}$$

Consequently, as N increases from zero, $1 - M$ increases, but beyond a certain point ($N \approx N_g$) $1 - m$ will start to decrease. As mixing proceeds, N_g will decrease and therefore for any given sample size N_g/N will decrease and so will $1 - m$. This is illustrated for successive times t_1, t_2, and t_3 in Fig. 4.36. A list of commonly used mixing indices to estimate mixture quality are given in Table 4.7.[63]

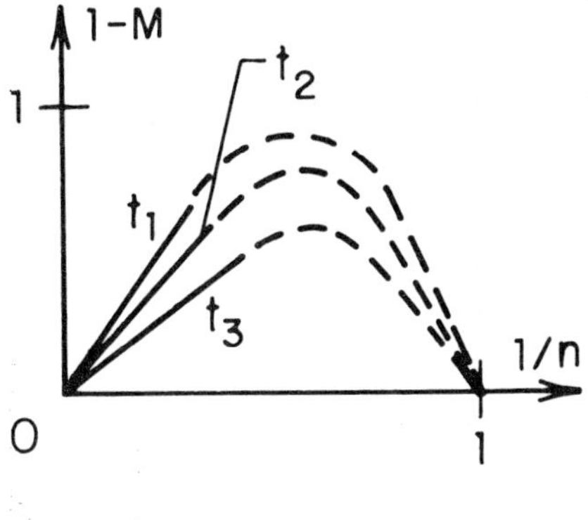

Fig. 4.36 Effect of sample size on analysis of convective mixing.

Table 4.7

COMMONLY USED MIXING INDICES[63]

Author	*Ref. no.*	*Mixing index*	*Change of index with improved mixing*
(1) Lacey, Weidenbaum, and Bonilla	48 62	$M = \frac{\sigma_r}{s}$	$<1 \longrightarrow 1$
(2) Kramer	54	$M = \frac{\sigma_o - s}{\sigma_o - \sigma_r}$	$0 \longrightarrow 1$
(3) Lacey	54	$M = \frac{\sigma_o^2 - s^2}{\sigma_o^2 - \sigma_r^2}$	$0 \longrightarrow 1$
(4) Ashton and Valentin	58	$M^2 = \frac{\log \sigma_o^2 - \log s^2}{\log \sigma_o^2 - \log \sigma_r^2}$	$0 \longrightarrow 1$

Table 4.7 *(cont'd.)*

(5) Poole, Taylor, and Wall	$M = \dfrac{c}{c_r}$ or $\dfrac{s}{\sigma_r}$	$\gg 1 \longrightarrow 1$
(6) Carley-Macauley, Rumpf, and Sommer	$M^2 = \dfrac{s^2 - \sigma_r^2}{(1 - 1/n)}$	$\gg 0 \longrightarrow 0$
(7) Gayle, Lacey, and Gary	$M = \dfrac{\chi_o^2 - \chi_r^2}{\chi_s^2 - \chi_r^2}$	$1 \longrightarrow 0$

where M = mixing index or variance due to mixing, s^* = observed between-sample composition standard deviation, σ_o = theoretical standard deviation in the unmixed state, σ_r = theoretical standard deviation of a random mixture, c, c_r = coefficients of variation of observed and random mixtures respectively, χ_o^2 = observed chi squared for mixture, χ_r^2 = expected chi squared for random mixture, χ_s^2 = expected chi squared for unmixed state.

Two Examples

The first simple example is given to illustrate the use of standard deviation in evaluating mixing efficiency in the case of noncohesive, coarse, binary particulate mixtures.[64] A mixer is examined in an effort to improve its efficiency. The process was one of mixing sand and lime in a spiral bladed "Ribbon mixer." This type of mixer is fitted with flights. The method of testing the sample was to neutralize with 5N HCl. Ideally the whole batch should have been neutralized by 10.8 cc of 5N HCl for a 10 g sample. The measure used was the standard deviation.

$$\text{Standard deviation} = \left[\frac{(\bar{X} - X)^2}{N}\right]^{1/2}$$

in which

$\bar{X} = 10.8$

X = observed ml HCl per sample, and

N = number of samples.

The results are reported in Fig. 4.37 and the numerical calculations are

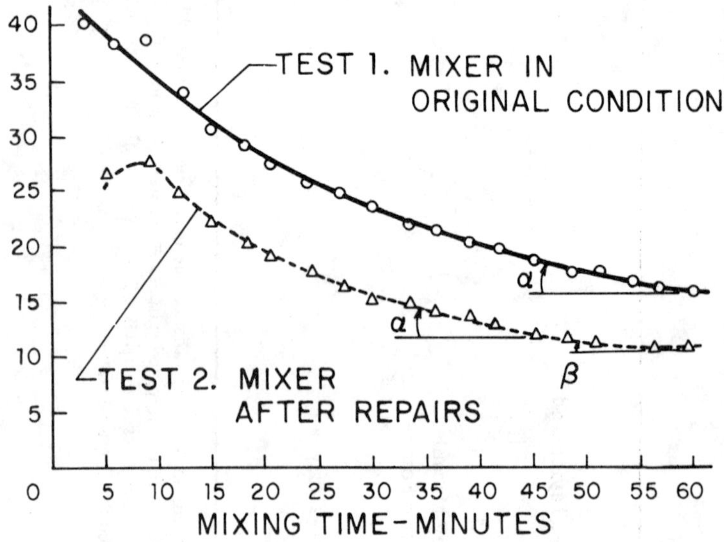

Fig. 4.37[64]

Table 4.8[64]

TESTS ON BATCH SAND, LIME MIXER

(avg. conc. 10.8 ml of 5N HCl)

Run No.	*Time after start* (min.)	*Test run No. 1*		*Test run No. 2*	
		Ml 5N HCl	σ	Ml 5N HCl	σ
1	3	51.3	40.4	10.1	0.7
2	6	47.2	38.4	47.3	25.8
3	9	3.1	38.8	38.7	26.5
4	12	50.1	33.9	24.3	23.9
5	15	7.8	30.4	17.8	21.6
6	18	34.3	29.4	14.7	19.8
7	21	19.1	27.4	15.6	18.4
8	24	20.8	25.8	13.8	17.2
9	27	25.8	24.9	7.8	16.3
10	30	17.1	23.6	6.2	15.5
11	33	12.1	22.6	4.7	14.9
12	36	11.2	21.8	8.9	14.3
13	39	13.3	20.8	11.5	13.7
14	42	12.8	20.0	10.8	13.2
15	45	14.2	19.3	12.2	12.8
16	48	10.1	18.7	14.2	12.4
17	51	5.9	18.2	10.7	12.0
18	54	10.6	17.7		
19	57	10.7	17.2	11.1	11.7
20	60	10.9	16.8	10.9	11.4

Note.– Only slide rule accuracy has been employed and in some cases decimals have been discarded.

given in Table 4.8. Results of run 1 were carried out and, after repairs, run 2 was completed. By comparing the slopes at A and B (α in each case) it is seen that the mixing time may be reduced from 60 minutes to 45 minutes.

In the second example the values of Chi square[65] were determined in order to follow the mixing of a three component mix. The sampling device collected three samples a,b, and c from four locations: 1, 2, 3, and 4.

$$\text{Chi square} = X^2 = \frac{(O - E)^2}{E} \tag{4.37}$$

in which

O = observed number of particles of a given color, and
E = expected number of particles of the same color.

Table 4.9

RESULTS OF THREE COMPONENT MIXING AFTER 23 REVOLUTIONS[65]

	Number of Particles			
Sample	Red	White	Blue	X^2
1a	3	9	18	6.10
1b	11	1	18	5.21
1c	10	5	15	0.28
2a	10	5	15	0.28
2b	11	5	14	0.68
2c	6	4	20	3.33
3a	17	2	11	10.84
3b	16	4	10	7.78
3c	13	6	11	2.84
4a	8	10	12	3.38
4b	7	7	16	0.68
4c	8	2	20	4.44
Expected Value =	9	6	15	ΣX^2 45.84

The results of run 1 are given in Table 4.9. The following illustrates the method of working out the values of chi^2:
For sample 1

O	3	9	18
$(O - E)$	−6	+3	+3
$(O - E)^2$	36	9	9
$\frac{(O - E)^2}{E}$	4	1.5	1.6

$$\frac{(O-E)^2}{E} = X^2 = 6.1$$

The value of X^2 for $t = 0$, the initial condition was calculated as follows. Thirty particles per sample (of 40 samples from each batch) were counted. If R and W are fixed, B is also fixed. Therefore, the number of degrees of freedom = 2 and the total for the 12 samples = 24; therefore $X^2_{t=0} = 24 \times 30 = 720$. The reduction of X^2 with mixing time is shown in Fig. 4.38.[65]

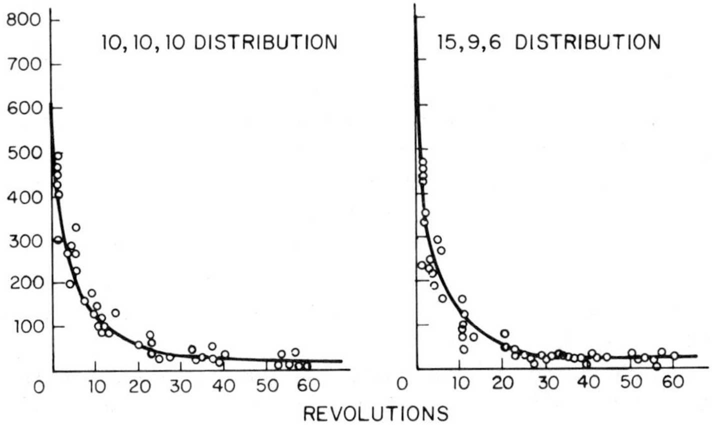

Fig. 4.38 Illustrating the use of chi^2 in following mixing performance.[65]

If $X^2 = \dfrac{NS^2}{\sigma^2}$ for the ideal mixture $X^2 = N = 1 + \nu$ (because $\nu = N - 1$) and 12 samples multiplied by 2 degrees of freedom each equal 24 which is equal to ν. Therefore, $N = 25 = X_I^2$.

By comparing this X^2 for the ideal mix with the actual experimental X^2 values, a comparative (fuzzy) measure of the degree of mixing is obtained. This is shown in Table 4.10.

Mixture Structure

In practice it is often necessary to have the individual particles in a mixture randomly distributed for example in multicomponent powder metal-

Table 4.10

FUZZY MEASURE OF THE DEGREE OF MIXING[65]

No. of mixer revs	0	2	5	11	23	35	55
X^2	720	330	300	90	46	21	30
Therefore $X_I^2/X^2 \cong$	0.03	0.08	0.10	0.28	0.56	1.02	0.82
R mix classification (comparative)[2]	0.1 very poor	0.1 vp	0.1 vp/p	0.1-0.3 p	0.3–0.7 fair	0.90 excel.	0.7–0.9 good

vp = very poor
p = poor

lurgy blends, and in solid rocket propellants a random mix is preferred. However, in the case of pharmaceutical tablets or fertilizers the mixing specification requires only that the deviations in composition from that specified should not exceed a given value. In many cases the degree of mixedness of the components in any individual tablet is unimportant. The first type of application requires some sort of spatial test in order to specify the nature of the mix. An interesting theoretical attempt has been made using chi^2 distribution with $2n$ degrees of freedom.[66] The method has been tested experimentally also.[67]

More recently, there have been attempts to assess the structure of particulate mixtures using two coefficients: the correlation coefficient $R(r)$[68] and the similarity coefficient S_{AB}.[69] These two coefficients are defined as

$$R(r) = \frac{(X_i - \bar{X})(X_{i+r} - \bar{X})}{\sigma^2}, \qquad -1 \leqslant R(r) \leqslant 1 \tag{4.38}$$

and

$$S_{AB} = \frac{\Sigma_r R(r)_A R(r)_B}{[\Sigma_r R(r)_A^2]^{1/2}\,[\Sigma_r R(r)_B^2]^{1/2}} \tag{4.39}$$

in which

r is a distance,
X_i is the composition at point i,
X_{i+r} is the composition at point r distant from point i,
$\bar{X}$ is the mean composition, and
σ^2 is the variance.

A and B refer to two different functions $R(r)$.

The function $R(r)$ plotted against r is called a Correlogram. An example is given in Fig. 4.39 for a model mixing experiment.[63]

The function $R(r)$ is a measure of mixture segregation and the function S_{AB} is a pattern recognition method which may be used to compare two different $R(r)$ functions. At the present time workers in the field are developing the appropriate techniques to enable these indices to be used.

Assessment of Mixers

During the past 20 years there has been a number of studies of the performance of batch type mixers for particulate solids[55] and more recently there have been some reports on continuous mixers.[70,71]

Batch mixers offer an advantage in that their capital costs may be lower than that of continuous mixers. Their operation is less complex because the loading is completed before the mixer is closed and operated. By comparison, continuous mixers must have their feed monitored constantly and there are problems associated with weighing on a continuous basis small quantities of components to be added to the process. Batch mixers are easier to handle for multicomponent mixing operations too. On the other hand, continuous mixers offer lower labor costs, less space, less inventory of particulate, and better mixing results.[71] Batch mixers that have been studies include the following:[55,72,73,74]

V mixer (twin shell),
rotating cube,
ribbon blender,
paddle type mixers,
conical mixers,
muller,
hammer mill, and
drum mixer.

It has been observed that design considerations often stem from the desire to facilitate the mixer manufacturers' problems rather than from the needs dictated by the mixing process itself.[75]

We have already seen earlier in this section that the standard deviation used to derive the mixing indices is a function both of the number of the particles in the sample and also the proportions of the components. The use of the thief-type sampler tends to predispose the investigator to study tumbling type mixers because this design is most readily accessed with this

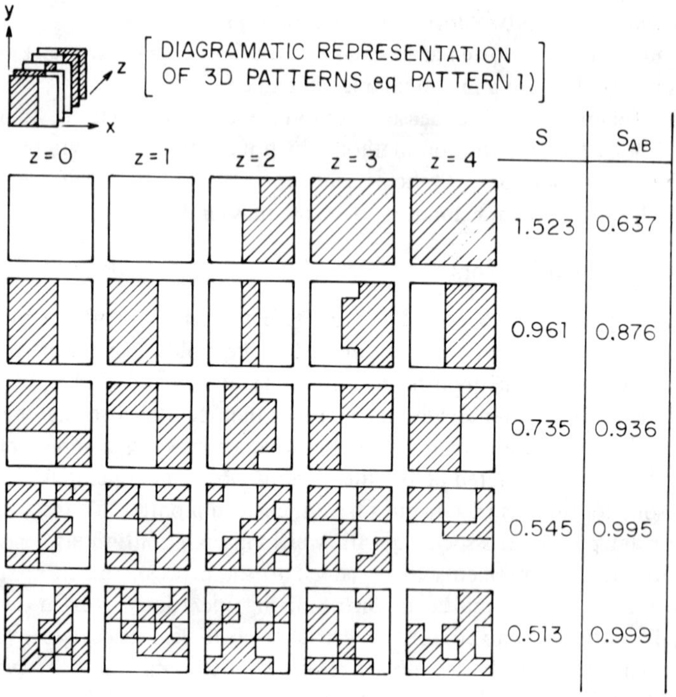

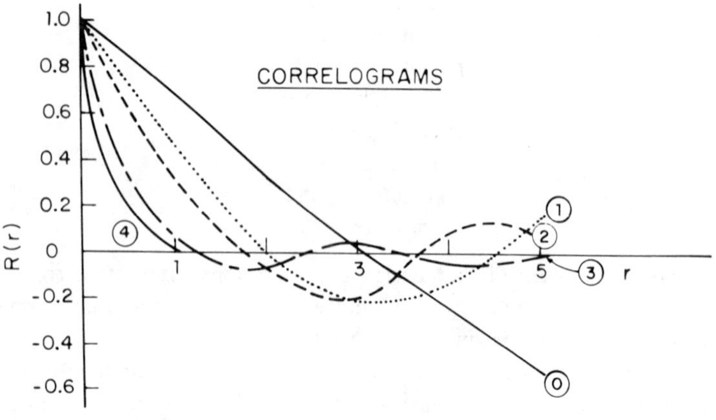

Fig. 4.39 Correlograms for model mixing experiment.[63]

type of sampler. The mixing operation in a batch mixer includes loading and unloading and frequently small laboratory-sized studies cannot be scaled up to mixers one or even two orders of magnitude larger.[55] In order to counter, to some extent, the above dissatisfactions, a technique has been used in which a discharge profile is determined.[55]

In the case of continuous mixers, it has been observed that some residence time of the feed in the mixer is necessary in order that back mixing may occur.[71] The number of studies on continuous mixing, however, is quite small. Some interesting results have been reported. A phenomenological approach has made it possible to predict with reasonable accuracy the quality of the mix coming from a mixer from a knowledge of the input and residence time.[71] Furthermore, a close agreement of actual variances with experimental variances has been obtained in the case of continuous mixing of a segregating particulate system.[76] Parameters used in these studies included the variance reduction ratio (V.R.R.) which is simply the ratio:

$$\text{V.R.R.} = \sigma_i^2/\sigma_o^2 \qquad (4.40)$$

in which

i indicates input, and

o indicates output.

Another parameter used is the correlation index I_c defined as follows:

$$I_c = R(r)e^{-r/\theta} \qquad (4.41)$$

in which $R(r)$ is the correlation coefficient and is $\text{cov}(x_t, x_{t+r})/\text{var}(x_t)$.

Some studies have also been reported of the patterns of segregation developed within a mixer.[77]

An interesting experimental study of the performance of dry continuous mixers has shown that the result obtained is very much dependent upon the properties of particulate to be mixed. In particular the flow properties were considered to be very significant.[70] It was found that some mixers are more suitable for certain mixes than others are.

Rate Theories

There has been a number of researches reported in the literature dealing with particulate mixing on the basis of an analog of diffusional mixing.[49,50] One of the first of these considered the process as being essentially an application of Fick's law for unidirectional diffusion. Thus the starting condition was that in which component A was topped by component B

with a dividing surface S between them (the interface). As mixing proceeds, it was envisaged that the area of this interface would increase.[78] Since S could not be measured directly, an indirect measure was computed as follows:[78]

$$x = \left(100 - \frac{n}{30} \cdot 100\right) \text{percent} \tag{4.42}$$

in which

$x/100$ is the fraction of system unmixed at any time, and

n is the number of samples of theoretical composition.

Some results from that research are shown in Fig. 4.40.

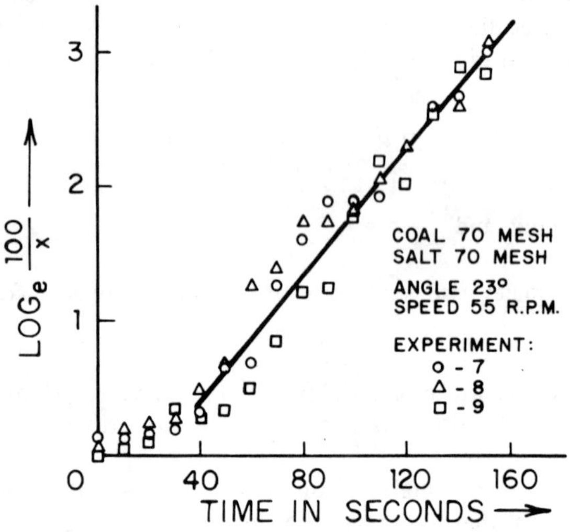

Fig. 4.40[78]

Further work by others has followed up this diffusional approach.[54,79,80,82] This work has been carried out on essentially monosized populations of particles such that the conclusion, "Micromixing by diffusion is the only mechanism operating in a simple barrel mixer, end loaded with ideal particles," has been reached.[81] However, when the populations under study differ in size, for example, demixing may occur.[59,80] It has been suggested that in these cases, "continuity" or "Kinematic" type wave equations might be more appropriate and useful.[59,83]

Mixing and Demixing

The first serious attempt to study the problem of concomitant mixing and demixing approached the problem presuming a modified kinetic relationship.[84]

The rate of attainment of randomness was assumed with time as shown in Fig. 4.41. This curve is independent of the alleged mechanism of mixing.

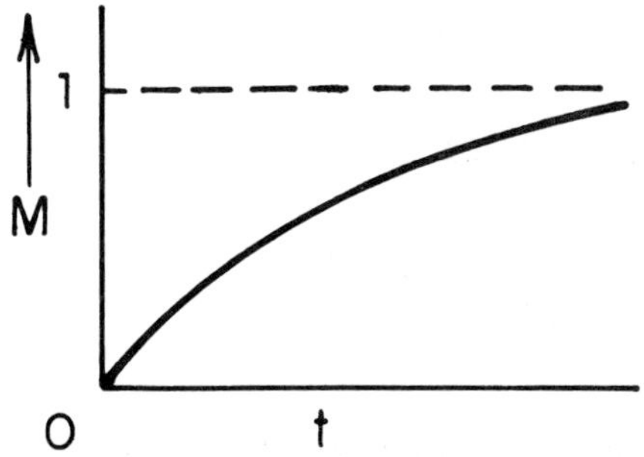

Fig. 4.41 Rate of attainment of complete mixing.

The mixing equation of the type $M = 1 - l^{-kt}$ fits the curve of Fig. 4.41. But it fits the limited number of results available over only a small range of conditions. It depends on the powder and upon the machine geometry.

It is assumed that M is related to the standard deviation as follows

$$M = 1 - \sigma/\sigma_o \qquad (4.43)$$

in which suffix o = unmixed material.

This relationship is not necessarily correct but its form is useful for developing concepts. According to this work two processes, mixing and demixing, are said to occur. Mixing is most rapid at the beginning of the whole process because any interchange of particles would tend to be between black and white.* In a well mixed particulate, there would be a much lower probability of B–W exchange as a B–B, W–W exchange tends to occur more often. Therefore,

$$\frac{dM}{dt} = A(1 - M),$$

*Black and white particles considered.

A = constant, and
M = degree of mixing.
Integration of this leads to Eq. (4.42).

While the particulate is unmixed the potential for demixing is small but once near "perfect" mixing is attained, the potential for demixing is a maximum if there is a difference between the particles which causes them to behave differently in a field of force (in this case gravity). Thus

$$\frac{dM}{dt} = -B\phi \tag{4.44}$$

in which
ϕ = the demixing potential, and
B = a constant.
Therefore, for the whole process total

$$\frac{dM}{dt} = A(1-M) - B\phi \tag{4.45}$$

By assuming two different starting configurations and following through with the appropriate set of calculations it was demonstrated that the degree of mixing varied with time to an extent that depended upon the method of loading: i.e., is the heavier material on top or underneath? After long mixing times, the value of M was found to equal

$$M = \text{Eff.}\left(1 - \frac{B^2}{A}\right) \tag{4.46}$$

in which Eff. is termed the intrinsic efficiency of the mixer.

The effect of loading, as outlined above, is shown in Fig. 4.42.

Using dimensional analysis it was shown that the constants A and B have the following make up:

$$A \alpha N \times \psi_1(D/d_m) \times \psi_2(N/N_c) \times \psi_3(f_m) \quad \text{and} \tag{4.47}$$

$$B \alpha N \times \psi_4(D/D_m) \times \psi_5(N/N_c) \times \psi_6(\Delta\rho/\rho_m) \times \psi_7(\Delta d/d_m) \tag{4.48}$$

in which
D is the mill diameter,
N is the speed of rotation,
N_c is the critical speed,
d_m is the particle mean diameter,

f_m is the particle friction,
ρ is the density, and
Δ indicates the difference between the two components.

It has been observed that *A, B,* and Eff. can be considered as fitting parameters and that, in effect, enough of the details are known to make investigations of this kind successful.

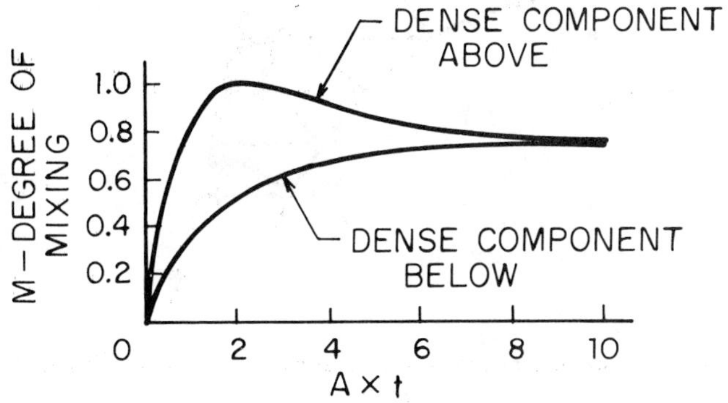

Fig. 4.42 The effect of loading of the heavier component in binary mixing in a barrel mixer.[84]

The Bridgewater Review[59]

1. It is necessary to fully consider the nature of the interparticle forces and to ascertain whether they are stronger or weaker than the applied force field. In general if the size is in excess of 100 microns, the cohesive forces are small and the mass can be treated as being "cohesionless."
2. Borrowing from soil mechanics permits one to consider simultaneously the shear stress, the normal stress and the density of the particulate in order to derive the critical state curve. This is illustrated in Fig. 4.43. This curve together with the elastic unloading relationships is considered to be a potentially fruitful road to follow in attempting to understand the fundamental behavior of a particulate in a mixer.
3. It is necessary to fully consider the shear velocity (this seems to increase the slope of the τ/σ curve, when increased) and also the effect of particle shape (more elongated particles are more difficult

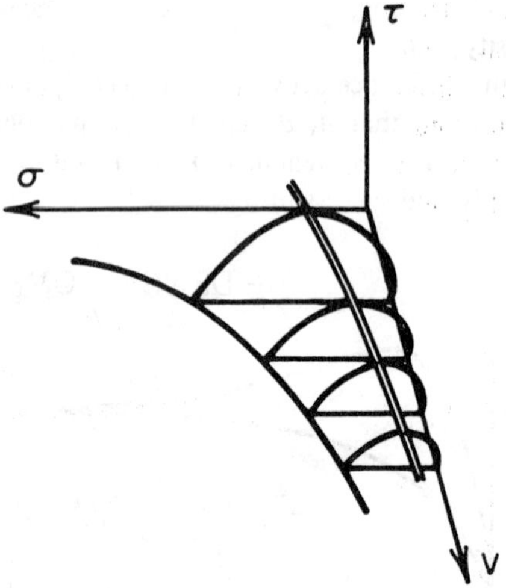

Fig. 4.43 Illustration of the critical state concept. The double line is the critical state line.[59]

to mix).

4. Many mixers have internal moving paddles with which the mixing process is assisted/achieved. Some studies have been reported of the movement of particulates over blades and the relationships between forces and system geometric properties.[85,86] It has been noted that there many be difficulties in building a dynamically similar model of a mixer[87] and much work remains to be done to clarify this problem.
5. Interparticle percolation has been studied with respect to the segregation occurring in the process of pouring particles onto a free surface. (However, it has also been shown that segregation within the body of the particulate due to percolation can also occur[88-91].) The total strain of the particulate is an important factor in determining the distance percolated by a particle through the bulk. Some interesting results are shown in Fig. 4.44.

In Fig. 4.44 particles of diameter d are percolating through particles of diameter D. The larger the strain, the greater the distance particles of diameter d percolate. This affords one explanation of mixing behavior of particulates that is actually observed. Thus, at the onset of mixing, there

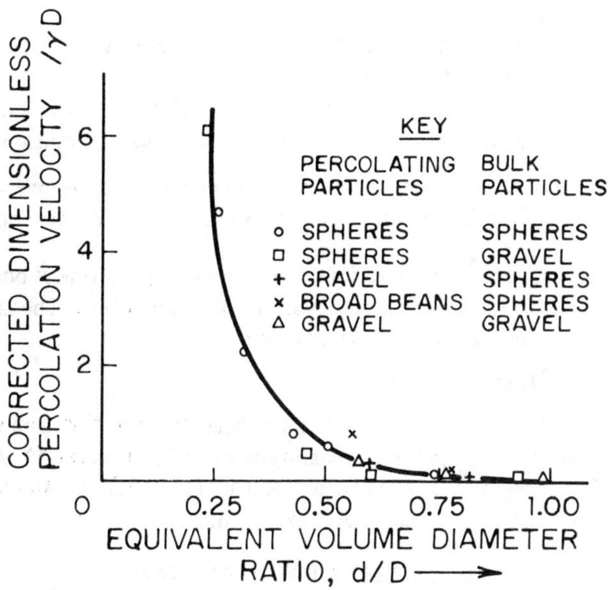

Fig. 4.44 Dimensionless percolation velocity vs. diameter ratio d/D, 1 = distance in strain γ.[59]

is a gradual improvement in the quality of the mix due to the intermixing of the different layers. However, the longer the mixing time, the more likely it is that percolation will lead to the gradual building up of more stable regions of the particulate in the mix (they will be more dense than the average). Failure zones tend to avoid these better packed regions, a random condition is not achieved, and a steady state system obtains.[92]

Random and Ordered Mixes

The classical approach to the study of mixtures has usually considered noncohesive particulates. Recently there has been an increased interest in systems showing cohesion. In contrast to so-called random mixing that is idealized in the case of cohesionless particulates, mixtures exhibiting cohesion are idealized in the concept of ordered mixing. This was first noted in 1976.* Some data reported therein is abstracted in Table 4.11.

According to Hersey,** mixtures may be divided on a basis of random,

*Yip, C.W., and Hersey, J.A. "Ordered Mixing." *Nature* 262, July 15, 1976, 202–203.

**Hersey, J. A. Tutorial on Mixing of Particulate Solids, Fan, L. T., Chairman, Fine Particle Society Annual Meeting, Philadelphia, May 1979.

ordered, and mixed.
Hersey believes that most mixtures may be of the mixed variety, i.e., both partially ordered and partially random.

A somewhat different approach represented by Harnby* uses a modified form of the Buslik equation (54a). This leads to the observations that

1. control of large diameter particles is necessary to control α_R^2 and
2. control of small diameter particles is necessary to control bonding forces.

These bonding forces have various origins including: physical bonding, moisture, electrostatic bonding, and Van der Waals bonding. The latter is believed to be important in ordered mixtures.

Modeling of Mixing Processes

Most recently considerable progress has been made in this area due to the treatment of mixing as a Markovian type stochastic process.** A book by L. T. Fan, Kansas State University and F. Lai, U.S.D.A., Manhattan, Kansas is expected shortly on this important topic.

4.6 SOLID-LIQUID MIXING

Whereas solid-solid mixes are neutral, in so far as they do not change until energy is introduced from without, solid-liquid, solid-gas, and also the majority of liquid-liquid mixtures are negative mixtures.

Unless energy is introduced into the system, the components segregate.

Suspension

The conditions necessary to achieve a bare suspension have been studied in the case of a variety of solids and fluids: sand and NaCl in H_2O, acetone, CCl_4, and K_2CO_3 solution.[93] Under Power (P) input conditions, specified by the Newton number,

$$\mathrm{Ne} = \frac{P}{l_L n^3 D} \tag{4.49}$$

in which

l_L = density of liquid,
n = rate of rotation of the stirrer, and
D = diameter of stirrer.

The effects of geometry and of materials properties have been related in

*Harnby, Tutorial on Mixing of Particulate Solids, L.T. Fan, Chairman, Fine Particle Society Annual Meeting, Philadelphia, May 1979.

**L.T. Fan, Tutorial on Mixing of Particulate Solids, Fine Particle Society Annual Meeting, Philadelphia, May 1979.

Table 4.11

ORDERED MIXING OF MICROFINE SALICYLIC ACID WITH COARSE SUCROSE CRYSTALS

Mean size fraction sucrose crystals (μm)	*Weight of fraction* (percent)	*Mean weight of salicylic acid/unit weight of sucrose* ($\times 10^{-4}$)	*Mean weight of salicylic acid/ unit weight of sucrose* (10^{-7} g cm^{-2})	*Average surface area of sucrose coated* (percent)
462.5	3.34	4.6	5.6	2.5
390.0	20.10	6.7	6.9	2.6
302.5	52.43	8.9	7.1	2.6
215.0	21.02	13.3	7.6	2.8
168.0	2.34	17.5	7.6	1.8
137.5	0.78	20.2	7.2	2.7

(see footnote p. 237 J. Hersey and also reference 201)

order to achieve a suspension of particles. If the ratio of T/D (where $T =$ Tank Diameter) is decreased to < 0.45, the particles spread more radially out along the base of the tank and pile up at the periphery. If T/D is > 0.45 the particles move inward and then up to the propeller. The number $S = f(1/B)$ is plotted against T/D for various values of T/C in which $C =$ the distance from the stirrer to the base of the tank. S is therefore a measure of efficiency.

Admixing of Gradually Increasing Quantities of Liquid

In the areas of ceramics, plastics compounding, fertilizer blending, concrete mixing and soil treatment, and lubricant blending in powder metallurgy a small quantity of liquid may be introduced into a large proportion of particulate solids. In dry solids mixing there is a range of rotational speeds below which the charge slips along the wall of the blender and above which projection occurs.[94] Mixing occurs as a result of tumbling and there is no necessity for the presence of shearing blades. But when mixing a small percent of liquid, the character of the mix changes as follows:

1. If a small quantity of liquid is added to a dry free flowing powder (e.g., H_2O to clay) the solid tends to form small pellets which resist disintegration and which are cushioned in the dry powder mass. The particulate is still essentially free flowing.
2. As more liquid is added, pelletizing continues until all free dry particles are agglomerated. Some of the pellets fracture during stirring.
3. At higher H_2O percents, pellets start to cohere and form clumps. This process is characterized by an increased plasticity of the material (c.f. potter's clay), so that the resistance to shear markedly increases.
4. At higher percents of H_2O added, the material becomes sticky and paste-like. It no longer fractures under load and it actually heals.
5. At the highest percentage H_2O content, the mass becomes free flowing once more.

Clearly the addition of a small quantity of dry powder to a particulate which already contains some liquid mixes with an efficiency which depends upon the proportion of liquid present as shown in Table 4.12.

Because of the changes in the consistency of the matrix material, the power requirement reaches a maximum for a water-clay ratio of 0.3 and this is the reason for the unusually rapid rate of mixing of this particular material. It should be noted that the initial rate of mixing is very rapid, and most of the time of mixing is taken up by a relatively inefficient pro-

Table 4.12[94]

EFFECT OF LIQUID PERCENT ON MIXING CHARACTERISTICS

Appearance	*Powder*	*Pellet & Powder*	*Pellet*	*Plastic*	*Paste*	*Liquid*
Rate of Homogenization	Rapid	Initially rapid then very slow	Very slow	Rapid	Very low	Rapid
Power required	Low	Low–Mod	Mod	High	Mod	Low
Shearing needed	Low	Small	Pellets have to be fractured	Large amount of shear required	Difficult to transmit	Yes

cess.[94]

Particle-Gas Mixtures

These are negative mixtures. Two major areas of interest are smoke clouds and dust clouds, of which the production of aerosols and dusting sprays is a specialized and important segment. Aerosols are fine sprays which are produced as is outlined in Chapter 3. The nature of the cohesive forces between particles which have to be overcome if the cloud is to be temporarily stabilized has been discussed.[95] The formation of dust clouds has been described by Bagnold in his classical work.[96]

4.7 INTERPARTICLE SEPARATION TECHNOLOGY

Most, if not all, processors of particulates regularly effect a separation of one fraction of the bulk solid from the remainder. This is the most simple level of interparticle separation. Reasons for doing this are many and varied. Some of the more obvious ones include: removing unwanted coarse particles (oversize); separating valuable constituents from an unwanted gangue; preparing different specifications of powders; obtaining differently shaped samples for laboratory study; removing magnetic particles from the bulk, etc. Of course, all particulate processors are bedevilled by the problem of the particulate effluent which they unleash on an unwilling host society. Although it may be an oversimplification one can differentiate the two groups in the following manner: interparticle separation is effected to enrich the particulate product; particle-fluid separation is effected to purify the fluid.

In this section separation processes involving dry separation of particles from each other are considered including mechanical methods (sieving) as well as magnetic, electrical, and gravity based separation. With the notable exception of sieving, for many years dry methods have been outweighed in their importance by wet methods.

Sieving

There is considerable variety in the specifications used in different countries and for this reason there is some effort being made to evolve an international sieve mesh standard. In general sizes of apertures range from 8 inches down to 400 mesh (~ 40 μ) and micromesh sieves (20, 10, 5, 2, and 1 μ). In the U.S. the ASTM system is widely used; in U.K., the BS 410 system; and in Germany the DIN 1171 system is used. The wire used in the manufacture of the mesh is carefully specified.

For example, with 200 mesh the specifications are:

aperture 75 μ, wire diameter 52 μ,
average aperture 70.4 $\rightarrow$ 79.6 μ,
$\not>$ 6 percent apertures lie outside 56 $\rightarrow$ 94, and
no apertures lie outside 42 $\rightarrow$ 108 μ.

The efficiency of sieving has been stated:[100]

$$\eta_N = 100\left[\frac{\begin{pmatrix}\text{Quantity of true coarse}\\ \text{particles in the coarser}\\ \text{collected fraction}\end{pmatrix}}{\begin{pmatrix}\text{Quantity of true coarse}\\ \text{particles in the feed}\\ \text{particles}\end{pmatrix}} - \frac{\begin{pmatrix}\text{Quantity of true fine}\\ \text{particles in the coarser}\\ \text{collected fraction}\end{pmatrix}}{\begin{pmatrix}\text{Quantity of true fine}\\ \text{particles in the feed}\\ \text{particles}\end{pmatrix}}\right] \tag{4.50}$$

which in Fig. 4.45[47] is

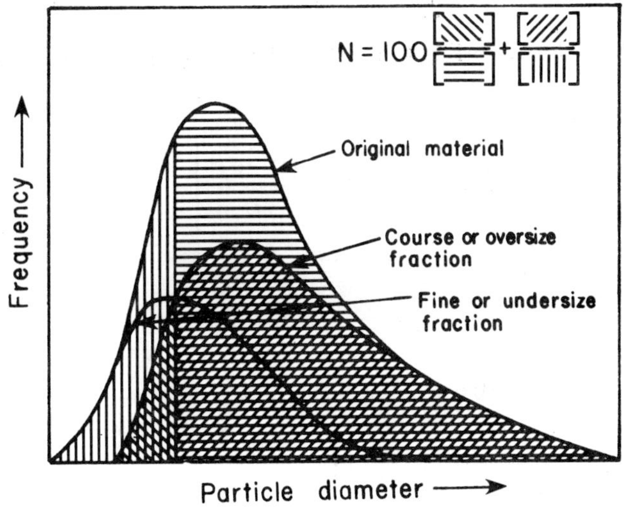

Fig. 4.45 Typical size-classification results.[47]

On an industrial scale 65 percent $< \eta_N <$ 75 percent and to increase this value further requires long sieves in order to effectively separate the coarse and the approximate sieve size fraction.

Batch Sieving. For batch sieving the weight of particles passing is proportional to the weight of appropriately sized particulate on the mesh (i.e., fit to pass), i.e.,

$$\frac{dW}{dt} = -KW$$

Therefore,

$$W = Me^{-Kt} \tag{4.51}$$

in which M = original weight on the screen.

*Continuous Sieving.** For continuous sieving,[101] the probability of a particle passing through aperture D is $(D-d)^2/D^2$. Therefore, the probability of the particle remaining on the sieve is $1 - (D - d)^2/D^2$. This is for one trial. If a particle has i trials at passage, then the fraction retained on the mesh is

$$r = \left[1 - \frac{(D-d)^2}{D^2}\right]^i \tag{4.52}$$

Eq. (4.51) is plotted in Fig. 4.46. (Note $[(D - d)/D] < 1$ for passage to occur.)

Taking logs

$$\ln r_i = i \ln \left[1 - \frac{(D-d)^2}{D}\right]$$

Therefore,

$$\ln r_i \cong -i\,\frac{(D-d)^2}{D}$$

If d is the size that is 50 percent retained (or passed), then

$$\ln 0.5 = -i\left(1 - \frac{d_{0.5}}{D}\right)^2$$

Therefore,

$$d_{0.5} = D - \left(\frac{.832D}{i^{1/2}}\right)$$

If the number of trials I per unit length of screen L = a constant, then

*Continuous sieving is often called screening.

$$i = IL$$

Therefore,

$$d_{0.5} = D - \frac{1}{\bar{L}} \frac{0.832D}{I^{1/2}} \tag{4.53}$$

in which I is a screening index and is a function of the system design, and D is ≡ screen aperture. Consequently, a plot of $d_{0.5}$ versus $1/L^{1/2}$ is a straight line.

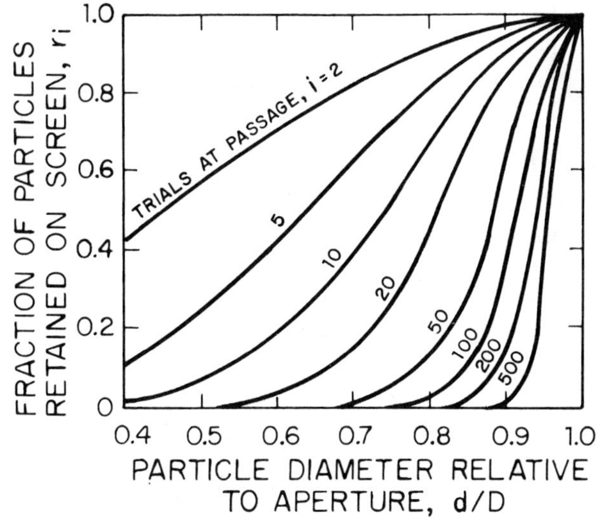

Fig. 4.46 Fractional retention of particles as a function of trials at passage.[102]

It has been shown that the sieving process may be considered as dividing into two regions as shown in Fig. 4.47.[103]

In region 1

$$P = at^b$$

in which

P = cumulative percent passing,

a = fraction passing per unit time,

b = ~ 1, and

t = sieving time.

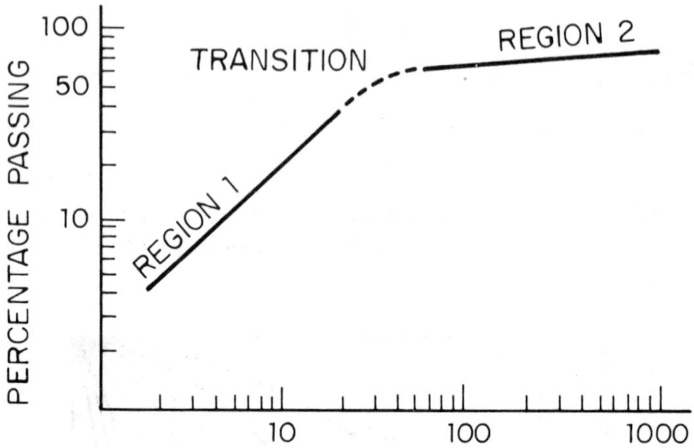

Fig. 4.47 The rate at which particles pass through a sieve[103] (reprinted with permission of American Society for Testing and Materials).

It was found that

$$a = f\left(\frac{W}{SA_o}, \frac{S}{d}\right)$$

Therefore,

$$\frac{aW}{A_o S} = C_1 \left(\frac{S}{K_s \bar{d}_m}\right) \frac{h}{\ln \sigma} \tag{4.54}$$

in which

C_1, h = constants,

σ = geometric standard deviation,

$K_S \bar{d}_m$ = linear function of geometric mass mean of the particle,

S = mesh aperture,

A_o = sieve operation area, and

W = total load on sieve.

In the second region the rate of passage decreases, so

$$\frac{dW}{dt} = \frac{P}{t \ln \sigma_g} \phi(z) \tag{4.55}$$

in which

$\phi(z)$ = normal probability function

and

$$P = \frac{dN}{Nd(\ln t)} = \ln \frac{\sigma_g}{\sigma_{gt}}$$

in which

N = number of particles able to pass.

Blinding During the Sieving Process

In the majority of conventional sieving processes the operative mechanisms involve low frequency (2 to 10 c.p.s.) shaking and/or banging actions. One major phenomenon which lowers sieving efficiency is the event known as blinding. In this event particles tend to block up and lodge in the mesh available for the sieving.

The process of blinding in batch sieving is a transient process and occurs in four distinct stages.[104]

1. In stage I a positive rate of blinding is obtained. The amount of blinding increases rapidly as the rate of passage of particles through the mesh increases.
2. Stage II of blinding occurs at a time in the sieving process at which the maximum amount of blinding occurs. The time to reach this peak of blinding depends upon the friction condition in the moving powder mass.[105] The more friction that is exhibited by the powder, the longer the time to reach this peak of blinding.
3. In stage III the rate of blinding decreases exponentially with time as the rate of passage of particles through the mesh diminishes.
4. In stage IV the rate of free passage through the mesh is negligible or nonexistent. The residual of particles is retained loosely on the surface of the mesh. Theoretically, the rate of blinding in stage IV is zero and the state is known as "hard blinding." It is this hard blinding which has traditionally been the cause for concern in its detrimental effect upon sieving efficiency.

The value of residual or hard blinding is strongly affected by both particle size and particle shape. These results, summarized in Fig. 4.48 show that:[105]

1. The level of residual blinding is markedly dependent upon mesh aperture below 100 microns. This is believed to indicate that the residual or hard blinding is due to high friction at the mesh-particle

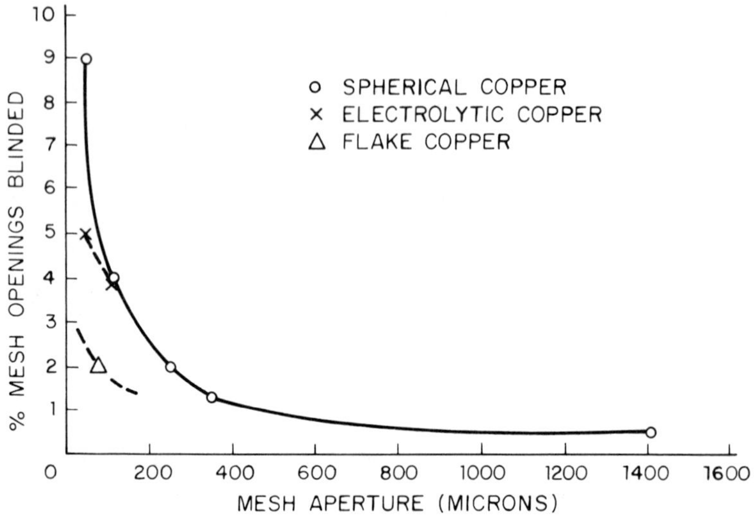

Fig. 4.48 Effect of particle size (shape or hard blinding).[105]

interface.

2. For a given size, irregular particles do not hard blind as much as spherical particles because they are more able to rotate and dislodge from the aperture.

The hypothesis that blinding is the controlling factor in the mechanics of sieving operations has been put forward and developed into a theory of sieving. The theory agrees with the experimental results very well.[106] The final equations derived relate the total amount of material per unit area of the sieve at any time, t, to the initial masses of undersized material and blinding material. This mathematical model contains three constants that vary only by machine. They are K, k, and k_1. The rate of passage of undersized material dW/dt is given by

$$\frac{dW}{dt} = -K(W - W_\infty)[1 - G(w_o - w)] \tag{4.56}$$

in which

K is used as a constant of proportionality and

$\frac{dW}{dt}$ is the rate of passage of undersized material through the sieve (screen),

W is the total mass per unit area on the sieve at any instant, t,
W_∞ is the total mass on the sieve at the true endpoint,
G is the area of sieve blinded by a unit mass of blinding material,
w_o is the mass of blinding material per unit area initially on the sieve, and
w is the mass of blinding material per unit area on the sieve at any instant, t.

The case of deblinding $\left(\frac{dw}{dt}\right)_o$ is given by

$$\left(\frac{dw}{dt}\right)_o = k_1 (w_o - w) \tag{4.57}$$

in which

$\left(\frac{dw}{dt}\right)_o$ is the rate of deblinding,
k_1 is the constant of proportionality, and
$w_o - w$ is the quantity of material trapped.

Also at any instant

$$\left(\frac{dw}{dt}\right)_i = -kw[1 - G(w_o - w)] \tag{4.58}$$

in which

i indicates the instantaneous rate of deblinding at any moment, and
k is termed the "diffusion coefficient."

Values of these constants were obtained for three machines: Rotap, Pascall, and Alpine. Their values are given in Table 4.13.[106]

Table 4.13

VALUES OF "DIFFUSION COEFFICIENTS" FOR PROPOSED DEBLINDING SIEVING MECHANISM[106]

	$k\ s^{-1}$	$k_1\ s^{-1}$	$K\ s^{-1}$
Rotap	45.71×10^{-4}	3.33×10^{-5}	11.43×10^{-3}
Pascell	1.79×10^{-4}	108.34×10^{-5}	8.97×10^{-3}
Alpine	36.60×10^{-4}	1.34×10^{-5}	11.44×10^{-3}

Eq. (4.55) through (4.57) are combined in one overall equation which relates W the total amount of material per unit area of the sieve at any time t as

$$W = W_\infty + (W_o - W_\infty) \left\{ \left[\frac{(1 - He^{-mt})}{1 - H} \right]^{-\frac{K}{k}} e^{-K\left(GJ + \frac{m}{2k} - \frac{b}{2k}\right)t} \right\} \quad (4.59)$$

in which

W_∞ is the total mass of material on a unit area of the sieve at the true end point value,

W_o is the initial mass of material on a unit area of sieve,

H is $\dfrac{2aW_o - b - m}{2aW_o + b + m}$,

J is $\dfrac{1}{G} - W_o$,

a is kG,

b is $k_1 + k - kGW_o$ and

c is $-k_1 W_o$.

The interesting point about this approach, quite apart from its inherent sense of realism, is that it offers a way of incorporating some of the characteristics of the particulate being sieved through the constant k. This had previously been attempted only once before and in a qualitative way only.[107]

Sieve Bends

These consist of a series of bars which are parallel, horizontal, and arranged on a concave frame. The feed is introduced at the top and the undersize is removed from between the bars. The range of particles treated is quite wide (from 8 mesh through 50 microns) and the particle size passing appears to be determined by the thickness of the layer deflected.[108]

The work of Nakajima and Whelen of the University of Queensland (Australia) as quoted by Reid* is of considerable interest. It appears that these authors have worked out an elegant way of deriving particle size and

*Grant, G., Reid, A.F., and Zuiderwyk, N. "Size and Simplified Shape Distribution of Ore Particles as Measured by Automated SEM," presented at the Fine Particle Society Annual Meeting in Philadelphia, May 1979. To be published in Proceedings by Industrial and Scientific Conference Management, Inc., Chicago, IL, 1979.

also certain particle shape parameters from sieving data. A basis of their approach is to say that the size of the particle equals the square root of the sum of the squares of the mesh aperture sides. This new work may open up renewed interest in sieving.

4.8 LABORATORY SEPARATION TECHNIQUES

In the past few years numerous techniques for the separation of particles from each other have been either proposed or experimentally demonstrated or both. In some cases, these techniques may be scaled up to a pilot plant stage and later to production units.

Mechanical Separation

For powder particles in which the size ranges from 100 microns on up, mechanical separation systems have been developed.[109–112] In one apparatus, a glass disc is rotated at a slight angle to the horizontal. A high percentage recovery of spheres (in excess of 99 percent) was reported in the case of a mixture of differently shaped glass particles.[113] A later report used an aluminum disc instead of a glass one. Again a good separation was reported and particularly in the case of 10/40# copper oxide up to 100 lb per day of material was separated.[114] It was found that the smaller the particle size and the more variety in the shapes to be separated, the slower the rate of separation. For example, using a device in which rolling was the predominant means of separating spheres from other shapes it was found that an average of only 3 g per 8 hours of operation of 100 micron alumina could be treated.[115] Other rolling devices have been developed for shape separation.[116]

Chromatographic Techniques

There are three distinct titles in this area:

Gel or permeation chromatography,[117]
Hydrodynamic chromatography,[118] and
Potential barrier chromatography.[119]

The principal of gel chromatography is related to separation by particle size by virtue of the differences between the movements of differently sized particles into and out of pores in a packed bed of granules. Thus the particles which are small enough to enter pores in the bed through Brownian motion take a longer average residence time than the other particles in the system.

In the case of hydrodynamic chromatography the basis of separation is also particle size but in this case pores are not required for the mechanism to operate. The principle involved is that the velocity profile between grains is nonuniform. Smaller particles are able to move closer to the surface of the grains and these therefore exhibit a lower than average velocity compared with the other particles in the population.

The proposed potential barrier chromatography promises to be much more selective towards particle properties than the other two methods enumerated above. Separation is expected to be possible on a basis of particle size and also on a basis of particle constitution including surface charge density, collision diameter of surface molecules, and Hamaker constant.[119] The principle involved concerns the Brownian movement of particles over a potential energy barrier in order to become physically attached to or detached from the surface of the grains. The origin of the potential energy barrier is a summation of the Born, double layer, and van der Waals forces.* These three forces are assumed to be capable of being lumped together into a boundary condition on the convective-diffusion equation. The motion of particles over this energy barrier is shown to be a first order reversible reaction and it is proposed that, by careful choice of column packing and the ionic strength of the carrier solution, the particle residence time can be closely controlled. Hence, the separation is affected.

Gravity Separators

These may be either absolute or relative settling types. The absolute type uses a fluid intermediate in density between the two particulates to be separated. (A suspension may have S.G. from 1.4 to 7.5 depending on composition.) The relative method uses Stokes Law of settling.

Flotation is a process in which bubbles are used to attach to and float out the required particulate from a prepared raw material.[101,120] The process finds wide application.

The first essential is the production of a stable (but not everlasting) froth by the use of "soap" addition coupled with gas bubbles. The most commonly used industrial frothers are pine oil and cresylic acid. The amount added is c.a. 0.0X x lb per ton of H_2O to give a 1-3 mm diameter bubble

*These three forces are treated in Chapter 7.

(this optimizes between stability and floating capacity). The soap has one end which is hydrophylic and the other end hydrophobic (see Fig. 4.49).

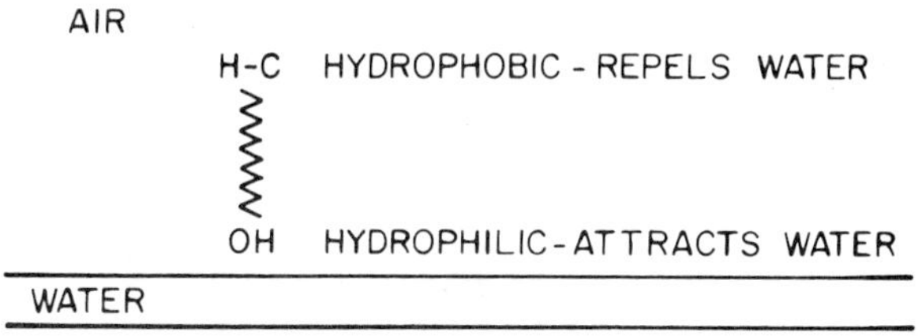

Fig. 4.49 Illustrating the principle of flotation.

The particle which is to be separated has to be (or made to be) possessed of a greater affinity for air than water. An air bubble which attaches onto such a particle then acts as a float. Many inorganic particulates must have their affinity for water reduced by selective adsorption of "collector" agents such as Xanthates. Unfortunately flocculation can occur and this must be minimized by agitation.

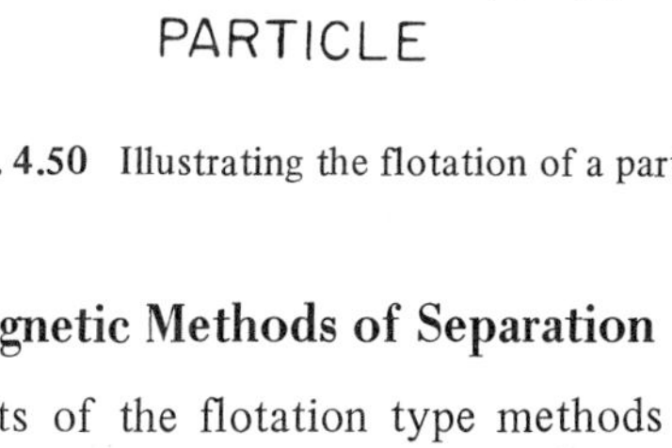

Fig. 4.50 Illustrating the flotation of a particle.

Electrical and Magnetic Methods of Separation

The developments of the flotation type methods of separation have overshadowed the application of other particle characteristics to interparticle separation. However, in order to reduce both water consumption and energy consumption (in drying) there is again some interest in electrostatic separation. Magnetic separation is of course currently attractive in relationship to clean up of the magnetic (inorganic) sulfur component in coal.[121]

Electrostatic Separation

This depends upon the differing ability of particulates to pick up an electric charge. This may be done by simply forcing the material into frictional contact with others. The tendency of different particulate materials to charge is shown in Table 4.14.[123] Those at the top of the column become positive with respect to those below.

Table 4.14[123]

FRICTIONAL ELECTRIC SERIES

Positive End	Serpentine
	Asbestos
	Topaz
	Mica
	Glass
	Calcite
	Barite
	Quartz
	Magnesium
	Lead
	Gypsum
	Zinc
	Beryl
	Sphalerite
	Magnetite
	Galena
	Pyrite
	Molybdenite
	Copper
	Antimony
	Stibnite
	Silver
	Silicon
	Sulfur
Negative End	Rubber

Electrostatic separation can also be effected by an applied electrostatic field in which conducting particles soon lose their acquired charge upon emergence from the direct influence of the field. A sophisticated application of this technique has been reported[124] in which the mineral in granular form (coal) was fluidized and subjected to electrostatic benefication.

Methods based upon the dielectric properties of particles to concentrate "lines of force" have also been developed. In these the particles, suspended in a medium of lesser dielectric constant, tend to move towards the region of highest field strength. A list of dielectric constants is given in Table 4.15.[123] The phenomenon is termed dielectrophoresis.

Magnetic Separation

Magnetic Separators are familiar in many plant operations in which the processor wishes to ensure the removal of tramp magnetic materials. Various designs of magnetic separators have been developed and are utilized. A major division occurs between wet and dry methods, the former being said to be more effective in regard to dispersion of the types of particle about to be magnetically separated.[47] On a basis of iron having a relative magnetic strength of 100, the data in Table 4.16 illustrates the broad spread of relative magnetic strengths for different particulate materials.[126] Of course, wet methods require magnetically strong particles and dry methods require free flowing particulate material.[47]

Ferromagnetic materials are not the only substances that can be separated. However, a high gradient field has been studied in respect to the separation of paramagnetic mineral particles.[125]

4.9 PARTICLE FLUID SEPARATION

Particle-fluid separation may include solids from liquids, solids from gases,* or droplets from gases. Separation systems make use of property differences such as:

1. barriers–size
2. density differences–settling process
3. electrical, magnetic, thermal differences
4. Other methods include: sonic energy to agglomerate particles in gas, chemical agents to agglomerate in a liquid, and a drying and

*Teoman Ariman, "Novel Concepts, Methods and Advanced Technology in Particle-Gas Separation," NSF/EPA Report, August 1978, ENG 77-02016.

evaporation process.

Table 4.15

VARIOUS DIELECTRIC CONSTANTS.[123]

Solid	*Dielectric Constant*
Anatase	31.0
Apatite	5.72
Arsenic	10.23
Bauxite	10.85
Beryl	5.73
Bismuth	>81
Borax	6.55
Calcite	6.36
Chromite	11.03
Copper	>81
Diamond	4.58
Dolomite	8.45
Galena	>81
Graphite	>81
Hornblende	7.37
Ilmenite	>33.7
Kaolinite	11.18
Lead chloride	39.06
Manganite	>81
Muscorite	10.00
Niccolite	<33.7
Potassium iodide	7.00
Quartz	6.53
Silver	>81
Silver chloride	14.13
Silver iodide	33.86
Sodium chloride	6.29
Stibnite	11.15
Sulfur	4.0
Talc	9.41
Tetrahedrite	≈ 81
Wolframite	12.51
Zincite	≈ 33
Zircon	6.09

Table 4.16

RELATIVE MAGNETIC ATTRACTABILITY.[126]

Substance	*Relative Attractability*
Iron (arbitrary standard)	100.00
Magnetite	40.18
Ilmenite	24.70
Pyrrhotite	6.69
Hematite	1.32
Zircon	1.01
Corundum	0.83
Manganite	0.52
Garnet	0.40
Quartz	0.37
Rutile	0.37
Molybdenite	0.23
Dolomite	0.22
Talc	0.15
Gypsum	0.12
Galena	0.04
Calcite	0.03
Witherite	0.02

Barrier Systems

Any arrangement which places a mechanical impediment to the flow of a solid particulate in a liquid is defined as a barrier system. A common method is to use vibrating screens. These are used to separate gravel, coal, and vegetable waste, to mention only a few. The angle of inclination is normally downward with the feed direction but if the degree of separation required is higher than that obtained by this means, the screen can be inclined in the other direction.

Liquid phase filtration uses many materials such as wire mesh, fabrics, and fixed particulate beds. Because the filter is continuously blinding up, the efficiency of filtration decreases with time. The driving force used for filtration is usually gravitational but it may also be either a pressure difference or centrifugal force.

The resistance to flow is related to the rate of fluid flow, thus

$$\text{driving force} = \text{rate} \times \text{resistance}$$

and the resistance can be written in terms of system characteristics, involving filtration cake depth.

$$R = \frac{\mu L}{KA} \tag{4.60}$$

in which

μ is a liquid viscosity,
L is cake depth,
A is area, and
K is the permeability, which is a function of porosity, specific surface, and *compressibility*.

Note: Because the cake is subjected to pressure during operation, some compression can occur and the cake porosity (ϵ) may change.

The pressure loss due to the laminar flow is

$$\frac{\Delta p}{L} = \frac{150\mu V_L(1-\epsilon)^2 Sv^2}{36\epsilon^3} \tag{4.61}$$

and the term $(1 - \epsilon)^2/\epsilon^3$ indicates that the porosity has a very big effect upon the filtration. Thus a mere ~25 percent change in ϵ can promote a tenfold change in $(1 - \epsilon)^2/\epsilon^3$ Fig. 4.51 shows a dimensionless curve which in effect characterizes the material being caked.[127] In the case of the cloth filters, plugging or blinding tends to increase the resistance of the membrane. It is due either to excessive pressure or to deflocculation as a result of high pressure shearing the agglomerates.

Gas Phase Filtration

Gas phase filtration is carried out in dust houses and bag houses. Particles from the very large to the very small can be removed from the gas and the process is extremely effective except for the removal of the last traces of fines. (Removing this remainder may be very important in the case of valuable or toxic dusts.) Because of the tendency to blind, especially with combined moisture present, the pressure drop can be excessive, necessitating frequent cleaning of the filter elements. Unfortunately these filter elements, if they are to be most economical, have to be for low temperatures.

However, by using varieties of composite materials for the filter media and also alloys for the higher extremes, filtration in the range from subzero

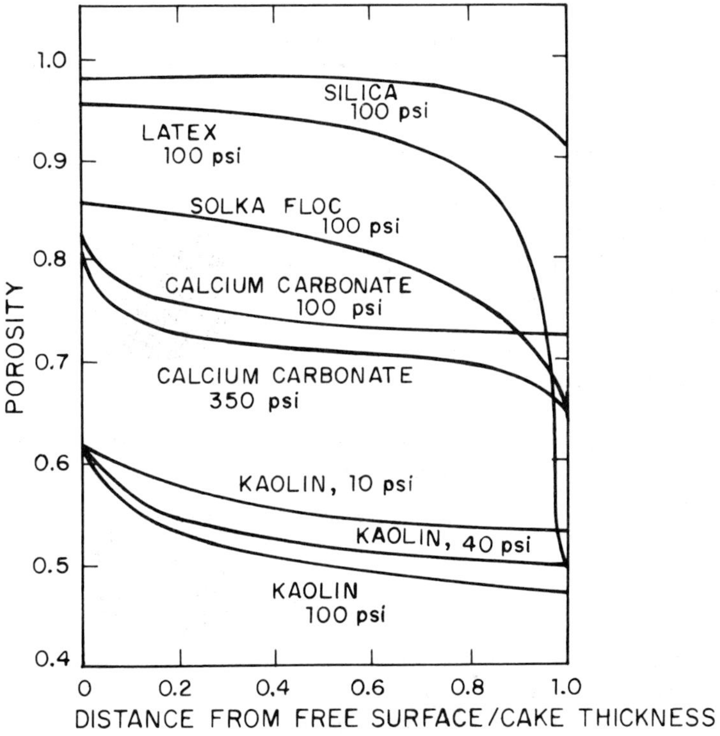

Fig. 4.51[127] Variation of cake porosity in relation to location in the cake.

through 750°C may be accomplished. Some of these types of media with their applications are illustrated in Table 4.17.[128]

Four stages of particle deposition on a filter medium are described:[129]

Stage 1. When the filter is new and clean, the incoming particles can deposit on the surfaces of the filter medium directly. This is a very brief stage of the whole process.

Stage 2. The continued deposition of particles occurs on free surfaces and also on previously deposited particles. These latter depositions lead to dendrite growths being formed. Examples of this process as it occurs are given in Fig. 4.52.[130] The range of assumed conditions is shown.

Stage 3. The dendrites grow, mutually intermesh, and coat each fiber.

Stage 4. If the adhesion forces are strong enough, the coatings on neigh-

Table 4.17

FILTER MEDIA, FACET[128]

Typical Applications			
1. Machine tool, farm implement, and materials handling equipment hydraulic systems. Automotive air cleaners, lubricating oil, and fuel line filters. Aircraft main fuel filters, lubricating oils, hydraulic system, and group support equipment (refueling) filters. Diesel fuel filters and precision instrument air line filters.	5–40 microns	-65°F to 365°F	Micropleat® 
2. Aircraft engine fuel filters. Hydraulic system filters. High temperature air filters.	3–40 microns	-65°F to 450°F	Microfil® 
3. Missile hydraulic supply and aircraft refueling filters. Petroleum industry filters. Air instrumentation filters, nitrogen, and gas filters.	Pleated cellulose sheets 0.5–10 microns	-65° to 275°F	Microedge® 
4. Aircraft and ground test equipment. Mobile fluid cleanup systems. Fuel injection systems.	Pleated sheets, bonded inorganic 5 microns up	-400° to 1000°F	Poromesh® 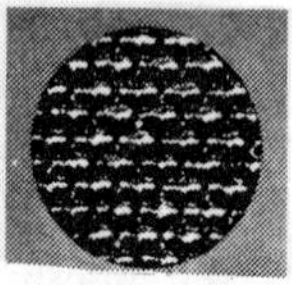

Table 4.17
(cont'd.)

Typical Applications	
5. Polymer processing filters. Chemical process filters, hot gas filters, spargers, flame arresters.	Resin impregnated washers 2 to 100 microns -400° to 1000°F Poroplate® 
6. Gasoline filters for automobiles, heaters, outboard motors, and stationary engines. Air cleaners for chain saws, stationary engines, air compressors, and air lines. Breather filters for hydraulic systems and power steering systems. Freon filters for air conditioning units.	Sintered/woven SS, Ni, Cu, etc. 40 microns -65° to 275°F Microben® Ribbons of resin impregnated cellulose 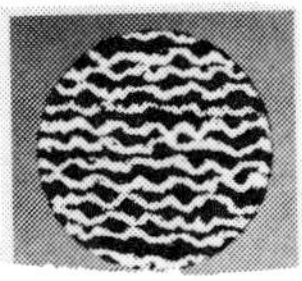
7. Missile ground test and support equipment. Airborne servo valves, gas diffusers, liquid oxygen filters, fuming nitric acid filters, helium and nitrogen filters. Turbine blades (with transpiration cooling), chemical and petroleum industry spargers, catalytic screens and corrosion-resistant filters.	Diffusion bonded wire mesh layers 5 to 1000 microns -400° to 1500°F Poroloy® Sintered/wound sheets SS, Ni, etc. 

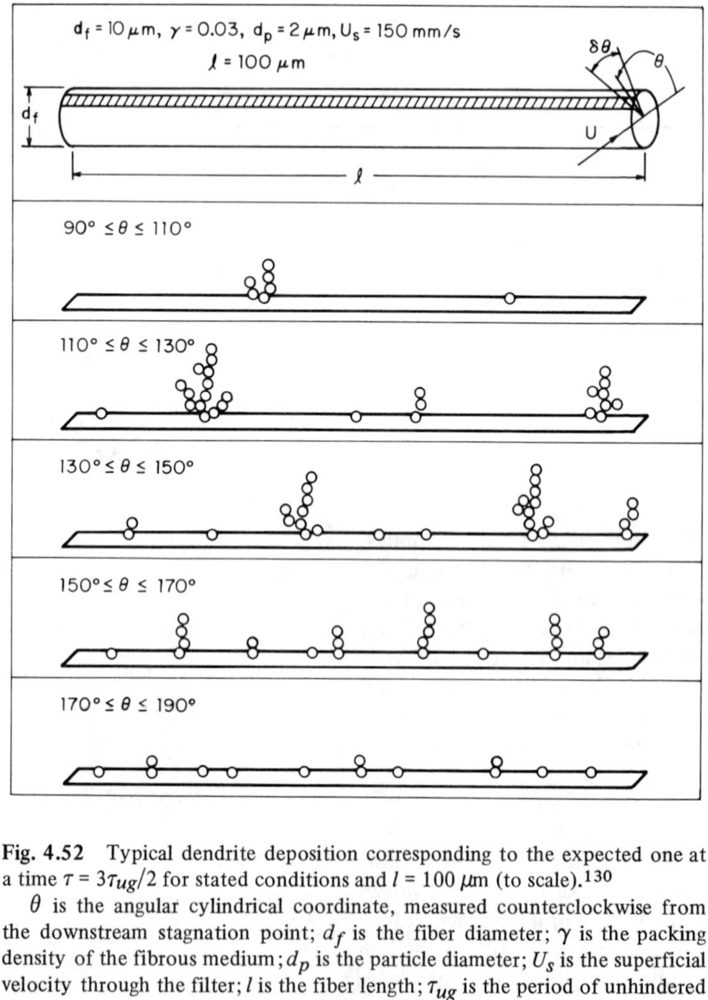

Fig. 4.52 Typical dendrite deposition corresponding to the expected one at a time $\tau = 3\tau_{ug}/2$ for stated conditions and $l = 100$ μm (to scale).[130]

θ is the angular cylindrical coordinate, measured counterclockwise from the downstream stagnation point; d_f is the fiber diameter; γ is the packing density of the fibrous medium; d_p is the particle diameter; U_s is the superficial velocity through the filter; l is the fiber length; τ_{ug} is the period of unhindered dendrite growth.

boring fibers eventually bridge the gap and form a cake which itself acts as a filter medium also. However, if the adhesion forces are less strong, the filter cake may diminish in size due to attrition from the bombardments of incoming particles and reentrained dendrites. Weaker adhesion promotes reentrainment on a continuing basis and there is little tendency for a filter cake to be formed.

Most theoretical treatments until recently have dealt with stage 1.[131] The development of a theoretical model for dendrite growth uses a convention for the idealization of dendrite structures as illustrated in Fig. 4.53[130] and leads to a prediction of the expected configuration and rate of growth of a dendrite as a function of θ and time.

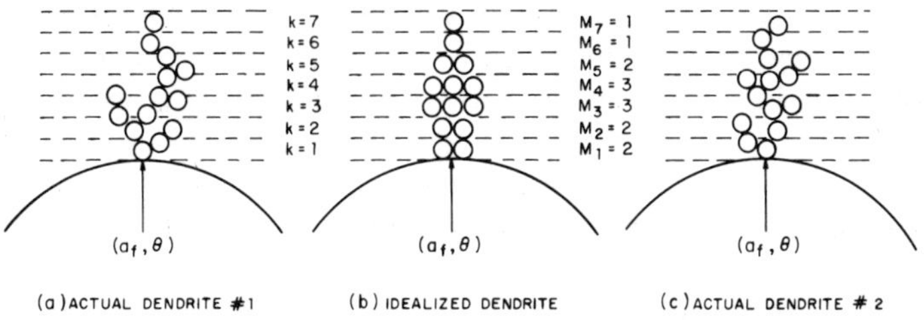

Fig. 4.53 Idealization of the dendrite configuration, (two-dimensional depiction). Different dendrites may correspond to the same idealized configuration.[130]

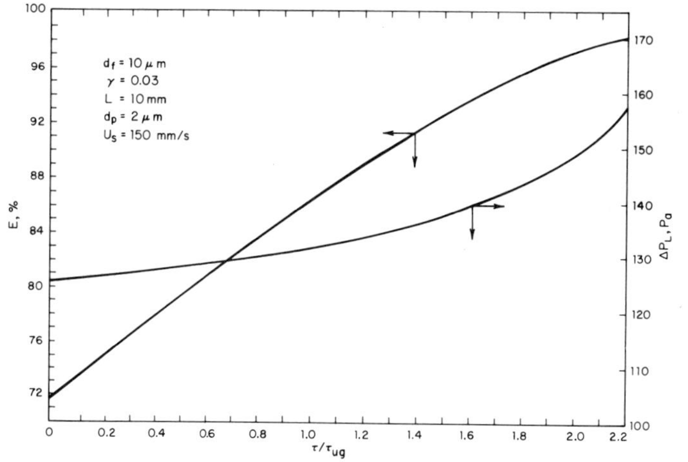

Fig. 4.54 Dependence of filter efficiency, E, and pressure drop across the filter, ΔP_L, on normalized time, τ/τ_{ug}, for given parameter values.[130]

Using this as a basis, the change of efficiency with time is calculated as shown in Fig. 4.54.[130] The treatment does not include inertial nor Brownian effects as it deals only with the case of pure interception. However, it is an interesting development based on an ingenious method for idealization of dendrite structure which may lead to a fully comprehensive model for predicting filter behavior.

Brownian motion becomes important in airborne particles. With $d <$ 0.5 μ, it imparts a relatively irregular motion to these very small particles and therefore in order to increase the probability of particle removal the gas flow has to be lowered. Raising the temperature tends to increase filtration efficiency. Like electrostatic charges reduce efficiency but an electrostatic charge between cloth and particulate helps to retain the blinded material.

This charge build-up on the fabric element may be either + or – and the materials can be listed in a triboelectric series.[132]

Gravity Sedimentation

Settling of this type is common in the metallurgical industries and particle settling from a gas is used in so-called settling chambers.

Factors which influence sedimentation include particle shape, size, and population density as well as system geometry, etc.

This is rationalized in the concept of resistance to flow (R), as

$$R = aLu\mu \tag{4.62}$$

in which

a = a shape factor,
L = a size factor,
u = the velocity, and
μ = the viscosity.

A particle accelerates until it reaches terminal velocity. The force F at this terminal velocity for a sphere is

$$F = \frac{\pi}{6} d^3 (\rho_p - \rho) \tag{4.63}$$

in which

d = particle diameter,
ρ_p = particle density, and
ρ = fluid density.

For a sphere $R = 3\pi du\mu$ and at terminal velocity $R = F$ and $u = u_t$. Therefore,

$$3\pi du_t\mu = \frac{\pi}{6} d^3 g(\rho_p - \rho)$$

and

$$u_t = \frac{d^2 g(\rho_p - \rho)}{18\mu} \tag{4.64}$$

This is Stokes's Law and applies if $Re < 0.1$.

In general the force on a particle moving in a fluid with velocity u is

$$F = \frac{\pi d^2 \rho u^2 Cd}{8} \tag{4.65}$$

When Cd is a drag coefficient, but, in the case of a gas in which the particles are similar in size to the mean free path of the gas molecules, a corrected velocity u_v is obtained

$$u_v = u_t \left(1 + \frac{2A\lambda}{d}\right) \tag{4.66}$$

Other factors which affect settling velocity are: the tendency to agglomerate (increase u), particle concentration (decreases in u if concentration is high), and when walls decrease u.

Sedimentation of Flocculants

In order to aid settling of fine particles, agglomeration can be achieved by the addition of various agents. Because of the loose, weak nature of the aggregate, the settled material may change substantially during the sedimentation process. This is shown in Fig. 4.55. The point where the sediment finally disappears is called the critical sedimentation point.

It is sometimes possible to control the shape of the flocs.[133] The information in Fig. 4.55 is of use in explaining the operation of thickeners and of clarifiers. In the former the aim of the process is to produce a material cake and in the latter process the aim is to produce a clear liquid. Both feed the suspension into a tank at the "suspension" level and both utilize a slow moving rake in order to remove the cake.

Centrifuges

Most centrifuges in use industrially are the continuous type dealing with liquid-particle suspensions. Gas phase centrifuges are of recent origin.

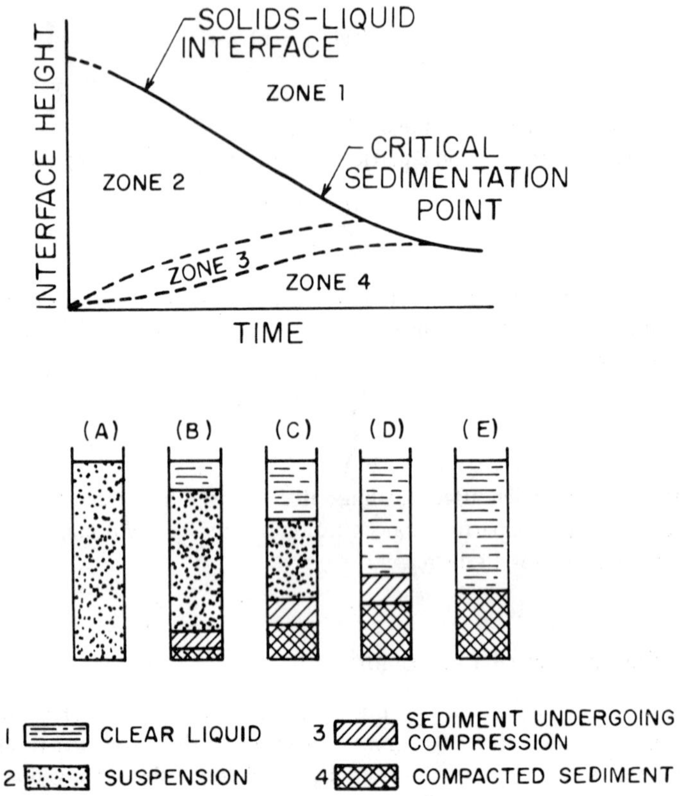

Fig. 4.55 Stages in the batch sedimentation of a flocculated suspension.[26,122]

Emission Sources and Performance of Various Air Filtering Devices

Table 4.18 gives a compilation of emission sources gleaned from the literature. Data on the particulate concentration, etc., is reported as mgm/liter, grains/scf, percent by weight (of the gas phase), and particle size information. References are also given.

A comparative summary of the performance of numerous devices is given in Table 4.19.[155]

Cyclones and Hydrocyclones

Gas phase cyclones have been in use for a long time but liquid phase hydrocyclones are of more recent application.

Cyclones are inertial separators which spin the polluted air at high velocities thus forcing the solid particles to settle out. Therefore, the force exerted on the pollutant particles is a centrifugal force. To effectively settle out particles in the diameter range of a few microns or below, extremely high velocities are required.* Furthermore, a cyclone settler could not usefully filter out submicron particles since this type of pollutant is essentially inertialess. However, cyclones are frequently used to eliminate large particles in the order of 10 microns or above diameter range.

The cyclone works in a manner as shown in Fig. 4.56. The tangential velocity of radius R in a cyclone of radius Ry is expressed by[165–167]

$$U_R = U_{TW}\left(\frac{Ry}{R}\right)^M \tag{4.67}$$

where

U_{TW} = velocity near wall, and

M = a constant ≈ 0.5.

The mechanism of the particle separation operates varyingly; therefore efficiencies cannot be calculated from theoretical considerations. Instead, a graphical method is used.

Scrubbers

These are low cost, highly efficient devices in which water is sprayed into the gas phase suspension. Impingement, Brownian motion, condensation, and agglomeration all play a part in the process and for efficient operation a fine, fast moving spray is desirable but not feasible because of interaction with the gas stream (entrainment); therefore an optimized situation is required in practice. The main disadvantage of scrubbers is that they, in effect, transform air/gas pollution into water pollution.

Cyclone Scrubber. The inlet enters tangentially at high velocity thus inducing a cyclone effect in the scrubber. The inlet is usually at the base of the unit and a water spray is admitted at the top. The particles impinge on the wetted walls and then into a hopper. This type of device is quite efficient for particles greater than 5 microns and has been used for smaller

*up to 3000 feet/s (1000 m/s)

sizes. But, the high velocities required necessitate large supply pressure heads and the overall assembly is complex.

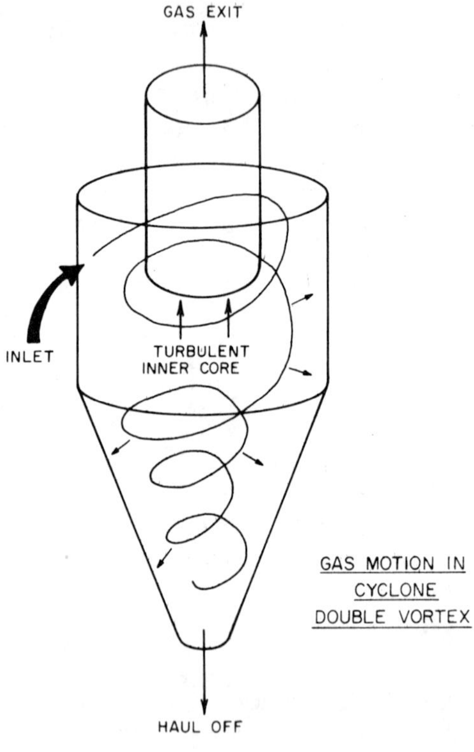

Fig. 4.56 Gas Cyclone.

Impingement Scrubbers. In this device particles settle out after interacting with water and striking an impingement plate. In the case of smaller particle sizes it is necessary to increase the gas velocity to effect a larger inertia of the solids. Also, high velocities are used to atomize the scrubbing liquor on the impingement plate. Because high velocities cause high pressure drops and require large amounts of energy, there is a practical limit to the smallest size of particles that can be efficiently scrubbed out.

Table 4.18

RANGE OF EMISSION SOURCES

Emission source	*Concentration* (mg/liter)	grain/ scf	percent by weight*	*Particle size*	*Reference*
Open hearth steel furnace (50 ton)	2.59	1.13	0.22	64.7 percent < 5 μ	134
Open hearth steel furnace	2.2	0.95	0.18	–	135
Electric arc steel furnace (50 ton)	1.2	0.54	0.10	71.9 percent < 5 μ	134
	5.0	2.2	0.42	–	135
Steel mill, blast furnace	19	8.5	1.6	–	135
Steel mill, sintering machine	3.9	1.7	0.32	–	135
Basic oxygen steel furnace	13	5.8	1.1	–	135
Steel mill, scarfing machine	1.1	0.5	0.095	–	135
Gray iron cupolas	2.0	0.88	0.17	65 percent > 20 μ	134
	1.2	0.525	0.105	mean size > 50 μ	136
Brass furnace	0.485	0.212	0.042	0.03–0.3 μ	134
Brass furnace, low-frequency induction furnace	0.11	0.05	0.01	–	137
Oil-fired tilting crucible (brass foundry)	3.0	1.3	0.25	–	137
Lead refining reverberatory furnace	6.6	2.9	0.55	0.3 μ	134

*weight of particles/weight of gas expressed as a percent.

Table 4.18 RANGE OF EMISSION SOURCES *(cont'd.)*

Emission source	*Concentration* (mg/liter)	grain/ scf	percent by weight*	*Particle size*	*Reference*
Lead refining blast furnace	up to 9	up to 4	0.8	60 percent > 20 μ	134
Aluminum works Soderberg furnace verticle bolts	3.0	1.3	0.25	–	138
Soderberg furnace horizontal bolts	0.14	0.06	0.01	–	138
Reverberatory aluminum sweating furnace	0.284	0.124	0.024	0.3 μ	134
Core oven, aluminum sweating furnace	0.480	0.205	0.039	64 percent < 10 μ	134 and 135
Reverberatory zinc sweating furnace	0.480	0.205	0.039	0.3 μ	134
Asphalt batch plants (vent line)	187	81.8	15.4	45 percent > 20 μ	134
Asphalt batch plants (drier)	85	37.2	7.0	70 percent > 20 μ	134
Mineral wool cupolas	2.94	1.28	0.24	75 percent > 45 μ	134
Mineral wool reverberatory furnace	0.71	0.31	0.06	60 percent > 44	134 and 135
Mineral wool blow chambers	0.222	0.097	0.02	90 percent fibers	134 and 135
Mineral wool curing ovens	0.50	0.22	0.04	90 percent fibers	134 and 135
Mineral wool coolers	0.08	0.035	0.007	90 percent fibers	134 and 135

*weight of particles/weight of gas expressed as a percent.

Table 4.18 RANGE OF EMISSION SOURCES *(cont'd.)*

Emission source	*Concentration* (mg/liter)	grain/scf	percent by weight*	*Particle size*	*Reference*
Dust losses from cyclones in grain cleaners (malted barley)	0.405	0.177	0.034	28 percent < 20 μ	134 and 135
Dust losses from cyclones in grain cleaners (Milo)	0.133	0.058	0.011	28 percent < 20 μ	134 and 135
Dust losses from cyclones in hammer mill (feed barley)	1.12	0.488	0.093	28 percent < 20 μ	134 and 135
Venting of asphalt saturator	0.95	0.416	0.079	–	134
Single-chamber incinerator (garbage)	3.2	1.4	0.27	60 percent > 20 μ	134 and 135
Multiple-chamber incinerator (garbage)	0.44	0.19	0.04	60 percent > 20 μ	134 and 135
Multiple-chamber incinerator (wood)	0.25	0.11	0.02	–	134
Incinerator	0.25	0.11	0.02	60 percent > 20 μ	134 and 135
Pathological-waste incinerator	0.162	0.071	0.013	–	134
Brakeshoe debounding in single chamber	3.4	1.5	0.28	–	134
Reclaiming electrical windings in single chamber	3.7	1.6	0.30	–	134
Wire reclamation with multiple-chamber retort furnace	37	16.2	3.08	–	134

*weight of particles/weight of gas expressed as a percent.

Table 4.18 RANGE OF EMISSION SOURCES *(cont'd.)*

Emission source	*Concentration* (mg/liter)	grain/ scf	percent by weight*	*Particle size*	*Reference*
Oil-fired steam generator	0.316	0.138	0.026	85 percent < 1 μ	134
Coal-fired power plant	7	3	0.6	–	139
Fuel oil combustion	0.073	0.032	0.006	0.4	140
Fluid catalytic cracking unit (petroleum refining)	0.37	0.137	0.026	–	134
Thermofor catalytic cracking unit (petroleum refining)	0.066	0.029	0.006	–	134
Contact sulfuric acid plant	0.32 6.3	0.14 2.75	0.03 0.52	– 98 percent < 3 μ	141 134 and 135
Chamber sulfuric acid plant	0.62	0.27	0.05	90 percent > 3 μ	141
Rotary frit smelter	0.48	0.210	0.04	45 percent < 5 μ	134 and 135
Coffee roaster	0.41	0.178	0.034	–	134
Direct-fired coffee roaster	0.44	0.193	0.037	–	142
Indirect-fired coffee roaster	0.29	0.127	0.024	–	142
Stoner and cooler coffee process	0.039	0.017	0.003	–	142
Meat smokehouse exhaust	0.26	0.115	0.022	–	134
Exhaust from fish mean drier	0.9	0.4	0.08	96 percent > 2 μ	134

*weight of particles/weight of gas expressed as a percent.

Table 4.18 RANGE OF EMISSION SOURCES *(cont'd.)*

Emission source	*Concentration*			*Particle size*	*Reference*
	(mg/liter)	grain/ scf	percent by weight*		
Oil furnace process of carbon black	48	21	4.0	–	143
Synthetic detergent process	7	3	0.6	50 percent > 500 μ	144
Rotary cement kiln dry process	15	6.4	1.2	47 percent < 10 μ	145
	16	7	1.3	mean size of 30 μ	136
Rotary cement kiln wet process	13	5.7	1.1	47 percent < 10 μ	145
	11	5	1	mean size of 30 μ	146
Verticle type pre-heater equipped cement kiln	34	15	2.8	mean size of 30 μ	146
Cotton gin	1.26	0.55	0.10	94.5 per-cent > 100 μ	147
Air drawn from mine shaft	0.0135	0.0059	0.0011	–	148
Single cylinder 4-cycle diesel engine exhaust	0.05	0.02	0.004	10–5000 m	149
Gasoline automobile engine exhaust 30 mph (lead con-centration)	0.00452	0.00192	0.00036	–	150

*weight of particles/weight of gas expressed as a percent.

Packed Bed Scrubber. Packed beds of granular solids are used as a filtering field. Using countercurrent flow the filtering efficiency of the unit is determined by the degree of packing of the filter beds. There are no moving parts and operating cost is relatively low. Pressure drops are low, low velocities are used, and finely packed beds may effectively remove 1–5 micron particles if the velocity is 0.8 ft/s or less. Because the unit tends to clog up due to the fine packing of the beds, application is restricted to gas with relatively low pollutant concentrations.

Table 4.19

EFFICIENCY OF DUST-ARRESTING EQUIPMENT AT VARIOUS PARTICLE SIZES[155]

Equipment	*Percentage efficiency at*		
	50 μ	5 μ	1 μ
Inertial collector	95	16	3
Medium-efficiency cyclone	94	27	8
Cellular cyclone	98	42	13
High-efficiency cyclone	96	73	27
Tubular cyclones	100	89	40
Jet-impingement scrubber	98	83	40
Irrigated cyclone	100	87	42
Self-induced-spray deduster	100	94	48
Spray tower	99	94	55
Fluidized-bed scrubber	> 99	98	58
Irrigated-target scrubber	100	97	80
Electrostatic precipitator	> 99	99	86
Disintegrator	100	98	91
Irrigated electrostatic precipitator	> 99	99	99
Annular-throat scrubber–low energy	100	>99	96
Venturi-scrubber- medium energy	100	>99	97
Annular-throat scrubber–medium energy	100	>99	97
Venturi-scrubber–high energy	100	>99	98
Shaker-type fabric filter	> 99	>99	99
Low-velocity bag filter	100	>99	99
Reverse-jet fabric filter	100	>99	99

Centrifugal Scrubber. The inlet stream of air and dust particles impinges on a set of rotating blades driven by mechanical means. A spray of water is added to the inlet to keep the blades wet and flush away the collected dust. Blade wear and corrosion is a problem and buildup on the rotary blades tends to throw the rotor out of balance. Power consumption is high, but the space required is minimal.

Orifice Scrubber. This device is also frequently known as a submerged nozzle scrubber. The polluted air stream passes through the submerged nozzle and is scrubbed by a water spray. The liquid is atomized by the kinetic energy of the gases. The dust laden water droplets are removed by a disengagement chamber supplied with baffles. The resulting sludge may be removed with a conveyor belt or some similar device (i.e., moving parts).

Corrosion and wear cause problems, and costs are higher if protective material is used.

An alternative device is the viscous filter. A conveyor belt traveling through the oil bath and the gas stream and moving perpendicular to the conveyor impinges on the belt so that the dust particles adhere to the oil. The dust settles as sludge when the belt again goes through the oil.

Jet Scrubber. A high velocity stream of water induces the gas flow due to a pressure difference created by the liquid flow. It is used when it is uneconomical to add a fan for the dust collection system and where a mist or an easily absorbed gas is to be removed from the gas stream. The atomized liquid droplets with absorbed dust and air are separated in a settling chamber.

Venturi-Type Scrubber. A high velocity gas stream induces the flow of water in the throat of a venturi. High efficiencies are obtained by impingement and extreme turbulence between the gas and water.

Flooded Disk Scrubber. The water spray is generated by directing a water stream on a rotating disk or drum. The gas flow then interacts with the liquid droplets.

Electrostatic Precipitation*

In this process the particulate becomes charged and separates in a manner dictated by the electrode configuration and potential. The gas phase is said to be relatively undisturbed and little or no pressure loss occurs. If the potential gradients between electrodes is $>$ 30 volts/cm a spark discharge occurs (in air) and short circuits the plates. Below 30 volts/cm a uniform electric field exists between the plates. If the system consists of

*see *Electrostatic Precipitation* by H.E. Rose and A.J. Wood, 2nd Edition, Constable, London, 1966.

an axial wire in a tube, the field is high near the core and electrical breakdown of the gas near the wire permits charging of the particulate to take place. This central breakdown region is called a corona and is visible as a blue glow. The electrons are discharged from the wire and migrate towards the collecting electrode (wire or plate). By colliding with gas molecules, other positive ions and electrons are produced. The swarm of electrons moves towards the lower field region where Mol + $e^- \rightarrow I^-$ occurs. The positive ions move towards the emitter wire, impact there, and produce new electrons. The overall result is that the space between electrodes is filled with unipolar ions (10^7-10^8 ions per cc) which effectively charge the particulate material passing through. This is termed a negative corona and is preferred to a positive corona. In this negative corona, breakdown occurs at the anode (positive) tube or plate but in a positive corona breakdown occurs near the wire and promotes spark discharge at comparatively low voltage.

The dust resistivity should lie between that of a good conductor and that of a good insulator (about 10^4-10^{10} ohm-cm). Conditioning agents such as moisture, acid mist or ammonia are sometimes added to reduce the particle resistivity so that they do not coat the electrodes and insulate them, thereby reducing the potential of the field.

Single-stage precipitators both ionize and filter the particles in one unit. The two-stage precipitators most commonly used consist of a first stage to ionize the gas and a second to filter the particles out. Pipe type precipitators are used for mists or water flushing, whereas plate types are for dry collection and large gas flows.

Advantages. Dust may be recovered dry and high efficiencies may be obtained even with small particles.

Low pressure drops and temperature drops are encountered even with high flow rates. There are few moving parts and they are designed to operate continuously. They may be used at high temperatures. Acids and tar mists which are hard to remove otherwise, may be collected, and the efficiency may be increased by increasing the size of the unit. There is a low power requirement, i.e., about 65 kilowatts for 500,000 cfm of air at about 95 percent efficiency.

Disadvantages. There is the production of ozone, a noxious gas. There is a high initial cost and precipitators are not easily adaptable to variable conditions (automatic voltage control may be used but optimum performance occurs under constant conditions). Some particles may not be collected due to their conductivity (the reciprocal of resistivity), and precipitators are not efficient for the gaseous phase. Precleaners may be needed to eliminate larger particles. Space requirements may be inhibitive if large efficiencies are required. Danger to personnel also occurs due to the high voltages. There is a lower limit to the particle size that can be efficiently collected (≈ 0.3 microns).

Mechanisms

Charging of a particle occurs by one of two mechanisms. Ion–particle collision in which the number of charges (n) acquired over time is expressed by:

$$n = \frac{E_c r^2}{e}\left(\frac{3\epsilon}{\epsilon + 2}\right)\left(\frac{t}{t + 1/\pi NeK}\right) \tag{4.68}$$

in which

r = particle radius,
E_c = the charging electric strength,
e = electron charge,
ϵ = dielectric constant of particulate,
N = ion density, and
K = ionic mobility.

In ion diffusion charging, which depends on the thermal activity of the ions and is therefore temperature sensitive,[152] the number of charges (n) occurring over time is expressed

$$n = \frac{rKT}{e^2} \ln\left(\frac{1 + \pi r U_r N e^2 t}{KT}\right) \tag{4.69}$$

in which

T = °Kelvin,
k = Boltzmann's constant, and
U_r = root mean square velocity.

It appears that when $d > 1\ \mu$, collision is the most important charging mechanism and when $d < 0.1\ \mu$ diffusion is the most important. In either case charging is essentially complete in 0.1 s and the process is relatively efficient. The efficiency (η) has been simply related to system characteristics as[151]

$$\eta = 1 - e\,\frac{-Au}{Q} \tag{4.70}$$

in which

A = plate area,
u = drift velocity, and
Q = gas flow rate.

This equation uses only a single U value and a summation of individual particle size fraction η values is usual.

Operation. Most of the unipolar ions do not attach to particles. Instead they go to the other electrode and produce a corona discharge current. If the deposit is conducting poorly, arcing occurs and operational voltage has to be reduced to prevent this fault continuing. Alternatively, moisture usually improves deposit conduction and H_2O injection is employed in many cases. It is called conditioning.[153]

Dielectrophoresis

Whereas the electrostatic precipitator uses a uniform electric field and uses electric power to charge the particulate, dielectrophoretic mechanisms utilize a nonuniform field and use the electric power to maintain this. No corona is necessary or obtained. The power requirement for dielectrophoresis is therefore slight as compared with normal electrostatic precipitation (less than 1 percent). Note that if the particles in the aerosol possess a natural charge, the separation affected is due to electrophoresis but, if the particles are uncharged, the separation must be by dielectrophoresis. In this process, the particle behaves as a dipole and interacts with others in the aerosol:

> . . . the dielectric forces acting on the particles are increased due to the interaction of many particles under the influence of the nonuniform field. . . and further. . . the field, free of corona discharge, which serves for the separation of the floating particles from the gases, has been disposed in a chamber through which the gas passes continuously at a low static pressure potential or remains stationary at ambient pressure. . .[154]

Systems employing this principle have been designed and built. In practice both electrophoresis and dielectrophoresis occur simultaneously if a portion of the particles in the aerosol have a natural charge. Deposition of the settled out particles appears to occur on both electrodes and suitable coverings, which may be removed.[155]

Other Methods

These include acoustic, thermal, and magnetic precipitators. At the present time, there is some interest in developing combined processes. For example, the use of a combined field of acoustic and electrostatic forces is expected to yield interesting results.[156]

4.10 COMPACTION OF PARTICULATE MATTER

Earlier in this chapter it was explained that for finely divided material to flow out of a container, the stress conditions within the arched surface have to reach a critical level. On the other hand, the solid mechanician normally does not want the particulate material of interest to move at all. His concern is therefore to identify the failure locus and then to operate well below it in practice.[168] In industrial practice of tabletting or compact production, as it is practiced in the powder metallurgy, pharmaceutical, and ceramics industry, the objective is to achieve a high degree of internal cohesion such that the "green" compact holds together during subsequent handling. In powder metallurgy, this involves sintering of the compact whereas in pharmacy the next step is dispensing followed by swallowing. Powder compaction is a good example of the compartimentalization that research and development work in powder science tends to follow in that those working in pharmacy tend to report their work to those in pharmacy, as do those in powder metallurgy and soil mechanics, and so on. There have been, however, greater attempts to interrelate information between the sciences over the years but progress is still very slow.

The aspects of importance in connection with the consolidation of particulates include: the structure of the particulate in the die or mold, interparticle friction, state of stress in the compact, mechanism of compaction, compaction equations, and ejection from the die.

Structure of Particulate Assembly

The early classic work on particle packing proposed that equally-sized spheres pack in one of four different ways leading to specific values of porosity as stated in Table 4.20.[157]

Table 4.20

SYSTEMATIC ASSEMBLAGES OF SPHERES[157]

Assemblages	*Porosity (percent)*	*Coordination number*
Cubic	47.6	6
Orthorhombic	39.5	8
Tetragonal	30.2	10
Rhombohedral	26.0	12

Nevertheless, the coordination number alone has been shown to be non-specific for a particular value of porosity.[158] The four types of packing have been amplified into nine variations each of which is derived by starting from a basic simple cubic cell upon which simple shearing processes are conducted.[159]

In practice, the set that is established by the particulate depends upon the environmental impulses that are transmitted to it during deposition and subsequently. That is illustrated in Table 4.21.[160] In addition, it is possible in practice to establish very high relative densities in particulate assemblies just by appropriate juggling of the distribution. This is shown in Table 4.22.[161] To date there appears to be no discernible effort to use this fact commercially.

Table 4.21

EFFECT OF DIFFERENT TREATMENTS ON THE PACKING OF SPHERES[160]

Material	*Treatment*	*Observed density (percent)*
0.38 cm dia. lead shot	Poured into glass beaker	55.3
	Shaken to maximum density	62.8
	Added in small quantities and tamped intermittently	64.1
0.32 cm dia. steel balls	Loose haphazard packing	60.1
	Dense haphazard packing	63.7
1/3 in. dia. ball bearings	One-dimensional vertical shaking	61
	Three-dimensional shaking	74

Table 4.22

EFFECT OF SPHERE GRADING ON POROSITY[161]

	Primary	*Secondary*	*Tertiary*	*Quaternary*	*Quinary*	*"Filler"*
Radius of sphere	*a*	0.414*a*	0.225*a*	0.175*a*	0.117*a*	very small
Relative no. of spheres	1	1	2	8	8	–
Wt. percent of spheres in final mixture	77.1	5.5	1.7	3.3	1.0	11.4
Percent voids in mixture	22.95	20.7	19.0	15.8	14.9	3.9

Internal Friction

Internal friction between particles is an important characteristic of particle masses. Its measurement has been somewhat clouded by the urge of a number of workers to demonstrate a significant relationship between angle of repose and some other property. Additionally, more than a few workers have failed to differentiate between this type of friction and the classic friction as understood to typify rubbing surfaces.[162] The Hausner Ratio[163] has been proposed as an indicator of the friction condition within a powder mass[164] and indeed, this ratio has been related to various angles of repose and flow rates, as shown in Fig. 4.57.[163] It is also related to the compaction behavior during the early stage of powder compaction, as shown in Fig. 4.58.[169]

Interest in the Hausner Ratio has prompted the investigation of an improved test for tap density[170] and also its application in ceramics processing.[171]

The angle of repose measured on the free-standing cone is symbolized by α. In the case of the rotating cylinder method the angle of repose is defined as the maximum angle which the surface plane of the powder made upon rotation of the container and is symbolized by α'. The angle between the surface of the powder mass and the horizontal, after the powder mass had collapsed as a result of further rotation is symbolized by β.[163]

Pressure Transmission

The main impediment to densification is not the interparticle friction

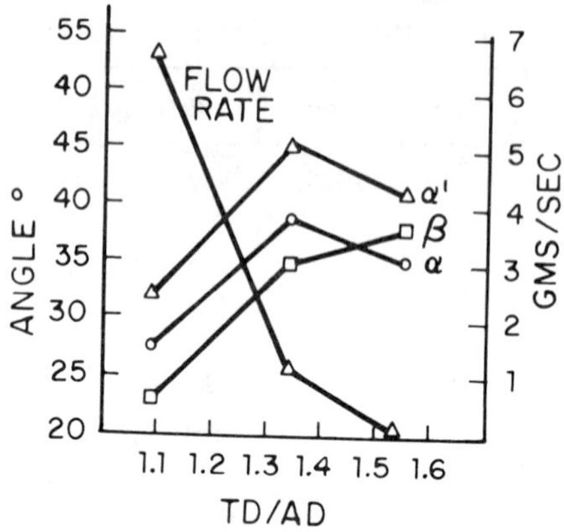

Fig. 4.57 Angles of repose and flow rates *versus* Hausner Ratio for three types of copper powder.[163]

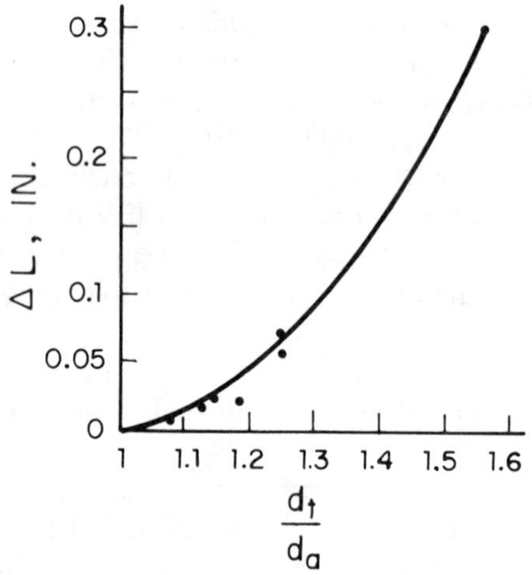

Fig. 4.58 Punch displacement vs. Hausner Ratio for copper and iron powders.[169]

but the friction between the die wall and the powder under compaction.[172] The effect upon density distribution of the compact is shown in Fig. 4.59.[173] This has also been demonstrated in the case of inorganics, etc.[174]

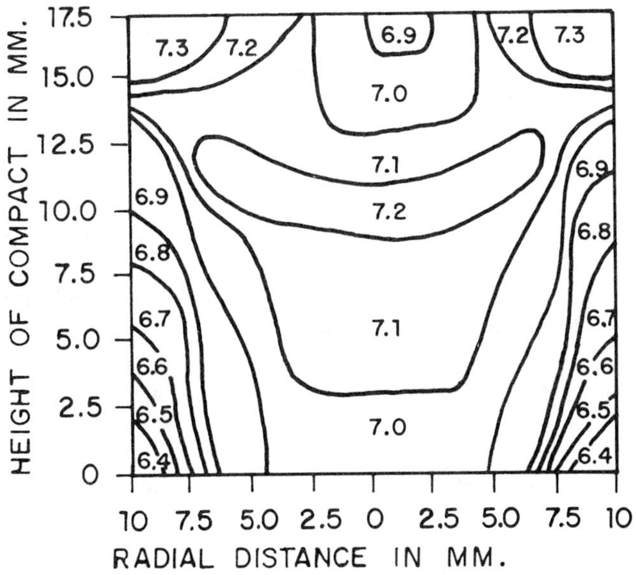

Fig. 4.59 Distribution of densities in a green nickel compact pressed at 103, 040 psi. Die diameter D is 0.79 in.; h/D, 0.87.[173] (© 1956, American Institute of Mining, Metallurgical and Petroleum Engineers, 345 E. 47th St., New York, NY, 10017).

The experimental technique of using a transducer as the bottom punch of a die set has been used to show that the wall friction effect depends upon the particle shape.[175]

A technical equation relating friction to compaction behavior has been studied:[176]

$$\sigma_d = 1 - \frac{16m\phi K}{\pi D^3 \rho}\sigma \tag{4.71}$$

in which

σ_d is the net specific pressure,
σ is the compressive stress,
m is the mass of powder,
ϕ is the coefficient of friction powder/wall,
D is the die diameter,
ρ is the density of mass, and

K is the coefficient of lateral pressure.

It has also been shown that in the case of metal powders, the effect of friction is not very large during the earlier stage of compaction when the relative density is lowest.[176]

State of Stress

A convenient apparatus for studying the state of stress in a particulate assembly under load has been described and is illustrated in Fig. 4.60. The sample is under hydrostatic stress within the rubber manifold shown. Under a given level of consolidation, (σ_c), the stress $\Delta\sigma_\nu$ (applied by the ram at the top) produces a corresponding strain ϵ_ν. The peak may be considered to be the failure point. The state of stress on a specific plane at angle θ to the major principal plane (i.e., plane on which σ_1 acts) is described by[177]

$$\sigma\theta = \frac{\sigma_1 + \sigma_3}{2} + \frac{\sigma_1 - \sigma_3}{2} \cos 2\theta \quad \text{, and} \tag{4.72}$$

$$\tau_\theta = \frac{\sigma_1 - \sigma_3}{2} \sin 2\theta \tag{4.73}$$

This pair of expressions is a straightforward application of Mohr Circle of Stress.

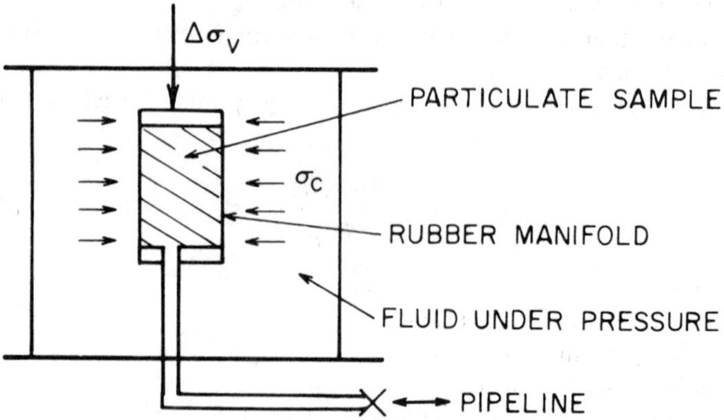

Fig. 4.60 Apparatus for imposition of the triaxial state of stress of a particulate.[177] Note that X inlet can be used to apply pressure or vacuum.

Mechanisms of Compaction and Compaction Equations

There has existed a fairly popular convention to divide the compaction processes occurring in a closed die into two stages: stage I, in which particle rearrangement processes are occurring, and stage II, which is usually termed the stage of bulk compression. The general idea behind this designation is that the particles move into lower energy positions under the action of the applied pressure, so they are not forced to deform plastically. This ideas seems to be supported by data which indicates that the friction condition in the moving powder mass is the controlling factor during this early stage. Results shown in Fig. 4.61 demonstrate this to be the case.[169]

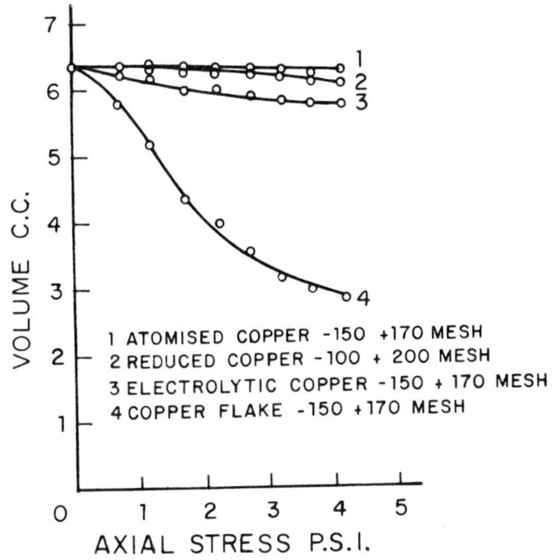

Fig. 4.61 Effect of particle shape on the volume–pressure curves for copper powder.[109]

However, this idyllic picture is really an oversimplification and in practice fragments tend to break off the individual particles, where they can, and this complicates the attempt to diagnose what is going on within the powder mass.[169,178,179,180] There are many equations relating the applied pressure to the volume of a particulate in closed die compaction and these have been extensively reviewed over the years.[183] A compilation of some of these equations is given in Table 4.23.[181]

Table 4.23

VARIOUS EQUATIONS DESCRIBING THE COMPACTION OF A POWDER MASS UNDER PRESSURE[181]

	Balshin	$\ln F = -c_1(V/V_\infty) + c_2$
	Smith	$\frac{1}{V} - \frac{1}{V_0} = c_3 P^{1/3}$
	Murray	$\ln\left(\frac{V}{V - V_\infty}\right) = c_4\left(\frac{V_\infty}{V - V_\infty}\right)^{1/3} + c_5 P$
	Ballhausen	$\ln\left(\frac{V_\infty}{V - V_\infty}\right) = c_6 P + \ln c_7$
	Konopicky	$\ln\left(\frac{V}{V_0 - V_\infty}\right) = c_8 P + \ln\left(\frac{V_0}{V_0 - V_\infty}\right)$
	Jones	$\ln P = -c_9\left(\frac{V}{V_\infty}\right)^2 + c_{10}$
*	Athy	$\frac{V - V_\infty}{V} = \frac{V_0 - V_\infty}{V_0} e^{-c_{11}P}$
	Nutting	$\ln\left(\frac{V_0}{V}\right) = c_{12}{}^{P^{c_{13}}}$
	Tanimoto	$\frac{V_0 - V}{V_0} = \frac{c_{14}P}{V_0} + \frac{c_{16}P}{P + c_{15}}$
	Terzaghi	$\frac{V - V_\infty}{V_\infty} = -c_{17}\ln(P + c_{18}) - c_{19}(P + c_{19}) - c_{20}P + c_{21}$

Table 4.23 *(cont'd.)*

Cooper	$\dfrac{V_0 - V}{V_0 - V_\infty} = c_{22} \cdot e^{-c_{23}/P} + c_{24} \cdot e^{-c_{25}/P}$
Gurnham	$P = c_{26} \cdot e^{c_{27}/V}$
Nishihara	$\ln\left(\dfrac{V_0}{V}\right) = -\left(\dfrac{P}{c_{28}}\right)^{1/c_{29}}$
Tsuwa	$\dfrac{V_0 - V}{V_0} = \dfrac{V_0 - V_\infty}{V_0} \cdot \dfrac{(1/c_{30})P}{1 + (1/c_{30})P}$
* Kawakita	$\dfrac{V_0 - V}{V_0} = \dfrac{c_{31} c_{32} P}{1 + c_{32} P}$

The equations that are said to be the most interesting are indicated in the table by an asterisk (*).[181] As an example of how these equations are constructed, the second one in the table may be derived as follows:

The relationship between applied pressure and volume is observed and analysis leads to the development of the following equation:[181]

$$C = \frac{V_o - V}{V_\infty} = \frac{abP}{1 + bP} \tag{4.74}$$

which may be rewritten as

$$\frac{P}{C} = \frac{1}{ab} + \frac{P}{a} \tag{4.75}$$

Thus by plotting P/C against P at large pressures one can back extrapolate to determine the intercept on the y ordinate (which is $1/ab$).

At very high pressures, the intercept is small compared with P and therefore the constant C, denoted by C_∞, is equal to a and is $(V_o - V_\infty)/V_o$. By combining this relationship with Eq. (4.74) one obtaines

$$(P + 1/b)(V - V_\infty) = \text{constant} \tag{4.76}$$

It is noted that this is similar to van der Waals' equation of state for gases and it is claimed that a is the initial porosity of the particulate in the die and that b is related to the resisting forces.[181]

A further development of this line of inquiry was to incorporate the effect of work hardening of plastically deformable (metal) particles by using the relationship[182]

$$\beta = \frac{b(1-a)}{a} \tag{4.77}$$

in conjunction with

$$Q = M\beta\gamma \tag{4.78}$$

in which

a and b are as before,

γ is the Meyer work hardening index, and

M is the Meyer hardness number (refer to "The Hardness of Metals" by Bowden and Tabor for a description of M).

The correlation shown in Table 4.24 shows that the hardness should be inversely proportional to the product[182] so that Q is a constant.

Ejection from the Die

The pressure required to eject the compact from the die is the function of many variables including the powder type, the amount, and the method of admixing the lubricant (see Table 4.25). Die wear directly affects ejection force and in turn this wear is dependent upon the type and hardness of the die steel, the punch/die clearance, and surface finish,[183] to mention only a few. Although it does not affect metal powder compacts very severely, ejection from a die can lead to the formation of laminar cracks due to the radial pressure resulting when the compact is being ejected from the die (see Fig. 4.62). This is especially severe when materials with high elastic moduli are pressed.

The basic cause of the friction between the die wall and the pressed

Table 4.24

RELATIONSHIP OF KAWAKITA CONSTANTS WITH POWDER PROPERTIES[182]

Powder type	*Kawakita constants*		*Initial porosity*	*Microhardness* (10 μm *diagonal*)	*Meyer index* γ	$Q = M \times \beta \cdot \gamma$
	a	*b*				
Sponge iron	0.697	0.235	0.688	168	1.70	29.2
Electrolytic copper	0.618	0.220	0.610	143	1.65	32.1
Atomized copper	0.676	0.284	0.663	115	1.85	29.0
Hydro-metallurgical nickel	0.554	0.147	0.540	148	1.76	30.8
Atomized 310 S/S	0.689	0.164	0.683	222	1.78	29.3
Atomized 410 L S/S	0.628	0.122	0.626	276	1.60	31.8

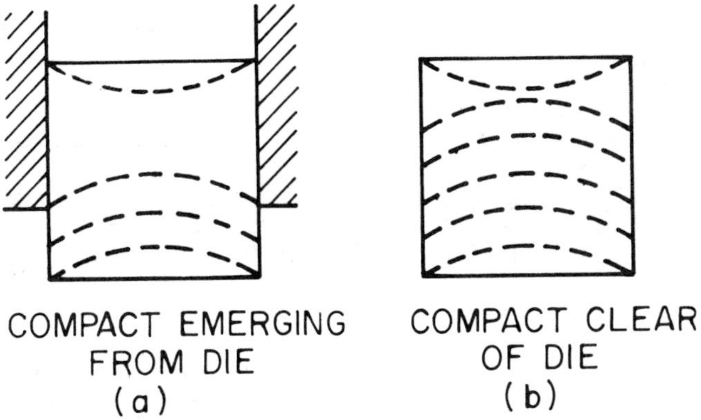

Fig. 4.62 Stages in production of laminar cracks by radial pressure.

material is formation of metallic or other junctions between them when the boundary lubrication breaks down. This is illustrated in Fig. 4.63.[162]

As a practical matter it is inconvenient to differentiate between the effect of friction during pressing and the effect of friction during ejection. A simple technical parameter $\bar{M}$ has been proposed to incorporate these

Table 4.25

EJECTIONS OF IRON POWDER COMPACTS

	Lubricant					
Powder type	*None*	*Benzene die wall*	*Benzene premixed*	*1% Stearic acid benzene*	*3% Stearic acid*	*3% Stearic acid benzene*
Atomized -150 to +170	$15.75 \frac{\text{in-lb f}}{\text{g}}$ $5.49 \frac{\text{g}}{\text{cc}}$ 4,505.0 psi	$14.55 \frac{\text{in-lb f}}{\text{g}}$ $5.39 \frac{\text{g}}{\text{cc}}$ 5,450.5 psi	$14.55 \frac{\text{in-lb f}}{\text{g}}$ $5.39 \frac{\text{g}}{\text{cc}}$ 4,594.4 psi	$6.00 \frac{\text{in-lb f}}{\text{g}}$ $5.81 \frac{\text{g}}{\text{cc}}$ 2,108.1 psi	$3.15 \frac{\text{in-lb f}}{\text{g}}$ $5.57 \frac{\text{g}}{\text{cc}}$ 991.0 psi	$3.06 \frac{\text{in-lb f}}{\text{g}}$ $5.93 \frac{\text{g}}{\text{cc}}$ 1,369.4 psi
Sponge -150 to +170	$11.46 \frac{\text{in-lb f}}{\text{g}}$ $5.45 \frac{\text{g}}{\text{cc}}$ 2,964.9 psi	$16.65 \frac{\text{in-lb f}}{\text{g}}$ $5.42 \frac{\text{g}}{\text{cc}}$ 4,774.7 psi	$15.30 \frac{\text{in-lb f}}{\text{g}}$ $5.57 \frac{\text{g}}{\text{cc}}$ 4,594.4 psi	$6.45 \frac{\text{in-lb f}}{\text{g}}$ $6.65 \frac{\text{g}}{\text{cc}}$ 1,972.9 psi	$4.67 \frac{\text{in-lb f}}{\text{g}}$ $5.73 \frac{\text{g}}{\text{cc}}$ 1,531.3 psi	$5.20 \frac{\text{in-lb f}}{\text{g}}$ $5.85 \frac{\text{g}}{\text{cc}}$ 1,639.6 psi
Electrolytic -150 to +170	$16.06 \frac{\text{in-lb f}}{\text{g}}$ $5.81 \frac{\text{g}}{\text{cc}}$ 4,504.5 psi	$10.35 \frac{\text{in-lb f}}{\text{g}}$ $5.57 \frac{\text{g}}{\text{cc}}$ 2,909.9 psi	$11.40 \frac{\text{in-lb f}}{\text{g}}$ $6.15 \frac{\text{g}}{\text{cc}}$ 3,513.5 psi	$3.60 \frac{\text{in-lb f}}{\text{g}}$ $6.25 \frac{\text{g}}{\text{cc}}$ 1,441.4 psi	$3.50 \frac{\text{in-lb f}}{\text{g}}$ $6.15 \frac{\text{g}}{\text{cc}}$ 1,306.3 psi	$3.87 \frac{\text{in-lb f}}{\text{g}}$ $6.15 \frac{\text{g}}{\text{cc}}$ 1,450.5 psi
Flake -150 to +170	$3.67 \frac{\text{in-lb f}}{\text{g}}$ $5.72 \frac{\text{g}}{\text{cc}}$ 1,621.6 psi	$3.00 \frac{\text{in-lb f}}{\text{g}}$ – 765.7 psi	$1.20 \frac{\text{in-lb f}}{\text{g}}$ – 1,081.1 psi	$1.00 \frac{\text{in-lb f}}{\text{g}}$ $6.06 \frac{\text{g}}{\text{cc}}$ 648.6 psi	$1.00 \frac{\text{in-lb f}}{\text{g}}$ $5.57 \frac{\text{g}}{\text{cc}}$ 405.4 psi	–

Work of ejection; density at ejection pressure.

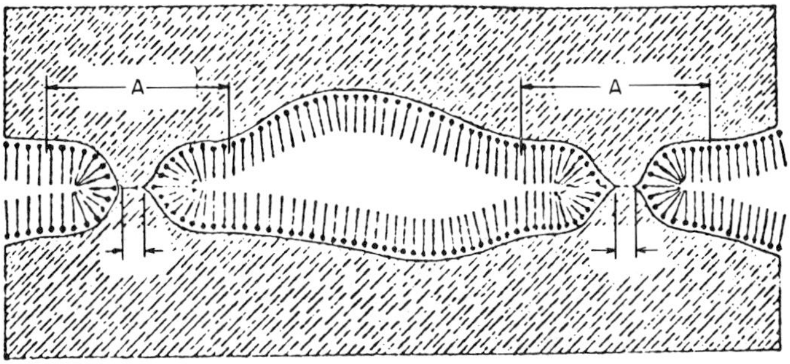

Fig. 4.63 Diagram showing mechanism of boundary lubrication. The lubricant film supports most of the load but over a small region of area aA metallic junctions are formed. These must be sheared together with the oil film when sliding occurs. The resistance to sliding may be expressed as the sum of two terms, the shear strength of the metallic junctions $s \times A$ and the shear strength of the lubricant film $s(1 - a)A$. For good lubricants the metallic junctions contribute little to the friction but are responsible for the wear observed.[162]

two effects into one number[184] in which $\bar{M}$ is the ratio of transmitted load per square inch of cross-sectional area divided by the ejection load per square inch of wall surface area. Clearly, an abnormally low value of $\bar{M}$ indicates either that the material being compacted has poor pressure transmission characteristics, poor ejection characteristics, or that both characteristics are poor under the conditions of testing. Conversely, if the testing conditions are altered in such a way as to increase $\bar{M}$, this indicates an improvement in either the pressure transmission characteristic, the ejection behavior, or both.

Despite the apparent utility of such a number, its application has not been taken up.

4.11 SINTERING

Sintering is a term which is commonly used to describe the heating of a particulate assembly in order to achieve interparticle bonding eventually leading to a polycrystalline aggregate of high density. Very often the theoretical density is desired. Understanding the complex and intricate processes that occur during sintering is one of the more intransigent problems facing materials science today. The technological and commercial importance of the process of sintering in metallurgy and ceramics in order to achieve industrially usable components is increasing year by year at a notable rate.

It is impossible to conduct anything but a most cursory review of the mechanisms of sintering. For more in depth treatment the reader is referred to a number of beginning references.[185,186,187] Fortunately, there is now an International Group for the Study of Sintering with its headquarters in Yugoslavia and a journal is published regularly.[188]

Over the last 30 years a great deal of effort has gone into basic and applied research on the sintering of finely divided matter (excluding baking processes from this discussion). The product from this activity is an immense quantity of literature that the beginner may find somewhat daunting. The approach taken in the following discussion has been made much easier by the introduction of the concept of the sintering diagram, which is discussed in a later section.

Capillarity

No specific sintering force is required to explain the phenomen of sintering. The forces responsible for the processes occurring are the same as the cohesive forces between atoms. Fig. 4.64 shows a cross section taken through a loosely piled assembly of copper particles sintered in order to develop neck growth.[189] One can observe the interparticle void located between three connected particles in the top left quadrant of the picture.

In order to develop necks between particles, some level of interparticle adhesion must develop. Note also that neck growth may be accompanied by densification of the powder mass or it may not be so accompanied.

In order to emphasize the almost baffling complexity of sintering, we restrict our discussion to pure one-component systems, idealized spherical particles, no applied stress, and stoichiometric compositions. Even so the phenomenon is complicated. For example, at least six clearly distinguishable mechanisms contribute to neck growth with or without bulk densification.[190] The possible routes for matter transportation are described in Table 4.26 and illustrated in Fig. 4.65.

The overall driving force is the reduction of the surface free energy of the system by reduction in the surface area of the material of the particulate mass. One should distinguish between the neck growth or sintering rate dx/dt and the shrinkage rate dy/dt. The latter is operative only when

Table 4.26

SIX TRANSPORT MECHANISMS IN SINTERING[190]

Mechanism No.	*Transport path*	*Source of matter**	*Sink of matter**
1	Surface diffusion	Surface	Neck
2	Lattice diffusion	Surface	Neck
3	Vapor transport	Surface	Neck
4	Boundary diffusion	Grain boundary	Neck
5	Lattice diffusion	Grain boundary	Neck
6	Lattice diffusion	Dislocations	Neck

*Sintering problems can be formulated in terms either of the flow of matter, or of the counterflow of vacancies. It is more convenient to focus on the flow of matter.

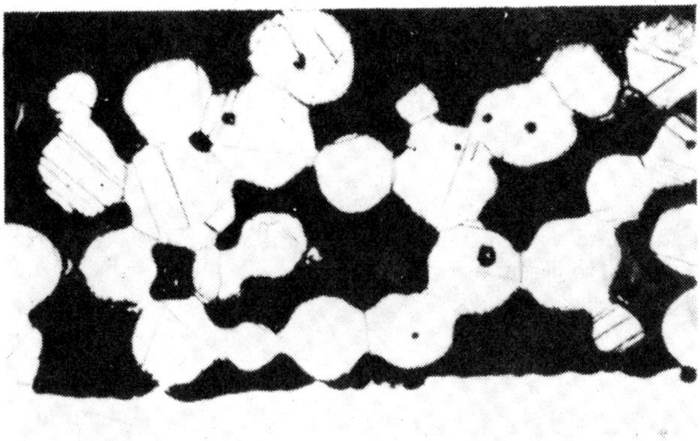

Fig. 4.64 Section through some loose sintered spherical copper powder, sintered at 1472°F for 6 h, illustrating neck growth and formation of grain boundaries in the original neck area (about 260 X).[189] (Courtesy Metal Powder Industries Federation).

the centers of the particles are approaching each other and this occurs either when matter is removed from the grain boundary region (mechanisms 4 and 5) or when matter is removed from the dislocations within the neck itself (mechanism 6).[190]

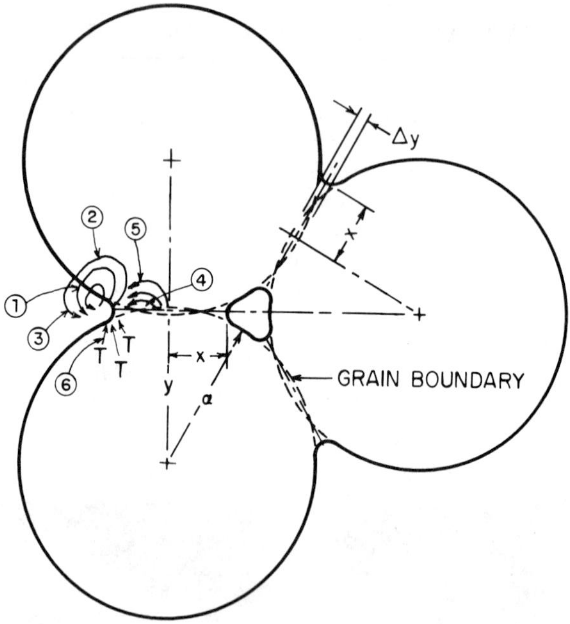

Fig. 4.65 Six alternative paths which permit diffusion-controlled sintering. All lead to neck growth. Only paths 4–6 cause the particle centers to translate together, and so permit densification.[190]

Sintering Diagrams

A very convenient way of collating information concerning the mechanisms of sintering as they operate in a particular system is by means of a sintering diagram.[190] One such for tungsten is shown in Fig. 4.66. The diagrams are constructed on the basis of calculations made from rate equations which describe the individual contribution that each mechanism makes to sintering (i.e., neck growth) or shrinkage in a given set of conditions.

Sintering diagrams identify the mechanism of sintering at a given temperature and also give the rate of operation. Note that the homologous temperature is plotted along the x axis (T_M is the melting point) and the normalized neck radius (i.e., x/a, in which a is the particle radius and x is the radius of the contact disc between two particles). The shaded portions

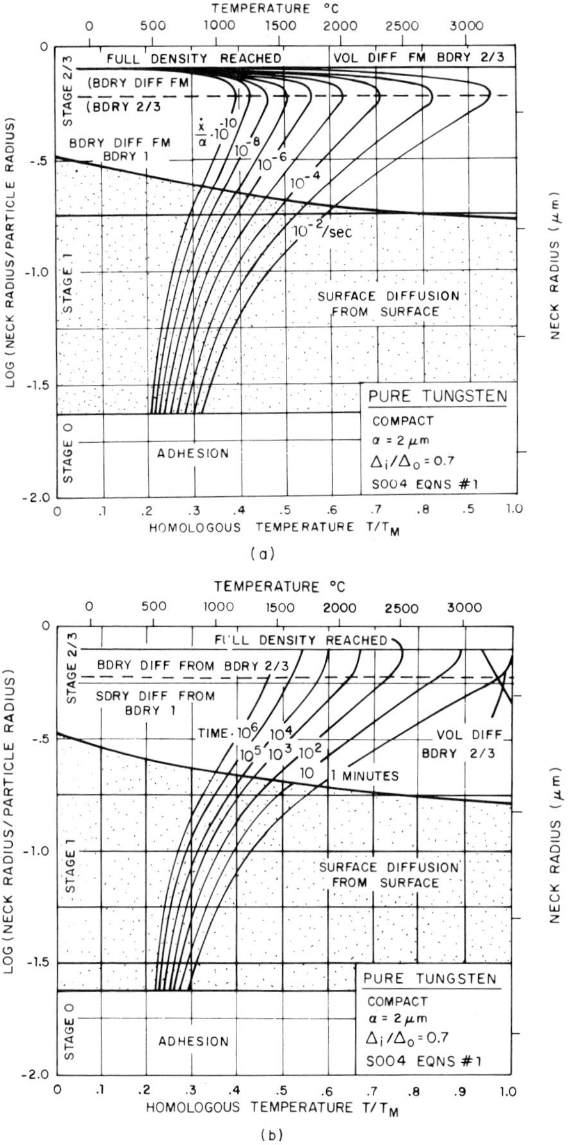

Fig. 4.66 (a) A sintering-rate diagram for tungsten. The fields indicate the dominant mechanism. In an unshaded field the dominant mechanism causes both neck-growth and densification: in a shaded field, only neck growth. The contours are lines of constant normalized sintering rate, $\dot{x}/a$; **(b)** A sintering-time diagram. It is identical with (a), except for the contours, which are now lines of constant time, t. This shows the neck size after a given time-temperature treatment.[190]

of the diagrams are regions within which neck growth without densification occurs. In the unshaded regions both occur. In Fig. 4.66(a), the contours of constant sintering rate are given (x/a) and in Fig. 4.66(b), contours of constant sintering time (to achieve a given neck size) are given.

Sintering diagrams thus encompass a great deal of information in a single comprehensive picture. They should prove to be extremely useful.

Four Stages of Sintering[190]

Stage 0 – Adhesion: When a particle comes into contact with another particle it is understood that an instantaneous neck formation occurs as a result of the interatomic forces acting between them. The specific effect depends upon the degree of cleanliness of the particle surfaces. Thus if the particle is clean, the effective surface energy is given by the clean surface approximation

$$\gamma_{eff} \overset{\text{clean}}{\approx} 2\gamma_s - \gamma_\beta \tag{4.79}$$

and, if the particle is dirty, the effective surface energy is given by the dirty surface approximation

$$\gamma_{eff} \overset{\text{dirty}}{\approx} \gamma_s/10 \tag{4.80}$$

in which

γ_{eff} is the effective surface energy,

γ_β is the grain boundary energy, and

γ_s is the surface free energy.

When two particles approach each other, one assumes that interatomic forces draw them closer together at about the velocity of sound leading to a neck formation where the maximum neck radius is given by the equation[191]

$$x \cong \left(\frac{\gamma_{eff}\, a^2}{G}\right)^{1/3} \tag{4.81}$$

in which G is the shear modulus.

To be realistic one must assume that the particle surface is in effect dirty and so use the dirty surface approximation. It follows then that the rate equation for neck growth is[190]

$$(\dot{x})_0 = \frac{ca^2}{x} \quad \text{for} \quad x < \left(\frac{\gamma_s a^2}{10G}\right)^{1/3} \quad \text{and}$$

$$(\dot{x})_0 = 0 \quad \text{for} \quad x \geqslant \left(\frac{\gamma_s a^2}{10G}\right)^{1/3} \tag{4.82}$$

Stage 1 – Diffusion Controlled Neck Growth: This becomes important above about 0.25 T_M. The total neck growth rate is simply determined by the sum of the contributions from the different mechanisms listed in Table 4.26. Thus

$$\dot{x}_1 = \sum_{i=1}^{6} \dot{x}_i \tag{4.83}$$

The individual contributions (refer to Table 4.26) are as follows:

Mechanism 1[192,193]

$$\dot{x}_1 = 2D_s\delta_s FK_1^3 \tag{4.84}$$

and

$$F = \frac{\gamma_s \Omega}{kT} \tag{4.85}$$

Mechanism 2[192,193]

$$\dot{x}_2 = 2D_v FK_1^2 \tag{4.86}$$

Mechanism 3[194]

$$\dot{x}_3 = P_v F\left(\frac{\Omega}{2\pi\Delta_O kT}\right)^{1/2} K_1 \tag{4.87}$$

Mechanism 4[195]

$$\dot{x}_4 = \frac{4D_B\delta_B FK_2^2}{x} \tag{4.88}$$

Mechanism 5[192]

$$\dot{x}_5 = 4D_v FK_2^2 \tag{4.89}$$

Mechanism 6[190]

$$\dot{x}_6 = \frac{4}{9} K_2 N x^2 D_v F \left(K_2 - \frac{3}{2} \frac{Gx}{\gamma_s a} \right) \tag{4.90}$$

In the above equations

K_1, K_2, K_3 are curvature differences which drive diffusion fluxes,
D_s is surface diffusion coefficient,
D_v is lattice diffusion coefficient,
D_B is grain boundary diffusion coefficient,
Ω is atom or molecular volume,
k is Boltzmann's constant (1.38×10^{-16} ergs/K; 138×10^{-23} J/K),
T is melting temperature (K),
T_M is melting temperature (K),
F is $\gamma_s \Omega / kT$ (typically of magnitude 10^{-6} cm),
δ_s is effective surface thickness,
δ_B is effective grain boundary thickness,
P_v is vapor pressure $[P_v = P_O \exp - (Q_{vap}/kT)]$,
Δ_O is theoretical density,
G is shear modulus, and
N is dislocation density.

Stage 2 – Boundary Diffusion from Sources on the Boundary and Stage 3 – Lattice Diffusion from Sources on the Boundary: These diffusions are the two important mechanisms in Stage 2 and Stage 3. Subtle differences between the two are ignored and it is assumed that they obey the same rate equations. As in Stage 1, the net sintering rate is given by the sum of the individual contributions for the individual mechanisms occurring during the process. Thus:

$$\dot{x}_{2,3} = \dot{x}_7 + \dot{x}_8 \tag{4.91}$$

In the above equation[190]

$$\dot{x}_7 = \frac{1}{16} D_B \delta_B F K_3^3 \left[\frac{1}{\log_e \left(\frac{x_f K_3}{2} \right) - \frac{3}{4}} \right] \tag{4.92}$$

and[190]

$$\dot{x}_8 = \frac{1}{16} x D_v F K_3^3 \left[\frac{1}{\log_e \left(\frac{x_f K_3}{2} \right) - \frac{3}{4}} \right] \tag{4.93}$$

in which

x_f is the final value of x when 100 percent density has been reached.

The above outline is extremely sketchy and the interested reader should refer to the original source referenced against each equation. The development of the sintering diagram is an elegant piece of work. It represents a sensible approach to the analysis of a transportation system.

Gas Pressure

In a real system of particulates, gas pressure opposes the tendency for sinter shrinkage. An approximate calculation involving use of the gas principle $P_1 V_1/T_1$ equals $P_2 V_2/T_2$ indicating that in the case of a closed spherical pore of diameter 44 μ, the pressure difference between the inside and outside of the pore is about 3 atmospheres pressure. This operation means that, if the pore is smaller than 44 μ diameter, it tends to shrink and if it is larger than 44 μ diameter it tends to grow.[196]

Sintering in the Presence of a Liquid Phase

This is quite a common industrial practice. It is carried out in order to:

1. achieve sintering which would be difficult or impossible in the solid (for example Cu-W contacts, Cu-Ag contacts);
2. achieve a much higher rate of sintering than normally obtained in all solid phase sintering (e.g., in the case of Cu-Fe compacts); and
3. to achieve a high finished density (some ceramic systems do this).

A set of data for the case of Cu-Fe is shown in Fig. 4.67. It can be seen that the process may be conveniently divided into stages:[197]

1. Rearrangement: this action occurs when the liquid phase melts and is simply due to the relative motion of individual particles relocating in lower energy positions in the particulate assembly.
2. Solution-Reprecipitation: this action occurs only when there is some solubility of the solid in the liquid phase. It is slower than the rearrangement stage.

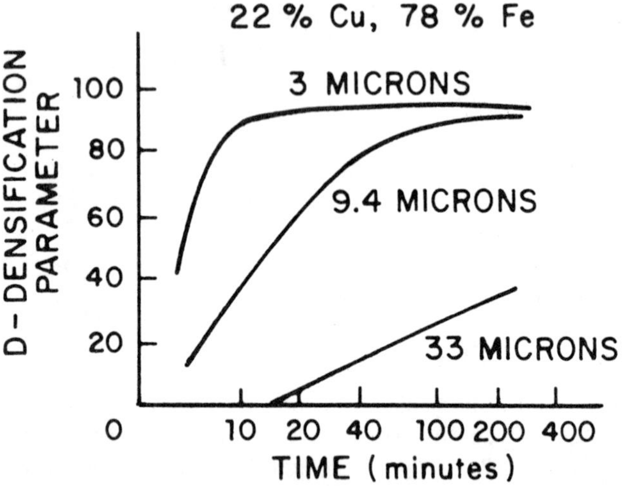

Fig. 4.67 Liquid phase sintering.

3. Coalescence: this action is in effect equivalent to normal solid state sintering.

The equation said to represent rearrangement is[197]

$$\frac{\Delta l}{l_o} = kr^{-1}\rho^{1+x} \tag{4.95}$$

in which

Δl is the change in length of the sample of initial length ρ_o,

r is the particle radius, and

$1 + x \cong 1$

The equation said to represent the solution-reprecipitation stage is given by[197]

$$\frac{\Delta l}{l_o} = k^1 r^{-4/3}\rho^{1/3} \tag{4.96}$$

A recent review of sintering theory is given in reference 198.

REFERENCES

1. Weisselberg, E. *Chemical Engineering Progress* 63, 11 (1967): 72–76.
2. Jenike, A. W. "Storage and Flow of Solids." *University of Utah Bulletin 123,* 1964, Univ. of Utah Engineering Experiment Station, Salt Lake City, Utah.
3. Reisner, W., and Rothe, M.V.E. *Bins and Bunkers for Handling Bulk Materials.* Trans. Tech. Publications. Rockport, Massachusetts, January 1971 (reprinted 1978). See also O. Molerus, *Powd. Tech.* 12 (1975): 259–275.
4. "Storage and Flow of Solids," AIChE Today Series, Kansas City, 1975.
5. Kvapil, R. *Aufber Tech.* 5 (1964): 139-144, 183–189.
6. Ashton, M. D., Cheng, D. C., Farley, R., and Valentin, F. H. H. *Rheol Acta* 4 (1965): 206–218.
7. Ashton, M. D., Farley, R., Valentin, F. H. H. *J. Sci. Instrum.* 41 (1964): 763-765.
8. Carr, J. F., and Walker, D. M. *Powder Technology* 1 (1967–68): 369–373.
9. Walker, D. M. *Powder Technology* 1 (1967): 228–236.
10. Williams, J. C., Birks, A. H., and Baattacharya, D. *Powder Technology* 4 (1970–71): 328-337.
11. Brown, R. L., and Hawkesly, P. G. W. *Fuel* 26 (1947): 159–173. See also Coughlin, R. W. *Powd. Tech.* 17 (1977): 65–71.
12. Bruff, W., and Jenike, A. W. *Powder Technology* 1 (1967–68): 252–256.
13. Brown, R. L. *Powders in Industry.* London: SCI Monograph 14, 1961, 150–164.
14. Shirai, T. *Chem. Eng. Japan* 16 (1952): 86–89.
15. Enge, T. A. ASME Publication. No. 68-MH-18.
16. Vand, V. *J. Phys. Coll. Chem.* 52 (1948): 300–304.
17. Link, J. M. *Bulk Materials Handling* 3, University of Pittsburgh, School of Engineering, edited by Hawk, M. C. 1975. pp. 289–313.
18. Durand, R., and Condolios, E. Le Journels d'Hydraulique, Grenoble, Soc. Hydrotech. de France, June, 1952.
19. Wasp, E. J. *Pipeline Engineer* 41, N12 (November 1969): 56–63.
20. Govier, G. W., and Aziz, K. *The Flow of Complex Mixtures in*

Pipes. New York: Van Nostrand Reinhold, 1972.

21. Metzner, A. B., and Reed, J. C. *AIChE J.* 1 (1955): 434–440.
22. Ismail, H. M. *Trans. ASCE* 117 (1952): 409–446.
23. Faddick, R. R. "A Mineral Slurry Data Bank." Report to U.S. Bureau of Mines, GO 110850, March, 1972.
24. Zandi, I., and Govatos, T. *J. of Hydr. Div.* ASCE 92, No. HY3, Proc. Paper 5244 (May 1967): 145–159.
25. Thomas, D. G. *AIChE J.* 8 (1962): 268.
26. Orr, C. Jr. *Particulate Technology.* New York: Macmillan, 1966.

    A perennial problem with pneumatic transport is that of erosion in bends, see for example: Mason, J.S., Smith, B.V. *Powd. Tech.* 6 (1972): 323; and Mills, D., and Mason, J. S. *Powd. Tech.* 17 (1977): 37–53.
27. Taylor, G. I. *Proc. Roy. Soc.* 138A (1932): 41–48.
28. Consolidated Engineering Co., 5960 Kansas St., Houston, Texas, Fluiflo® pump.
29. Continental Screw Conveyor Co.
30. Cambelt International Corp.
31. The Young Industries, Inc. Young Industries Rotary Valve Bulletin # 220–201. Muncy, Pennsylvania 17756.
32. Buhler-Miag, Inc., P.O. Box 9497, 1110 Xenium Lane, Minneapolis, MN 55440, Bulher-Miag diverter valve, Bulletin PH-DV-01.
33. Schwarts, C. E., and Smith, J. M. *Ind. Eng. Chem.* 45 (1953): 1214–1215.
34. Bernard, R. A., and Wilhelm, R. H. *Chem. Eng. Progr.* 46 (1950): 233–244.
35. McHenry, K. W., and Wilhelm, R. H. *AIChE* 3 (1957): 83–91.
36. Latinen, G. A. "Mechanism of Mixing in Fixed and Fluidised Beds of Solids." Ph.D. thesis, Princeton University, 1954.
37. Ebach, E. A., and White, R. R. *AIChE* 4 (1958): 161–169.
38. Cairns, E. J., and Prausnitz, J. M. *Chem. Eng. Sci.* 12 (1960): 20–34.
39. Guinness, R. C. *Chem. Eng. Prog.* 49, 113 (1953).
40. Wilhelm, R. H. *Research* 3, 161 (1950).
41. Zenz, F. A. *Fluidized Beds.* write to Dr. Zenz for a copy, Chem. Engg., Manhattan College, N.Y., N.Y. 10 . A very interesting series of studies of large scale fluidized beds is available. See for example, Whitehead, A.B., Dent, D.C., McAdam, J. C. H. *Powd. Tech.* 18 (1977): 231–237, Pt. V. See also *Powd. Tech.* 14 (1976): 61 (Pt. III) and *Powd. Tech.* 15 (1976): 77 (Pt. IV).

42. Wen, C. Y., and Hasinger, R. F. *AIChE J.* 6 (1960): 220–226.
43. Bakker, P. J., and Heertjis, P. M. *Chem. Eng. Sci.* 12, 266 (1960). For some recent work on effect of interparticle forces on the expansion of a gas fired fluidized bed see Mutser, S.M.P., and Rietma, K. *Powd. Tech.* 18 (1977): 239–248. For a study of particle mixing in a gas fluidized bed see Rowe, P. N. and Nienow, A. W. *Powd. Tech.* 15 (1976): 141–147.
44. Rone, et al. *Nature* 195 (1962): 278–279.
45. Dynapore, T. M. Michigan Dynamics Division of AMBAC Industries, Inc., Subsidiary of United Technologies Corp., 32400 Ford Road, Garden City, MI.
46. Dalla Valle, J. M. *Micromeritics.* New York: Pitman, 1948.
47. Orr, C., Jr. *Particulate Technology.* New York: Macmillan, 1966.
48. Lacey, P. M. C. *Trans. Inst. Chem. Eng.* 21 (1943): 53–59.
49. Hausner, H.H. and Beddow, J.K. *Mixing of Powders.* Bibliography published by Hoeganaes Corp., Riverton, NJ 08077, 1972.
50. Cooke, M. H., Stephens, D. J., and Bridgewater, J. *Powder Technology* 15, 1 (1976): 1–20. For recent work in Japan, see Yano, T., Sato, M., and Toroshita, K. *Powd. Tech.* 20 9–14.
51. *Powder Technology* 15, 2, edited by Williams, J. C., U. Bradford, U.K,. November/December 1976.
52. Hersey, J. A. *J. Soc. Cosmet. Chem.* 31 (1970): 259–269.
53. Hersey, J. A. *Powder Technology* 15, 1 (1976): 149–153.
54. Lacey, P. M. C. *J. App. Chem.* 4 (May 1954): 257–268.
54(a). This is called the Buslik equation after its author in Bull Am. Soc. Test. Mat. 66 (1950): 165, AZT.
55. Harnby, N. *Powder Technology* 1 (1967): 94–102.
56. Brown, R. L. *J. Inst. Fuels* 13 (1939): 15.
57. Williams, J. C. *Fuel Soc. J. Un. Sheffield* 14 (1963): 29.
58. Ashton, M. D., and Valentin, F. H. H. *Trans. Inst. Chem. Eng.* (London) 44, 1966, T166. A study of particle velocities has been reported. See Oko, K., Walewender, W. P., and Fan, L. T. *Powd. Tech.* 18 (1977): 171–178.
59. Bridgewater, J. *Powder Technology* 15, 2 (1976): 215–236.
60. Danckwerts, P. V. *Research* 6 (1953): 355–361.
61. Ure, Unpublished work, reported by Lacey, 1943 (see Reference 48).
62. Weidenbaum, S. S., and Bonilla, C. F. *Chem. Eng. Prog.* 51, 1 (1955): 27J–36J.
63. Scholfield, C. *Powder Technology* 15, 2 (November 1976): 169–180.
64. Bannister, H. *Chem. and Proc. Eng.* 40 (1959): 53–54.

65. Gayle, J. B., Lacey, O. L., and Gary, J. H. *Ind. and Eng. Chem.* 50, 9 (1958): 1279–1282.
66. Shinnar, R., and Naor, P. O. *Chem. Eng. Sci.* 15 (1961) 220–229.
67. Shinnar, R., Kattar, A., and Steg, I. *Chem. Eng. Sci.* 18 (1963): 677–683.
68. Danckwerts, P. V. *App. Sci. Research* A3 (1952): 279.
69. Horwitz, L. P., and Shelton, G. L. *Proc. Inst. Radio Eng., NY.* 49 (1961): 175.
70. Harwood, C. F., Walanski, K., Luebcke, E., and Swanstrom, C. *Powder Technology* 11 (1975): 289–296.
71. Williams, J. C. *Powder Technology* 15, 2 (1976): 237–243.
72. Gray, J. B. *Chem. Eng. Progress* 53, 1 (1957): 25J–32J.
73. Roseman, B., and Donald, M. B. *Brit. Chem. Engr.* 7, 11 (1962): 823–827.
74. Adams, J. F. E., and Baker, A. G. *Trans. Inst. Chem. Engr.* 34 (1956): 91–107.
75. Brown, C. O. *Ind. Eng. Chem.* 57A (1950): 42.
76. Williams, J. C., and Richardson, unpublished work, reported in reference 71.
77. Sugimot, M., Endoh, K., and Tanaka, K. *Kagaku Kogaku* 31 (1967): 145.
78. Coulson, J. M., and Maitra, N. K. *Ind. Chemist.* (February 1950): 55–60.
79. Lloyd, P. J., and Yeung, P. C. M. *Chem and Proc. Engr.* (October 1967): 57–61.
80. Donald, M. B., and Roseman, B. *Brit. Chem. Eng.* 7, 10 (October 1962): 749–753.
81. Cahn, D. S., and Fuerstenau, D. W. *Powder Technology* 1 (1967): 174–182.
82. Hogg, R., Chan, D. S., Healey, T. W., and Fuerstenau, D. W. *Chem. Eng. Sci.* 21 (1966): 1025.
83. Wallis, G. B. *One Dimensional Two Phase Flow.* New York: McGraw-Hill, 1969.
84. Rose, H. E. *Tran. Inst. Chem. Eng.* 37 (1959): 47–64.
85. Novosad, J., and Standart, G. *Collect, Czech. Chem. Commun.* 30 (1965): 3247.
86. Bagster, D. F., and Bridgewater, J. *Powder Technology* 3 (1969–70): 323.
87. Bridgewater, J., Bagster, D. F., Chen, S. F., and Hallem, J. H. *Powder Technology* 2 (1968/69): 198.
88. Bridgewater, J., Sharp, N. W., and Stocker, D. C. *Trans. Inst. Chem. Eng.* 47 (1969): T114.

89. Bridgewater, J., and Ingram, N. D. *Trans. Inst. Chem. Eng.* 49 (1971): 163.
90. Masliyah, J., and Bridgewater, J. *Trans. Inst. Chem. Eng.* 52 (1974): 31.
91. Bridgewater, J., and Scott, A. M. *Trans. Inst. Chem. Eng.* 52 (1974): 317.
92. Bridgewater, J. *Powder Technology* 5 (1971/72): 257.
93. Zweitering, T. N. *Chem. Eng. Sci.* 8 (1958): 244–253.
94. Michaels, A. S., and Puzinauskas, V. *Chem. Eng. Prog.* 50, 12 (1954): 604–614. The dispersion of powders in liquids is reported about in Parfitt, G. D. *Powd. Tech.* 17 (1977): 157–162.
95. Corn, M. *J. Air Poll. Control Assoc.* 11 (1961): 523–528 and 566–575.
96. Bagnold, R. *The Physics of Blown Sands and Desert Dunes.* (this book was written many years ago. It deals with the saltation of sand particles in prevailing desert winds. I understand that Bagnold was made an FRS on the strength of this work. The reader cannot fail to be impressed with the good judgement of the Royal Society. The book is a masterpiece).
97. Reynolds, O. Papers on Mechanical and Physical Subjects, Cambridge, London, 1900.
98. Ergun, S. *Anal Chem.* 23 (1951): 151–156. See also Propster, M., and Szekely, J. *Powd. Tech.* (1977): 123–138 for a study of porosity in multilayered solids.
99. Baron, J. *Chem Eng Prog.* 48 (1952): 118–124.
100. Newton, H. W., and Newton, W. H. *Rock Prod.* 35 (1932): 26–30.
101. Gaudin, A. M. *Principals of Mineral Dressing.* New York: McGraw Hill, 1955.
102. Miwa, S. *Chem. Eng. (Japan)* 24 (1960): 151.
103. Whitby, K. J. *ASTM Special Technical Publication* 234 (1958): 3–23.
104. Nichols, G. V., Hess, H. L., and Beddow, J. K. *Powder Technology* 3 (1969): 57–59.
105. Roberts, T. A., and Beddow, J. K. *Powder Technology* 2, 2 (1968): 121–124.
106. Rose, H. E. *Journal of the Japanese Society of Powder and Powder Metallurgy.* 22, 2 (1975): 55–70.
107. Nichols, G. V., Hess, H. L., and Beddow, J. K. "Filtration as a Part of the Mechanism of Sieving." paper presented at International Sympsium on Powder Technology, ASTM, Chicago, 1968.

108. Elkson, J. C., and Ehinger, G. A. *Chem. Eng. Prog.* 59 (January 1963): 76–80.
109. U.S. Patent 1,190,926; July 11, 1916, Lotozky, A.
110. U.S. Patent 871,536; November 19, 1907, Thompson, E.
111. U.S. Patent 897,489; September 1, 1908, Prinz, F.
112. U.S. Patent 1,291,278; January 14, 1919, Ulrich, J.
113. Riley, G. S. *Powder Technology* 2 (1969): 315–319.
114. Klar, E. *Powder Technology* 3 (1969/70): 313–314.
115. Waldie, R. *Powder Technology* 7 (1973): 244–246.
116. Glenzen, W. H. and Ludwick, J. C. *J. Sed. Petrol.* 33 (1963): 23–40.
117. Determan, H. *Gel Chromatography.* New York: Springer Verlag, 1968.
118. Dimarzio, E. A., and Guttman, C. M. *J. Pol. Sci.* B7 (1969): 267.
119. Ruckenstein, E., and Prieve, D. C. *AIChE J.* 22, 2 (March 1976): 276–283.
120. Gaudin, A. M. *Flotation.* 2nd ed. New York: McGraw-Hill, 1957.
121. Workshop on Novel Concepts, Methods, and Advanced Technology in Particulate-Gas Separation, April 1977, NSF, U Notre Dame, session on High Gradient Magnetic Field in Particulate-Gas Separation, Chairman BYH Liu.
122. Comings, F. W., Pruiss, C. E., and De Bord, C. *Ind. Eng. Chem.* 46 (1954): 1164–1172.
123. Ralston, O. C. *Electrostatic Separation of Mixed Granular Solids.* New York: Elsevier. 1961.
124. Bergougnon, M. A., Incoulet, I. I., Anderson, J., and Parobek, L. International Conference on Powder and Bulk Solids Handling, Chicago, May 1976, Session 8, Paper 6, London: Powder Advisory Center, U.K.
125. Dobby, G., and Finch, J. A. ibid, Session 8, Paper 4.
126. *Encyclopedia of Chemical Technology,* Vol 8, Eds. Kirk, R. E., and Othmer, D. F., p. 625, New York: Wiley–Interscience, 1952.
127. Tiller, F. M., and Cooper, H. R. *AIChE* 8 (1962): 446.
128. Filter Facet Media, Bulletin FDP 48R 9/75, Facet Prod. Div., Facet Enterprises, 434 W. 12 Mile Rd., Madison Heights, MI 48071.
129. Payatakes, A. C. *Powder Technology* 14 (1967b): 267.
130. Payatakes, A. C. to appear in *AIChE* 23 (1977): 192–202.
131. Davies, C. N. *Air Filtration.* New York: Academic Press, 1973.
132. Frederick, E. R. *Chem. Eng.* 68 (1961): 110.

133. Farnard, J. R., Smith, H. M., and Puddington, I. E. *Can. J. Chem. Eng.* 39 (1961): 94–97.
134. Danielson, J. A. *Air Pollution Engineering Manual,* PHS, Cincinnati, Ohio, National Center of Air Pollution Control, 1967.
135. Duprey, R. L. Compilation of Air Pollutant Emission Factors, (Bureau of Disease Prevention and Environmental Control, Durham, North Carolina, National Center for Air Pollution Control) 1968.
136. Sterling, Morton, "Current State and Future Prospects–Foundry Air Pollution Control," Proceedings of the National Conference on Air Pollution, PHS, Washington, Division of Air Pollution, 1966.
137. Cleary, G. J. M., and Palmer, D. G. *Australasian Eng.,* Sydney (August 1967): 55–56, 58.
138. Böhlen, B. *Chem. Eng.* CE257–61 (September 1968): 221.
139. Cuffe, S. T., and Gerstle, R. W. "Emissions from Coal-Fired Power Plants." A Comprehensive Summary, PHS, Cincinnati, Ohio, National Center for Air Pollution Control (presented at Amer. Ind. Hyg. Assoc. Meeting, Houston, Texas, May 1965) 1967.
140. Smith, L. S. "Atmospheric Emissions from Fuel Oil Combustion." PHS, Cincinnati, Ohio, Division of Air Pollution, 1962.
141. Manufacturing Chemists' Association, Inc., "Atmospheric Emissions From Sulfuric Acid Manufacturing Process." PHS, Cincinnati, Ohio, Division of Air Pollution, 1965.
142. Partee, F. "Air Pollution in the Coffee Roasting Industry." PHS, Cincinnati, Ohio, Division of Air Pollution, 1966.
143. Drogrin, I. *Journal of the Air Pollution Control Association* 18, 4 (April 1968): 216–228.
144. Phelps, A. H., Jr. *Journal of the Air Pollution Control Association* 17, 8 (August 1967): 505–507.
145. Kriechelt, T. E., Kemnitz, D. A., and Cuffe, S. T. "Atmospheric Emissions from the Manufacture of Portland Cement." PHS, Cincinnati, Ohio, Bureau of Disease Prevention and Environmental Control, 1967.
146. Doherty, R. E. "Current Status and Future Prospects–Cement Mill Air Pollution Control." Proceedings of the National Conference on Air Pollution, PHS, Washington, Division of Air Pollution, 1966.
147. Cuffe, S. T., and Knudson, James C. "Considerations for Determining Acceptable Ambient and Source Concentrations for Particulates from Cotton Gins," and "Control and Disposal of Cotton Ginning Wastes." PHS, Cincinnati, Ohio, Bureau of

Disease Prevention and Environmental Control, 1967.
148. Uskov, V. I. *Testing of An Electrostatic Filter, Air Pollution in Mines; Theory, Hazards, and Control.* trans. by N. Kaner, Akademiia Nauk SSSR Institut Gornogo Dela., 1966.
149. Frey, J. W. and Corn, M. *Amer. Ind. Hyg. Assoc. Journal* 28, 5 (September–October 1967): 468–478.
150. "Performance Evaluation of an Electrostatic Particle Precipitator for Gasoline-Engine Exhaust." Chemical & Physical R.D. and Motor R.D. Programs, National Air Pollution Control Administration, 1968.
151. White, H. J. *Ind. Eng. Chem.* 37 (1955): 932–939.
152. White, H. J. *Trans.* AIEE 70 (1951): 1186–1191.
153. White, H. J. *Chem. Eng. Prog.* 52 (1956): 247.
154. U.S. Patent by N. Gothard. This may be compared with the German Patent 571159 by Hahn, C. in which the particle laden gas is under pressure.
155. Gothard, N., Gothard Industries, Arlington, Texas, personal communication, 1979.
156. Scholz, P. D., personal communication.
157. Graton, L. C., and Frazer, H. J. *J. Geol.* 43 (1935): 785.
158. Gray, W. A. *The Packing of Solid Particles.* London: Chapman and Hall, 1968.
159. Smalley, I. J. *Powder Technology* 4 (1971): 69.
160. James, P. J. *Powder Metallurgy International* 4, 2 (1972): 82–85; 4, 3 (1972): 145–149; 4, 4 (1972) (reviews).
161. Horsfield, H. T. *J. Soc. Chem. Ind.* 53 (1934): 107.
162. Bowden, F. P., and Tabor, D. *The Friction and Lubrication of Solids.* Pt. 1, London: Oxford University Press, 1954.
163. Grey, R. O., and Beddow, J. K. *Powder Technology* 2 (1968/69): 323–326.
164. Hausner, H. H. *Intl. J. Powder Met.* 3, 4 (1967): 7–13.
165. ter Linden, A. J. *J. Proc. Inst. Mech. Engrs.* 160 (1949): 233–240.
166. Rietema, K. *Chem. Eng. Sci.* 15 (1961): 303–309.
167. Stairmand, C. J. *Trans. Inst. Chem. Eng.* 29 (1951): 356–383.
168. Kane, H. personal communication, Materials Engineering, Univ. of Iowa, 1979.
169. Kostelnik, M. C., Kludt, F. H., Beddow, J. K. *Int. J. Powd. Met.* 4 (4) (1968): 19–28.
170. Kostelnik, M. C., and Beddow, J. K. *Modern Developments in Powder Metallurgy.* vol. 4, Processes, pp. 29–48, ed. by Hausner, H. H., New York: Pelnum, 1971.
171. Katz, R. N. in *Treatise on Materials Science and Technology.* 9,

pp. 35–49, "Ceramic Fabrication Processes." ed. by Wang, F.Y. New York: Academic Press, 1976.
172. Sheinhartz, I., McCulloguh, H. M. , and Zambrow, J. L. *J. Metals* 6 (1954):
173. Kuzynski, G. C., and Zaplatynskyi, I. *J. Metals* 8 (1956): 215. (©1956, American Institute of Mining, Metallurgical and Petroleum Engineers, 345 E. 47th St., New York, N.Y. 10017)
174. Train, D. *Trans. Inst. Chem. Engr.* 35 (1957): 258.
175. Beddow, J. K. *Intl. J. Powd. Met.* 4, 1 (1968): 27–29.
176. Kunin, N. F., and Yurchenko, B. D. *Poroshkovaya Metallurgiya.* 8 (1968): 15.
177. Kane, H., personal communication, Materials Engineering, Univ. of Iowa, 1979.
178. Felter, E. J. *J. Am. Ceram. Soc.* 44, 8 (1961): 381–385.
179. Beddow, J. K., *Intl. J. Powd. Met.* 9, 4 (1973): 127–131.
180. Train, D., and Lewis, C. J. *Trans. Inst. Chem. Engrs.* 40 (1962): 235.
181. Kawakita, K., and Ludde, K-H. *Powd. Tech.* 4 (1970/71): 61–68.
182. Brackpool, J.L. U. Technology, Loughborough, M. Tech. thesis.
183. Bockstiegel, G., and Svensson, O. *Modern Developments in Powder Metallurgy.* vol. 4, Processes, ed. by Hausner, H. H., pp. 87–114, New York: Plenum Press, 1971.
184. Beddow, J. K., Sadjack, R., McNalley, R., and Nasta, M. D. *Intl. J. Powd. Met.* 6, 2 (1970): 13–23.
185. Kuzynski, G. C. et al. eds. *Sintering and Related Phenomena.* New York: Gordon & Breach, 1967.
186. Kingery, W. D. et al. eds. *Kinetics of High Temperature Processes.* MIT Press, 1959.
187. Gray, T. J., and Frichette, V. D. *Kinetics of Reactions in Ionic Systems.* New York: Plenum Press, 1969.
188. Science of Sintering, a journal with the text in English and abstract in Russian. Published by International Institute for Science of Sintering, Belgrade, Yugoslavia.
189. Hirschhorn, Joel. *Introduction to Powder Metallurgy,* Metal Powder Industries Federation, N.J., 1969, p. 160.
190. Ashby, M. F. *Acta Metallurgica* 22 (March 1974): 275–289.
191. Easterling, K. E., and Tholen, A. R. *Physics of Sintering.* special issue, 1971: p. 77. also *Met. Sci. J.* 4 (1970): 130. also *Acta Met.* 20 (1972): 1001.
192. Wilson, T. L., and Shewman, P. G. *Trans AIME.* 236 (1966): 48.
193. Kuzynski, G. C. *Trans AIME.* 185 (1949): 169.

194. Kingery, W. D., and Berg, M. *J. App. Phys.* 26 (1955): 1205.
195. Johnson, D. L. *J. App. Phys.* 40 (1969): 192.
196. Lenel, F. V., lecture notes, Powder Metallurgy Short Courses, UCLA, 1967.
197. Kingery, W. D. *J. App. Phys.* 30 (1959): 301.
198. Kuzynski, G. C. *Science of Sintering* 9, 3 (1977): 243–264.
199. Dodge, D. W., Metzner, A. B. "Turbulent Flow of Non-Newtonian Systems," *AIChE* 5 (1959): 189–204.
200. Dynapac Mfg., Inc., P.O. Box 368, Stanhope, NJ 07874, refer to Dynapac Air Stimulator Systems, Technical Data, Bulletin # SA-060-0575.
201. Yip, C. W. and Hersey, J. A., *Nature,* 262 (1976): 202–203.

Chapter 5

DESCRIPTION OF PARTICULATE ASSEMBLIES

5.1 DESCRIPTION OF PARTICLE SETS

Fine particle sets contain myriads of individual particles and the description of the characteristics or properties of these sets must be expressed in probabilistic statistics if data are available. This chapter presents an account of some of the statistical methods used in fine particle science and technology to describe fine particle sets. In addition, an introductory discussion, following the justification of the term "particle set," examines some of the effects of particle size and shape (where these are known) upon the properties of particles and their sets. With some knowledge of single particles, their formation, production, handling, and processing, it is now possible to begin to comprehend particle sets as the common form in which fine particles are found.

Definition of a Particle Set

A standard definition of a powder in a dictionary reads as follows: "any dry substance in the form of fine dust-like particles produced by crushing, grinding, etc."[1] Despite the vagueness of this definition it does convey to the reader two essential characteristics of a powder: the minute size of the particles and their origin. Of course, the same definition is too exclusive, as it implies that the predominant size is that of dust-like particles and furthermore it also implies that powders originate only due to mechanical action on a solid. In practice powders contain particles covering a very wide range of sizes, as was shown in Table 2.1. And also, in addition to mechanically produced powders, there are particles formed by electro-deposition, chemical deposition, atomizing, fluid dispersion, and vapor deposition.

A more inclusive definition of a powder is "a set of particles of matter in some environment."[2] This carries with it two interesting and interconnected implications: the concept of the powder being a 'set' of particles leads one to expect that the response of the powder to external stimuli can be treated in terms of the response of a group of particles. Secondly, it serves to remind us that a powder 'exists' in either a liquid or a gas, usually. Perhaps the most important objective in the scientific treatment of powders is a direct result of viewing powders as a "set" of particles: in seeking a description of the powder it is necessary to identify the set in terms that can be measured and to generalize the measurements in terms of convenient statistical formulation.

In Chapter 2 some of the important aspects of individual particles were discussed. It was observed that particle size, shape, and surface area are the three characteristics most often of concern. It can come as no surprise therefore to learn that these same three characteristics are primary in the case of sets of particles called powders. The most well-developed methods of description are those concerned with size distributions (mean diameters, histograms, variance, skewness, etc.); less developed are those means of description concerned with surface areas of particle sets; least developed are means of descriptions of shape and shapes distributions of particle sets. One important aspect of a powder in its environment deserves to be specially emphasized: the environment of the members of the set consists not only of an enveloping fluid but also of particle to particle contacts. From this it follows that the particles in a powder are assembled into a sort of structure, albeit one not so easily described as are the structures of solids.* It is in deference to these patterns of particles in a powder that the heading of this section is entitled "Description of Particle Sets" rather than a simple phrase such as "description of powders." A second and equally important reason for using the term particle set is that the treatment is not specifically limited to "powder" particles. For example, a distribution curve can be used to represent a mass of measurement data on aerosol particles just as on gravel. If the treatment in the discussion specifically pertains to one type of particle, it will be noted.

*For those interested in statistical descriptions of the structure of a set of particles in the form of a bed, K. Gotoh has done some interesting work. See, for example, K. Gotoh in "Advanced Particulate Morphology," CRC Press, Eds. Beddow, J. K. and Meloy, T. P., 1979.

5.2 PROPERTIES OF PARTICLES AND PARTICLE SETS AS INFLUENCED BY VARIATIONS IN PARTICLE SIZE AND SHAPE

Many examples were discussed in the Introduction and also in Chapter 2 illustrating the wide varieties of fine particles produced and the capabilities of different industrial processors to handle and utilize these variable and varying substances. In order to demonstrate the practical importance of describing fine particle sets in terms of the three particle characteristics of size, shape, and surface area, some of the influences that variations in particle size and shape have on the properties of particle sets, expressed in terms of the fundamentals of physics, chemistry and mechanics are cursorily examined.

Surface Area. A single drop of water of 1 cm diameter has a surface area of ca. 3×10^{-4} m^2. The same drop of water broken up into droplets of 0.1 micron diameter has a surface area of 30 m^2. This large surface area component in powders accounts for many interesting phenomena including their high chemical reactivity, even explosivity; their higher toxicity relative to the bulk solid; and the faster sintering rates in ceramics and powder metallurgy. A general chart for relating particle diameter, number of atoms contained therein and particle weight is shown in Fig. 5.1.[3]

Electrostatic Charge. Usually small particles acquire electrons due to a variety of alternate processes including–electron exchange mechanisms, spray electrification, contact potential, contact separation electrification, and ion diffusion in gases. In any specific dust (or aerosol) the net charge may be small but the charge distribution within the set of particles can vary from showing an excess of electrons on a proportion of the particles to showing a deficiency of electrons on another proportion of the particles. Some examples of natural charges on fine particulates (or aerosols) are given in Table 5.1. The numerical limit for a dry air environment is about 1.6×10^{10} electrons per cm^2 of surface (8 esu/cm^2) and as is shown in Fig. 5.2, this serves as an upper limit for a wide variety of particles.[5] This predilection of fine powders to acquire charge is a disadvantage if clumping occurs. This interferes with many procedures such as sedimentation, sieving on very fine sieves, and flow characteristics. However, it offers some considerable advantages with respect to electrostatic type precipitators in which electrons are pumped onto the particles in order to encourage their motion through an electric field in a desired direction.

Light Scattering. Light is transmitted through a homogeneous disper-

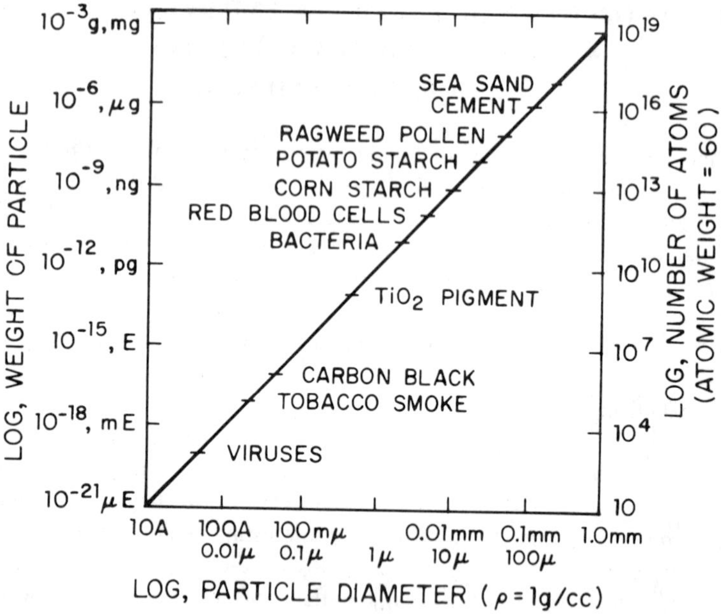

Fig. 5.1 Relationship between particle size, weight and number of atoms.[3]

Table 5.1

NATURAL CHARGES ON AEROSOLS

Material	*Specific charge* (esu/g) Positive	Negative	*Charge distribution* Percent Positive	Percent Negative
Fly ash	1.9×10^4	2.1×10^4	31	26
Gypsum dust	1.6	1.6	44	50
Copper smelter dust	0.2	0.4	40	50
Lead fume	0.003	0.003	25	25
Laboratory oil fume	0	0	0	0

Based on White, H.S. *Industrial Electrostatic Precipitation.* 1963, Reading: Addison-Wesley, p. 64.

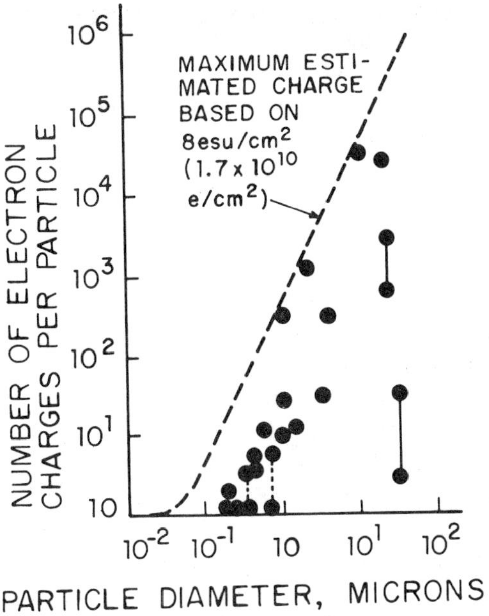

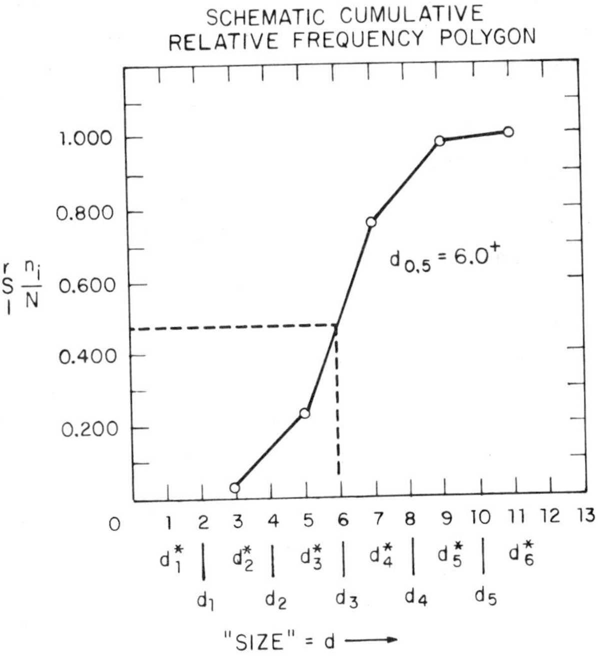

Fig. 5.2 Measured and theoretical single particle charge.[5]

sion of particles to an extent which depends upon the particle size according to the Lambert-Beer law

$$I = I_o \exp(-abcd) \tag{5.1}$$

in which

I_o = incident light intensity,
I = emergent light intensity (parallel),
a = particle extinction coefficient (dimensionless),
b = length of path through dispersoid (cm),
c = concentration of particles (number/cm^3), and
d = particle projected area, cm^2.

The attenuation may be due to both scattering and absorption. The extent of each may be observed by noting the attenuation normal to the main beam and at various other angles. The attenuation which is directly observed is termed the extinction. This light scattering property of particulates is used in a variety of size determination methods.[9,10] On the debit side it is responsible for the poor visibility through fogs and smogs.[11]

*Specific Gravity.** It is an elementary observation that the specific gravity of a particulate is less than that of the solid from which it is formed. How much less can be observed from the data shown in Table 5.2. There is a considerable variance in these specific gravity figures apparently depending upon the methods used to prepare the samples.[5] For example, there is variation in the loose (i.e., apparent) density to an extent that depends upon the treatment to which the packing is subjected.[12] Also there is a variation in the tap density depending upon the method used to mechanically agitate the powder.[13]

Adsorption has already been discussed in detail in Chapter 2. Because adsorption is a surface phenomenon, the large surface area of powders bestows on them large adsorptive power. Thus the higher the specific surface of the powder (surface area per unit volume) the greater is the capacity for adsorption. This adsorptive capability is used in gas adsorption methods for determining powder surface area[14] and also for measuring pore size distribution.[15] Yet another important set of analytical techniques is available in chromatography. There are three basic techniques involving powders: gas liquid partition chromatography, gas-solid chromatography, and liquid-solid chromatography. One useful application of the ability of

*Note: In general, the more complex the particle shape, the larger the surface area, the smaller the specific gravity, the more the light scattering capability and the larger the number of electric charges that can be retained on the surface.

powders to adsorb from solution is exemplified by the clarification of wines.[11]

Table 5.2

PARTICLE DENSITIES OF AGGLOMERATES[a]

Material	*Floc density* (g/cm³)	*True density* (g/cm³)
Silver	0.94	10.5
Mercury	1.70	13.6
Cadmium oxide	0.51	6.5
Magnesium oxide	0.35	3.6
Mercuric chloride	1.27	5.4
Arsenic trioxide	0.91	3.7
Lead monoxide	0.62	9.4
Antimony trioxide	0.63	5.6
Aluminum oxide	0.18	3.7
Stannic oxide	0.25	6.7

[a]From Whytlaw-Gray and Patterson[63]

Cohesion and Adhesion refer respectively to forces acting on particles of the same composition and particles of different composition. In general, the force of adhesion between two particles is

$$F = 2\pi r\sigma \tag{5.2}$$

in which σ is the surface tension and is the surface energy of the water-air interface and r is the radius of two equally sized particles.[16] On a convention force-distance diagram this relationship gives a curve with a minimum at a distance of an atomic diameter. The Van der Waals attractive force being opposed by Born repulsive forces.[17]

In the case of the presence of a liquid phase, iono-electrostatic forces come into play which, unlike Van der Waals forces, fall off rather steeply with distance (see curves 2, 3, 4 in Fig. 5.3). These effects can promote the development of two quite distinct adhesion force levels: for example, in curve 3 of Fig. 5.3, at large interparticle distances, there is a small attractive force F_{min}. Only when an applied force F_{max} overcomes the force barrier does the adhesion force increase to F'_{min}, and this at a much smaller interparticle distance.[17]

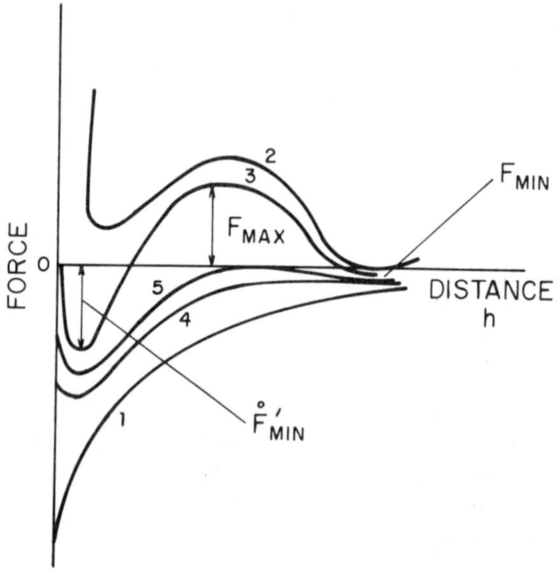

Fig. 5.3 Force diagrams for interaction of two solid spheres in a liquid.

The case in which there is a liquid meniscus separating the powder particles is much more complicated to solve analytically but appears to lead to similar solutions as indicated in Fig. 5.4. The hatched area indicates that in this region the adhesion is vanishingly small. This effect appears to be a major difference between power particle adhesion in the presence of a liquid and powder particle adhesion in the presence of meniscus films.[17]

Chemical Reactivity. The basic requirement for a chemical reaction is that there be chemical contact and this naturally leads one to expect that powders should be particularly reactive when compared with the same material in bulk form. This is borne out in the case of burning by the burning time square law[18]

$$t = Kd_o^2 \tag{5.3}$$

in which

t is the time to complete the burning process,

d_o is the initial particle diameter,
K is the burning constant, and
the atmosphere is infinite. If the oxygen supply is restricted, this condition can be factored into the equation.

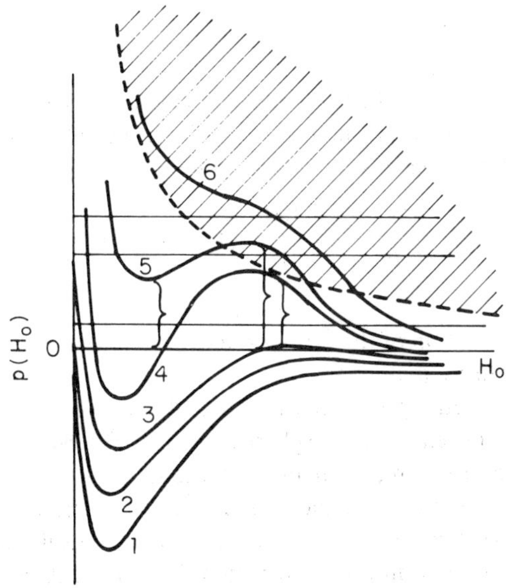

Fig. 5.4 Force diagrams for interaction of two spheres joined by a liquid meniscus.[17]

The reactivity of aerosols can be divided into two considerations: whether a steady state has been achieved or whether reactions vary in time and/or space. In the case of a steady state condition, one can assume that either the reaction rate is controlled by processes within the particle (diffusion, reaction) or that it is controlled by processes in the gas phase. Thus if the internal particle environment controls, then[19]

$$R = c_i(D_i k_r)^{1/2} \tag{5.4}$$

in which
R is the rate of reaction of the dissolved gas with the particle (first order),
D is the diffusion coefficient,
k is the first order rate constant, and
i indicates internal.

If the gas environment controls, then[19]

$$R = \frac{c_e D_e}{r} \tag{5.5}$$

in which

r is the particle radius, and
e indicates external.

In nonsteady state conditions, empirical equations are often resorted to.[20] An equation representing ammonia reacting with sulfuric acid drops when internal diffusion controls the reaction rate is

$$\frac{-d[A]}{dt} \quad k_r \frac{3[B]}{r} (1 - FZ)[A] \tag{5.6}$$

in which

Z is the fraction of the drop reacted,
F is a variable multiplier,
$[A]$ is the reactive gas concentration, and
$[B]$ is the concentration of spherical aerosol particles.

As was observed before, all of these relationships are developed from the most simple assumption of particle sphericity. Variations in shape (and also in surface texture) serve to complicate analytical treatments. The usual procedure is to introduce variations in shape as shape factors. These are discussed later in this chapter.

Motion of Particulates can occur due to a variety of field forces and also due to fluid flows of one sort of another. This wide range of influences and effects is shown in Table 5.3. As an approximate generalization it can be said that the smaller the particulate size, the more important do the nongravity field forces become.

Resistance to motion in a fluid is characterized by the well-known Stoke's law

$$F = 3\pi\gamma V d \tag{5.7}$$

in which

γ is the viscosity of the fluid in g/cm-s, and
d is particle diameter.

The origin of this fluid resistance is the stresses in the fluid ahead as the particle cleaves out its path. For simplicity, a spherical shape is assumed.

As the particle velocity (or its size) increases, inertial effects become more important. As shown in Fig. 5.5 as the Reynolds number increases

Table 5.3[5]

INFLUENCES ON PARTICLE MOTION

A. *Field forces*
- Gravitational
- Electrical
  - Coulombic
  - Image
  - Dielectrophoretic
  - Induced
  - Space charge
- Magnetic

B. *Partial inertia and fluid mechanics*
- Drag
- Centrifugal or vortex flow
- Inertial (fluid flow around submerged object)
- Inertial (particle trajectory relative to fluid)
- Turbulent (convective transport of fluid)
- Shear gradient
- Coriolis

C. *Stochastic processes*
- Diffusion (due to concentration gradient of particle phase)
- Diffusiophoresis (due to concentration gradient of molecular species in gas)
- Thermal gradient

D. *Other factors*
- Photophoretic (photon pressure gradient)
- Sonic (alternating fluid pressure gradient)

to a high value, eddies appear downstream of the particle, eventually separating to produce a wake.[21] This series of events is summarized in the relationship between F and fluid velocity

$$F = \rho C A V^2/2 \, , \tag{5.8}$$

in which

ρ is the fluid density in g/cc,
C is the drag coefficient,

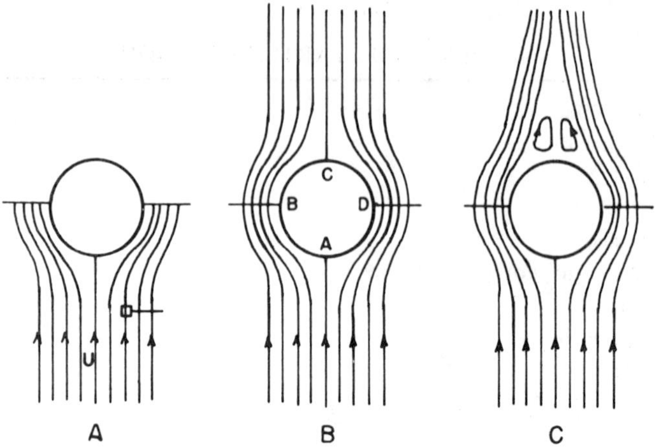

Fig. 5.5 Concepts of fluid resistance. (A) Inertial resistance due to acceleration of fluid elements to one side. (B) Ideal fluid. (C) Real fluid at velocity giving rise to turbulent wake (From *Applied Hydro and Aeromechanics,* Prandit and Tietjens, 1934, with permission of McGraw-Hill Book Co.).[21]

A is the projected area of the particle normal to its motion, and
V is the velocity of the particle relative to the fluid.

The same series of events is graphically represented in Fig. 5.6 as a plot of C, the drag coefficient versus Reynolds number Re where Re equals $dV\rho/\mu$. The points on the curve of Fig. 5.6 represent experimental results. One can summarize all of this by saying that, at low relative velocities, fluid resistance is the main force on the particle. As the relative velocity increases, fluid inertia (turbulence and wake formation) is the main force on the particle.

There are many areas of concern in connection with particle motion all the way from patterns of motion of dust clouds in the stratosphere (and even beyond) to the very personal inquiry as to the motion of particles in our lungs. For the moment this study is restricted to five specific topics of particle motion which are briefly discussed below:

a. gravitational sedimentation,
b. motion in a centrifugal field,
c. motion in an electrostatic field,
d. motion in a nonuniform gas flow, and
e. motion in a thermal gradient.

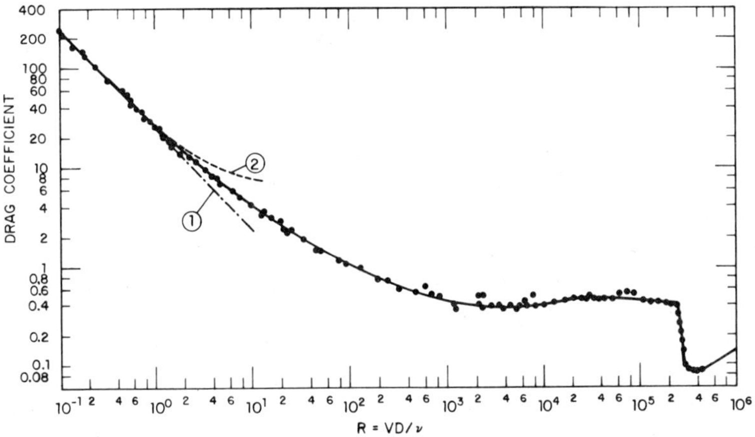

Fig. 5.6 Drag coefficient as a function of Reynolds number for spheres[22] (1) Stoke's law; (2) Oseen's approximation, $C_R = (24/Re)\ [1 + (0.187/Re)]$ (From *Boundary Layer Theory,* by Schlicting, 1960, used with permission of McGraw-Hill Book Co.).

In a still fluid, a particle soon reaches a steady terminal velocity, typical of its 'size' (assumed in the first instance to be spherical in shape). This fact is the basis of the design of gravity sedimentation methods which are widely used in ore beneficiation, waste treatment practice (wet methods), and in the partial cleansing of gases in many industrial applications. Sedimentation is used in a wide variety of test methods for size determination in which particle diameters are related to particle velocity. Correction factors are applied for particle shape variation, particle slip (which occurs when they are so small that they begin to slip between the molecules of the suspending fluid and settle out at a higher velocity than expected), wall effects, fluid viscosity, and others. The experimental methods may be divided into two classes depending upon the way in which the data is obtained: incremental methods and cumulative methods. Elutriation methods are also used in which the grading of the particles is carried out in an upward moving current of air.

When a suspending fluid is constrained to move in a circular path (as in a cyclone) the particles therein experience a centrifugal force which moves the particles to the outer wall of the cyclone to an extent that depends upon their diameter

$$V_{rad} = \frac{d^2 \rho}{18\mu} \cdot \frac{V_{tan}^2}{R} \tag{5.9}$$

in which

V is the particle velocity in radial or tangential direction, and

R is the radius of turn of the path described by the constrained fluid.

The ratio of centrifugal force to gravitational force is termed the separation factor which in dust collection cyclones is not more than 200 (as compared with 5000 in a laboratory centrifuge).

The motion of a particle in an electrostatic field may be influenced in an unexpected manner if the particle possesses a natural charge. However, even if the particle is not charged, it can be polarized by an electric field and, if the field is nonuniform, the resultant force can cause the particle to move.[23] Similarly, a charged particle can create an electric field (image force) when in the vicinity of a polarizable object[24] with a resultant effect on its motion. An important application of motion in an electrostatic field is that of precipitators. As shown in Table 5.4 above 0.2 microns, particles are charged mostly by ion bombardment. Below this size, the major mode of particle charging is via ion diffusion. Because the collecting velocity expressed as

Table 5.4[62]

NUMBER OF CHARGES ACQUIRED BY PARTICLES[a]

Particle diameter (μm)	*Field charging Exposure time* (s) 0.01	0.1	1	∞[b]	*Diffusion charging Exposure time* (s) 0.01	0.1	1	10
0.2	0.7	2	2.4	2.5	3	7	11	15
2.0	72	200	244	250	70	110	150	190
20.0	7200	20,000	24,400	25,000	1100	1500	1900	2300

[a]From Lowe and Lucas[62]
[b]Limiting charge.

$$V = (Ene/3\pi\mu d)C \tag{5.10}$$

in which

E is the electric field strength locally,
n is the number of charges e on the particle, and
C is the slip correction factor,[24]

is proportional to n, the efficiency of these precipitation devices is lower the smaller the particulate dispersion passing through.

Fig. 5.7 illustrates the origin of the separation effected in *inertial impaction devices.* The circled numbers indicate the velocity pattern in the jet. The collection efficiency is controlled by the sharpness of curvature of the fluid stream and also by the velocity of the jet. This is illustrated in Fig. 5.8 for various values of S/W (when S is the distance from the jet to the impaction surface and W is the jet width) in the case of a round jet. Note that in such situations it is usual to use the dimensionless Stokes number (Stk) which is a dimensionless inertial parameter relating the particles stopping distance to some characteristic length of the system. Usually this is

$$Stk = \rho C d^2 U / 9 W \mu \tag{5.11}$$

in which U is the average velocity of the particle in the stream and the other symbols are as defined for Eq. (5.8), et seq.

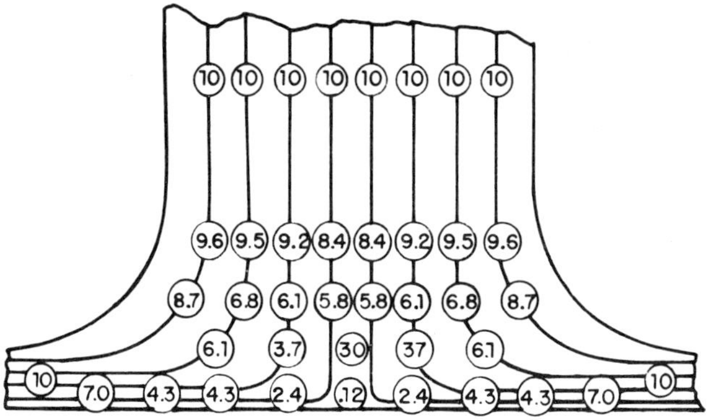

Fig. 5.7 Impingement of a free jet on a flat plate (From Jorgensen, Courtesy Buffalo Forge Co.).[64]

Motion in thermal gradient moves from the hot surface to the cold surface due to the molecular momenta being larger in the case of those coming from the hot surface. The velocity of a specific particle is directly proportional to the thermal gradient

$$V = -\frac{3}{2}\frac{C\mu}{\rho T}\frac{k_f}{2 + k_p}\frac{\Delta T}{\Delta x} \tag{5.12}$$

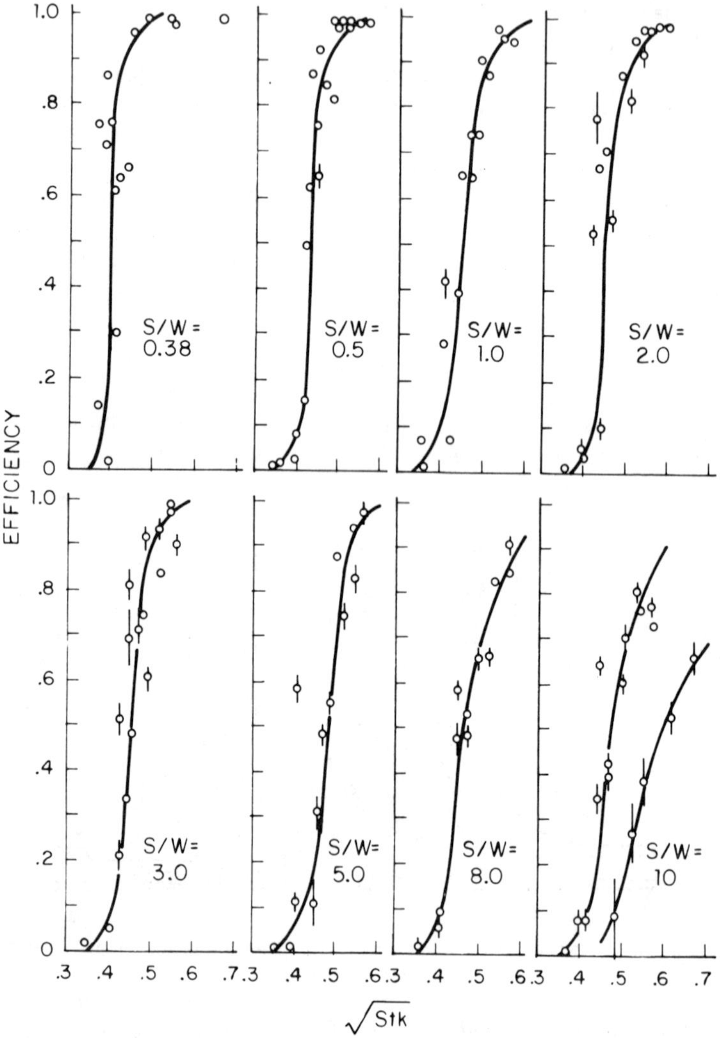

Fig. 5.8 Experimental collection efficiency curves for impaction from round jets (Courtesy of British Occupational Hygiene Society).[61]

in which k indicates thermal conductivities.

The relative importance of these various forces is shown in Fig. 5.9 where it is clear that reduction of particle size has a dramatic effect on the magnitude of all of the forces considered.

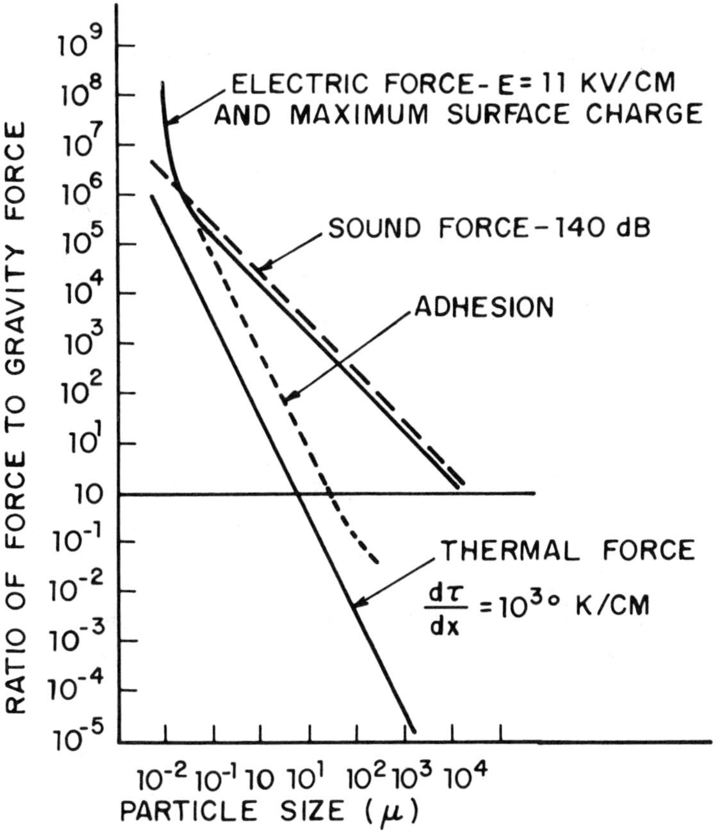

Fig. 5.9 Electrical and other forces that can be exerted on aerosol particles. Courtesy of Academic Press, New York.[7]

Particle interactions have been ignored thus far for the sake of analytical simplicity. However, if the concentration of particles is such that they interact with each other in response to external stimulus, then the physical phenomena become observably different from this first simple picture. For example the set of particles tends to behave more and more as a unit as in the case of the "base surge" effect which accompanies large explosions. In these events, large quantities of particulate material settle out around the main explosion stem.[11] On a smaller scale, in particle size analysis the topic is treated in Chapter 5. On a larger scale in industrial processing, and at higher particulate concentrations, the subject becomes a province of fluidized beds.

Brownian motion is caused by the collision of atoms or molecules with a portion of matter in motion. When the matter is extremely small, the motion it describes is erratic (the drunken walk) but in the case of larger particles, the sharp changes in direction are smoothed out and the path may be described as a tortuous curve. The effect is familiar to us all as the motes dancing in the sunbeam. One area in which it is particularly important is that of the behavior of aerosol particles in filters and lung passages and the like. In these environments, the larger particles are captured by the surfaces by impaction and the smaller ones are entrained by diffusion.

5.3 FUNDAMENTAL STATISTICAL CONCEPTS*

Some basic statistical concepts are fundamental to a generalized development of the methods of "Small Particle Statics." The diagrams and examples do not contain actual particle size data. The synthetic sample data are used in order to permit a simplified presentation of calculation methods.

The basic foundation of statistical analysis is the mathematics of probability.[25,26] Methods of statistical analysis are techniques used to aid interpretation in the analysis of data obtained by repetitive observation. In characterizing a particulate system observations would in general be made on a sample of particles removed from the system. The system is the *population* or total mass of powder whose characteristics are in question. The *sample* is the portion of the population on which measurements are made in order to obtain *estimates* of the characteristics of the population. In order that the sample reliably represent the population, it should be a *random sample;* that is, be chosen so that it has the same probability of being chosen as any other sample. The *statistics* which characterize the population are the variates (numerical quantities) or functions of the variates obtained from the observations made on the sample.

* I am indebted to Homer D. Lewis for permission to freely use and quote his superlative report (see Ref. 2).

Consider a hypothetical system (population) of perfect sodium chloride crystals containing $N = 10$ crystals or particles. Assume that the exact length, (linear dimension d) of the major cube diagonal of each particle as given in Table 5.5 is known.

Table 5.5[2]

HYPOTHETICAL "SIZE" DISTRIBUTION OF $N = 10$ NaCl CRYSTALS

Number of particles	*"Size" = cube diagonal*
n	d_i (microns)
2	4
6	6
2	8

"Size" is defined for this particular system as the dimension d = cube diagonal.

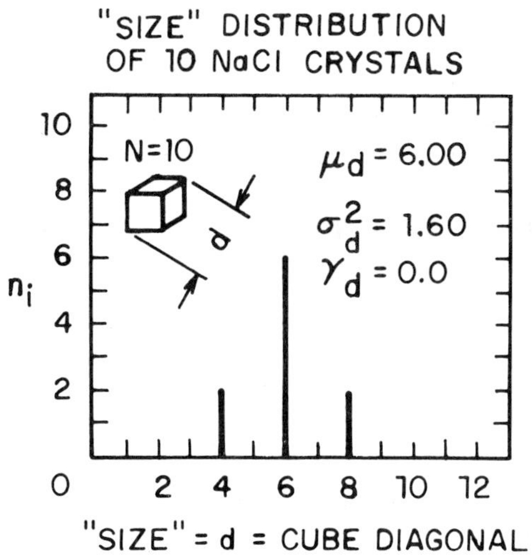

Fig. 5.10 Description of particle "size" distribution by line chart.[2]

One method to describe the particle size distribution is to draw a line chart as shown in Fig. 5.10, but such a chart would not provide a convenient quantitative description of a population consisting of many particles of varied "size." The population distribution of the 10 NaCl crystals could be adequately described by three parameters, the mean μ_d, variance σ_d^2, and skewness γ_d. These parameters are defined as

Mean:
$$\mu_d = \sum_1^N d_i/N \tag{5.13}$$

Variance:
$$\sigma_d^2 = \sum_1^N (d_i - \mu_d)^2/N \tag{5.14}$$

and

Skewness:
$$\gamma_d = \sum_1^N (d_i - \mu_d)^3/N \tag{5.15}$$

For the system of NaCl crystals,

$$\mu_d = (4+4 + 6+6+6+6+6+6 + 8+8)/10 = 6.00,$$

$$\sigma_d^2 = [(4-6)^2 + (4-6)^2 + (8-6)^2 + (8-6)^2]/10 = 1.60, \text{ and}$$

$$\gamma_d = [(4-6)^3 + (4-6)^3 + (8-6)^3 + (8-6)^3]/10 = 0.0.$$

These parameters are also respectively the first moment about the origin, second moment about the mean, and third moment about the mean of the distribution of d.

These parameters are considered respectively as measures of central value, dispersion or variability, and asymmetry about the mean for the population distribution of d. Although, for our purpose, these parameters adequately describe a distribution, it can be argued in general that a distribution is not quantitatively defined unless its moment generating function (i.e., moments of all order) is known.[27] Notice that $\gamma_d = 0$ for the symmetric distribution.

In practice, the population distribution is never known. Therefore consider a more "realistic" example of a sample of $N = 1000$ particles taken from a population (particualte system) of 10^9 perfect NaCl crystals. The N particles in the sample yield many different measured values of size plotted as d = cube diagonal as listed in Table 5.6.

Because the sample contains a large number of particles having varied values of d, presentation and interpretation of the sample data is simplified by grouping or classification into I "size intervals ($d_i^* < d \leqslant d_{i+1}^*$) in which

Table 5.6[2]

SCHEMATIC PARTICLE "SIZE" DATA
FOR SAMPLE OF NaCl CRYSTALS

Interval end point value	*Midpoint value*	n_i	n_i/N
d_i^*	d_i		
1			
	2	6	0.006
3			
	4	224	0.224
5			
	6	540	0.540
7			
	8	220	0.220
9			
	10	10	0.010
11			

$$N = \sum_{1}^{I} n_i = 1000$$

d_i values are the midpoint values of each of the i "size" intervals, $i = 1, 2, 3, \ldots, I$. The midpoint values are defined as

$$d_i = \frac{d_i^* + d_{i+1}^*}{2} \tag{5.16}$$

The values of n_i are the *frequency* or number of particles observed in the ith interval (d_i^*, d_{i+1}^*) and the values n_i/N, when $N = \sum_{1}^{I} n_i$, are the *relative frequency* of occurrence of particles in the ith interval. The observed n_i/N values give an estimate of the probability of occurrence of particles in the population having "size" d in the range $d_i^* < d \leqslant d_{i+1}^*$. In a real analysis of $N = 1000$ particles (or observations), the number of "size" intervals would be greater than five. Cochran[28] provides some useful rules for selecting a proper value of I (number of intervals) with respect to N. These are summarized in Table 5.7.

Methods for graphical display of the data of Table 5.7 are illustrated in Fig. 5.11 and 5.12. A plot of n_i/N vs. d_i^* values as shown by the solid line is a *histogram* or *sample distribution* of d. From this type of graph the

Table 5.7[2]

N	200	400	600	800	1000	1500	2000
I	12	20	24	27	30	35	39

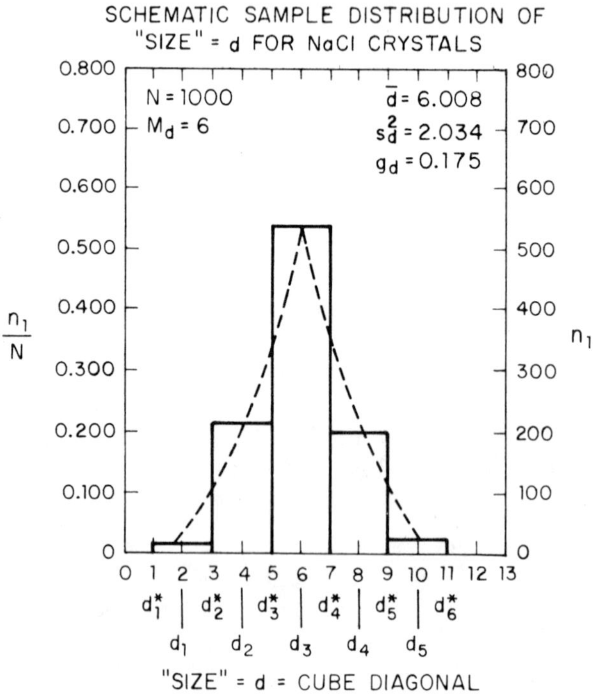

Fig. 5.11 Graphical display of Table 5.6 data. Histogram and relative frequency polygon.[2]

sample M_d, an estimator of the population mode or most probable value of d, can be obtained. M_d is the value of d having the greatest observed value of n_i or can be approximated as d_i of the "size" interval having the greatest value of n_i/N. A plot of n_i/N vs. d_i values, as shown by the dashed line of Fig. 5.11, is called the *relative frequency polygon.*

A plot of the values $\overset{r}{\underset{1}{S}}\, n_i/N$ vs d_r^* would produce the cumulative relative frequency polygon of Fig. 5.12. The values $\overset{r}{\underset{1}{S}}\, n_i/N$ are the total fraction of

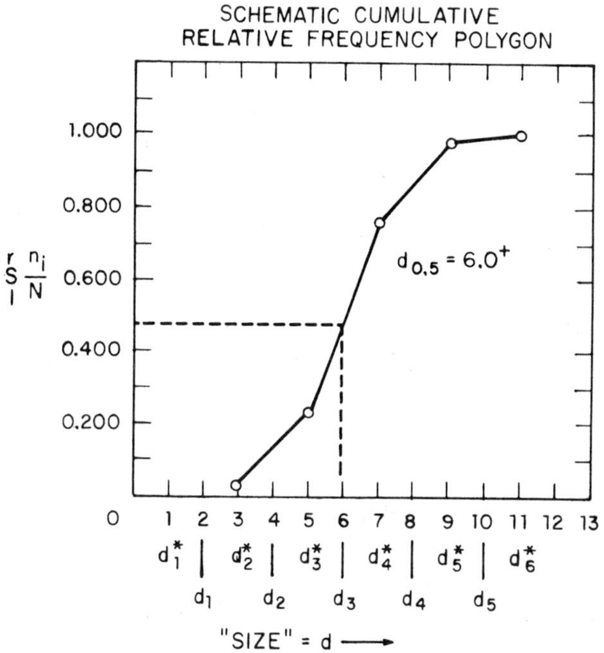

Fig. 5.12 Cumulative relative frequency polygon plot of Table 5.6 data.[2]

occurrence of particles through the rth interval where $i = 1, 2, 3, 4, \ldots, I$. From this graph, estimates can be obtained of the probability of occurrence of particles of "size" less than any particular "size" d in the population distribution. A statistic called the median of d, another measure of central value, can also be obtained. The median is the 0.5 quantile value of d or $d_{0.5} = 6.0^{+}$ for the data of Fig. 5.12.

Values of *mean, variance,* and *skewness* of the sample distribution of d cannot be obtained directly from either of these graphs. These sample statistics are defined as follows:

Mean:
$$\bar{d} = \sum_{1}^{I} \frac{n_i}{N} d_i \tag{5.17}$$

Variance:
$$s_d^2 = \sum_{1}^{I} \frac{n_i(d_i - \bar{d})^2}{N - 1} \tag{5.18}$$

Skewness:

$$g_d = \sum_{1}^{I} \frac{n_i}{N} (d_i - \bar{d})^3 \qquad (5.19)$$

These statistical formulas are related to the population parameters in the following way: $\bar{d} \stackrel{\bullet}{\approx} \mu_d$, $s_d^2 \stackrel{\bullet}{\approx} \sigma_d^2$, and $g_d \stackrel{\bullet}{\approx} \gamma_d$, where $\stackrel{\bullet}{\approx}$ is read as "is an estimator of." The term $N - 1$ is used in the denominator of (5.18) to make s_d^2 an unbiased estimator of σ_d^2.[25] However, because the sample size N should be large, i.e., > 500 in "small particle statistics," s_d^2 can be defined as $s_d^2 \approx \sum_1^I (n_i/N)(d_i - \bar{d})^2$. The values of the sample statistics, for the data of Table 5.5, are listed in Fig. 5.8. Notice the + value of g_d. This positive skewness indicates the sample distribution is asymmetric with $M_d < d_{05} < \bar{d}$; however, it is not proof that $\gamma_d \neq 0$. Negative skewness would indicate $\bar{d} < d_{0.5} < M_d$.

5.4 MEAN DIAMETERS

The size distribution may be represented by a discrete function—a histogram as shown in Fig. 5.11. If the number of particles involved is large and if the size interval is small, the distribution may be considered to be continuous. The "mean diameter" of such a set is expressed by

$$\bar{D} = \Sigma DP(D) \qquad (5.20)$$

For different fields of application it is useful to specify different mean diameters.[30] Thus in general

$$\bar{D}^p = \Sigma D^p P(D)$$

and

$$\bar{D}^q = \Sigma D^q P(D)$$

Therefore

$$\frac{\bar{D}^q}{\bar{D}^p} = \frac{\Sigma D^q P(D)}{\Sigma D^p P(D)} \qquad (5.21)$$

The order of $p + q$ specifies the particular mean diameter which is being investigated as shown in Table 5.8.

The development of the mean of these various $\bar{D}$'s is illustrated with the following example calculations on 245 particles measured in the optical

Table 5.8

VARIOUS MEAN DIAMETERS AND THEIR NOMENCLATURE[29]

Order $(p+q)$	p	q	*Name of* $\bar{D}$ *symbol*		*Where used*	*Other symbols*	*Alternative nomenclature*	
1	0	1	Linear	$\bar{D}_l$	Comparison, evaporation	$\bar{d}_{10}$	$x_{NL} = \frac{\Sigma\, dL}{\Sigma\, dN} = \frac{\Sigma x\, dN}{\Sigma\, dN}$	Number, length mean diameter
2	0	2	Surface	$\bar{D}_s$	Absorption	$\bar{d}_{20}$	$x_{NS} = \frac{\Sigma\, dS}{\Sigma\, dN} = \frac{\Sigma x^2\, dN}{\Sigma\, dN}$	Number, surface mean diameter
3	0	3	Volume, mass	$\bar{D}_v, \bar{D}_m$	Comparison, hydrology atomizing	$\bar{d}_{30}$	$x_{NV} = \frac{\Sigma\, dV}{\Sigma\, dN}^{1/3} = \frac{\Sigma x^3\, dN}{\Sigma\, dN}^{1/3}$	Number, volume mean diameter
3	1	2	Surface-diameter	$\bar{D}_{s-D}$	Absorption	$\bar{d}_{21}$	$x_{LS} = \frac{\Sigma\, dS}{\Sigma\, dL} = \frac{\Sigma x^2\, dN}{\Sigma x\, dN}$	Length, surface mean diameter
4	1	3	Volume-diameter	$\bar{D}_{v-D}$	Evaporation, molecular diffusion	$\bar{d}_{31}$	$x_{LV} = \frac{\Sigma\, dV}{\Sigma\, dL}^{1/3} = \frac{\Sigma x^3\, dN}{\Sigma x\, dN}^{1/3}$	Length, volume mean diameter
5	2	3	Sauter	$\bar{D}_{\text{Saut}}$	Efficiency studies	$\bar{d}_{32}$	$x_{SV} = \frac{\Sigma\, dV}{\Sigma\, dS} = \frac{\Sigma x^3\, dN}{\Sigma x^2\, dN}$	Surface, volume mean diameter
7	3	4	DeBroukere	$\bar{D}_{DeB}$	Combustion, equilibrium	$\bar{d}_{43}$	$x_{VM} = \frac{\Sigma\, dM}{\Sigma\, dV} = \frac{\Sigma x^4\, dN}{\Sigma x^3\, dN}$	Volume, moment mean diameter
							$x_{WM} = \frac{\Sigma\, dM}{\Sigma\, dW} = \frac{\Sigma x\, dW}{\Sigma\, dW} = \frac{\Sigma x^4\, dN}{\Sigma x^3\, dN}$	Weight, moment mean diameter

Table 5.9

SAMPLE DATA FOR PURPOSES OF ILLUSTRATION OR CALCULATIONS[12]

Mean of size group D microns	*Number of particles in each size n*	$n\bar{D}$	$n\bar{D}^2$	$n\bar{D}^3$	$n\bar{D}^4$
2.5	2	5.0	12.5	30.3	75.6
7.5	10	75.0	562.5	4218.8	31640.6
12.5	56	700.0	8750.0	109375.0	1367187.5
17.5	82	1435.0	25112.5	439488.8	7691053.1
22.5	35	787.5	171718.8	398671.9	8970117.2
27.5	22	605.0	16647.5	457806.3	12589672.0
32.5	26	845.0	27462.5	892531.3	29007265.6
37.5	7	262.5	9843.8	369140.6	13842773.4
42.5	5	212.5	9031.3	383828.1	16312696.2
	$\Sigma_n = 245$	$\Sigma_{n\bar{D}} = 5127.5$	$\Sigma_{n\bar{D}^2} = 115{,}141.4$	$\Sigma_{n\bar{D}^3} = 3{,}055{,}091.1$	$\Sigma_{n\bar{D}^4} = 89{,}782{,}481.2$

microscope. The dimension $\bar{D}$ is measured in a *consistent* manner. The raw data is given in Table 5.9. Using this data and the definitions in Table 5.8 the following is calculated:

1. *The linear mean diameter* $\bar{D}_{\text{linear}}$ (or $\bar{d}_{10}$) is

$$\bar{D}_l = \frac{\Sigma n\bar{D}}{\Sigma n}$$

$$= \frac{5127.5}{245} = 20.9\,\mu\,\bar{D}_{\text{linear}} \,.$$

2. *The surface mean diameter* $\bar{D}_{\text{surface}}$ (or $\bar{d}_{20}$) is

$$\Sigma n = \frac{\Sigma n\bar{D}^2}{\bar{D}_s^2}$$

So

$$\bar{D}_s^2 = \frac{\Sigma n\bar{D}^2}{\Sigma n}$$

$$\bar{D}_s = \frac{\sqrt{\Sigma n\bar{D}^2}}{\Sigma n}$$

$$= \frac{\sqrt{115141.4}}{245}$$

So

$$\bar{D}_s = 21.7\,\mu\,.$$

Therefore $[\pi\bar{D}_s^2 \times \Sigma n]$ = total surface area of the spray.

3. *The volume mean diameter* $\bar{D}_{\text{volume}}$ (or $\bar{d}_{30}$) is

$$\Sigma n = \frac{\Sigma n\bar{D}^3}{\bar{D}_v^3}$$

Therefore

$$\bar{D}_v^3 = \frac{\Sigma n\bar{D}^3}{\Sigma n}$$

or

$$\bar{D}_v = \frac{\sqrt[3]{\Sigma n\bar{D}^3}}{\Sigma n}$$

Therefore

$$\bar{D}_v = \frac{\sqrt[3]{3055091.1}}{245}$$

$$= 23.0\ \mu$$

Therefore $[\pi\bar{D}_v^3 \times \Sigma n]$ = total volume of spray.

4. *The Sauter diameter* $\bar{D}_{\text{Sauter}}$ (or $\bar{d}_{32}$) is the volume-surface diameter which if known allows one to calculate the surface area of a particular volume of fluid converted to spray.

$$\bar{D}_{\text{Saut}} = \frac{\Sigma n\bar{D}^3}{\Sigma n\bar{D}^2}$$

$$= \frac{3055091.1}{115141.4}$$

$$= 26.5\ \mu$$

Thus 100 cc of spray with $\bar{D}_{\text{Saut}} = 26.5\ \mu$ has a surface area of

$$\frac{100}{26.5 \times 10^{-4}} \approx 3.6 \times 10^4\ \text{cm}^2$$

5.5 SHAPE FACTORS

Particle size measurements differ depending upon the equipment used to carry out the measurements and upon the shape characteristics of the powder particles. The effect of the equipment is illustrated in Fig. 5.13 which shows "size" defined by three typical measurement methods. Notice that sedimentation methods measure occurrence of particles as weight, w_i', rather than as frequency, n_i. The difficulty in unique definition of "size" has an attendant problem in the definition of particle shape.

As illustrated in Table 5.10 particle "size" can be defined as a linear dimension, an area, volume, or weight, and for a given shape, the concepts of "size" are related by a shape factor. That is, a general shape factor C_k

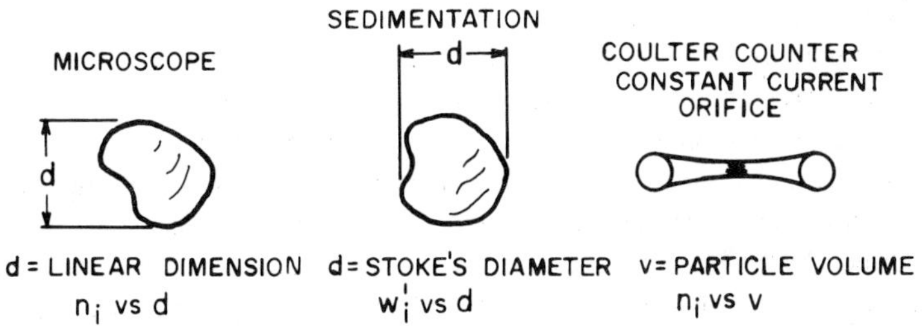

Fig. 5.13 Examples of the relationship between definitions of "size" and the measuring device.[2]

Table 5.10

PARTICLE "SIZE" DEFINITION

"Size" definitions	*Particle shapes*			
	Sphere	*Irregular*	*Regular cubic*	*Regular parallelepiped*
$d =$	Diameter	Maximum dimension	Edge	Edge or length
$v = C_3 d^3 =$	$\frac{\pi}{6} d^3$	$C_3 d^3$	d^3	$\frac{L}{d} d^3$
$a = C_2 d^2 =$	πd^2	$C_2 d^2$	$6d^2$	$2 + \left(\frac{4L}{d}\right) d^2$
$w =$	ρv	ρv	ρv	ρv

relates the kth power of the linear dimension d to some "geometrical" particle property. In general, property

$$Q = \alpha_r d^r$$

For example, $v = C_3 d^3$ and C_3 is a volume shape factor, or $A = C_2 d^2$ and

C_2 is the surface area shape factor.*

This preliminary discussion indicates that shape factors can be defined in many ways, therefore, careful definition of the meaning of any specific shape factor must be made with respect to the application of the term.

In general two distinct types of shape factor have been distinguished: those based upon geometrical characteristics and those based upon aerodynamic characteristics. Finally, shape factors relate to the powder and should not be applied in the case of individual particles, because they are statistical quantities.[24]

Geometrical Shape Factors

The Volume Shape Factor C_3 is shown to be independent of particle diameter in Fig. 5.14.[30] Furthermore it has been shown that C_3 is independent of diameter in the case of coal particles in the range of 3-76 μ[31] and in the case of coal , UO_2, ThO_2, and quartz particles in the respirable size range.[32,33] Some values of C_3 for a variety of particulates and based upon a number of different types of diameters are given in Table 5.11. Theoretical volume shape factors are given in Table 5.12. By comparison with Table 5.11 it appears that actual particulates approximate a regular shape with an axial ratio of 5.[24] It has been observed that C_3 depends upon the method of grinding used to produce the powder in the case of quartz materials.[34,35]

The Surface Shape Factor C_2 is also independent of particulate diameter as shown in Fig. 5.15.[30] A variety of C_2 values is given in Table 5.13. The ratio of C_2/C_3, which is termed the specific surface factor[10] is shown for quartz of various sizes in Table 5.14. This factor is obtained as follows:

The surface area S of n particles is

$$\Sigma\, nC_2d^2$$

---

*Notice the volume and surface shape factors for the idealized cubic particle. Recall the example discussed previously of measurement of cube diagonal as "size" d. In this case, the volume shape factor $C_3 = \dfrac{1}{3\sqrt{3}}$ and the surface area shape factor is $C_2 = 2$. Suppose the elongated "particle" in Table 5.10 has the measure d. An "acicularity ratio" can be defined as L/d. The volume shape factor would also be L/d since $v = (L/d)d^3$. In general, the assumption is usually made that an "average" or equivalent shape, hence C_3 and C_2, can be defined for the particles in a powder, no matter how irregular the particles.

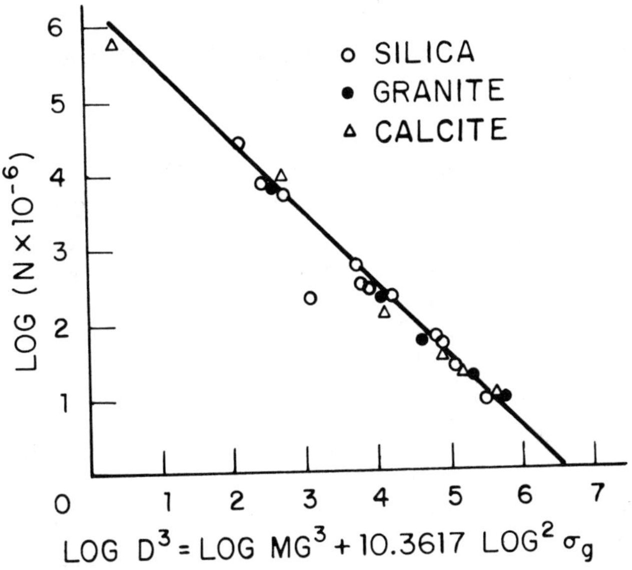

Fig. 5.14 Number of particles per gram as a function of particle diameter[30] D is the diameter of average mass, M_g is the count median diameter, and σ_g is the geometric standard deviation. Courtesy of The Franklin Institute.

The volume V of n particles is

$$\Sigma\, nC_3 d^3$$

The ratio S/V is

$$\frac{C_2}{C_3}\,\frac{\Sigma nd^2}{\Sigma nd^3}$$

Therefore

$$\frac{C_2}{C_3} \text{ is equal to } d_{32}\,\frac{S}{V}\rho \tag{5.22}$$

in which

ρ is the particle density, and
d_{32} is the Sauter mean diameter.

Table 5.11

VOLUME SHAPE FACTORS FOR SEVERAL SUBSTANCES[24]

Substance	Volume shape factor based on: Projected area diameter	Martin's diameter	Feret's diameter	Orientation
Portland cement	0.48[a]	0.48	0.28	stable
Glass	0.35[a]	0.35	0.16	stable
Quartz	0.31[a]	0.31	0.15	stable
Silica	0.27[b]	–	0.14	–
Calcite	0.27[b]	–	0.14	–
Granite	0.27[b]	–	0.14	–
Quartz	0.31[b]	–	0.16	–
Coal	0.21	–	–	stable
Coal	0.25	–	–	?
Quartz	0.21	–	–	?
Quartz	0.29	–	–	not quite random
Quartz	0.34	–	–	random
Coal (Pittsburgh)	0.29[c]	–	–	?
Silica	0.26[c]	–	–	?
Mica	0.10[c]	–	–	?
Fly ash	0.61[c]	–	–	?

[a] Assuming $D_M \approx D_P$

[b] Estimated from D_F value using $D_F/D_P = 1.2$

[c] Volume measured by Coulter counter

Aerodynamic Shape Factors[24]

In general the resistance force on a single particle can be represented by[36–38]

$$F_R = 3\pi\eta d U K_R \tag{5.23}$$

in which

K_R is the aerodynamic shape factor, and

U is the relative velocity.

When d is assumed to be equal to (particle surface area/π)$^{1/2}$, K_R can be calculated for a variety of isometric shapes as shown in Table 5.15.[24] For

Table 5.12

THEORETICAL VOLUME SHAPE FACTORS[24]

Particle shape	*Volume shape factors related to D_P at random orientation for*			
	Isometric particles	Regular shapes of axial ratio 2	5	10
Sphere	0.52			
Cube octahedron	0.45			
Octahedron	0.41			
Cube	0.38			
Tetrahedron	0.29			
Parallelepiped (rect × sq)		0.35	0.29	0.20
Oblate spheroid		0.46	0.26	0.14
Prolate spheroid		0.47	0.33	0.24
Cylinder		0.40	0.30	0.23

Table 5.13

PROJECTED AREA DIAMETER SURFACE SHAPE FACTORS[24]

Substance	*Orientation*	*Size range* (μm)	*Method of surface area measurement*	C_2
Quartz	stable	2–6	Light extinction	2.2
Quartz	?	0.2–10	N_2 adsorption	2.9–3.8
Quartz	stable	0.6–1.8	N_2 adsorption	5.2
Quartz	random	0.6–1.8	N_2 adsorption	5.2
Silica (vitreous)	stable	1.2–4.3	N_2 adsorption	4.7
Silica (tridymite)	stable	0.7–5.2	N_2 adsorption	4.8
Diamond	stable	3–6	Light extinction	1.8
Coal (bituminous)	stable	3.5–9	Light extinction	1.9
Coal (bituminous)	random	0.6–4.3	N_2 adsorption	15.6
Coal (anthracite)	stable	2–10	Light extinction	2.4

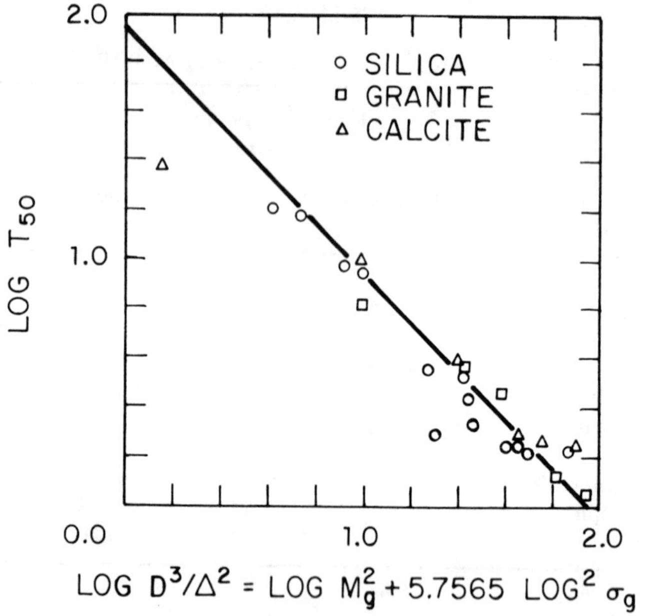

Fig. 5.15 Relative scattered light intensity (∝ area) as a function of particle diameter.[30] D is the diameter of particle of average mass, Δ is the diameter of particle of average surface, M_g is the count median diameter, and σ_g is the geometric standard deviation *(Courtesy of The Franklin Institute)*.

Table 5.14

SOME MEASURED VALUES OF C_2C_3 FOR QUARTZ[24]

Microscope measurement	*Size range*	*Method of surface area measurement*	C_2C_3
D_P	0.2–10 μm	N_2 adsorption	14–18
D_P	0.1–2 μm	N_2 adsorption	34.6
D_F	0.05–0.38 cm	Permeability	6.3

K_R values for other shapes, including chains and clusters consult Reference 24. When the value of d used is the projected area diameter, the shape factor K_{RP} is termed the projected area resistance shape factor and some values of this for various shapes are shown in Table 5.16 and Table 5.17.

Table 5.15

RESISTANCE FORCE RELATIONSHIPS FOR ISOMETRIC PARTICLES[24]

Shape	K_R [a]	$C_R \cdot Re$ [b]	Re
Sphere	1.00	24.0	⩽0.07
Cube-octahedron	0.99	22.9	⩽0.07
Cube	0.98	25.6	⩽0.07
Cube	0.94	24.3	⩽0.1
Cube	0.97	25.4	⩽0.1
Cube-octahedron	0.98	22.8	⩽0.01
Octahedron	0.98	19.4	⩽0.01
Cube	0.97	25.1	⩽0.01
Tetrahedron	0.96	17.0	⩽0.01

[a] $K_R = F_R/3\eta U(\pi S)^{1/2}$; F_R is the experimental resistance force.
[b] $C_R \cdot Re = 8F_R/\eta PU$

When the d value used corresponds to the diameter of a sphere of equal volume, χ is termed the equivalent volume diameter resistance shape factor. Some values of χ are given in Tables 5.16 and 5.17.[24] For K values for a fiber, consult Reference 24 and Table 5.18.

It should be noted that, in the case of the definitions of sphericity and circularity,[39] the aerodynamic shape factors are related in the following way:

$$\text{Sphericity is } \theta = \frac{\text{surface of equivalent volume sphere}}{\text{actual surface of particle}} = \frac{K_{RP}}{\chi}^{2} \tag{5.24}$$

$$\text{Circularity is } \phi = \frac{\begin{array}{c}\text{circumference of circle of area}\\ \text{equal to particle's projected area}\end{array}}{\text{actual perimeter of particle}} = \frac{D_p}{D_F} \tag{5.25}$$

Shape Coefficients and Factors

The various shape factors discussed previously ignore considerations of particle proportions in so far as they can be differentiated from the geo-

Table 5.16

AERODYNAMIC SHAPE FACTORS[a]
FOR PARTICLES OF REGULAR SHAPE[24]

Particle shape	*Aspect ratio*	$\frac{C_{3p}}{K_{RP}}$	K_{RP}	χ
Sphere		0.52	1.0	1.00
Cube octahedron		0.47	0.99	1.03
Octahedron		0.42	0.98	1.06
Tetrahedron		0.30	0.96	1.17
Parallelepiped	0.25	0.27	1.07	1.30
Parallelepiped	4.0	0.24	1.12	1.40
Oblate spheroid	2	0.46	1.00	1.05
Oblate spheroid	5	0.26	0.98	1.24
Oblate spheroid	10	0.15	0.97	1.50
Prolate spheroid	2	0.46	1.01	1.05
Prolate spheroid	5	0.30	1.09	1.27
Prolate spheroid	10	0.19	1.23	1.60

[a] For negligible slip.

metric properties of the particle.[40] The following particle characteristics have been specified and/or derived:

L, B, and T = the limiting dimensions of a particle in decreasing order of magnitude (see Fig. 5.16);

$$n = \text{elongation} = L/B; \tag{5.26}$$

$$m = \text{flatness} = B/T; \tag{5.27}$$

$$\alpha_a = \text{area ratio} = \pi d_a^2/4LB; \tag{5.28}$$

$$p_r = \text{prismoidal ratio} = \text{mean thickness}/T; \tag{5.29}$$

$$\alpha_{v,a} = \text{volume coefficient} = \text{volume of particle}/d_a^3; \tag{5.30}$$

r = rugosity coefficient = perimeter of particle profile, including minor irregularities and corrugations divided by perimeter of smooth curve circumscribing particle profile; and (5.31)

$$\phi_r = \text{circularity including effect of rugosity} = \phi/r\,. \quad (5.32)$$

The volume coefficient $\alpha_{v,a}$ has been shown as[41,42]

$$\alpha_{v,a} = \frac{\pi}{8}^{3/2} \left(\frac{p_r}{\alpha_a^{1/2}}\right)\left(\frac{1}{mn^{1/2}}\right) \quad (5.33)$$

A chart which graphically relates m, n, and the shape factor F (when F equals n/m) is shown in Fig. 5.17.[43]

Table 5.17

AERODYNAMIC SHAPE FACTORS OF VARIOUS SUBSTANCES[24]

Substance	*Range of* D_P, (μm)	C_{3p}/K_{RP}	C_{3p}	K_{RP}	χ
Coal	0.56–4.27	0.26±0.02	0.38±0.03	1.50±0.17	1.88±0.16
	5–15	0.29±0.06[a]	–	–	–
	2.5–12.5	0.27±0.02[a]	–	–	–
(mines)	2.7–25.6	0.27±0.07	–	–	–
(lab)	3.0–29.6	0.19±0.06	–	–	–
	>4	0.28	0.25±0.01	0.9	1.15
Glass	2–10	0.24±0.02	–	–	–
Quartz	0.65–1.85	0.24±0.02	0.35±0.04	1.43±0.2	1.84±0.22
	>4	0.23	0.21±0.01	0.91	1.23
	2–8	0.19±0.06[a]	–	–	–
China clay	2–8	0.20±0.05[a]	–	–	–
Rock (mine)	3.5–12	0.21±0.05	–	–	–
UO_2	0.21–0.63	0.40±0.02	0.34±0.03	0.85±0.27	1.11±0.07
	0.63–1.68	0.27±0.03	0.34±0.06	1.23±0.17	1.60±0.12
	0.21–1.68	0.36±0.06	0.34±0.04	0.95±0.21	1.24±0.24
ThO_2	0.23–0.68	0.31±0.02	0.23±0.03	0.75±0.06	1.19±0.03
	0.68–3.38	0.21±0.05	0.23±0.05	1.14±0.27	1.70±0.32
	0.23–3.38	0.26±0.06	0.23±0.04	0.93±0.27	1.42±0.33
Cotton	2.4–19	0.18±0.02[b]	–	–	–
Asbestos	2.1–25.5	0.17±0.03[b]	–	–	–

[a] Aggregates were not included.
[b] Mixed particles and fibers.

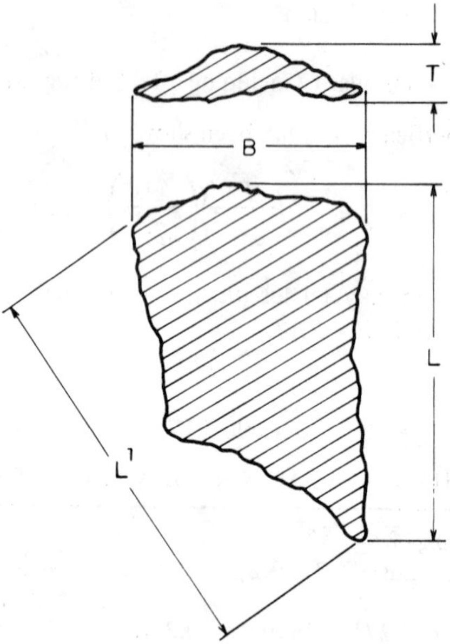

Fig. 5.16 *T, B, L* and *L'* in the Heywood method.[66]

Table 5.18

AERODYNAMIC SHAPE FACTORS FOR FIBERS[a] [24]

Fiber type	*Diameter* (μm)	β = *Length/ diameter*	D_A/D_P	C_{3p}/K_{RP}	K_{RP}
Glass	1.5–8	≅ 2–309	$3.12/\sqrt{\Psi}$	$2.04/\Psi$	$0.385\beta\Psi$
Asbestos					
Amosite	0.6–3	–	3.5	2.56	0.31β
Amosite	< 1	4–200	$2.18\beta^{0.116}$	$0.99\beta^{0.232}$	$0.79\beta^{0.768}$
Crocidolite	0.8–4	–	3.0	1.88	0.42β
Crocidolite	< 1	4–150	$2.19\beta^{0.17}$	$\beta^{0.342}$	$0.78\beta^{0.658}$
Chrysotile	0.8–4	–	2.5	1.31	0.61β

[a] $\Psi = [1 + (0.5/\beta)]^{4.4}$.

[b] Values not corrected for Reynolds numbers.

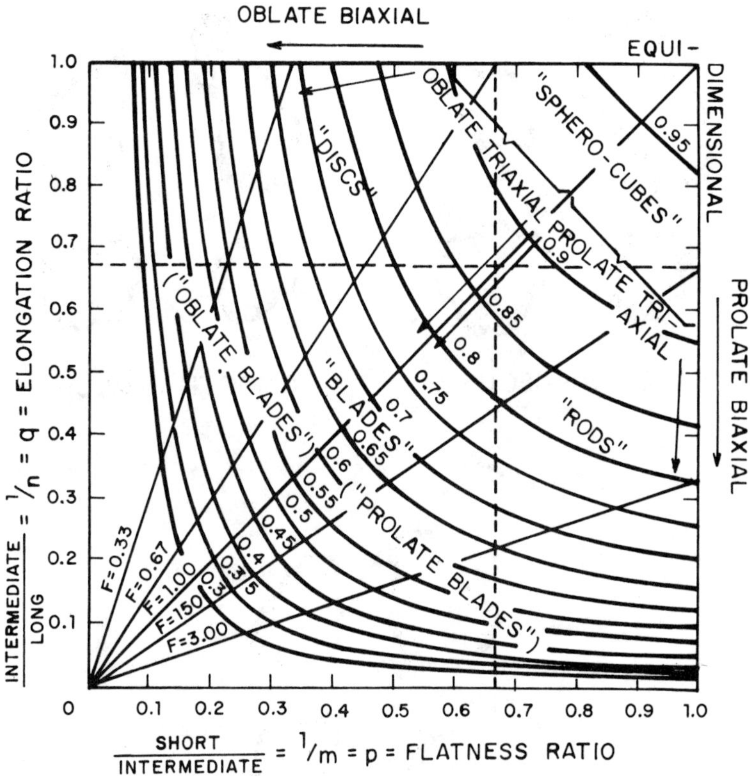

$$\theta = \frac{12.8\sqrt[3]{p^2 q}}{1 + p(1+q) + 6\sqrt{1 + p^2(1+q^2)}} \qquad F = \frac{p}{q}$$

Fig. 5.17 Chart for converting the parameters *T, L,* and *B* into sphericity θ and shape factor *F* where $m = B/T$ and $n = L/B$ (see Fig. 3.16) (after Aschenbrenner, 1956; reproduced with permission from *J. Sediment. Petrol.*).[65]

The utility of the shape coefficients enumerated above is illustrated in the case of the shape analysis of lunar fines sample 12056.72.[40] The three distinct shape types abstracted from the sample are shown in Fig. 5.18. Detailed particle profiles (plan and elevation) are shown in Fig. 5.19. The enumerated shape coefficients in the case of the three particulate types are given in Table 5.19.

A set of results taken from another source[35] shows that in the case of crushed glass (a noncrystalline material that is brittle in response to the crushing forces on it), $1/m$ the flatness ratio, $1/n$ the elongation ratio, and

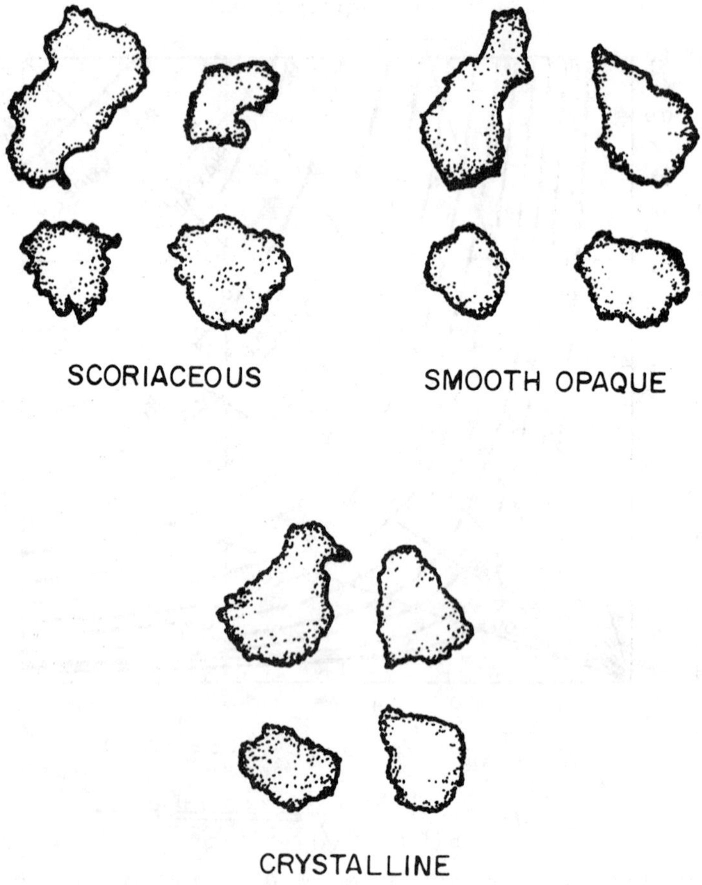

Fig. 5.18 Typical shapes of 700 μm diameter particles.[40]

$\alpha_{v,a}$ the volume coefficient all vary with the size variation of the crushed product. This is shown in Fig. 5.20.[35]

5.6 DISTRIBUTION FUNCTION AND FUNCTIONAL MODEL*

If the measurements of d_i^* recorded in Table 5.6 and Fig. 5.11 could be

*The author is indebted to Homer D. Lewis for permission to freely use and quote his superlative report (see Ref. 2).

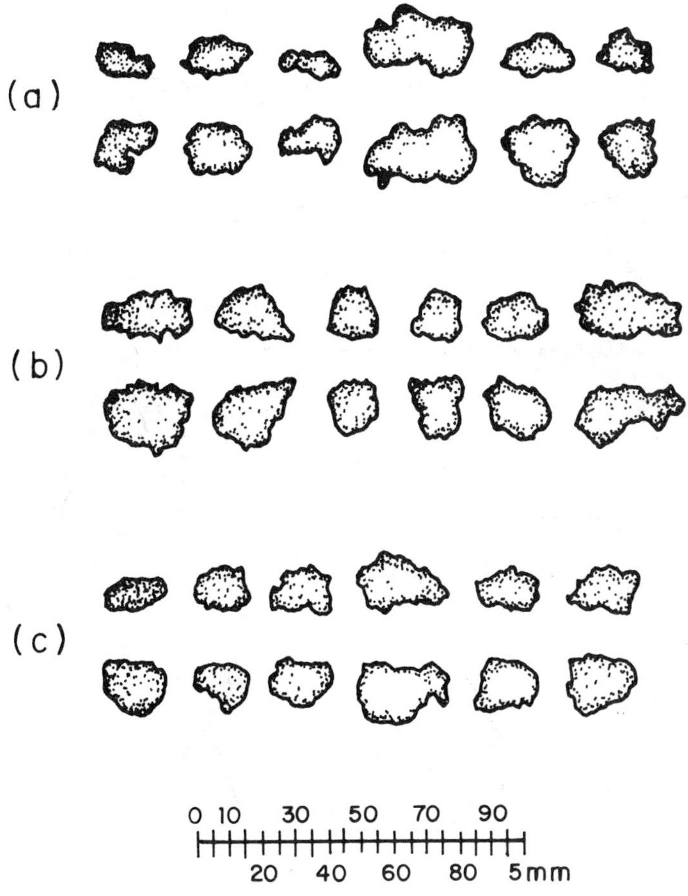

Fig. 5.19 Particle profiles; elevation and plan views (upper and lower rows respectively) (a) Scoriaceous agglomerates (b) Smooth opaque particles (c) Crystalline or transparent particles.[40]

made with accuracy and precision, it would be possible to construct a histogram having many more "size" intervals, and, in the limit, the values of d would be continuous. If some function $h(d)$ could be found that approximated the histogram, $h(d)$ would be a mathematical model for the distribution of d. Such a function is called the *probability density* or *density function.* The density function must have the following properties:

$$h(d) \geqslant 0$$

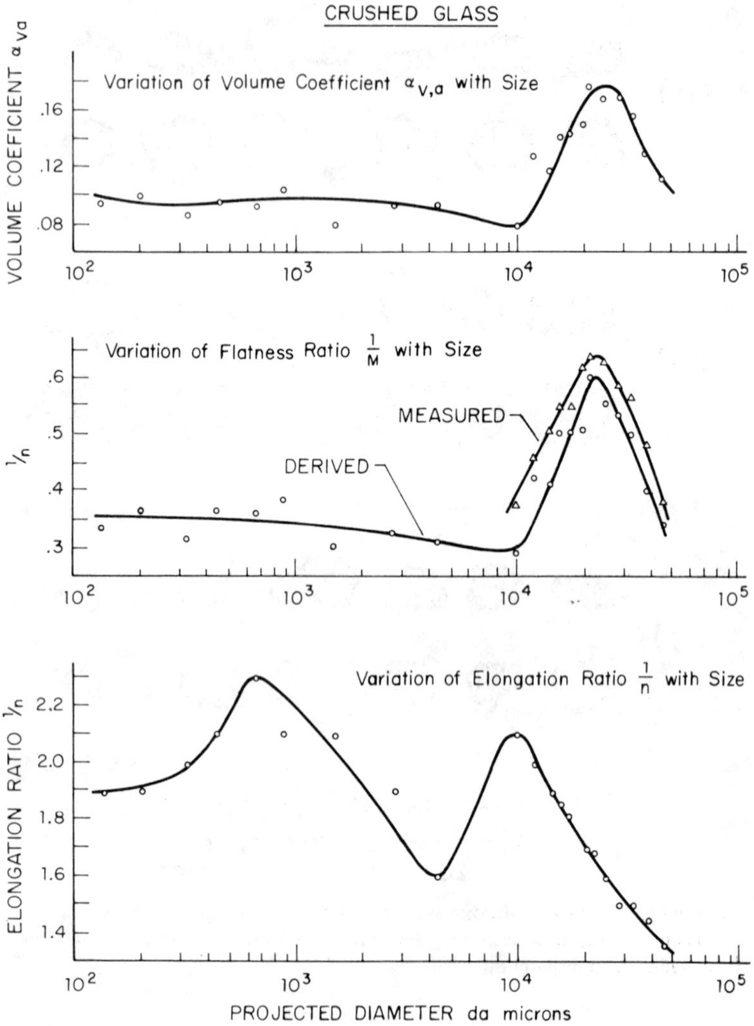

Fig. 5.20 Crushed glass.[35]

$$\int_{-\infty}^{\infty} h(d)\, dd = 1$$

and

$$Pr[d_i^* < d < d_{i+1}^*] = \int_{d_i^*}^{d_{i+1}^*} h(d)\, dd \qquad (5.34)$$

Eq. (5.34) indicates that integration of $h(d)$ over any particular interval gives the probability that the variable d has values in that interval. The use of a functional model to represent statistical data similar to the data discussed implies that

$$\frac{n_i}{N}\,(d_i^*, d_{i+1}^*) = \int_{d_i^*}^{d_{i+1}^*} h(d)\, dd$$

That is, integration of the function over the interval (d_i^*, d_{i+1}^*) gives a fraction approximately equal to n_i/N observed in the same sample data interval as indicated in Figs. 5.21(a) and 5.21(b).

Further, as illustrated in Figs. 5.21(c) and 5.21(d)

$$\overset{r}{\underset{1}{S}} \frac{n_i}{N} \approx H(d) = \int_{-\infty}^{d_{r+1}^*} h(d)\, dd \qquad (5.35)$$

for cumulative data. The histogram and $h(d)$ are drawn to indicate that the model need not be symmetric about some central value.

It is assumed that the normal or Gaussian density function

Table 5.19

SHAPE COEFFICIENTS MEASURED ON 30 PARTICLES 700 μm PROJECTED AREA DIAMETER[40]

Shape coefficient	*Scoriaceous*	*Particle type smooth opaque*	*Crystalline*
n	1.33	1.38	1.32
m	1.37	1.17	1.20
α_a	0.70	0.72	0.74
$\alpha_{v,a}$	0.32	0.35	0.34
r	1.10	1.04	1.04
ϕ^*	0.93	0.93	0.93
$\phi_r{}^*$	0.84	0.89	0.90

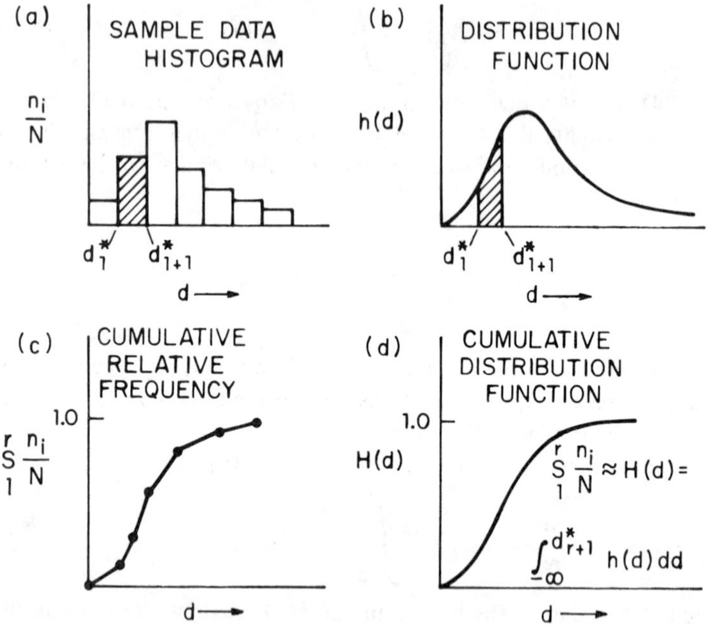

Fig. 5.21 Functional representation of sample data.[2]

$$p(d) = \frac{1}{\sqrt{2\pi}\sigma_d} \exp - \frac{(d - \mu_d)^2}{2\sigma_d^2}$$

having mean μ_d, variance σ_d^2, and $\gamma_d = 0$, represents the data of Fig. 5.11. A particular advantage of using the normal function as the model is that the function is completely defined by μ_d and σ_d^2, hence the sample statistics $\hat{\mu}_d$ and $\hat{\sigma}_d^2$ are the only estimators required to characterize the size distribution of the population. A plot of the cumulative form of the sample data, Fig. 5.12 on normal probability paper[44] would give a straight line, from which first approximations of the sample statistics $\hat{\mu}_d$ and $\hat{\sigma}_d^2$ can be obtained as shown in Fig. 5.22. Probability paper is ruled such that a plot of

$$P(d) = \int_{-\infty}^{d^*_{r+1}} p(d)\, dd$$

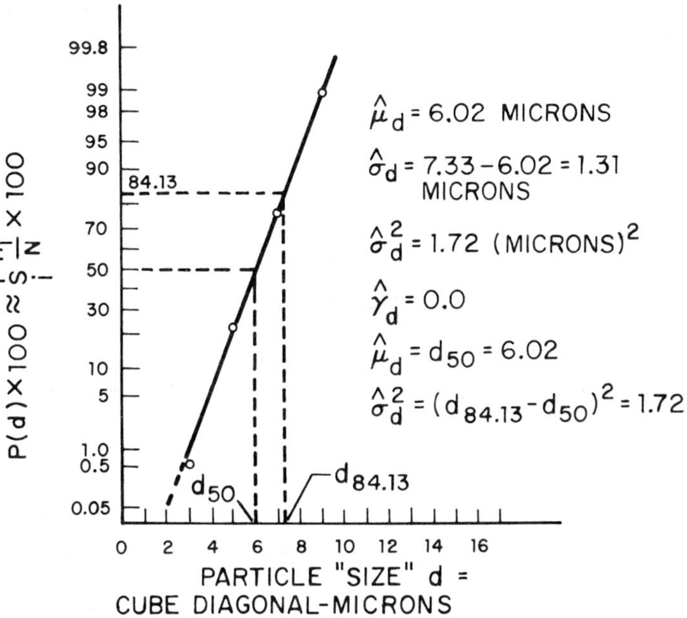

Fig. 5.22 Probability plot of Table 5.6 data.[2]

on the ordinate vs d on the abscissa gives a straight line. Because $p(d)$ is normal and symmetric about μ_d, the mean, median, and mode are identical, and μ_d is defined such that

$$\int_{-\infty}^{\mu_d} p(d)\, dd = 0.5$$

Notice that because the ordinate values, Fig. 5.22, are usually labeled as $P(d) \times 100$, the value of $\hat{\mu}_d = d_{0.5} = M_d$ is the 50th percentile value of d, or d_{50}. The variance $\hat{\sigma}_d^2$ is computed as $(d_{84.13} - d_{50})^3$ since $P(d)$ in the interval $[-\infty, (\mu_d + \sigma_d)]$ is

$$P(d)\,[-\infty, (\mu_d + \sigma_d)] = \int_{-\infty}^{(\mu_d+\sigma_d)=d} p(d)\, dd = 0.8413$$

It is emphasized that the estimate of σ_d^2, $(d_{0.8413} - d_{0.50})^2 = \hat{\sigma}_d^2$ is a property of $p(d)$. This estimate is *only* valid for the normal function and can be made for our data *only* because in plotting on normal probability paper it is *assumed* that the sample data are normally distributed.

To illustrate the origin of the estimate $\hat{\sigma}_d = (d_{0.8413} - d_{0.50})$, consider calculating the value of $P(d)$ in the interval $[(\mu_d - \sigma_d), (\mu_d + \sigma_d)]$. By changing of variable,

$$t = \frac{d - \mu_d}{\sigma_d} \qquad \text{and} \qquad \sigma_d dt = dd$$

then

$$P(t) = \frac{1}{\sqrt{2\pi}} \int_{-1}^{+1} e^{-\frac{t^2}{2}} \, dt$$

$P(t)$ can be expanded as a power series, and it is easily shown that $P(t)$ $(-1, +1) = P(d)\,[(\mu_d - \sigma_d), (\mu_d + \sigma_d)] = 0.68249$. Therefore, in the interval $(\mu_d, \sigma_d), P(d) = 0.34125$, and

$$\int_{-\infty}^{\mu_d + \sigma_d} p(d)\, dd = 0.8413$$

This is, in general, the method by which tables of the normal function are devised.[45,46]

The values of $\hat{\mu}_d$ and $\hat{\sigma}_d^2$ obtained from the model $p(d)$, or values of statistics obtained from any functional model, should be obtained from an objective fitting procedure such as Probit[47] for cumulative data or minimum chi squared[48] for interval data.

5.7 TEST OF STATISTICAL HYPOTHESES (STATISTICAL INFERENCE)*

A statistical hypothesis is the statement of a theory applied to some population. For example, in the case of the "sample data" plotted in Fig. 5.22, the hypothesis is that the data are from a normal population having certain values of μ_d and σ_d^2. A complete test of the hypothesis requires first comparison of the sample data to theoretical values obtained from a hypothetical normally distributed population, and then comparison of the sample mean and variance to the hypothetical mean and variance.

To illustrate the basic idea of testing a statistical hypothesis, the hypothesis that the sample data of Table 5.6 are obtained from a normal popu-

*The author is indebted to Homer D. Lewis for permission to freely use and quote his superlative report (see Ref. 2).

lation, i.e., the population is a normal distribution, will be tested.

The basic procedure for the test of the hypothesis is:[49]

1. state the hypothesis.
2. select a level of significance for the test.
3. compute the appropriate test statistic from experimental data.
4. determine from tabulated theoretical values of the distribution of the test statistic the range of values which will cause acceptance or rejection of the hypothesis.
5. accept or reject the hypothesis on the basis of the computed value of the test statistic.

The test of the hypothesis that the data of Table 5.6 are from a normal population is a chi squared[48,49] goodness of fit test.

Choose a 5 percent level of significance, α, for the test. The choice of $\alpha = 5$ percent implies acceptance of a 0.05 probability of rejecting the hypothesis even if it is true.

The test statistic $\hat{\chi}^2$ has a χ^2 distribution with $(k - 1) - p$ degrees of freedom. The test consists of comparing the observed frequencies, n_i, in the intervals (d_i^*, d_{i+1}^*) of Table 5.6 or Fig. 5.11 with theoretical frequencies expected in the same interval. The test statistic is calculated as

$$\hat{\chi}^2 = \sum_{1}^{I=k} \frac{(n_i - F_i)^2}{F_i} \tag{5.36}$$

in which F_i = theoretical values of n_i, and the number of degrees of freedom $= (k - 1) - p$, in which p = number of parameters estimated for the population. For this example $p = 2$ and the number of degrees of freedom $= (5 - 1) - 2 = 2$.

The normal curve which was *fitted* to the data has the equation

$$N\,p(d) \approx \frac{N}{\sqrt{2\pi}\hat{\sigma}_d} \exp\left[-\frac{(d - \hat{\mu}_d)^2}{2\hat{\sigma}_d^2}\right] \tag{5.37}$$

F_i values are calculated as

$$F_i = \int_{d_i^*}^{d_{i+1}^*} p(d)\,dd$$

which are obtained from normal distribution tables[53] of $z = (d - \mu_d)/\sigma_d$. The computations for χ^2 are illustrated in Table 5.20. Since $\alpha = 0.05$ is chosen, the value $\hat{\chi}^2$ is compared with the theoretical 95th percentile value

Table 5.20[2]

d_i^*	$\dfrac{d_i^* - \hat{\mu}_d}{\hat{\sigma}_d}$	d_i	$P(d)$	F_i	n_i	$\dfrac{(n_i - F_i)^2}{F_i}$
1	−3.82		0			
		2		11	6	2.273
3	−2.30		0.0107			
		4		210	224	0.933
5	−0.77		0.2207			
		6		555	540	0.405
7	+0.76		0.7763			
		8		212	220	0.302
9	+2.28		0.9887			
		10		11	10	0.091
11	+3.81		0.9999			

$N = 1000$

$\hat{\mu}_d = 6.01$

$\hat{\sigma}_d = \sqrt{\hat{\sigma}_d^2} = 1.31$

$$\sum_{1}^{5} \frac{(n_i - F_i)^2}{F_i} = \hat{\chi}^2 = 4.00$$

of χ^2. For a chi-squared test, if $\hat{\chi}^2 < \chi^2$, the hypothesis is not rejected. The value of $\chi^2_{0.95}(2) = 5.99$. Since $\hat{\chi}^2 = 4.00 < 5.99$ the conclusion drawn from the test is that there is no reason to reject the hypothesis that the population distribution is normal. Basic procedures for testing hypotheses concerning the mean and variance are the same as those outlined earlier, however, different test statistics must be used.[49]

5.8 PARTICLE SIZE DATA TYPES I AND II*

Particle "size" data obtained from all possible methods of "size" measurement can be classified into two general types[55] as illustrated in Fig. 5.23. Type I data are sample distribution data, and diagrams of the type shown are histograms. Type II data are sample distribution data only if $k = 0$. Diagrams of this type are called *moment relationships.* In the case of Coulter counter data, "size" is defined as $C_3 d^3 = v$, when v is measured, and fraction of occurrence in the various "size" intervals is measured as n_i/N. These data are called type (3,I) data where 3 denotes the power that

*The author is indebted to Homer D. Lewis for permission to freely use and quote his superlative report (see Ref. 2).

d is raised to. Sedimentation methods measure "size" as d equals equivalent Stokes diameter, and fraction of occurrence as w_i'/W or weight fraction. The weight of particles in an interval (d_i^*, d_{i+1}^*) can be estimated as

$$w_i' \approx \rho C_3 n_i d_i^3 \tag{5.38}$$

and weight fraction as,

$$\frac{w_i'}{W} \approx \frac{\rho C_3 n_i d_i^3}{\sum_1^I \rho C_3 n_i d_i^3} \tag{5.39}$$

hence sedimentation data is of the general Type II or data type (3,II). Data obtained from microscopic count can be considered either Type I or II.

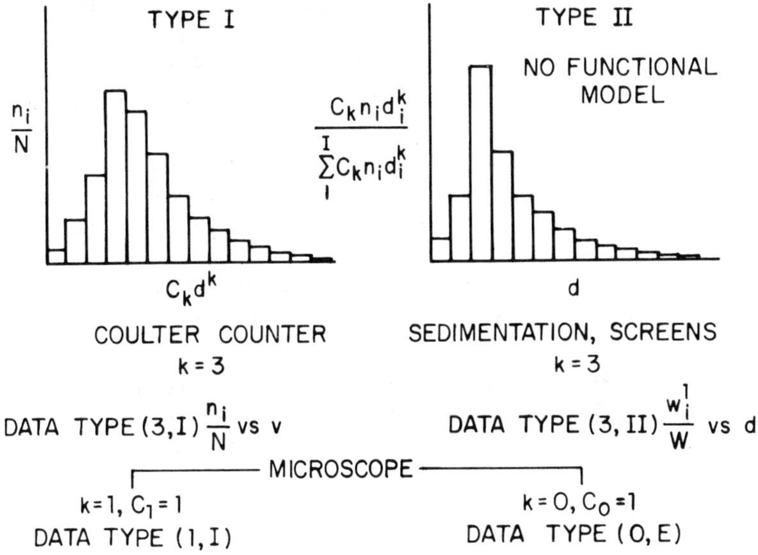

Fig. 5.23 The two general types of sample data.[2]

5.9 CALCULATION OF SAMPLE STATISTICS AND DATA COMPARISON

The Finite Interval Model

If a functional model representing particle "size" sample data is not recognized, sample statistics can be computed in a manner analogous to the methods previously described for calculating mean, variance, and skewness. The definitions of these statistics are given in Table 5.21. The statistics in the first and last columns are estimators of the three parameters of a population distribution, however, those in the middle are related to the distribution parameters only through the approximation $w_i' \approx \rho C_3 n_i d_i^3$. Nevertheless, they are *sample statistics* and do define the sample data. If an instrument measuring weight fraction data were consistently used, it would certainly be legitimate to use these sample statistics for both sample characterization *and* sample comparison. As a matter of fact, the statistics computed directly from the sample give the *better basis for sample comparison* than, for example a $\bar{d}_3$ value computed from Coulter Counter data.

It is possible that there may be a special requirement to compare a $\bar{d}$ value from optical count data with a $\bar{d}$ value obtained from weight fraction or Coulter Counter data. These calculations should be based on the proper definition of midpoint values, endpoint values, and the approximation $w_i' \approx \rho C_3 n_i d_i^3$.

Fig. 5.24 illustrates the effect of transforming, (0, I) data to (3, I) data. The variable endpoint values are transformed by use of the definition $v_i^* = C_3 d_i^{*3}$. Notice the distortion of the interval width on transforming the variable to v.

Fig. 5.25 illustrates the effect of converting relative frequency to weight fraction. Notice that w_i'/W in a particular interval is greater than n_i/N for values of $d > 1$.

The general conversions and transformations for Types I and II data have been developed in detail in Reference 50. To illustrate the method, consider computing $\bar{d}$ from the weight fraction and Coulter Counter data.

In computing $\bar{d}$ from Coulter Counter data, the endpoint transformations are used.

$$d_i^* = \frac{v_i^*}{C_3}^{1/3} \tag{5.40}$$

and the midpoint definition

$$d_i = \frac{\left(\frac{v_i^*}{C_3}\right)^{1/3} + \left(\frac{v_{i+1}^*}{C_3}\right)^{1/3}}{2} \tag{5.41}$$

Table 5.21

	Optical	*Sediment*	*Coulter C.*
Mean:	$\bar{d} = \sum_{1}^{I} \frac{n_i}{N} d_i$	$\bar{d}_3 = \sum_{1}^{I} \frac{w_i'}{W} d_i$	$\bar{v} = \sum_{1}^{I} \frac{n_i}{N} v_i$
Variance:	$s_d^2 = \sum_{1}^{I} \frac{n_i(d_i - \bar{d})^2}{N-1}$	$s_{d_3}^2 = \sum_{1}^{I} \frac{w_i'}{W} (d_i - \bar{d})^2$	$s_v^2 = \sum_{1}^{I} \frac{n_i(v_i - \bar{v})^2}{N-1}$
Skewness:	$g_d = \sum_{1}^{I} \frac{n_i}{N} (d_i - \bar{d})^3$	$g_{d_3} = \sum_{1}^{I} \frac{w_i'}{W} (d_i - \bar{d})^3$	$g_v = \sum_{1}^{I} \frac{n_i}{N} (v_i - \bar{v})^3$

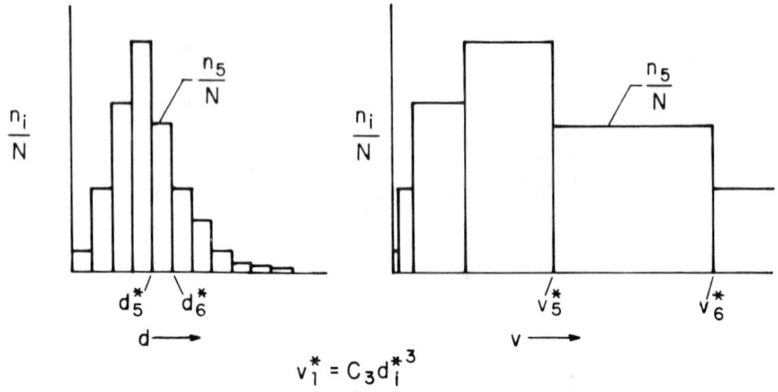

Fig. 5.24 The effect of Type I transforms.[2]

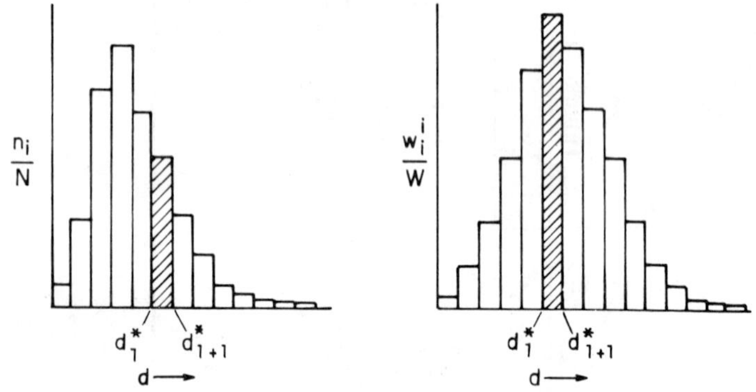

Fig. 5.25 The effect of Type II transforms.[2]

and $\bar{d}$ is calculated as,

$$\bar{d} \approx \sum_{1}^{I} \frac{n_i\left[\left(\frac{v_i^*}{C_3}\right)^{1/3} + \left(\frac{v_{i+1}^*}{C_3}\right)^{1/3}\right]}{2N} \tag{5.42}$$

The calculation of $\bar{d}$ from weight fraction data is slightly more complex. The fraction of occurrence in a given "size" interval is converted by use of the approximations,

$$w_i' \approx \rho C_3 n_i d_i^3$$

and

$$\frac{w_i'}{W} \approx \frac{\rho C_3 n_i d_i^3}{\sum_1^I \rho C_3 n_i d_i^3}$$

assuming C_3 and ρ constant for all intervals.

Since

$$\frac{w_i'}{W} \approx \frac{n_i d_i^3}{\sum_1^I n_i d_i^3}$$

then

$$\frac{\frac{w_i'}{W}}{d_i^3} \approx \frac{n_i}{W}$$

and $\bar{d}$ may be written,

$$\bar{d} \approx \frac{\sum_1^I \frac{w_i}{W}\frac{d_i}{d_i^3}}{\sum_1^I \frac{w_i'}{W}\frac{1}{d_i^3}} \tag{5.43}$$

In Eq. (5.43) the assumption is made that the midpoint value d_i represents the diameter of a particle of average weight in each of the I intervals. This is not necessarily true. In 1910 J. Mellor suggested the following definitions to estimate the average weight $\bar{w}_i$ of a particle in the ith interval:

$$w_i' \approx \int_{d_i^*}^{d_{i+1}^*} \rho C_3 y^3 \, dy$$

then,

$$\bar{w}_i = \frac{\rho C_3 \displaystyle\int_{d_i^*}^{d_{i+1}^*} y^3\, dy}{\displaystyle\int_{d_i^*}^{d_{i+1}^*} dy}$$

and

$$\bar{w}_i = \frac{\rho C_3 \left({d_{i+1}^*}^4 - {d_i^*}^4\right)}{4\,(d_{i+1}^* - d_i^*)}$$

A value of d, related to $\bar{w}_i$ can be defined such that,

$$d_{Mi}^3 = \frac{\bar{w}_i}{\rho C_3}$$

Then

$$d_{Mi} = \left(\frac{{d_{i+1}^*}^2 + {d_i^*}^2\, d_{i+1}^* + d_i^*}{4}\right)^{1/3}$$

and the expression for $\bar{d}$ (from weight fraction data) can be rewritten as

$$\bar{d} = \frac{\displaystyle\sum_1^I \frac{w_i'}{W} \frac{d_i}{d_{Mi}^3}}{\displaystyle\sum_1^I \frac{w_i'}{W} \frac{1}{d_{Mi}^3}} \tag{5.44}$$

By expanding the expressions in Table 5.21 for variance and skewness in d_3 and ν, the same transformations and conversions to get s_d^2 and g_d from these data can be made.

5.10 SUMMARY OF THE METHODS FOR FINITE INTERVAL DATA

The *data* or *statistics computed from data* from different types of measuring devices cannot be directly compared. Sedimentation data or

C.C. can be compared with microscope count data by using proper conversions of fraction of occurrence or transformation of "size" intervals. However, for data comparisons, wherever possible statistics computed directly from sample data should be compared. One should be especially careful of converting n_i/N data to w'_i/W. Computer experiments have indicated[50] a sample of $N = 2500$ may not be adequate.

The Log Normal Model

Many different types of functions have been used to represent particle "size" data. Perhaps the most useful is the log normal function. Development of the use of the log normal function began with the work of Galton and Macallister[51] in 1879. Their work led to the use of the log normal function by Hazen[52] in 1914; Kapteyn[55] in 1916; Wightman[34] 1921–1924; Loveland and Trevelli,[55] 1927; Hatch and Choate,[30] and many others.

Histograms of particle "size" data type (0, II) or (1, I) are in general positively skewed. If the sample distribution variable is defined as

$$x = \ln d \tag{5.45}$$

when $\ln = \log_e$, i.e., "size" is defined as $x = \ln d$, the normal density function,

$$p(x) = \frac{1}{\sqrt{2\pi}\,\sigma_x} \exp\left[-\frac{(x-\mu_x)^2}{2\sigma_x^2}\right] \tag{5.46}$$

is often a good approximation of the sample data. The use of $p(x)$ as a model implies the same approximation method as Eq. (5.35). That is,

$$\frac{n_i}{N}(x_i^*, x_{i+1}^*) \approx \int_{x_i^*}^{x_{i+1}^*} p(x)\,dx \tag{5.47}$$

and the distribution variable is $x = \ln d$. The effect of transforming the variable is illustrated by Fig. 5.26. Notice $p(x)$ is normal, having identical values of mean, median, and mode. The transformation implies the population is log normal in d, and normal in $x = \ln d$.

The use of the log normal function as a model has three basic advantages:

1. the sample data are defined by only two statistics.
2. various data forms are related through functional relationships

among their sample statistics; and

3. "geometric properties" such as particle surface area and volume can be estimated as simple exponential functions.

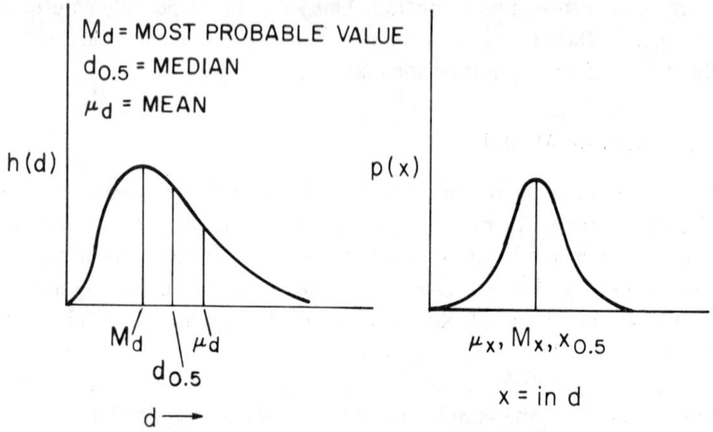

Fig. 5.26 Illustrating the relation between a log normal density function and its normal transform.[2]

5.11 THE GENERAL TYPES OF LOG NORMAL FUNCTIONS*

The various forms of the log normal function can be classified as one of two types analogous to Types I and II data forms, as shown in Fig. 5.27. Consider Type II first and define $f_k(x)$ in terms of $p(x)$, using the approximation

$$\frac{\dfrac{n_i}{N}\, d_i^k}{\sum_1^I \dfrac{n_i}{N}\, d_i^k} \approx \int_{x_i^*}^{x_{i+1}^*} f_k(x)\, dx \tag{5.48}$$

Since

*The author is indebted to Homer D. Lewis for permission to freely use and quote his superlative report (see Ref. 2).

$$x = \ln d, \qquad d^k = e^{kx}$$

and

$$\frac{n_i d_i^k}{\sum_1^I n_i d_i^k} \approx \frac{\int_{x_i^*}^{x_{i+1}^*} e^{kx} p(x)\, dx}{\int_{-\infty}^{\infty} e^{kx} p(x)\, dx} \tag{5.49}$$

After multiplying the exponential terms, completing the square in the integrands, and rearranging constant terms, the expression[30] is written as

$$\frac{n_i d_i^k}{\sum_1^I n_i d_i^k} \qquad \int_{x_i^*}^{x_{i+1}^*} f_k(x)\, dx$$

$$= \frac{\exp k\mu_x + \dfrac{k^2}{2}\sigma_x^2 \sqrt{2\pi}\ \sigma_x}{\exp k\mu_x + \dfrac{k^2}{2}\sigma_x^2 \sqrt{2\pi}\ \sigma_x} \times \frac{\int_{x_i^*}^{x_{i+1}^*} \exp\left[-\dfrac{\left(x-(\mu_x+k\sigma_x^2)\right)^2}{2\sigma_x^2}\right] dx}{\int_{-\infty}^{\infty} \exp\left[-\dfrac{\left(x-(\mu_x+k\sigma_x^2)\right)^2}{2\sigma_x^2}\right] dx} \tag{5.50}$$

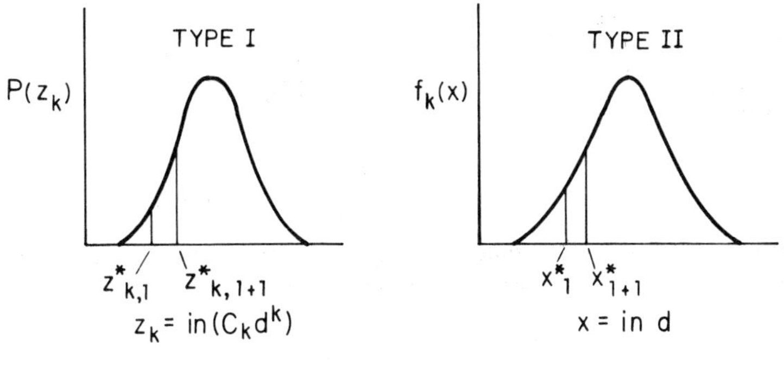

Fig. 5.27 Schematic representation of the general form of Type I and Type II log normal density functions.[2]

This demonstrates the following properties of the function representing Type II data:

1. $\int_{-\infty}^{\infty} f_k(x)\, dx = 1$, this condition is necessary if the integral is to have properties of a density function;
2. $$\int_{-\infty}^{\infty} d^k p(x)\, dx = \exp\left[k\mu_x + \frac{k^2}{2}\sigma_x^2\right] \qquad (5.51)$$
   is the value of the kth moment of the log normal function, $p(x)$;
3. $$f_k(x) = \frac{e^{kx}\, p(x)}{\exp\left[k\mu_x + \frac{k^2}{2}\sigma_x^2\right]} \qquad (5.52)$$
   is the general definition of a kth moment function analogous to the normal density function; and
4. the relationships among the parameters of Type II functions are
   $$\mu_{xk} = \mu_x + k\sigma_x^2 \qquad (5.53)$$
   and
   $$\sigma_{xk}^2 \equiv \sigma_x^2 \qquad (5.54)$$

Now the relationships among the parameters of the Type I functions will be briefly developed. If $z_k = \ln (C_k\, d^k)$, the variable x in Eq. (5.46) can be transformed by use of the expressions $z_k = \ln C_k + kx$, and $dz_k = k\, dx$ to obtain the approximation

$$\frac{n_i}{N}\ \frac{1}{\sqrt{2\pi}\, k\sigma_x} \int_{z^*_{k,i}}^{z^*_{k,i+1}} \exp - \frac{[z_k - (\ln C_k + k\mu_x)]^2}{2k^2\sigma_x^2}\, dz_k$$

$$= \int_{z^*_{k,i}}^{z^*_{k,i+1}} p(z_k)\, dz_k \qquad (5.55)$$

The form of Eq. (5.55) indicates $p(z_k)$ is a normal function of z_k. The relationships among the parameters of the Type I functions are,

$$\mu_{z_k} = \ln C_k + k_{\mu_x} \qquad (5.56)$$

and

$$\sigma_{z_k}^2 = k^2 \sigma_x^2 \tag{5.57}$$

5.12 COMPARISON OF SAMPLE STATISTICS–LOG NORMAL MODEL*

The utility of these relationships is illustrated by Fig. 5.28 and 5.29 which schematically demonstrate the method for obtaining sample statistics from the log probability plot of three typical forms of sample data. These plots are analogous to Fig. 5.22. It is emphasized that the assumption has been made that $\log_e$ "size" is the normally distributed variable. The values of the variances are properties of the appropriate density function as discussed for Fig. 5.22. For example $\hat{\sigma}_x^2 = (x_{0.8413} - x_{0.5})^2$ has no direct relationship to the variance of the sample distribution of d, in fact $\hat{\sigma}_d^2$ is an exponential function of $\hat{\mu}_x$ and $\hat{\sigma}_x^2$ as will be seen later.

It is also emphasized that values of mean and variance should be obtained from a fitting routine such as Probit or Chi-Squared Minimum.

The sample statistics $\hat{\mu}_x$, $\hat{\sigma}_x^2$, $\hat{\mu}_{x3}$, $\hat{\sigma}_{x3}^2$, $\hat{\mu}_{z_3}$, and $\hat{\sigma}_{z_3}^2$ are the logical statistics to use for sample characterization comparison. The relationships among the distribution parameters given in Eqs. (5.53), (5.54), (5.56), and (5.57) are the basis for comparing the sample statistics obtained from different data forms, i.e., measurement methods. The sample statistics are related as follows:

$$\hat{\mu}_{x3} \approx \hat{\mu}_x + 3\hat{\sigma}_x^2 \tag{5.58}$$

$$\hat{\sigma}_{x3}^2 \approx \hat{\sigma}_x^2 \tag{5.59}$$

$$\hat{\mu}_{z_3} \approx \ln C_3 + 3\hat{\mu}_x \tag{5.60}$$

and

$$\hat{\sigma}_{z_3}^2 \approx 9\hat{\mu}_x^2 \tag{5.61}$$

If for some reason it is necessary to examine the statistics of the distribution of "size" other than ln "size," the expressions of Table 5.21 are used as the basic definitions with the proper function approximating the fraction of occurrence. For example, to compute the values of $\bar{d}$, s_d^2, and g_d, the approximation

*The author is indebted to Homer D. Lewis for permission to freely use and quote his superlative report (see Ref. 2).

$$\frac{n_i}{N} \approx \int_{x_i^*}^{x_{i+1}^*} p(x)\,dx$$

is used, the expressions of Table 5.21 are expanded where necessary, and the following estimators are obtained.

$$\bar{d} \approx \exp\,[\hat{\mu}_x + 0.5\hat{\sigma}_x^2] \tag{5.62}$$

$$s_d^2 \approx \exp\,[2\hat{\mu}_x + 2\hat{\sigma}_x^2] - \bar{d}^2 \tag{5.63}$$

and

$$g_d \approx \exp\,[3\hat{\mu}_x + 4.5\hat{\sigma}_x^2] - 3\exp\,[3\hat{\mu}_x + 2.5\hat{\sigma}_x^2] + 2\bar{d}^3 \tag{5.64}$$

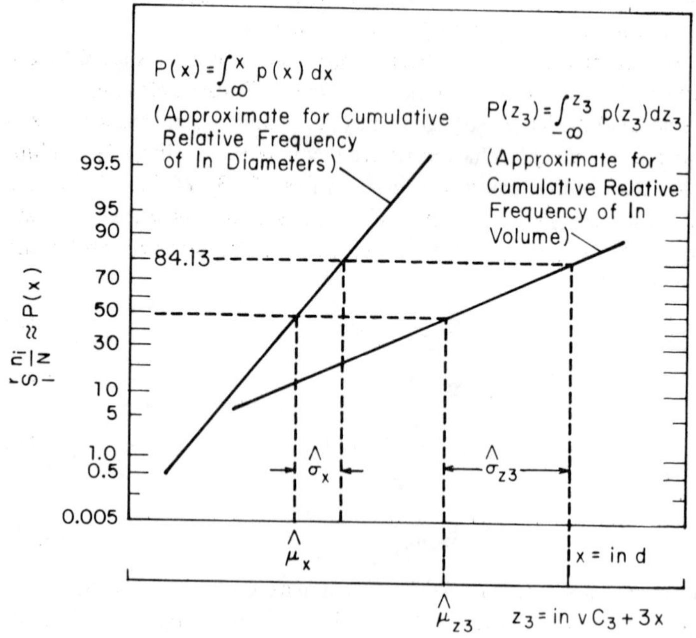

Fig. 5.28 Log probability plot–Type I.

The mode and median are obtained as

$$M_d \approx \exp\,[\hat{\mu}_x - \hat{\sigma}_x^2] \tag{5.65}$$

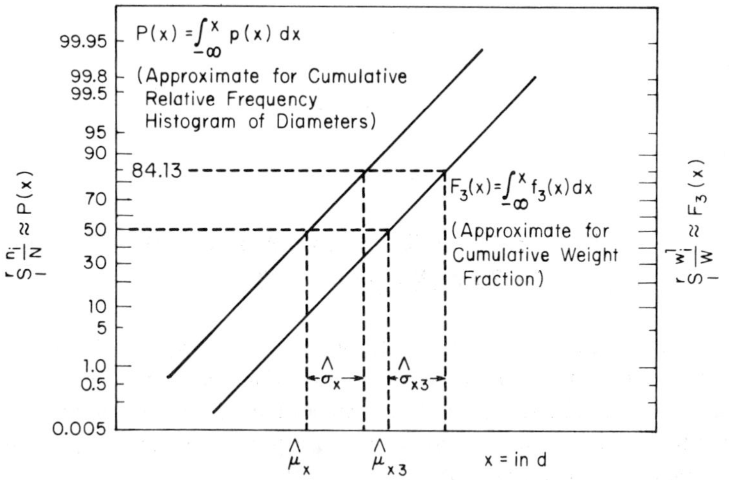

Fig. 5.29 Log probability plot–Type II.

and

$$d_{0.50} \approx e^{\hat{\mu}_x} \tag{5.66}$$

By using Eqs. (5.58), (5.59), (5.60), and (5.61), these statistics can be obtained from any data form. For example, an estimate of $\hat{\mu}_d \approx \bar{d}$ from Coulter Counter data is obtained as

$$\bar{d} \approx \left(\frac{1}{C_3} \exp \hat{\mu}_{z_3} + \frac{0.5}{3} \hat{\sigma}^2_{z_3} \right)^{1/3}$$

and from weight fraction data,

$$\bar{d} \approx \exp \hat{\mu}_{x3} - 2.5\hat{\sigma}^2_{x3}$$

Expressions for $\bar{d}_3$, $s^2_{d_3}$, g_{d_3}, $\bar{\nu}$, s^2_{ν}, and g_{ν} are obtained in the same manner as are Eqs. (5.62), (5.63), and (5.64). For example, in the $\bar{d}_3$ exp $\hat{\mu}_{x3} + 0.5\hat{\sigma}^2_{x3}$ and $\bar{\nu}$ exp $\hat{\mu}_{z_3} + 0.5\hat{\sigma}^2_{z_3}$ discussion, inequalities of the following type are apparent:

$$\bar{d} \neq e^{\hat{\mu}}x$$

$$s^2_d \neq e^{\hat{\sigma}^2}x$$

$$\bar{d}_3 \neq \bar{d}$$

$$s_{d_3}^2 \neq s_d^2$$

and

$$\bar{d} \neq \frac{\bar{v}}{C_3}^{1/3}$$

5.13 SURFACE AREA AND SPECIFIC SURFACE CALCULATIONS*

Specific surface is one of the characteristics of a powder often considered in attempts at process behavior correlation. S_W can be measured by adsorption techniques (BET)[56] or estimated from the particle "size" distribution. Such estimates assume nonadsorptive "smooth shell particles." Comparison of calculated values of S_W with BET values can give some insight into the character of the particle surface.

To illustrate the computation of S_W as a sample statistic,[50] consider calculating S_W from weight fraction data for the finite interval model and from microscopic count data for the log normal model. For the finite interval model the approximation $a_i' \approx C_2 n_i d_i^2$ is used to define S_W as

$$S_W \approx \frac{\sum_1^I a_i'}{W} = \frac{C_2}{\rho C_3} \sum_1^I \frac{w_i' d_i^2}{W d_{Mi}^3} \tag{5.67}$$

Notice that this expression implies a knowlege of C_2 and C_3. A better estimate of S_W is obtained if a value representing the diameter of a particle having "average" surface area in the interval is used rather than d_i in the numerator of Eq. (5.67). This value of d is defined as[50]

$$d_{Li} = \left[\frac{(d_{i+1}^*)^2 + (d_{i+1}^*)(d_i^*) + (d_i^*)^2}{3}\right]^{1/2} \tag{5.68}$$

Calculation of S_W by this method is laborious if I is large enough to adequately define the data. If appropriate, the use of the log normal model greatly facilitates the calculation. For example, from the microscopic count (0, II) data and with the definition of the general moment function,[54]

*The author is indebted to Homer D. Lewis for permission to freely use and quote his superlative report (see Ref. 2).

the approximation of S_W can be written as

$$S_W \approx \frac{NC_2}{\rho NC_3} \frac{\int_{-\infty}^{\infty} d^2 p(x)\, dx}{\int_{-\infty}^{\infty} d^3 p(x)\, dx} = \frac{C_2}{\rho C_3} \exp\left[-(\hat{\mu}_x + 2.5\hat{\sigma}_x^2)\right] \quad (5.69)$$

Notice that Eq. (5.69) shows that $S_W \not\approx C_2/\rho C_3 \bar{d}$. From the relationships among sample statistics for various data forms, S_W can be easily estimated from any type of data as illustrated by Table 5.22.[50]

5.14 OTHER DISTRIBUTIONS

The normal function often has a small particle tail and is not symmetrical about the mean. This was once thought to be due to inaccuracies in small particle size determination. It is now known to be a true effect. The normal distribution becomes less skewed and more or less symmetrical when the log normal representation is used. In this distribution, the variables defining the population are the log geometric standard deviation:

$$= \frac{\left[\Sigma (n \log \bar{D} - \log \bar{D}_G)^2\right]^{1/2}}{\Sigma n} \quad (5.70)$$

and the log mean diameter:

$$= \frac{\Sigma\, n \log \bar{D}}{\Sigma\, n} \quad (5.71)$$

One of the problems associated with normal and log-normal type plots is that the points deviate from the straight line at the two tails of the distribution. Fortunately in most cases the error is small because the number of particles involved is small. However, not all populations can be so readily collated and other distributions have been described in the literature.

The Hatch-Choate Distribution[57]

The Hatch-Choate Distribution[57] for an assymetrical population, combining the established relationship for D with the log-normal distribution gives:

Table 5.22[2]

FUNCTIONAL ESTIMATES OF PARTICLE SURFACE AREA, VOLUME, AND WEIGHT

Property Estimated	Measuring Devices		
	Microscope (relative frequency of d)	*Micromerograph, Sedibal, etc. (weight fraction of d)*	*Coulter Counter (relative frequency of v)*
N	Measured	$\frac{W}{\rho C_3} \exp[-(3\hat{\mu}_{x_3} - 4.5\,\hat{\sigma}_x^2)]\ 10^{-12}$	Measured
$A(\text{cm}^2)$	$NC_2 \exp(2\hat{\mu}_x + 2\hat{\sigma}_x^2)\ 10^{-8}$	$\frac{WC_2}{\rho C_3} \exp[-(\hat{\mu}_{x_3} - .5\,\hat{\sigma}_x^2)]\ 10^4$	$\frac{NC_2}{C_3^{2/3}} \exp\left[\frac{2}{9}(3\hat{\mu}_{z_3} + \hat{\sigma}_{z_3}^2)\right]\ 10^{-8}$
$V(\text{cm}^3)$	$NC_3 \exp(3\hat{\mu}_x + 4.5\,\hat{\sigma}_x^2)\ 10^{-12}$	$\frac{W}{\rho}$	$N \exp(\hat{\mu}_{z_3} + .5\,\hat{\sigma}_{z_3}^2)\ 10^{-12}$
$W(\text{g})$	$\rho NC_3 \exp(3\hat{\mu}_x + 4.5\,\hat{\sigma}_x^2)\ 10^{-12}$	Measured	$\rho N \exp(\hat{\mu}_{z_3} + .5\,\hat{\sigma}_{z_3}^2)\ 10^{-12}$
$s_v(\text{cm}^{-1})$	$\frac{C_2}{C_3} \exp[-(\hat{\mu}_x + 2.5\,\hat{\sigma}_x^2)]$	$\frac{C_2}{C_3} \exp[-(\hat{\mu}_{x_3} - .5\,\hat{\sigma}_x^2)]\ 10^4$	$\frac{C_2}{C_3^{2/3}} \exp\left[-\frac{1}{18}(6\hat{\mu}_{z_3} + 5\hat{\sigma}_{z_3}^2)\right] 10^4$
$s_w(\text{cm}^2/\text{g})$	$\frac{C_2}{\rho C_3} \exp[-(\hat{\mu}_x + 2.5\,\hat{\sigma}_x^2)]\ 10^4$	$\frac{C_2}{\rho C_3} \exp[-(\hat{\mu}_{x_3} - .5\,\hat{\sigma}_x^2)]\ 10^4$	$\frac{C_2}{\rho C_3^{2/3}} \exp\left[-\frac{1}{18}(6\hat{\mu}_{z_3} + 5\hat{\sigma}_{z_3}^2)\right] 10^4$

Units: d in microns, ρ in g/cm^3

$$\begin{aligned}
\log \bar{D} &= \log \bar{D}_G + 1.151 \log_G^2 \\
\log \bar{D}_s^2 &= \log \bar{D}_G^2 + 4.605 \log_G^2 \\
\log \bar{D}_v^3 &= \log \bar{D}_G^3 + 10.362 \log_G^2
\end{aligned} \tag{5.72}$$

Thus once the geometric mean diameter and the geometric standard deviation are known, the diameters which approximate the average diameter of the particles (D), the diameter of the particle of average surface area (D_s), and the diameter of the particle of average volume (D_v) can be calculated.

The Rosin-Rammler Distribution[58]

This expression was developed to plot the grinding products of friable materials such as coal:

$$n = 100\, e^{(-d/c)^b} \tag{5.73}$$

in which n is the percent of particles with the diameter more than d, c is called the Rosin-Rammler mean, and b is the dispersion constant. If b is high, the particles are closely grouped. If small, a wide range of particle sizes exists. A log-log plot gives a straight line when log $100/n$ versus d on log-log paper is plotted. The Rosin-Rammler mean c (d_{RR} on graph) is the diameter above which lies 36.8 percent of the spray volume. Therefore, log $100/36.8 = 0.43 = c$ having a value of ~ 190. From the graph b, the slope of the line, can be determined (Fig. 5.30).[59]

The Nukiyama-Tanasawa Distribution[60]

$$\log (1/d^2 \times N/\Delta d) = \log K_1 - (K_2 d)^{K_3}/2.303 \tag{5.74}$$

in which

- d is the droplet diameter,
- N is the number of droplets in the diameter range d,
- K_1 is the constant calculated from the equation,
- K_2 is the slope of line $\times$ 2.303 = constant, and
- K_3 is the constant determined experimentally for each nozzle.

To use this equation, count the droplets in two size ranges and by substitution obtain a complete distribution of droplet sizes.

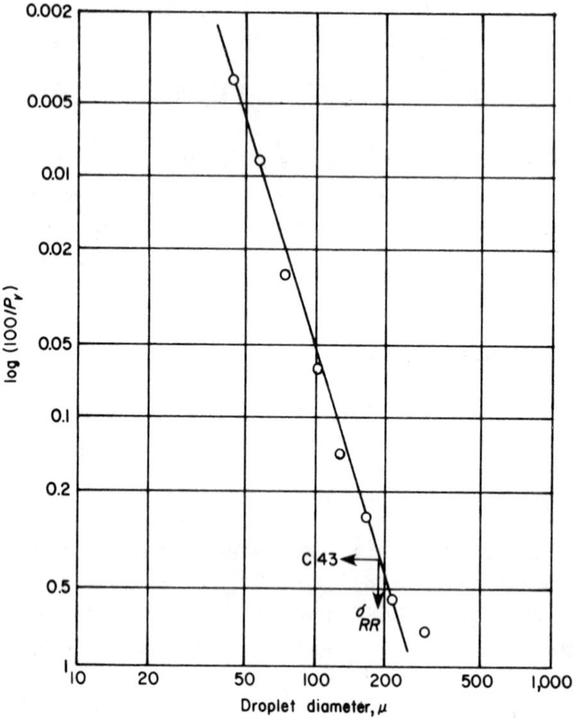

Fig. 5.30 Fan-spray droplet distribution plotted in accordance with Eq. (5.73) [After R. P. Fraser and P. Eisenklam, "Liquid Atomisation and the Drop Size of Sprays," *Trans. Inst. Chem. Engrs.* (London), 34 (1956): 304.[59]

5.15 CHAPTER NOTATIONS AND DEFINITIONS

A Partial List of Symbols and Definitions[2]

Powder (particulate material). A set of particles of matter in some environment.
Particle. A unit of matter having a defined boundary and whose "size" and "shape" depend on interatomic or intermolecular bonding forces and the process of formation.
Population. A population of powder particles is the main or total mass of powder whose properties or characteristics are in question.
Population parameters. Constants which define the population distribution.
Sample. A portion of the population upon which measurements are performed in order to obtain *estimates* of the characteristics of the population.
Statistics. The variates (numerical quantities) or functions of the variates obtained from measurements made on a sample.
Random sample. A sample chosen by a method such that it has the same probability of being selected as any other sample.
d. A linear particle dimension characteristically measurable for all particles in the population of particles.
k. An integer = 0, 1, 2, 3, . . . , determined by a particular defined geometric property of the particle.
C_k. Shape Factor; a coefficient constant relating the kth power of d to a geometric property of the particle (e.g., for $k = 2$, spherical particles, $C_2 = \pi$, and $C_2 d^2$ = surface area of the particle).

(cont'd.)
Weight Factor. ρC_3 = volume shape factor multiplied by particle density.
a. The surface area of an individual particle.
v. The volume of an individual particle.
W. The weight of an individual particle.
"Size." A measured particle characteristic that may be a linear dimension, d; an area, a; a volume, v; or a weight, w.
$d_i^* < d_{i+1}^*, a_i^* < a_{i+1}^*, v_i^* < v_{i+1}^*, w_i^* < w_{i+1}^*$. The range of sizes in the ith "size" intervals; $i = 1, 2, 3, \ldots, I$, when I = total number of intervals.
d_i, a_i, v_i, w_i. The "midpoint" values of the respective i^{th} "size" intervals.
a_i', v_i', w_i' . The total area, volume, or weight respectively in the ith "size" interval.
n_i . The *frequency* (i.e., number) of particles occurring in the ith "size" interval.
N . $\sum_1^I n_i = n_1 + n_2 + n_3 \ldots n_I$: Total number of particles in I size intervals.
$\frac{n_i}{N}$. Relative frequency of occurrence of particles in the ith "size" interval. (Note: $\Sigma \frac{n_i}{N} = 1$)

(cont'd.)
$A = \sum_1^I a_i' \approx \sum_1^I n_i C_2 d_i^2 .$ The total surface area of all the particles in I size intervals.
$\frac{a_i'}{A}$. The relative surface area or surface area fraction of particles in the ith "size" interval.
$V = \sum_1^I v_i' \approx \sum_1^I n_i C_3 d_i^3$. The total volume of all the particles in I "size" intervals.
$\frac{v_i'}{V}$. The relative volume or volume fraction of particles in the ith size interval.
ρ . Particle density, weight/unit volume.
$W = \sum_1^I w_i'$. The total weight of all the particles in I "size" intervals. $W \approx \sum_1^I \rho n_i C_3 d_i^3$
$\frac{w_i'}{W}$. Relative weight or weight fraction of particles in the ith "size" interval.
$\underset{i=1}{\overset{r}{S}} \frac{C_k n_i d_i^k}{\sum_1^I C_k n_i d_i^k}$. The general expression for the cumulative fraction less than "size" for a fractile diagram, or cumulative frequency polygon, where r refers to the rth "size" interval through which the partial sum is computed. For example, if $k = 0$, then $\underset{1}{\overset{r}{S}} \frac{n_k}{N}$

(cont'd.)

is the partial sum of all values of n_i/N of a relative frequency histogram for $i = 1, 2, 3, \ldots, r$, and r may take any value of $i = 1, 2, \ldots, I$. The series of partial sums,

$$\frac{n_i}{N}, \frac{n_i + n_2}{N}, \frac{n_1 + n_2 + n_3}{N}, \ldots, \mathop{S}_{1}^{I} \frac{n_i}{N} = 1\,,$$

gives the fraction less than "size" for the fractile diagram.

$$m'_k = \frac{\sum_1^I n_i d_i^k}{N}\,.$$

The kth moment about the origin of the frequency histogram of d.

$$m'_1 = \bar{d} = \frac{\sum_1^I n_i d_i}{N}\,.$$

The first moment about the origin of the frequency histogram of d, the arithmetic mean of d.

$$\bar{d}_k = \frac{\sum_1^I n_i d_i^{(k+1)}}{\sum_1^I n_i d_i^k}\,.$$

An average dimension of sample particles based on a kth moment relationship. (When $k = 0$,

$$\bar{d}_o = \bar{d} = \frac{\sum_1^I n_i d_i}{N}$$

and $\bar{d}$ is the arithmetic mean which is identically equal to m'_1.)

$$m_k = \frac{\sum_1^I n_i (d_i - \bar{d})^k}{N}\,.$$

The kth moment about the mean of the frequency histogram of d.

(cont'd.)

$$m_2 = s^2 = \frac{\sum_1^I n_i(d_i - \bar{d})^2}{N}.$$

The second moment about the mean of the sample distribution (frequency histogram) of d, the sample variance of d. For s^2 to be an unbiased estimate of the population variance, the denominator must be $N - 1$.

$$s_{d_k}^2 = \frac{\sum_1^I n_i d_i^k (d_i - \bar{d}_k)^2}{\sum_1^I n_i d_i^k}.$$

The measure of sample dispersion for the kth moment relationship. $s_{d_k}^2$ is called the sample variance of d for the kth moment relationship.

$$\bar{a} = \frac{\sum_1^I n_i a_i}{N} \approx \frac{\sum_1^I n_i C_2 d_i^2}{N}.$$

Average individual surface area of particles in the sample.

$$\bar{v} = \frac{\sum_1^I n_i v_i}{N} \approx \frac{\sum_1^I n_i C_3 d_i^3}{N}.$$

Average individual volume of particles in the sample.

$$\bar{w} = \rho \bar{v} \approx \frac{\sum_1^I n_i \rho C_3 d_i^3}{N}.$$

Average individual weight of particles in the sample.

$$s_a^2 = \frac{\sum_1^I n_i (a_i - \bar{a})^2}{N - 1}.$$

Sample variance of particle surface areas.

(cont'd.)

$$s_v^2 = \frac{\sum_1^I n_i(v_i - \bar{v})^2}{N-1}.$$

Sample variance of particle volumes.

$$s_w^2 = \rho^2 s_v^2 \approx \frac{\sum_1^I n_i(\rho C_3 d_i^3 - \bar{w})^2}{N-1}.$$

Sample variance of particle weights.

$$p(h) = \frac{1}{\sqrt{2\pi}\,\sigma_h} e^{-\frac{(h-\mu_h)^2}{2\sigma_h^2}}.$$

The definition of the probability density of a normally distributed population of the random variable h, having mean μ_h, and variance σ_h^2.

$$g_{d_k} = \frac{\sum_1^I n_i d_i^k (d_i - \bar{d}_k)^3}{\sum_1^I n_i d_i^k}.$$

The basic definition of sample skewness.

$$\Pr(h_0 < h < h_1) = \frac{1}{\sqrt{2\pi}\,\sigma_h} \int_{h_0}^{h_1} e^{-\frac{(h-\mu_h)^2}{2\sigma_h^2}}\, dh.$$

The probability that h is found in the interval (h_0, h_1).

$x = \ln d.$

The definition of x; $\ln d$ is the natural logarithm of d.

μ_x.

The mean value of a normal distribution of the population of x.

(cont'd.)

σ_x^2 .	The variance of a normal distribution of the population of x.
$z_k = \ln C_k d^k$.	The definition of z_k .
μ_{z_k} .	The mean value of a normal distribution of the population of z_k.
$\sigma_{z_k}^2$.	The variance of a normal distribution of the population of z_k.
$f_k(x) = \dfrac{e^{kx} e^{-\frac{(x-\mu_x)^2}{2\sigma_x^2}}}{\sqrt{2\pi}\,\sigma_x \exp\left(k\mu_x + \dfrac{k^2\sigma_x^2}{2}\right)}$	The definition of the kth moment relationship for the log normal function.
μ_{xk} .	The first moment about the origin of $f_k(x)$. This parameter is analogous to μ_x but is not a "mean" in the strict technical sense.
σ_{xk}^2 .	The second moment about μ_{xk}. This parameter happens to be identically equal to σ_x^2; however, it is not a variance in the strict technical sense.
$\approx$.	Approximately equal to.
$\dot{\approx}$.	Estimates or is estimated by.
	Designator of a sample statistic for samples represented by a functional model. (Example: $\bar{d}$ $\hat{\mu}_d$ μ_d ; read as $\bar{d}$ is approximately equal to $\hat{\mu}_d$ which estimates μ_d , the mean of the population of d.)

REFERENCES

1. Websters New World Dictionary of the American Language, World Publishing, New York, 1968.
2. Lewis, Homer D., "Small Particle Statistics." USAEC Informal Report LA-5254-MS, 1973 (A superb report, freely quoted in this chapter by permission of the author).
3. *The Particle Atlas,* Ed. 2, McCrone, Walter C. and Delly, John Gustav. Vols I–IV, Ann Arbor Science Publishers, Inc. 1973, 1979.
4. White, H. S. *Industrial Electrostatic Precipitation.* Reading: Addison-Wesley, 1963.
5. Billings, C. E., First, M. W., and Silverman, L. *Particle Size Analysis in Industrial Hygiene.* New York: Academic Press, 1971.
6. Green, H. L., and Lane, W.R. *Particulate Clouds–Dusts, Smokes, and Mists.* 2nd ed., Princeton, D. Van Nostrand, 1964.
7. Whitby, K. J., and Liu, B. H. Y. *Aerosol Science.* ed. Davies, C. N., New York: Academic Press, 1966.
8. DallaValle, J. M., Orr, C., and Hinkle, B. L. *Brit. J. App. Phys. Supplement* 3 (1954): 5198.
9. Jelinek, Z. *Particle Size Analysis.* New York: Halstead Press, 1974.
10. Allen, T. *Particle Size Measurement.* London: Chapman & Hall, 1968.
11. Cadle, R. D. *Particle Size.* New York: Reinhold, 1965 (for a more recent reference in hindered settling, see Davies, L., Dollimore, D., Sharpe, J. H. *Powder Technology* 13 (1976): 123.
12. DalleValle, J. M. *Micromeritics.* New York: Pitman, 1948.
13. Kostelnick, M. and Beddow, J. K. *Modern Developments in Powder Metallurgy,* ed. Hausner, H. H. pp. 28–48, New York: Plenum, 1972.
14. Atkins, J. H. *Analyt. Chem.* 36 (1964): 579.
15. Shuil, C. G. *J Am. Chem. Soc.* 70 (1948): 1405.
16. Bradley, R. *Phil. Mag.* 13 (1932): 853.
17. Derjaguin, B. V. *Powders in Industry.* S.C.I. Monograph 14, London. (1961): 102–113.
18. Eissenhigh, R. H., and Falls, I. *Physical Chemistry of Aerosols.* Discussion 30, Faraday Society, London, 1961.
19. Johnson, H. F., and Caughanowr, D. R. *Ind. Eng. Chem.* 50 (1958): 1169.

20. Cadle, R. D., and Robbins, R. C., See (18).
21. Prandlt, L. L., and Tietjens, O. G. *Applied Hydro and Aeromechanics.* New York: McGraw Hill, 1934.
22. Schlichting, H. *Boundary Layer Theory,* New York: McGraw Hill, 1960.
23. Abraham, M., and Becker, R. *Electricity and Magnetism.* New York: Harner, 1932.
24. Mercer, T. H. *Aerosol Technology in Hazard Evaluation.* New York: Academic Press, 1973.
25. Hoel, P. G. *Introduction to Mathematical Statistics.* New York: Wiley.
26. Brownlee, K. *Statistical Theory and Methodology in Science and Engineering.* New York: Wiley, 1960.
27. Dixon and Massey. *Particle Size Measurement.* London: Chapman & Hall, 1968.
28. Cochran, W. G. *Annals of Math. Stat.,* vol. 23, 3, 1952.
29. Mugele and Evans, *Ind. Eng. Chem.* 36, 4 (1951): 1317–1324.
30. Hatche, T., and Choate, S. *J. Franklin Inst.* 207 (1929): 369–387.
31. Robbins, W.H.M. *Brit. J. App. Phys. Suppl.* 3, S82-S, 85 (1954).
32. Kotrappa, P. *Aerosol Sci.* 2 (1971): 353–359.
33. Kotrappa, P. *In Assessment of Airborn Particles.* Thomas, Ill. 1972, ed. Merar, et al.
34. Steinhartz, A. R. *Trans. Soc. Chem. Ind.* 65 (1946): 314–320.
35. Bird, K. E. "The Assessment of Particle Shape and Its Influence on Sizing Analysis." M.S. Thesis, U. London, U.K., 1966.
36. Gurel, S., Ward, S. G., and Whitmore, R. L. *Brit. J. Appl. Phys.* 6 (1955): 83–87.
37. Meorow, J. A., Malaika, J. *Trans. Am. Geophys.* vol. 31 (1950): 74–82.
38. Heiss, J. F., and Coull, J. *Chem. Eng. Prog.* 48 (1952): 133–140.
39. Wadell, H. *J. Franklin Institute* 217 (1934): 459–490.
40. Heywood, H. *Proc. Second Lunar Sci. Conf.* 3 (1971): 1489–2001, Plenum Press.
41. Heywood, H. *Chem. Ind.* (London) 56 (1937): 149–154.
42. Heywood, H. *J. Pharm. Pharmac.* 15 (1963): 565–735.
43. Lees, *Sedimentology,* 3 (1964) 2–21.
44. Hald, A. *Statistical Theory with Engineering Applications.* New York: Wiley, 1952.
45. *Handbook of Chemistry and Physics*, Ed. Robert C. Weast 59th ed., Boca Raton, Florida: CRC Inc., 1978.
46. Hald, A. *Statistical Tables and Formulae.* New York: Wiley, 1952.

47. Finney, D. J. *Probit Analysis.* London: Cambridge U. Press, 1947.
48. Cramer, H. *Mathematical Methods of Statistics.* New Jersey: Princeton U. Press, 1951.
49. Dixon and Massey, *Introduction to Statistical Analysis.* second ed., New York: McGraw Hill.
50. Lewis, Homer D., and Goldman, A., Report LA 3656, Los Alamos, USAEC, 1968.
51. Galton, F., and MacAllister, D. *Proc. Royal Soc.* 29 (1879): 365–367.
52. Hazen, A. *Trans. ASCE* 77 (1914): 1549–1550.
53. Kapteyn and Van Uhren, *Skew Frequency Curves in Biology and Statistics.* Gronigen, 1916.
54. Wightman, et al. *J. Phys. Chem.* 25 (1921): 181, 561; 27 (1923): 1, 141, 466; 28 (1924): 529.
55. Loveland, R., and Trevelli, A. *J. Franklin Institute* Aug–Sept, 1927.
56. Gregg, S., and Sing, K. *Absorption, Surface and Porosity.* New York: Academic Press, 1967.
57. Herdan, G. *Small Particle Statistics.* Amsterdam: Elsevier, 1953.
58. Rosin, P., and Rammler, E. *J. Inst. Fuel* 7 (1933): 29–36.
59. Fraser, R. P., and Eisenklam, P. *Trans. Inst. Chem. Engrs.* 34, 304 (1956).
60. Nukiyama, S., and Tanasawa, Y. *Trans. Soc. Mech. Engrs. (Japan)* 5 (1939): 1–4.
61. Mercer, T., and Stafford, R. G. "Impaction From Round Jets," *Ann. Occup. Hyg.* 12 (1969): 41–48.
62. Lowe, H. J., and Lucas, D. H. The Physics of Electrostatic Precipitation, *Brit. J. Appl Phys.* Suppl. 2, 1952.
63. Whytlaw-Gray, R., and Patterson, H.S. *Smoke,* London: Edward Arnold and Co., 1932.
64. Jorgensen, R. *Fan Engineering,* 7th Edition, Buffalo Forge Co., Buffalo, NY.
65. Aschenbrenner, *J. of Sedimentary Petrology,* 1956.
66. Heywood, H. *Trans. Inst. Chem. Engr* 25 (1947): 14–24.

Chapter 6

FINE PARTICLE CHARACTERIZATION

6.1 FROM PAST TO FUTURE

In a sense, the four main chapters up to this point have reviewed what is past. This chapter, however, deals with the future of particle science. The size and shape characterization of fine particles may be summarized in the term morphology. This is the topic of this chapter, in essence, the geometrical properties of fine particles. Physical-chemical properties of fine particle are considered separately in Chapter 7. This chapter points towards the future in that there are currently significant advances being made in the morphological characterization of fine particles. If these advances bear the fruit of their promise, more rigorous scientific relationships will gradually supercede much of the empiricism in this field. The topics covered in this chapter include:

principles and fundamentals of the methods for particle size determination,
principles of shape determination methods,
pattern recognition and particle characterization,
design of a system for particle shape analysis,
feature extraction,
particle signature and the Meloy equations,
property representation,
stereology, and
deterministic, statistical, and fuzzy classifiers.

A primary ideal of fine particle science is to describe in detail the physical, chemical, electrical. and magnetic properties of the myriads of particles of a representative sample; and to condense/compress this multitudinous set of information into a small set of informational items which succinctly, precisely, and adequately describe the original population.

This ideal represents a gigantic problem of information collection and condensation which has hardly been tackled at all yet. In fact, one might even go so far as to say that it is only recently that this problem has even been appreciated. The upshot of this lack of a system for collecting and processing information is that, at the present time, size determination methods are mainly quality control tools rather than means for assessing properties in a rigorous scientific manner. Until recent time, shape assessment methods have been totally inadequate. Many other particulate material properties are, on the whole, inadequately understood from a fundamental point of view.

In the previous chapter the group of statistical descriptors commonly used to summarize information about fine particles, particularly information concerning the "size" distribution of particles in a powder was discussed. These descriptors permit the observer to condense a large amount of information about size and thus permit these characteristics to be related to the observed behavior of the particle set in response to various environmental stimuli. While this procedure may be adequate for some purposes, it is certainly insufficient for use in fundamental studies, on two grounds. First, the use of "size" alone is clearly incomplete as a description of the physical characteristics of a particle set. It is incomplete even for the description of the geometric properties of a particle set because it ignores morphic properties of the particles. Second, the characterization of particle sets is not rigorous. For example, the concept of size is well understood but, when the testing has to be done, the precise meaning of the emergent data is unclear. Moreover, no one method for evaluating "size" is able to relate some size parameter of a spherical particle to that for a highly "irregular" particle. For example, it is an easy matter to decide how to take microscopic diameter measures for some spherical particle material, but what should the rule be in the case of the "irregular" particle material? Who is to decide these rules, assuming, that is, that they can indeed be formulated at all? Even more confusion abounds when one considers the morphological properties of fine particles. We are all familiar with terms such as spherical, dendritic, angular, rough, irregular, oval, disc-like, ellipsoidal, and so on and so forth. In the first place, we rather loosely bandy about terms which describe two-dimensional characteristics and other terms which describe three-dimensional characteristics without ever adequately differentiating them from each other. In fact, whether two-dimensional shape descriptors are adequate to characterize three-dimensional shape has never been fully discussed. It appears to be a difficult problem that awaits formulation. Secondly, because of the inherent practical difficulties of obtaining a complete shape (or morphological) description of a sample of particles, there has been a tendency to limit both the scope and application of the descriptors used. Thirdly, in those cases in which both size and shape characteristics of a particle set are determined, they tend to be quoted rather as a shopping list than as locating a point in some type of hyperspace. To summarize, one can say that, in both size and shape analysis, there is inadequate initial information with the consequence that any condensation of information fails to adequately or precisely describe the set of fine particle properties being sought. In addition, there is inadequate utilization of appropriate and available methods for data handling.

6.2 SIZE ANALYSIS AND SAMPLING

Although size analysis has been taken to include the determination of shape by some workers,[1] in this chapter, as far as is necessary, the two characteristics have been separated from each other and from consideration of the methods of determining surface area of powders. As is shown in Table 6.1 there are numerous categories of particles size analysis methods covering sizes all the way from fractions of an inch (using sieving riddles, which are not included in the Table) down to extremely small parts of a micron. The analysis of sets of fine particles is possible in the case of the majority of the available methods in the form of smoke, powder, suspension, and sol. A fairly detailed procedural chart for size analysis of sets of particulate materials[2] is shown in Fig. 6.1. This chart may be used in conjunction with Table 6.1 as aids in clarifying possible strategies in the case of specific particle size anlysis problems.

Sampling

The general philosophical approach to size analysis was discussed in Section 5 of Chapter 2 under the heading of Size Classification Strategies.

Methods which are used for the sampling of particulate materials may be conveniently divided into two groups: those for sampling powders and those for sampling dispersions.

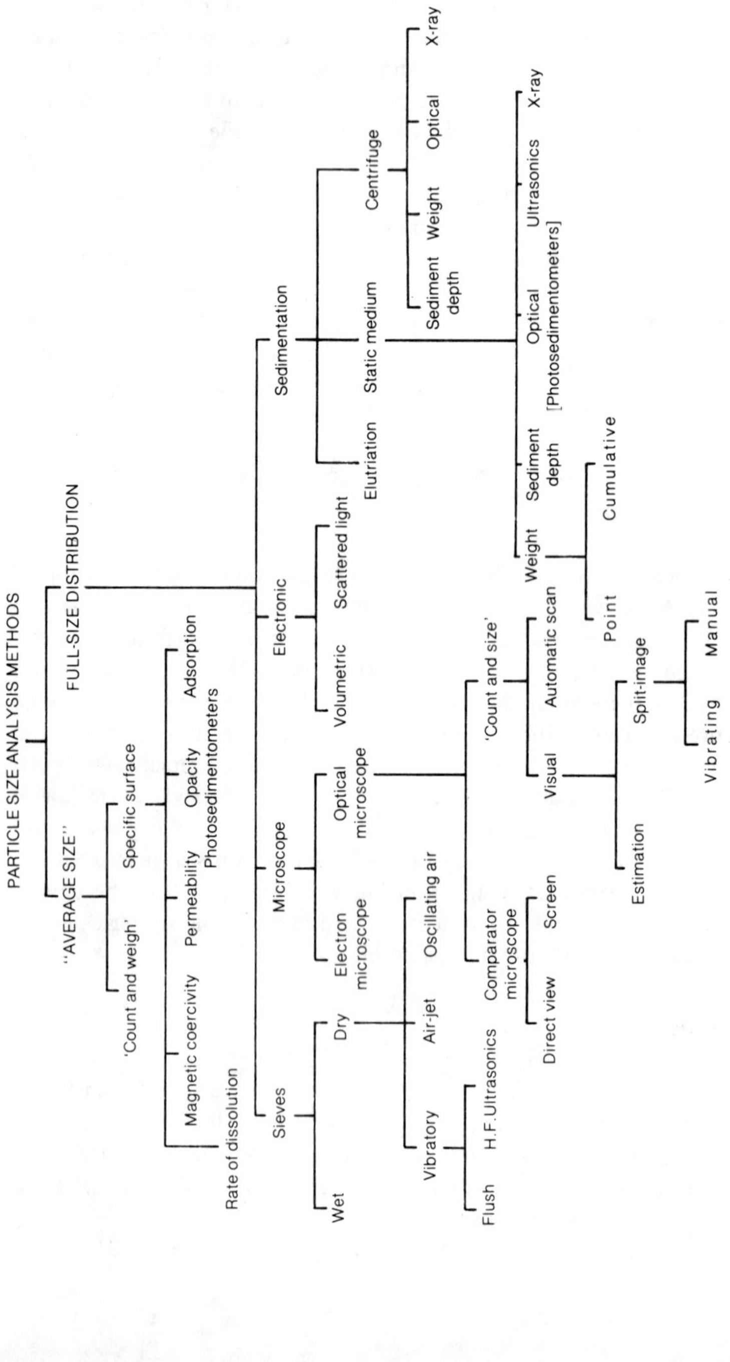

Fig. 6.1 Classification of methods of particle size analysis (after W. Hurst and G. Spence).[129]

Table 6.1

SUITABILITY OF PARTICLE SIZE ANALYSIS METHODS[1]

Method	*Range* (μm)	*Gas-liquid (foam)*	*Gas-solid (porous material)*	*Liquid-gas (fog)*	*Liquid-liquid (emulsion)*	*Liquid-solid (solid emulsion)*	*Solid-gas (smoke, powder)*	*Solid, liquid (suspension, sol)*	*Solid-solid (solid mixture)*	{*Continuous phase*} {*Disperse phase*}
Microscopy	0.4–100	x		x	x	x	x	x	x	
Ultramicroscopy	0.01–2			x	x	x	x	x	x	
Electron microscopy	0.0005–5						x	x	x	
Light-scattering	0.1–30			x	x	x	x	x	x	
X-ray analysis	0.005–1						x			
Sieve classification	> 40						x	x		
Ultrafiltration	> 0.002							x		
Sedimentation	1–50			x			x	x		
Centrifugation	0.01–1							x		
Ultracentrifugation	0.0005–0.01							x		
Elutriation	1–100						x	x		
Diffusiometry	< 0.1							x		
Osmometry	< 0.1							x		
Gel permeation chromatography	< 0.1									
Viscometry	two dimensions smaller than 0.0001				x			x		
Permeametry	0.5–100		x				x			
Adsorption	0.002–50		x				x	x		
Conductometry	> 0.2							x		
Radiometry	> 0.002			x	x		x	x		
						*			*	

Note: These two combinations marked thus * can be studied if the medium is transparent.

Powder Sampling

Large tonnages of powders are transported in boxcars, trucks, and sundry other containers. The difficulty of obtaining laboratory-sized samples that actually represent the bulk should not be underestimated and indeed the specifications shown in Table 6.2 testify to the seriousness of the difficulty. When the bulk sample has been obtained, a laboratory-sized sample is taken either by momentarily and repeatedly interrupting the flow of powder to obtain incremental samples or by splitting the stream of powder using any one of the devices shown in Fig. 6.2. It has been observed that a most useful sampler is the spinning riffler, if properly used,[2] as is shown in Table 6.3. Once the laboratory-sized sample is in hand, repetition of the previous splitting procedures yields the desired analysis sample (for really precise work a microsplitter is used all the way through to the analysis sample[3]). Alternatively, the skilled analyst may use a manual quartering and remixing procedure.[4] In all of these methods, the aim is to attempt to reduce the magnitude of error by substituting mechanical means in place of the sampler's manual procedures. This quite laudable aim is never fully realized and in truth there is no substitute for the experienced sampler. While this is true at every stage of sample reduction, it is most clearly true at the bulk sampling stage when shovels, sampling thieves, and the like are standard equipment. In sampling operations, the analyst/sampler must be aware of the potential difficulties due to segregation in the powder mass. It has been observed that the factors influencing the tendency of fine particles to segregate in binary component systems in the case of a simple filling of a cylindrical container from a conical funnel include:[5] the particle characteristics including diameter ratio, shape, density, and percent fines in powder; the system characteristics, including the height of drop into the container, the die and nozzle diameters, and the rate of filling.

One important aspect of the mechanism of segregation appears to be one of filtration or percolation of fines down through the moving larger particles. This produces an inner mound of fines. The major variables are the diameter ratio (see Fig. 6.3), the percent fines in the mixture (see Fig. 6.4), and also the rate of fill—the more rapid the filling, the less time for segregation to occur.[5] In practice, a variety of vibrational impulses may be imparted accidently or deliberately to the powder mass before or after it is sampled. Segregation in a vibrating, transient bed[6] and in a vibrating container[7,8] have been reported. In general, it cannot be assumed that vibration increases segregation as it may actually reduce it due to a mixing effect.

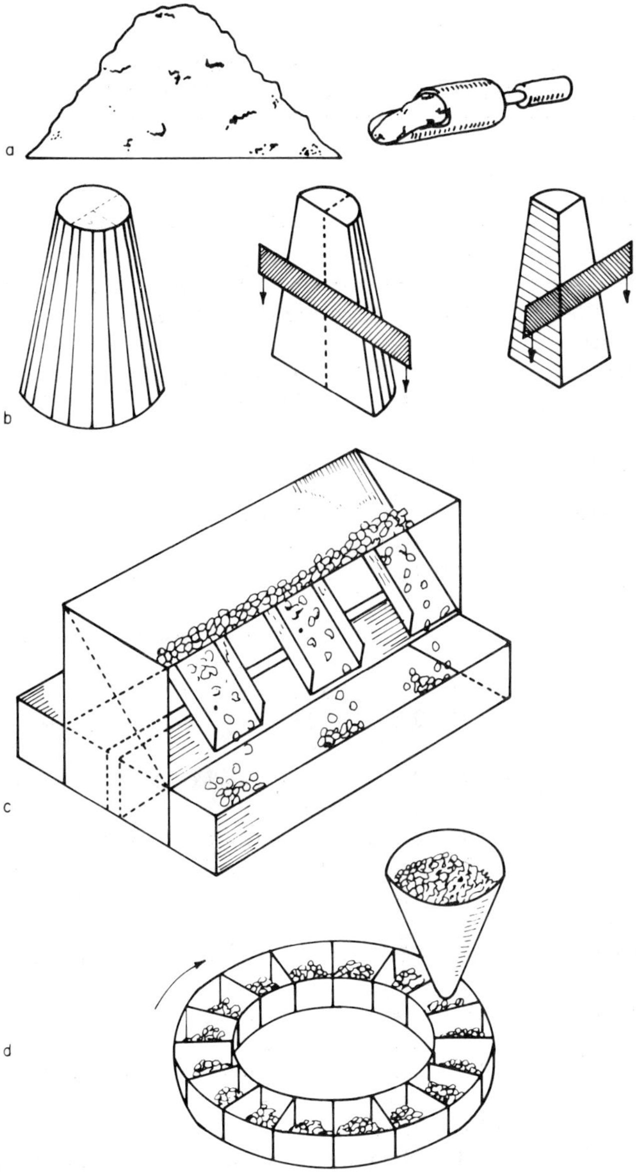

Fig. 6.2 An illustration of five sampling techniques: (a) scoop sampling from a heap, (b) cone and quartering, (c) chute splitter, (d) spinning riffler, and (e) table sampler.[2]

Table 6.2

LIST OF ASTM PUBLISHED STANDARDS ON SAMPLING OF PARTICULATE MATERIALS

Material	ASTM *Desig-nation*	*Title of Standard*
Aggregates	D 75	Methods of Sampling Stone, Slag, Gravel, Sand, and Stone Block for Use as Highway Materials
Asbestos fiber. . .	D 2590	Method of Sampling Asbestos Fiber for Testing
Bituminous materials	D 140	Methods of Sampling Bituminous Materials
	D 979	Methods of Sampling Bituminous Paving Mixtures
Calcium chloride .	D 345	Methods of Sampling and Testing Calcium Chloride
Carbon black . . .	D 1799	Method for Sampling Packaged Shipments of Carbon Black
	D 1900	Method for Sampling Bulk Shipments of Carbon Black
Casein.	D 1697	Method of Sampling Casein and Similar Protein Materials
Ceramic clays . . .	C 322	Method of Sampling Ceramic Whiteware Clays
Coal	D 197	Method of Sampling and Fineness Test of Powdered Coal
	D 492	Method of Sampling Coals Classified According to Ash Content
	D 2013	Method of Preparing Coal Samples for Analysis
	D 2234	Methods for Mechanical Sampling of Coal
Coke.	D 346	Method of Sampling Coke for Analysis
Lime and limestone	C 50	Methods of Sampling, Inspection, Packing, and Marking of Lime and Limestone Products
Magnesium.	C 237	Method of Sampling Magnesium Oxychloride Compositions and Ingredients
Metal powders . .	B 215	Methods of Sampling Finished Lots of Metal Powders
Plastics	D 1898	Recommended Practice for Sampling of Plastics
Soap powders. . .	D 460	Methods for Sampling and Chemical Analysis of Soaps and Soap Products

(Courtesy ASTM, 1916 Race St., Philadelphia, PA)

Table 6.2

LIST OF ASTM PUBLISHED STANDARDS ON SAMPLING OF PARTICULATE MATERIALS

Material	*ASTM Designation*	*Title of Standard*
Sorptive mineral products	C 431	Methods for Sampling and Evaluation of Sorptive Mineral Products Used as Floor Absorbents
Statistical probabilities	E 105	Recommended Practice for Probability Sampling of Materials
	E 122	Recommended Practice for Choice of Sample Size to Estimate the Average Quality of a Lot or Process
	E 141	Recommended Practice for Acceptance of Evidence Based on the Results of Probability Sampling
Aggregates	D 75	Methods of Sampling Stone, Slag, Gravel, Sand, and Stone Block for Use as Highway Materials
Asbestos fiber	D 2590	Method of Sampling Asbestos Fiber for Testing
Bituminous materials	D 140	Methods of Sampling Bituminous Materials
	D 979	Methods of Sampling Bituminous Paving Mixtures
Calcium chloride	D 345	Methods of Sampling and Testing Calcium Chloride
Carbon black	D 1799	Method for Sampling Packaged Shipments of Carbon Black
	D 1900	Method for Sampling Bulk Shipments of Carbon Black
Casein	D 1697	Method for Sampling Casein and Similar Protein Materials
Ceramic clays	C 322	Method of Sampling Ceramic Whiteware Clays

(Courtesy ASTM, 1916 Race St., Philadelphia, PA)

Table 6.2 *(cont'd.)*

Material	*ASTM Designation*	*Title of Standard*
Coal	D 197	Method of Sampling and Fineness Test of Powdered Coal
	D 492	Method of Sampling Coals Classified According to Ash Content
	D 2013	Method of Preparing Coal Samples for Analysis
	D 2234	Methods for Mechanical Sampling of Coal
Coke	D 346	Method of Sampling Coke for Analysis
Lime and limestone	C 50	Methods of Sampling, Inspection, Packing, and Marking of Lime and Limestone Products
Magnesium	C 237	Method of Sampling Magnesium Oxychloride Compositions and Ingredients
Metal powders	B 215	Methods of Sampling Finished Lots of Metal Powders
Plastics	D 1898	Recommended Practice for Sampling of Plastics
Soap powders	D 460	Methods for Sampling and Chemical Analysis of Soaps and Soap Products
Sorptive mineral products	C 431	Methods for Sampling and Evaluation of Sorptive Mineral Products Used as Floor Absorbents
Statistical probabilities	E 105	Recommended Practice for Probability Sampling of Materials
	E 122	Recommended Practice for Choice of Sample Size to Estimate the Average Quality of a Lot or Process
	E 141	Recommended Practice for Acceptance of Evidence Based on the Results of Probability Sampling

(Courtesy ASTM, 1916 Race St., Philadelphia, PA)

Table 6.3

COMPARISON OF SAMPLING METHODS[2] –
SIX OPERATORS, STANDARD DEVIATION (σ)

	Average standard deviation $\sigma_A = \sqrt{(\Sigma\sigma^2/6)}$	*Maximum* σ	*Minimum* σ
Spinning riffler	0.642	0.85	0.40
Chute splitter	1.02	1.04	0.34
Cone and quartering	1.15	1.80	0.61

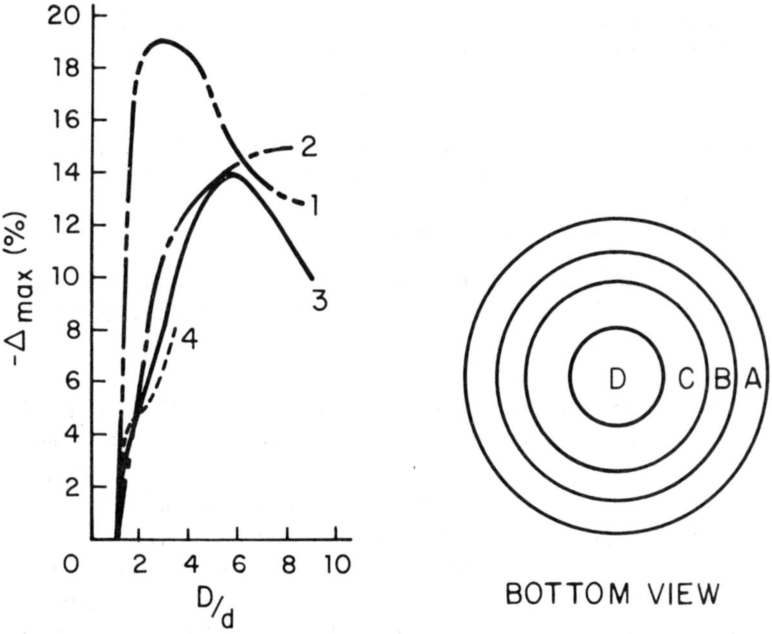

Fig. 6.3 Maximum Segregation.[5] $-\Delta_{max}$ vs D/d for Zone A of Radial Sampler. Curve 1, $D = 2000\ \mu$; Curve 2, $D = 841\ \mu$; Curve 3, $D = 250\ \mu$; Curve 4, $D = 149\ \mu$.

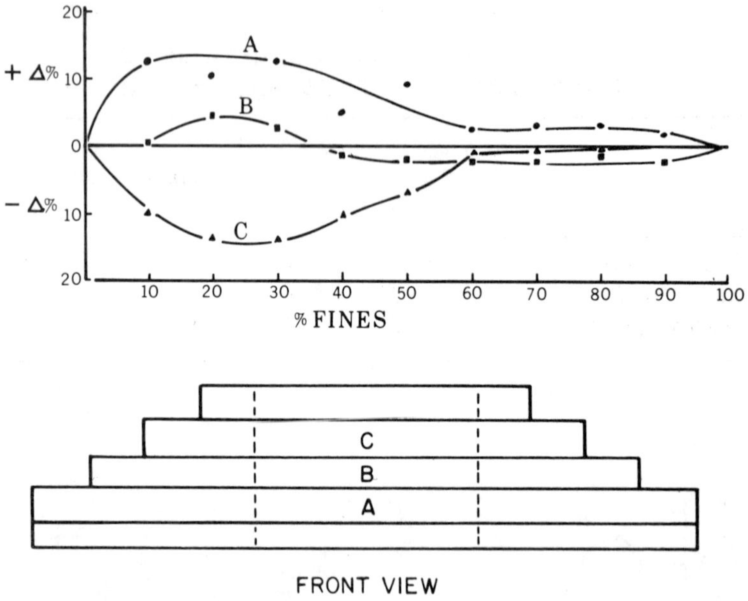

Fig. 6.4 Segregation.[5] Δ percent vs percent fines in mixture for zones A, B, and C in vertical sampling device (D = 2000 μ, d = 841 μ).

Sampling of Fluid–Solid Systems

In the case of liquid/solid particulate systems, a number of elementary precautions should be observed in order to insure a representative sample:[1] agitate the liquid prior to sampling, avoid evaporation of liquid, and avoid agglomeration and formation of secondary particles.

In the case of gaseous dispersions, it is much more difficult to avoid some segregation-type effect during a sampling operation, as will be clear below.

The methods for samplings of gaseous dispersions include at least four categories: sampling methods which do not classify the sample, sampling methods which classify according to size, respirable activity samplers which classify and prepare a respirable sample, and isokinetic sampling. Methods which do not classify the sample include: settling, impingement, or impaction, thermal precipitation, electrostatic precipitation, and filter samplers. The principles upon which these methods are based have been touched on briefly in Chapter 3.

Nonclassifying Sampling Methods

Dynamic settling chambers which use a slow moving aerosol are unsuitable for the collection of particles below approximately 5 μ, but they can be used as a presampler to remove particles which are too large to be of hygienic or other significance. The control of the dispersion flow means that the observer can note the volume of air from which the sample of particles is retained. This is not the situation in the case of the various methods of "dust fall" collection which are used from time to time (such as the collection on a sticky slide in the open, for example). As the title implies, static settling methods utilize a simple design in which the aerosol is retained in a closed vessel of suitable form. One major problem with this method is that the long time of settling may permit agglomeration, and also may lead to curtailment of settling period with consequent loss of the finest particles.

Impingement settlers include both wet and dry high velocity jet impactors of standard size and also smaller midget impingers. Two of the important problems with these instruments are the size and number separation in the dust ribbon sample obtained and the tendency for deflocculation. The former necessitates the very careful examination of samples and the latter tends to produce a finer sample than the original aerosol unless the jet velocity is reduced.[9]

Thermal precipitators move the aerosol at a measured rate through a steep thermal gradient. The drawbacks of this method include heat effects on the particles and separation according to size over the deposit.

Electrostatic precipitators also tend to size-separate, especially if a central wire is used. A point source electrostatic precipitator minimizes this problem. One precaution should be mentioned—operators should use the precipitators only in nonexplosive atmospheres.

Filter samplers are of three types: those with soluble porous beds, membrane filters, and fibrous filter beds. The soluble beds use soluble powders such as sugar and volatile materials such as salicylic acid. Polystyrene fiber filters are also used. One disadvantage with the use of these filters in conjunction with size determination is that the powder particles may be affected by the heat in the case of volatile beds, and by agglomeration tendencies in the case of the soluble beds.

Membrane filters are thin films of dried porous cellulose based gels. The films are 150 μ thick and contain closely controlled pore sizes available in the range from 8 μ to 0.01 μ. One advantage of their use in microscopy is that the filter material has the same refractive index as the immersion oil. When the oil is placed on the deposit, the filter disappears from view, greatly facilitating observation.

There are many paper and fiber bed filters which are used to collect samples for size analysis purposes (trade names including Whatman, Milipore, Gelman, and AEC). Depending upon the material of construction, the filters can be used at different levels of elevated temperature; for example, paper for room temperature work, glass fibers for intermediate temperatures (500°C), and ceramics for elevated temperature work (1100°C).

Classifying Sampling Methods

Methods which classify the particles into size groups include the cascade impactors and the winnowing samplers. The cascade impactors consist of a series of connected chambers (from four through six, depending upon the model). A final stage filter may also be incorporated in the design. The aerosol is drawn through the sampler which increases in velocity as it enters each successive chamber of the instrument. This increase is brought about by the successive reduction of the areas of the input jets into the chambers. This contrived increase in velocity increases the probability of impingement of finer particles in the end chambers. The particles are impinged onto slides for the convenience of the observer. The device has a number of advantages. For instance, once the calibration is completed, evaluation of additional samples of the same material can be done simply by weighing the deposit on each chamber slide, also, it can be used for droplets as well as for particles. A serious disadvantage is that from deposition on the chamber walls up to 30 percent of the required sample may be lost. Yet another problem is that there is no sharp cut-off in the portions of the population sampled in each chamber. In most cases this presents no more of a problem than other classifiers, but it leads to erroneous results for monosize aerosols.

Several winnowing type samplers have been designed, the basic principle of which is the use of a rotating cone to generate centrifugal force (the aerosol spectrometer can develop up to 30,000 g), thus greatly increasing the settling velocities of the particles onto the sampling surfaces. A more complete account of the sampling devices is contained in a review.[14] Fig. 6.5 shows the ways in which particles of increasing fineness enter into the respiratory system. Note that the finer particles penetrate to the deepest recesses of the lungs. Respirable activity samplers are designed with the intent that they collect a sample which represents in size distribution the probability of deposition in the alveolar regions of the lungs. The respirable fraction has been defined in a number of studies[10–12] based upon the probability of deposition. An alternative method has been developed based upon the combined probabilites of deposition and retention for a long period of time.[13] Two main groups of respirable samplers are elutriators and cyclones. Specially designed impingement samplers have also been developed. For a comprehensive review see reference 14. The importance of isokinetic sampling is illustrated in Fig. 6.6. The on-stream sampling of aerosols has to be carried out so that the velocity in the sample tube matches the mainstream velocity. Otherwise, if the sample tube velocity is less than the stream velocity, smaller particles fail to enter the tube and are swealed away in the mainstream. Conversely, if the tube velocity is too high, an excess of fines is drawn into it. Anisokinetic sampling therefore, distorts the size distribution with respect to the expected distribution. Care has to be taken in order to insure that the trajectories in the probe are continuous with those in the mainstream in order to avoid anisokinetic sampling. The observer must also take note of segregation of particles in the mainstream so as to avoid atypical samples.[15]

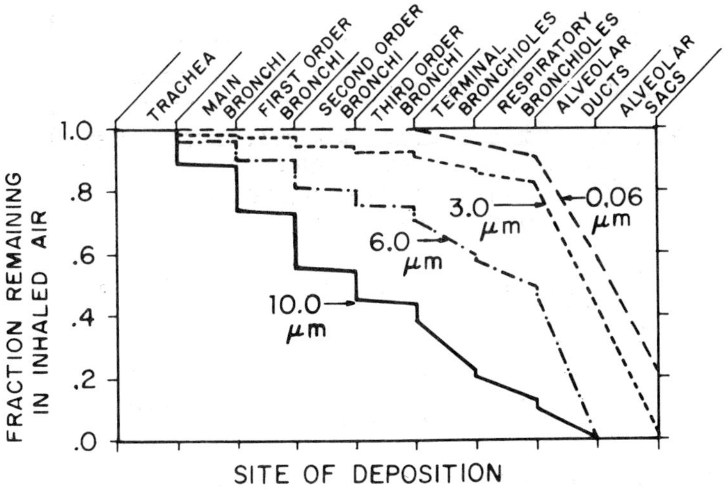

Fig. 6.5 Lung deposition, during inspiration, of particles that enter the trachea. Abrupt changes represent impaction at airway entrance. Other changes represent diffusion and sedimentation losses. Losses in the alveolar sacs represent changes with time rather than distance traversed.[14]

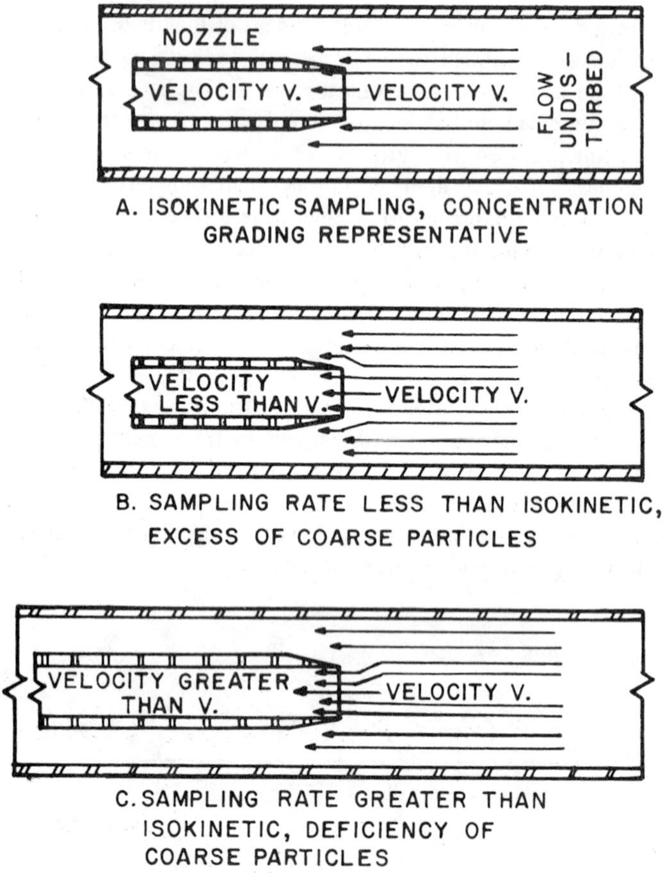

Fig. 6.6 Streamlines around a sampling probe (from Stairmand).[124]

6.3 FUNDAMENTALS OF THE METHODS FOR DETERMINING PARTICLE SIZE

Although there are well in excess of a dozen groups of methods for determining particle size, it is possible to discern some common principles of operation and some distinguishing principles as follows:

1. There is a number of methods which are fundamentally static and observational in character. Examples of these include light microscopy,

electron microscopy, and X-ray methods. Although there may be some actual motion of the particles in some versions of the above, these motions are made for sampling reasons, usually.

2. The majority of size analysis methods require that the particle set move. Its motion is observed and deductions as to its size characteristics are made. For example, in sieving, the powder particles are forced to move and their differing responses are ascribed directly to size differences. Sedimentation, centrifuging, and viscometry are additional examples of methods which are fundamentally dynamic and deductive in character.

There is a number of detailed accounts of particle size analysis methods available in book form.[1,2] It is therefore unnecessary to itemize the practical aspects of these methods herein. However, the specific principles or theoretical foundations of each method are discussed. For an example from the literature of the typical test data generated in the laboratory, see reference 1.

Static Methods

Optical Methods. The particle, when magnified in the microscope, must be at least as large as the resolving power of the human eye. The resolving power of a microscope equals $1/d$ when

$$d = \frac{0.5\,\lambda}{\gamma \sin \alpha} \tag{6.1}$$

in which

λ is the light wavelength,
$\gamma \sin \alpha$ is the numerical aperture,
α is the angle subtended at the lens, and
γ is the refractive index of the medium between the lens and the object observed.

Substitution of typical figures in Eq. (6.1) yields a d value of approximately 0.5 μ. Because the resolving power of the human eye is about 200 μ, obviously it is necessary to magnify the object up to a level at which the human eye can resolve the image. This means that the magnification required is 200/0.5 or about 400X.

Enlargements of photomicrographs are commonly used to make it easier for the observer to work on the information in the picture, because the human observer is more comfortable when not being forced to work at the limit of resolution. Empty magnification is said to be employed when particles below the resolution of the microscope are magnified and appear as diffused circles. An example from the literature is given in Table 6.4 in the case of size analysis of sand.[16]

Table 6.4

SIZE ANALYSIS OF SAND[16]

Size class (μm)	% w/w distribution	cumulative
0.4–0.8	0.01	0.01
0.8–1.6	0.06	0.07
1.6–2.4	0.16	0.23
2.4–3.0	0.33	0.56
3.0–4.0	0.67	1.23
4.0–8.0	1.12	2.35
8.0–12.0	9.42	11.77
12.0–16.0	7.29	18.06
16.0–20.0	6.56	24.62
20.0–30.0	7.49	32.11
30.0–40.0	20.22	52.33
40.0–60.0	11.81	69.14
60.0–90.0	19.46	83.60
90.0–120.0	14.57	98.17
over 120.0	1.83	100.00

In the case of the electron microscope, the resolving power is two orders of magnitude higher than that of the optical microscope. However, there are some problems associated with the use of the more powerful instrument. For example, the intensity of the electron beam can promote thermally induced changes in the sample, especially organic ones. This problem is intensified if a great deal of searching for a suitable field has to be done. A thin sputtered layer of Al or some other suitable metal may be used (coarse dispersions) to stabilize the sample and in this case some of the damaging effects of the heating by the electron beam can be ameliorated. For best results, organic samples may be embedded in polymethylmethacrylate. As in the case of the optical methods, it is usual to record the results on photographic paper and take measurements off these. An example of a histogram for the electron microscope sizing of coal dust particles is shown in Fig. 6.7.[17]

Size class (μm)	*Number of particles*	*Size class* (μm)	*Number of particles*
0.23–0.32	48	0.90–1.30	85
0.32–0.45	50	1.30–1.80	64
0.45–0.64	64	1.80–2.60	35
0.64–0.90	65	2.60–3.60	13

Fig. 6.7 Coal dust particles "sized" in electron microscope.[17] (Reproduced by permission of Health and Safety Executive, Steel City House, West Street, Sheffield SI ZGQ, England.)

The phenomenon of light scattering is variously used for the characterization of sizes of particulates. For example:

1. molecular weight determination,[18] ($M = 1.27 \times 10^6$);
2. particle size of synthetic latex,[19] ($L = 1700$ A); and
3. particle size of emulsion polymers,[20] ($d = 0.4\ \mu$).

The starting relationship in all of these methods is that due to Rayleigh[21] who assumed that the particles scattering the light are optically isotropic and nondispersed, and have a radius smaller than the wavelength of the incident light. The basic idea of the treatment is that the oscillating electrical field of the incident light causes the particle to behave as a linear electrical oscillator.

The Mie theory[22] is an extension of the Rayleigh theory to larger particles (equal or larger than the wavelength of the incident radiation). In these cases, the wavelength of the scattered light is a resultant of the light from all of the many points of the surface. The summation is quite complex.

According to the Rayleigh theory, the total amount of light scattered is a function of the particle volume, V; radiation wavelength, λ; and the refractive index (m^2):

$$s = s\left[V^2, \frac{1}{\lambda^4}, (m^2)\right] \tag{6.2}$$

The intensity of the scattered radiation is a function of s, the distance from the particle to the observation point D_{obs} and a function of the angle between the direction of the scattered light and the reverse direction of the incident light (γ):

$$I = I\left[s, \frac{1}{D_{obs}}, (\gamma)\right] \tag{6.3}$$

Practical examples are given in reference 1 and 18-20.

If X-rays are used instead of light scattering, two methods may be utilized: either the method of low angle scattering or the method due to Scherrer. The latter involves the use of conventional Debye-Scherrer type X-ray instruments. For further details and calculated examples, refer to reference 17.

Dynamic Methods

Perhaps the most popular method of size determination is sieve analysis. The principle of operation of the method is to arrange for a vertically stacked set of sieves to be moved about in an abrupt but regular manner. The top sieve contains the original weighed portion of powder. By a series of abrupt motions, the particles gradually move down the stack and finally locate in the appropriate sieve. Or, in the case of the finest particles of powder, they pass through all of the sieves in the stack and come to rest on the metal pan below.

Because different countries have different wire mesh standards, it is difficult to say anything thought-provoking about them (see Table 6.5 for the Tylor, ASTM, IMM, and DIN standards, respectively), except that a $(2)^{1/2}$ relationship between openings of successive screens is often looked for.

Many laboratories use the ROTAP machine to accomplish the steady disturbance that the sieving method requires. However, other machines are available and have been tested. Table 6.6 shows the results of the Broersma and Seger-Cramer machines being used to test a sand sample in a trial with the ROTAP model. Details of the three machines are given in Table 6.7 also. Numerous studies of sieving mechanisms have been reported in the literature from time to time (see, for example, references 23 and 24). An interesting development has been an interest in blinding. This occurs when a particle either permanently or temporarily blocks a mesh opening. Some effects of particle size and shape upon permanent blinding have been reported[25] and more recently, a new theory of sieving based upon the

Table 6.5

SETS OF STANDARD SIEVES

Tylor		ASTM		IMM		DIN
Hole size						
in.	mm	in.	mm	in.	mm	mm
						0.060
						0.075
						0.090
						0.100
						0.12
				0.0055	0.139	
						0.15
				0.0062	0.157	
0.0069	0.175					
		0.0070	0.177			
				0.0071	0.180	
						0.20
0.0082	0.208					
		0.0083	0.210			
				0.0083	0.211	
0.0097	0.246					
		0.0098	0.250			0.25
				0.0100	0.254	
0.0116	0.295					
		0.0117	0.30			0.3
				0.012	0.317	
		0.0138	0.35			
0.014	0.351					
						0.4
0.016	0.417					
		0.0165	0.42			
				0.017	0.421	
0.020	0.495					
		0.020	0.50			0.5
0.023	0.589					
		0.023	0.59			
						0.6
				0.025	0.635	
0.028	0.701					

Table 6.5 *(cont'd.)*

Tylor		ASTM		IMM		DIN
Hole size						
in.	mm	in.	mm	in.	mm	mm
		0.28	0.71			
						0.75
				0.031	0.792	
0.033	0.833					
			0.840			
0.039	0.991					
		0.039	1.00			1.0
				0.042	1.056	
0.046	0.168					
		0.047	1.19			
						1.2
				0.050	1.270	
0.055	1.397					
		0.056	1.41			
						1.5
				0.062	1.574	
0.065	1.651					
		0.066	1.68			
0.078	1.981					
		0.079	2.0			2.0
0.093	2.362					
		0.094	2.38			
						2.5
				0.100	2.54	
0.110	2.794					
		0.111	2.83			3.0
0.131	3.327					
		0.132	3.36			
0.156	3.962					
		0.157	4.00			4.0
0.185	4.699					
		0.187	4.76			
						5.0
0.221	5.613					
		0.223	5.66			6.0
0.263	6.680					

Table 6.5 *(cont'd.)*

Tylor		ASTM		IMM		DIN
Hole size						
in.	mm	in.	mm	in.	mm	mm
		0.265	6.73			
0.312	7.925					
		0.315	8.00			
0.371	9.423					
0.441	11.20					
0.525	13.33					
0.624	15.85					
0.742	18.85					

Table 6.6 [125]

COMPARISON OF SCREENING ANALYSES ON BROERSMA, SEGER-CRAMER, AND TYLER ROTAP MACHINES

Sample IV (sand)

Screen gauge	*Sifting percentage*		
	B	S-C	T-R
mm	%	%	%
1.00	11.2	11.3	11.2
0.85	0.3	0.2	0.3
0.72	3.6	3.7	3.3
0.60	7.4	7.1	7.1
0.50	10.8	10.3	10.6
0.42	15.4	–	15.9
0.35	15.4	–	15.4
0.30	10.1	9.8	9.9
0.25	15.4	15.6	15.4
0.21	5.8	6.3	5.9
0.175	3.5	4.0	3.8
< 0.175	1.1	0.9	0.9

Sifting period of Broersma machine 7½ minutes, increasing rotation speed of the sieves from 0.67 ... 1.33 s^{-1} (40 ... 80 rev/min); sifting period of Seger-Cramer machine six minutes, at 3.3 oscillations per second (200 oscillations per minute); and sifting period of Tyler ROTAP machine 15 min at 2.5 s^{-1} (150 rev/min) of the sieves.

(temporary) blinding mechanism has been proposed.[26] Normal sieving practice goes down to 400 mesh, but in order to reach 10 microns or below, it is necessary to use microsieves and often wet sieving is practiced in these cases.[24]

Table 6.7

TECHNICAL AND FINANCIAL COMPARISON OF THE THREE SCREEN ANALYSIS MACHINES WITH STANDARD TYLER SIEVES
(Diameter: 200 mm, Screen Distance: 50 mm)[125]

Model	Broersma*	Seger-Cramer	Tyler ROTAP
Number of sieves	12	5	6
Sifting time for sand	7½ min	2 min	2 min
Screen analysis results	Identical and reproducible for all three machines		
Motion	Centerline of sieves describes a conical surface, of which the axis includes an acute angle with the vertical; cone pointed downwards.	Centerline of sieves describes pendulum motion in a vertical plane; sieves rotate about their centerline.	Centerline of sieves describes a straight cylinder, with a vertical axis.
Auxiliary motion	Small hammer per sieve knocks radially on screen, in a free pendulum motion in phase with the rotation.	Rather heavy driven hammer per sieve knocks radially on sieve wall.	A heavy driven hammer knocks vertically on the nest of sieves; one can put rubber disks in sieves.
Power	0.07 kW (0.1 hp)	0.14 kW (0.2 hp)	0.26 kW (0.35 hp)
Noise	None	Much	Much
Foundation	None	Concrete	Heavy cast iron frame
Relative cost	1	2.3	1.6

*Small model has sieves with a diameter of 100 mm, a screen distance of 25 mm, and a repulsion motor of 0.06 kW (1/12 hp).

Another widely used mechanical method is that of gel permeation chromatography. In this method, a solution of the substance whose molecular weight distribution is sought is passed through a column of discrete spherical gel particles in the range of 100–2000 microns. Those molecules too large to enter the pores in the gel bed pass out through the system. The molecules remaining distribute themselves variously according to their relative size and the end result is a curve relating molecular size to elution volume.

The most well-known gravity method of size determination is that of sedimentation, the principles of which have been dealt with in Chapter 5. Various items of equipment may be utilized all the way from sedimentation balances to a simple Andreasen pipette. The essence of the technique is to mix a low volume percent of the particulate material in a suitable suspension medium, to place the whole in a constant temperature enclosure, and to time the rate at which the suspended solids settle out. A set of data for diamond powder in ethanol is given in Table 6.8. A diagram of the Andreaen pipette is given in Fig. 6.8. The interested reader should consult appropriate references for practical details of the method.

Table 6.8[1]

DIAMOND POWDER IN ETHANOL,
ANDREASON PIPETTE DATA.

d (μm)	*Sedimentation time*, s	*Sedimentation distance*, cm	*Sediment** % w/w
>20			6.2
<20	267	12	93.8
<10	915	10.3	57.4
< 5	3070	8.6	31.0
< 2.5	9860	6.9	14.0

*The last column is a numerical representation of the cumulative curve.

Numerous correction factors are necessary in the straight sedimentation methods. Centrifuging methods are used in the case of particles which are too small to separate out "voluntarily," and are still too small for diffusion to be an appreciable factor. Sedimentation in a centrifugal field is also subject to the same influences as is sedimentation in a gravitational field and corrections are therefore also required. These include: correction for concentration of the suspended phase, correction for particle shape and

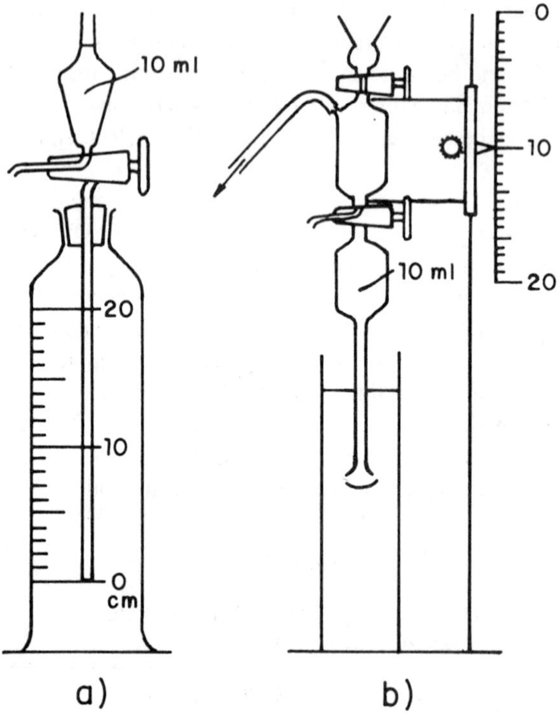

Fig. 6.8 Two modifications of the Andreasen sedimentation pipette.[1]

orientation, the solvation effect and the electroviscosity effect.[1] In the case of particulate materials in which the size is extremely small (such as sols and molecular colloids), the investigator must use the ultracentrifuge. The regular centrifuge rotates at 4000 rpm, whereas the ultracentrifuge rotates at 60,000 rpm. Even so, in the latter it is necessary to take into account diffusional effects because of the extreme fineness of the particle sizes involved. A practical example of the centrifugation method is given for the size determination of an organic pigment (in the range 0.1-2 μ) and for the use of the ultracentrifuge in the case of molecular weight determinations on viscose rayon nitrate, glycogen, and cotton cellulose, respectively.[1]

Yet another sedimentation method is one of determining the rate of sedimentation in a moving stream in which the stream flow rate is successively adjusted in order to divide the sample into fractions (those particles that are too small to settle are carried off in the stream).

Methods of size determination concerned with principles of osmometry are divided into two groups—those which derive from application of Fick's laws of diffusion, and those which are related to the measurement of osmotic pressure. The diffusional methods work on the basis of the equalization of chemical potential in the case of a pure solvent and a solution being placed in contact. A color change or an alteration in refractive index are measured and related to particle size. The method has been reported in the determination of the degree of aggregation of benzopurpurine (in aqueous solution) produced by the addition of NaCl. The methods of osmometry depend upon an empirical relationship between the osmotic pressure and concentration ratio which contains a number of virial coefficients which are themselves complicated relationships between solvent molecular weight, entropy, and heat of dilution, solvent and solute densities, and other factors.

The so-called translational methods include viscometry and permeametric methods, of which three shall be mentioned. Viscosity of a linear molecular colloid can be determined in a capillary viscometer and the values obtained are then related to the molecular weight of the substance. Gas permeametry is based upon developments of the Kozeny relationship for the process of water penetration through sand beds. A standard compact of the powder under study is made and incorporated in the appropriate equipment (for example, the Carmen apparatus), wherein it is subjected to a passage of a known volume of air under standard conditions. The time of passage is noted and substitution in a derived form of the Kozeny relationship leads to the elucidation of the specific surface of the powder.

The transient flow method involves the flow of gas from a container into an evacuated vessel, via a bed of the particulate matter under test. The method is such that only the flow delay (the time required to attain a stable gas flow) need be measured and substituted in a standard equation for the specific velocity of Knudsenian flow. The flow of a liquid through a porous layer of particles may be used as the dynamic process of concern (in conjunction with a modified form of the Kozeny equation). It should be noted that the permeametric methods are used for the determination of specific surface properties of particulates.

6.4 PRINCIPLES OF SHAPE DETERMINATION METHODS

Particle shape may be briefly defined as the pattern of relationships among all of the points of the surface of a particle. This definition is common to both earlier approaches and more recent work. Earlier workers tended to measure one specific feature of the particle (its ratio of maximum to minimum radius; ratio of perimeter to area; ratio of Martin to Feret diameter) and to use this as a representation of the shape. More recently there has been a concerted effort to measure a large number of sample points of the particle (at the present time these are taken from particle profiles) and to represent this sample in mathematical ways (for example using Fourier representation). The shape can then be represented by sets of appropriate coefficients handled in various ways (see references 60, 96). The principle underlying this approach is that thorough sampling of the profile of a particle collects all of the information concerning the particle morphology. This information is then condensed and manipulated in various ways in order to develop sets of significant features, such that these features represent the original pattern as faithfully as is possible. The eventual aim of this research movement is to relate the morphological features to the properties and origins of the particle sets.

The remainder of this chapter is focused on these new developments. In order to aid in our understanding of the suitability of the various methods for shape assessment, of their advantages and their disadvantages with respect to each other, and also to aid in our developing a structured consensus in the general area of particle shape assessment, the various schemes have been reviewed and classified.

Particle shape is a fundamental characteristic of a fine particle set and therefore of powders. During the last almost four decades there has been a steady level of interest in the problems associated with assessing the shapes of individual powder particles.[27] The reported literature is widely scattered and not always easy to track down and obtain. Since the advent of the computer, a major interest has developed in the general area of pattern recognition. Problems which have been studied in this area include: meteorological cloud patterns,[28] spark chamber images,[29] chromosome shapes,[30] letters of the alphabet,[31] networks,[32] simple geometric shapes.[33,34] Three recent texts about pattern recognition are also available.[35,36,61] However, many of the earlier methods for assessing particle shape do not necessarily involve the use of the computer.

In Fig. 2.40 the various methods reported in the literature for particle shape analysis have been arranged into a four-class scheme. Each of the four classes is further subdivided into what have been termed categories. The four general classes are as follows:

Class 1. The assessment of specific characteristics of individual particles.
Class 2. Measurements of bulk properties of powders.
Class 3. Shape representation by various mathematical techniques.
Class 4. The use of words to convey shape characteristics.

The first two deal with the means of observation. The third transforms a digital image into alternate mathematical form. The last deals with means of description. The categories within these various classes are:

Categories of Class 1. The assessment of shape on an individual particle basis may be divided into three categories. In the first of these the particle shape is related to distances measured between tangents parallel to the particle contour. In the second category, a variety of standard shape outlines is used as a shape comparator. In the third category lengths of specific types of intercepts are measured and used to characterize particle shape.

Categories of Class 2. There are as many categories in this class as there are different methods used to measure bulk properties of the powder mass. For instance, the Beddow method (the use of tap-apparent density ratio), the flow rate method, the bounce method, etc.

Categories of Class 3. At this time there are four categories of generating shapes: matrix mapping, polynomial generation, Fourier series analysis, and syntactic methods.

Categories of Class 4. This class may be conveniently divided into four categories: the use of shape group, the use of essentially separate verbal definitions, the analysis of the informational content of words and the use of morphology classes.

The following notation has been adopted for easy referral to the chart. The first integer denotes the class and the second integer denotes the category. Thus (1.3) indicates that particle shape is assessed on individual particles by methods involving the measurements of lengths of intercepts.

Distances Between Parallel Tangents Class (1.1)

Heywood[37] defines the three limiting dimensions of an irregular particle as the thickness (T), breadth (B), and length (L). The thickness (T) is the minimum distance between two parallel planes tangential to the opposite surfaces of the particle. The breadth (B) is the minimum distance between the two parallel planes which are perpendicular to the planes defining thickness and are tangential to opposite surfaces of the particle. The length (L) is the distance between two parallel planes which are perpendicular both to those planes defining thickness and breadth and which are also tangential to opposite surfaces of the particle. The scheme of this method is outlined in Fig. 5.16. Note that (L) does not necessarily define the longest dimension of the particle, which in Fig. 5.16 is denoted by L'.

There are two other methods of defining dimensions due to Krumbein[38] and Lees,[39,40] respectively. In Krumbein's system, the longest dimension L' is selected to define the length of the particle, and the breadth is then measured perpendicular to L'. Lees method is somewhat similar to this in that he defines (L) as the absolute longest dimension of the particle. However, he further defines an intermediate diameter as the longest dimension in the cross section of the particle, and the thickness is defined as the widest dimension of the particle in its narrowest plane. The problem with this method is that the measured dimensions are not necessarily mutually perpendicular nor orthographic. Heywood's method has the advantage over the other two in that it is less complicated, and it has led to the establishment of relationships between shape ratios and sieving characteristics of powders.[41] The two ratios are defined as the flatness ratio, $m = B/T$ and the elongation ratio, $n = L/B$. If d is the projected area diameter (the diameter of the circle having the same area as the projection of the particle when viewed in its most stable position), and A is the sieve aperture, Heywood found that

$$\frac{d}{A} = 0.98\left(\frac{2nm^2}{m^2 + 1}\right)^{1/2} \tag{6.4}$$

Some specific values of d/A are tabulated in Table 6.9.

Standard Shape Comparisons Class (1.2)

A method of comparing the particle with the characteristics of an equivalent sphere[41] leads to the definition of a surface shape coefficient f and a volume shape coefficient k:

$$\text{Surface area of the particle} = fd^2$$

and

$$\text{Volume of the particle} = kd^3$$

in which d is the projected area diameter.

Wadel[42,43] conceived of two shape factors, sphericity and roundness

Table 6.9[65]

VALUES OF d/A FOR VARIOUS VALUES OF $m + n$

Elongation ratio n	*Flatness ratio*					
	1.0	1.2	1.5	2.0	2.5	3.0
1.0	1.00	1.06	1.14	1.23	1.28	1.32
1.2	1.09	1.17	1.25	1.35	1.41	1.44
1.5	1.22	1.30	1.39	1.51	1.57	1.61
2.0	1.41	1.51	1.61	1.74	1.82	1.87
2.5	1.57	1.68	1.80	1.95	2.03	2.08
3.0	1.73	1.84	1.97	2.13	2.23	2.28

$$\text{sphericity} = 4.84\frac{k}{f}^{2/3} \qquad (6.5)$$

$$\text{roundness} = \text{mean of } \frac{\text{radius of inscribed circle}}{\text{radii of all corners and edges}} \qquad (6.6)$$

These two concepts have been represented in the form of charts for convenient assessment, as have other factors such as angularity and the serration index. Appropriate references are given in Table 6.10.

Table 6.10

TYPES OF SHAPE COMPARATOR CHARTS AND THEIR SOURCE

Author	*Chart*	*Reference number*
Rittenhouse	Sphericity	123
Krumbein	Roundness	38
Lees	Angularity	39, 40
Mackey	Ellipsoidal factor	55
Chaplin	Ten factors	57

The rugosity coefficient is an attempt to assess the roughness of the particle surface. It is defined as the perimeter of the particle profile, including minor irregularities and corrugations, divided by the perimeter of a smooth curve circumscribing the particle profile.[44]

An interesting use of these many shape assessment factors is involved in the reported particle shape characteristics of lunar fines sample 12057, 72.[44] The shadow replicas of three typical particle profiles are given in Figs. 5.18 and 5.19. The particle shape "factors" are reported in Table 5.19.

Hausner[45] proposed a method of assessing particle shape in which the particle under examination, is compared with an enveloping rectangle of minimum area. If the rectangle length is a and its breadth is b, three characteristics are defined:

$$x = a/b \quad \text{– the elongation factor,}$$
$$y = A/ab \quad \text{– the bulkiness factor, and}$$
$$z = C^2/12.6A \quad \text{– the surface factor,}$$

in which

A is the projected area of the particle and

C is the circumference of a spherical particle.

Hausner selected a population of shapes and calculated the three shape factors for each shape as shown in Table 6.11. Medalia[46] has made a most interesting adaptation of Hausner's method to the study of the shapes of carbon black aggregates. In essence, the method involves the development of two dimensionless shape factors termed the anisometry and the bulkiness. The object under study is represented as an ellipsoid with equivalent radii of gyration about the central principal axis. The anisometry of the object is then taken as that of the ellipsoid which is L_a/L_b. The bulkiness factor is determined from a comparison of the volume of the ellipsoid with that of the particle itself. Fig. 6.9 shows the procedure as applied to a typical aggregate. Another variable of the approach uses a standardized template of ellipses as comparators.[47]

Length of Intercepts Class (1.3)

Church[48] proposed the use of the ratio of the expected values of Martin's diameter and Feret's diameter (see Fig. 6.10) as a shape factor for a population of elliptical particles. He investigated the utility of this method for a monosize population and suggested possible methods for extending the technique to the study of other shapes and three dimensions. Cole[49] has developed a range of instruments for carrying out image analysis, one of

Table 6.11

HAUSNER SHAPE COEFFICIENTS FOR TYPICAL SHAPES[45]

	Particle Characteristics		
Shape	Elongation x	Bulk y	Surface z
I	1.00	0.78	1.00
II	1.00	0.78	1.98
III	1.00	0.71	2.04
IV	1.34	0.76	1.89
V	2.23	0.72	1.77
VI	1.97	0.54	2.20
VII	1.60	0.74	1.40
VIII	3.60	0.52	7.50
IX	4.00	0.83	1.89
X	1.40	0.31	2.80
XI	1.60	0.50	1.84

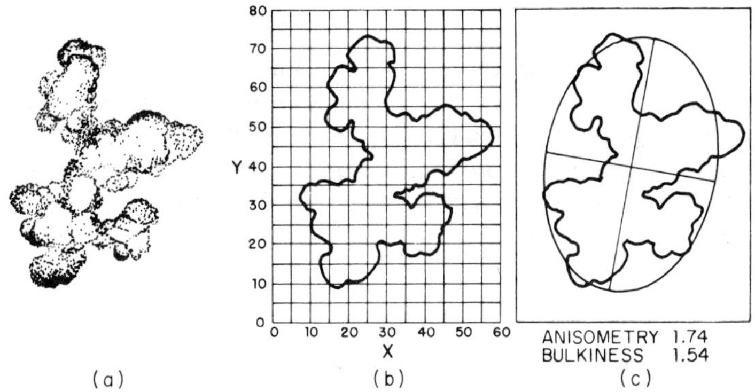

Fig. 6.9 (a) Carbon black aggregate, (b) Contour of aggregate traced for manual digitizing, and (c) Silhouette of aggregate with central principal axes and radius-equivalent ellipse.[46]

these being the Quantimet 720. This is capable of carrying out measurements of Martin's and Feret's diameter, particle periphery, longest dimension, and area in combination with a number count. The basic geometric classifications are shown in Fig. 6.11. The system is capable of a very high rate of measurement—less than one second per field of view has been reported.

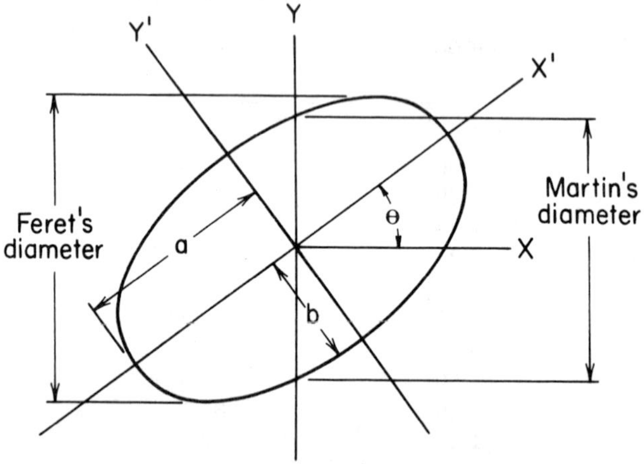

Fig. 6.10 Martin and Feret diameters for an elliptical particle.[48]

Chalkley et al.[50] investigated the use of a needle throwing method to estimate the surface to volume ratios for monosize particles. In this method, if l is the length of the needle, c the number of times the needle cuts a profile, and h the number of times the needle ends lie within the profiles then lh/c = profiles area/perimeter, and so, in three dimensions, $lh/c = 4$ volume/surface of the particles. Fig. 6.12 illustrates the results obtained with this method.

Ratio Apparent/Tap Density Class (2.1)

Beddow[51] proposed the use of this ratio as a shape parameter for metal powder particles and demonstrated its relationship to other bulk powder properties,[52] sieving behavior,[53] and compaction behavior.[54] The usefulness of the method relies on the rapid establishment of a limiting value of tap density in the Rotap density test. Although this value is established in the case of metal powders (see Fig. 6.13), it is not established in the case of nonmetallic powders (see Fig. 6.14).

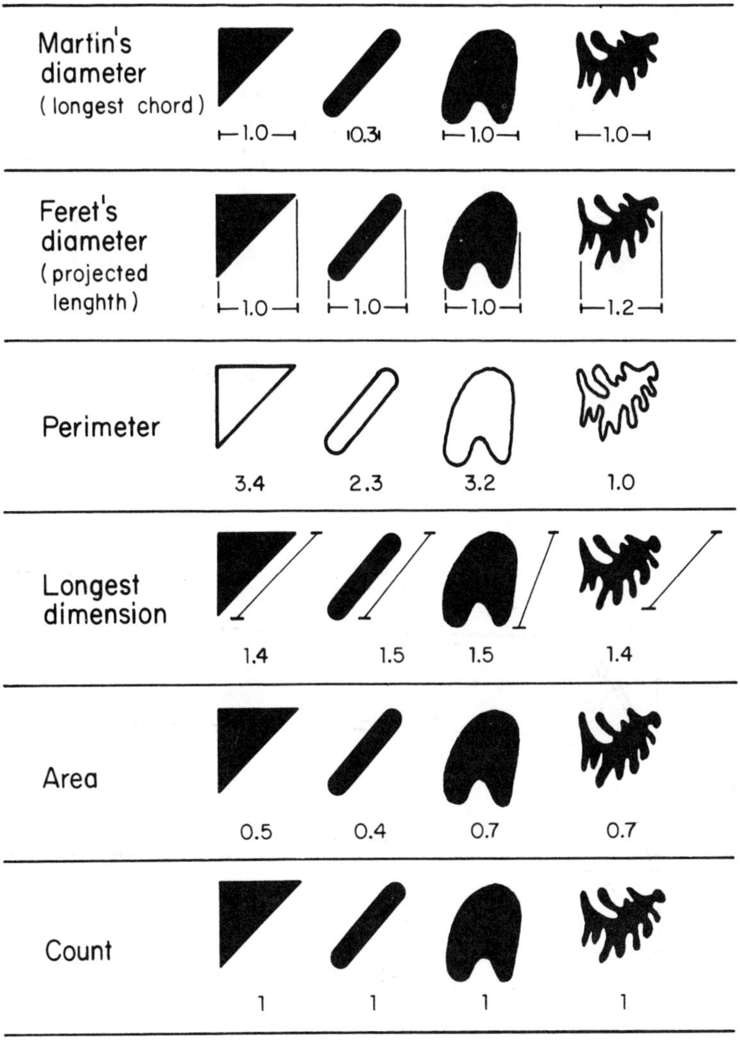

Fig. 6.11 Illustrating the potential of Cole's method.[49]

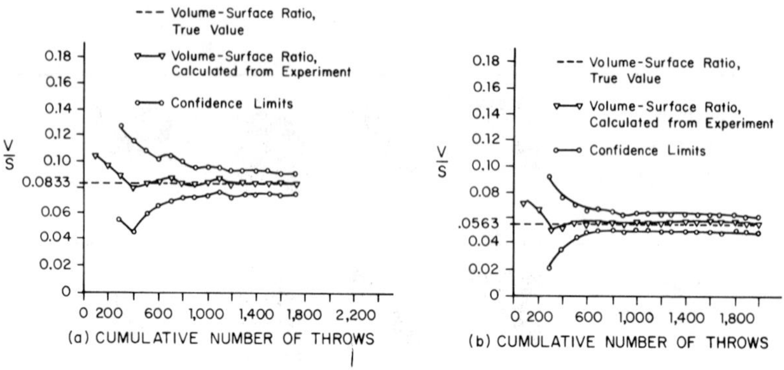

Fig. 6.12 Pin throwing experiments.[50]

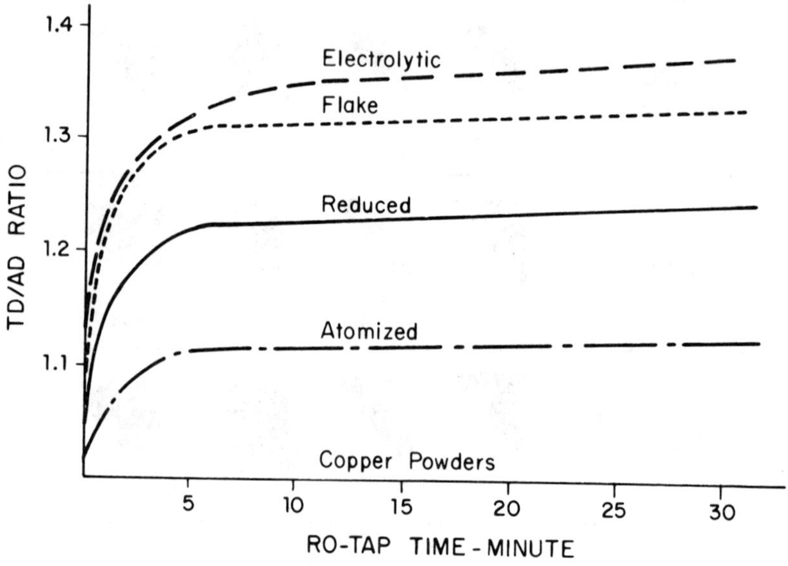

Fig. 6.13 *TD/AD* for metal powders.[54]

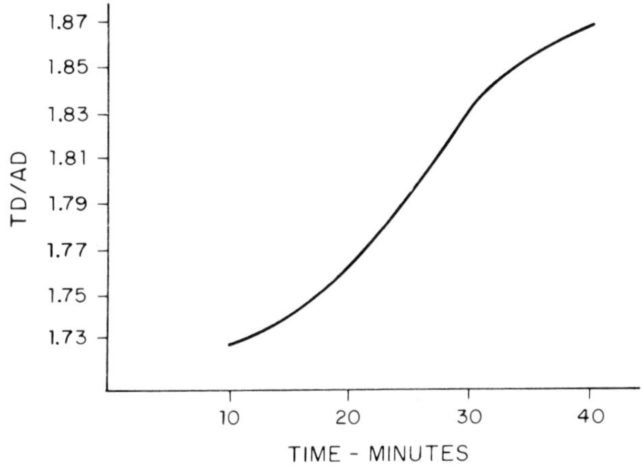

Fig. 6.14 *TD/AD* for light magnesium oxide.

Other Bulk Methods Class (2.2)

The porosity of a powder mass under a given degree of compaction has been used as an indicator of shape.[55,56] The use of bulk properties such as flow rate, permeability, and bouncing has also been reported.[57] A modification of the latter method has been used to separate particles of different shape.[58]

Fourier Series Class (3.1)

A procedure for applying a Fourier Analysis-Synthesis technique for particle shape-profile analysis and classification has been developed.[59,81] Using this procedure it is possible to obtain both an analytical set of coefficients which are in themselves a shape classification tool and to approximate the original profile by feeding the coefficients back into the computer program and generating a shape profile mathematically. Comparison between the computer generated shape profile and the original enables the experimenter to quickly establish the requirements for additional analytical measurements. Fig. 6.15 shows a representative set of regenerates for a range of points per particle profile 12, 24, 48, and 100 and a range of coefficients 3, 5, 10, 15, and 20.

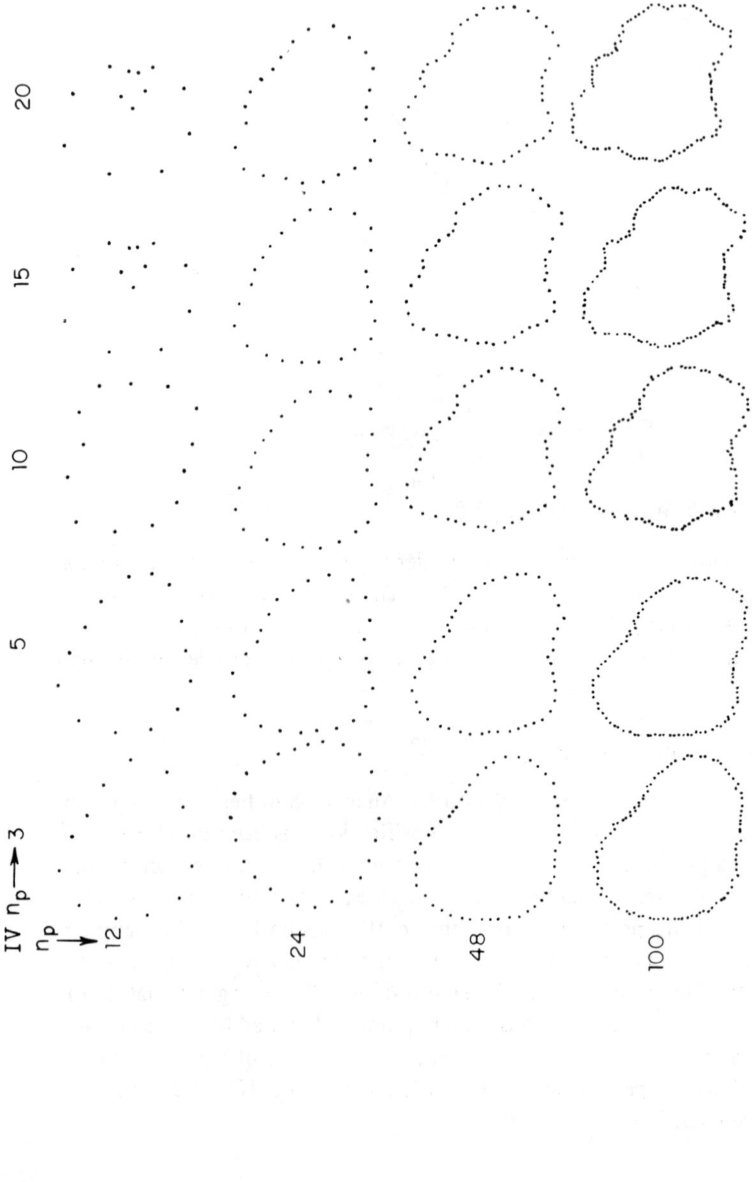

Fig. 6.15 Fourier analysis-synthesis regenerated profiles of shape IV of Table 6.11[59]; n_p = number of sample points on perimeter and n_k = number of Fourier coefficients.

Syntactic Methods Class (3.2)

A language consists of strings of characters called terminal symbols which are arranged into certain structures by the operation of the rules of the grammar of the language.[61] A visual pattern can be recognized if the proper terminal symbols and appropriate rules for the pattern generation have been used. For further information the references given in the introduction should be consulted.[28–36]

Other Methods of Generating Shapes Class (3.3)

The technique of obtaining the periphery of the particle in the form of a polynomial equation has been tried, but suffers from the disadvantage that if the closeness of fit is too demanding, the calculated trace curve starts to deviate widely from the actual contour of the particle. This is an illustration of Gibbs Phenomenon and demonstrates the limitation of this method. (Note also that the Fourier method is also subject to the same limitation.) Theoretically the shape of a particle can be represented by matrix mapping. Furthermore, it should be possible to design a set of photocells in conjunction with a fiber-optical lens systems to provide an observation tool with which to examine microsections or profiles of particles under study. The matrix obtained from one observation would consist of a set of zeros and ones, either digit representing the location of the profile. The form of the data would be conveniently designed for computer processing.

Shape Groups Class (4.1)

The observer naturally attempts to use words to designate the shape of the particles under observation, and various authors have chosen arbitrary ratios of B/L and T/B to designate these words. The methods of Zingg,[62] Rosslein[63] and British Standard 812/1960 are shown in Fig. 6.16. Heywood[64] developed the concept of the volume coefficient for an equidimensional particle k_e defined as

$$k_e = km(n)^{1/2} \tag{6.7}$$

He was able to relate k_e to the shape groups shown in Table 6.12. Bird[65] studied the variation of m with size for crushed glass. Fig. 6.17 shows that both calculated and measured values of $1/m$ agree closely, thus illustrating the utility of Heywood's concept of k_e.

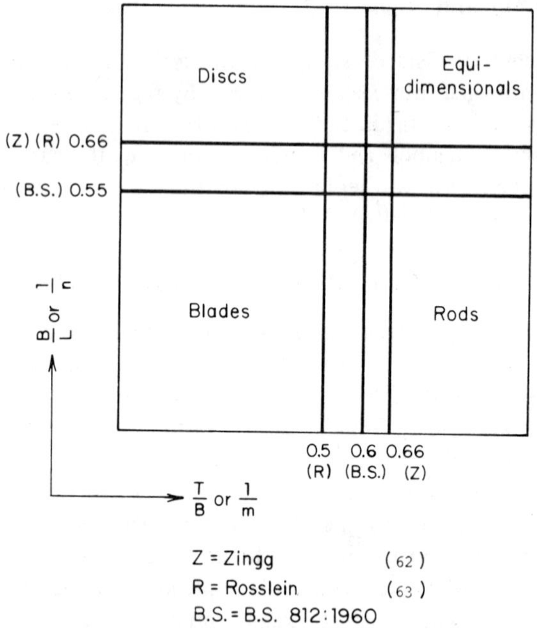

Fig. 6.16 Three-dimensional shape groups.[62,63]

Table 6.12

ILLUSTRATING THE COMBINED USE OF CLASS 1 AND CLASS 4 METHODS[65]

Shape group	k_e
Angular	
Tetrahedral	0.38
Prismoidal	0.47
Subangular	0.51
Rounded	0.54

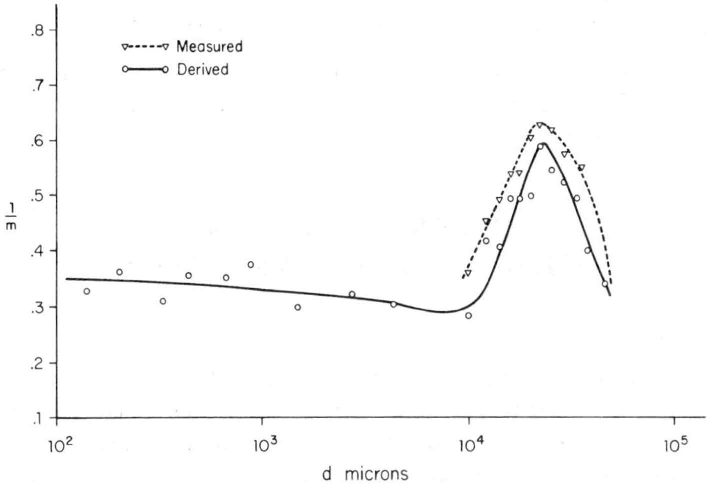

Fig. 6.17 $1/m$ versus d from Bird's result.[65]

Other Methods Using Words

A fairly comprehensive glossary of descriptive terms for particle shapes is given in British Standard 2955.[66] One approach which has not so far been attempted is to try to use the concepts of information theory to measure the informational content of words used to describe shape. If this were possible, it would afford a way of assessing the utility of verbal descriptions by comparing the amount of information obtained with the actual effort expended in obtaining it.

Riley and Magnuson[67] have observed that it may not be sufficient to describe the shape of individual particles in order to specify the important behavior of a representative population. For example, they differentiated between at least four types of shape characteristics in the case of refractory powders:

Type A particles with hollows/porosity in them,
Type B dense particles,
Type C agglomerated particles, and
Type D hollow particles, equivalent to agglomerated pores.

However, not all of these characteristics are appropriately measured in conventional shape analysis—for example, the Type D particle would only be recognized after sectioning.

6.5 PATTERN RECOGNITION AND PARTICULATE CHARACTERIZATION[60]

Introduction

It was noted at the beginning of this chapter that there exists a formidable problem of information collection and condensation (or compression) in fine particle characterization. The field of pattern recognition is concerned with the modelling of pattern classes and the formulation of decision rules (via various alternate criteria) in order to render the decision-making process as to the classification of a particular set of data (the pattern) a feasible and relatively reproducible and reliable process. The study and utilization of the principles and methods of pattern recognition appears to be fairly important, therefore, in the area of fine particle characterization.

At the present time there are some 30 methods of particle shape analysis which have been either developed or proposed.[27] However, until recently there has been little interaction between the area of pattern recognition and particle shape analysis with the consequence that particle shape analysis methods do not generally utilize pattern recognition principles.

The essential elements in pattern recognition are summarized as follows:

1. There must be a choice, for comparison purposes.
2. The criteria are assumptions that some decision rule can be based upon a specific measurement regime. Some criteria include:
   a. distance in the pattern space is the unit of discrimination,
   b. clustering in the pattern space is the unit of discrimination,
   c. expected value of loss or gain is the unit of discrimination,
   d. probability of failure is the unit of discrimination, and
   e. double criteria can be used: minimum cost of misclassification coupled with the lowest probability of error is the unit of discrimination.

3. The observer collects information and divides it into relevant and irrelevant categories. The relevant information may be either deterministic or indeterministic.
4. The abstracted information is then subjected to the iterational procedures of an approprite algorithm. The algorithms use either gradients (which yield directions for change) or size (which deal with systems of equations and indicate the amount of change required to improve the classification.
5. The iterations should lead to convergence in order that successful termination of the procedure shall be achieved.

The following discussion briefly outlines five of the methods for pattern recognition and attempts to show how these can be made use of in particle shape analysis. The five basic classification methods of pattern recognition dealt with here are:

1. classification by distance functions.
2. classification by likelihood functions.
3. classification by trainable classifiers–deterministic methods.
4. classification by trainable classifers–statistical methods.
5. classification by syntactic methods.

The following discussion leads to the expectation that various methods of pattern recognition are extremely useful in the search for improved methods of particle shape analysis. In particular, it can be expected that they will complement those methods of analysis already in use.

Classification Methods

The main objective of a classification process is to yield decisions as to the pattern class into which a specific pattern can be classified. The first step in the classification process is to collect appropriate data measured from the patterns to be recognized. Each element of this data describes a characteristic of the pattern. These data elements can be arranged in the form of a measurement or pattern vector.

The second step in the process is to classify (i.e., group into appropriate sets) the pattern vectors. The methods considered here pertain to the second step.

1. Pattern Classification by Distance Functions

A. The Concept. Assume that the classification process should be capable of recognizing M different pattern classes, denoted by $w_1, w_2, w_3, \ldots, w_M$. The recognition problem now reduces to that of choosing the best alternative out of the M possible alternatives that will approximate as closely as possible to the pattern X under consideration.

B. The Operational Procedure. See Fig. 6.18. If $D_i < D_j$, then

$$X\{W_i\} \tag{6.8}$$

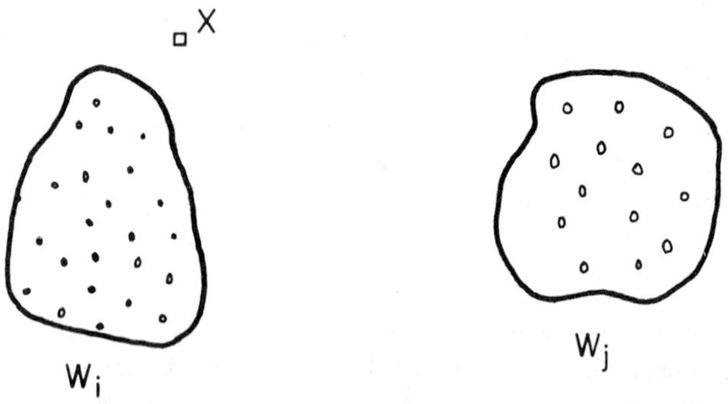

Fig. 6.18 Patterns classifiable by proximity concept.[60]

C. A Practical Example. Many differing shape factors have been reported in the literature which attempt to represent the shape properties of particles. As an illustration, two particular examples–anisometry and bulkiness,[46] are selected. As shown in Fig. 6.19, the centrum of a circular type population is represented by point 1 and that of an elliptical population is represented by point 2. The measured values of the two variables for each particle examined are then plotted on the same graph. The respective distance functions can then be measured and the individual particles assigned to their appropriate category.

2. Pattern Classification by Likelihood Functions

A. The Concept. Pattern classification by likelihood functions uses statistical concepts to derive the decision rules that minimize the probability of committing classification errors. In the case of classes W_1, W_2, there are $P(X|W_1), P(X|W_2)$.

B. Procedure. The objective is to minimize error of misclassification. Select sample patterns are assigned to class W_J which produces loss expectation R_J when

$$R_J = \sum_{I=1}^{M} P(W_1/X), L_{IJ} \qquad (6.9)$$

L_{IJ} is the loss if X is assigned to W_J when it should really be assigned to W_1. Similarly, R is deduced for other selected values of J. The pattern is correctly assigned when R_J is a minimum for $J = 1, \ldots, M$. This scheme is illustrated in Fig. 6.20.

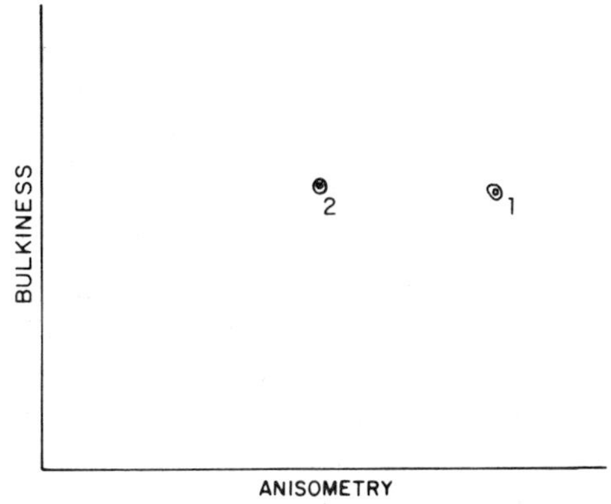

Fig. 6.19 Illustration of the use of distance functions.[60]

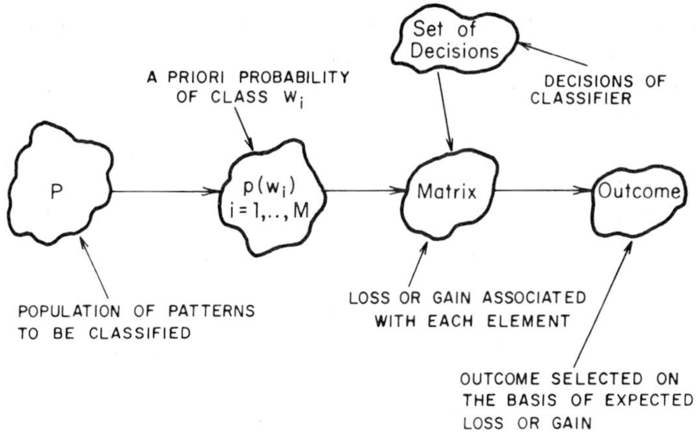

Fig. 6.20 Conceptual framework for pattern classification by likelihood functions.[60]

C. A Practical Application. As an example of the potential use of this method of shape classification, the Fourier analysis–synthesis method is selected. This technique transforms a set of measured (X, Y) coordinates of the particle profile into a matrix of Fourier coefficients. It is known that in the case of simple numeral shapes, the observer can correlate the characteristics of the matrix with those of the number which it represents.[70] Assume that the first five or so member coefficients in a given set may be representative of some feature of the particle and the remaining $(n - 5)$ coefficients may be representative of another feature of the particle. Using automatic image measuring systems[71] it should be possible to establish probability density functions of coefficients (and their associated properties) with a view to standardizing measures for both features.

3. Trainable Pattern Classifiers–The Deterministic Approach

A. The Concept. In the trainable pattern classifiers method the decision functions are generated by means of iterative, "learning" algorithms as opposed to the direct calculation in the two methods described earlier. As the name implies, the deterministic approach does not make any assumptions regarding the statistical properties of pattern classes. The principle of this method could be illustrated through a simple example of the perceptron algorithm which is based on this method.

B. Procedure. Given two training sets of patterns belonging to two different pattern classes, this algorithm choses an arbitrary initial decision function. If the decision function fails to classify the corresponding training pattern correctly, the coefficients of the decision function are modified using a constant correction increment. This process is repeated until satisfactory classification is attained.

C. A Practical Example. As an example of the possible application of this approach, consider gas and water atomized powders. It can be conceived that the gas atomized powder particle characteristics might be divisible into two groups: smooth spheres and smooth ellipsoids. Similarly, the water atomized powder particles might be divisible into two groups: rounded irregular and elongated irregular. As shown in Fig. 6.21, these shape groups could be delineated with comparatively few approximations.

4. Trainable Pattern Classifiers–The Statistical Approach

A. The Concept. The statistical approach of trainable pattern classifiers

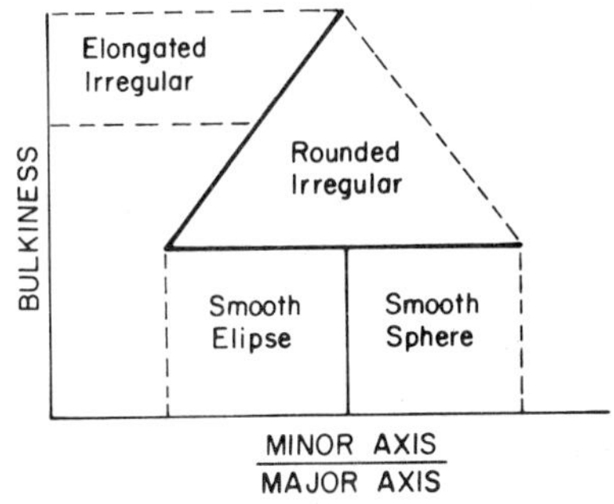

Fig. 6.21 A tentative representation of shape classes of atomized powders (this does not claim to be accurate; it is for purposes of illustration only).[60]

takes into account the statistical properties of pattern classes in determining the decision function. As in the method of likelihood functions, this method obtains the optimal decision function by minimizing the probability of misclassification of a pattern (i.e., minimizing error). In addition, the learning effect is incorporated through the use of a suitable probability function in the decision function.

B. A Practical Application. Whereas the deterministic TPC, as exemplified above, is what one might call a conceptual tool, the statistical trainable pattern classifier is an analytical tool. If automatic particle shape measuring and data handling devices are available, the observer can determine just how much error is associated with a particular set of already defined shape groups by using optimizing techniques in conjunction with probability density functions of the measured data. These procedures can therefore assist the observer in modifying the set of shape groups in order to make them more in line with reality and consequently more useful.

5. Syntactic Pattern Recognition

A. The Concept. In essence, this method attempts to establish the rules with which a specific pattern can be developed from a given set of pattern primitives. This is by way of analogy with formal linguistics[72] in which a specific pattern (the language) can be developed from the primitives (the letters, words) by the operation of a set of rules.

The entity of concern is called a grammar G, which generates a language $L(G)$. In general, a grammar is defined as a fourtuple,

$$G = \{V_T, V_N, S, P\} \tag{6.10}$$

in which

V_T indicates the terminal symbols, seen in the case of the visual pattern. They are also called primitives;
V_N indicates the nonterminal symbols;
S indicates the starting symbol(s); and
P indicates the production rules. These explain how to proceed from the starting symbol and finally how to produce the terminal symbols in the appropriate structure.

Fig. 6.22 illustrates the meaning of these terms.

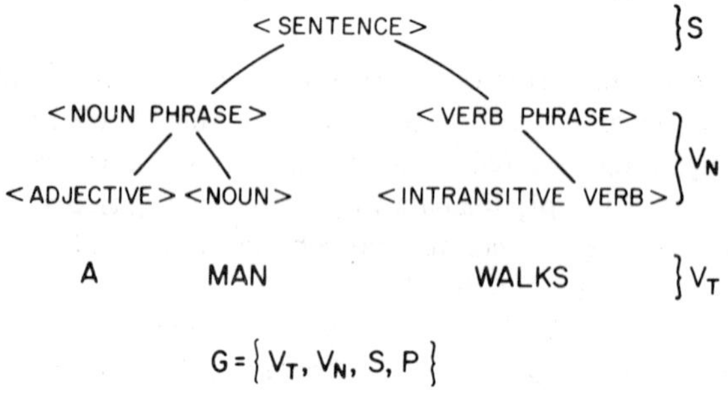

Fig. 6.22 An example of a Fourtuple grammar.[60]

B. A Practical Application. Consider the PDL* grammar with rules shown in Fig. 6.23. The grammer is fourtuple as follows:

$$G = \{V_N, V_T, S, P\} \tag{6.11}$$

$$V_N = \{S, A_1, A_2, A_3, A_4\} \tag{6.12}$$

$$V_T = \{a \nearrow \ b \searrow \ c \rightarrow \ d \downarrow\}. \tag{6.13}$$

The production rules are charted:

Step

$$P:\ S \rightarrow d + A_1 \qquad 1 \tag{6.14}$$

*PDL = Picture description language.

$$A_1 \rightarrow c + A_2 \qquad 2$$

$$A_2 \rightarrow {}^{-}d{*}A_3 \qquad 3$$

$$A_3 \rightarrow a + A_4 \qquad 4$$

$$A_4 \rightarrow b \qquad 5$$

Note: ^{-}d indicates $d\uparrow$ and $S \rightarrow d + A_1$ should be read as S *is replaced by* $d + A_1$. Applying the rules generates the following productions:

i.e., a pentagon has been constructed.

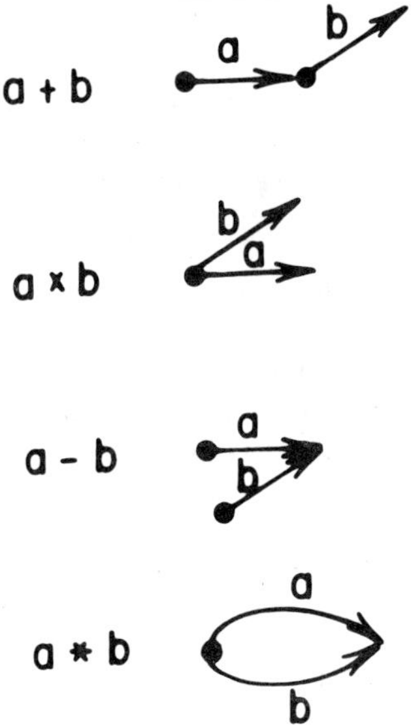

Fig. 6.23 PDL rules.[60]

6.6 ON THE DESIGN OF A SYSTEM FOR PARTICLE SHAPE ANALYSIS

A Tentative Outline of a System

The primary purpose of a system of particle shape analysis is to enable an observer to unequivocally identify the shape or shapes of a set of particles in a way which is acceptable to other observers and then to permit communication of this information to other interested observers so that they too can understand the identification. The proposed system consists of a series of steps, each of which constitutes an action or set of actions intended to provide information such that the observer can decide whether a practicable sufficiency of information has been obtained or whether more is needed. The system proceeds from the most general description to more and more specific identifications. Clearly, the first step must supply for the most gross requirements enough information and no more, yet provide for a satisfactory overview for those whose observation must be more acute. Then, later steps develop more minute information for those observers with special needs and interests. At each step, identification can be exact within known limits of preciseness and communication can be effected.

Step 1 satisfies the major reason for an observer's wanting particle shape data in the first place. This is the necessity to determine and to communicate particle shape data unequivocally to another observer. For this purpose, the first step should have the following characteristics:

1. it should be logical.
2. it should require a minimum of standardized equipment and/or facilities,
3. it should demand only a readily attainable level of skill, and
4. the bulk of the necessary activity should be confined to little more than general observation, decision making, and a recording of the result (or results).

It follows from the above that although Step 1 may be complex internally, the observer need not understand these details in order to use it effectively.

Dispatch in carrying out an examination of a set of particles must be possible if Step 1, with its attendant technique(s) is to be fully utilized in the powders field.

Step 2 of the proposed system of shape analysis is a judgment step which comprises three operations.

First, the observer has to decide if the information obtained from Step 1 is sufficient for the purpose in hand. This is not a simple, clear-cut decision to make because, even though there may be a need for additional information, the extra effort may not be worth the expected results.

Second, the observer has to be able to ascertain the cost of making more particle shape measurements in the case of each of the available methods. (There are some 30 or more such methods.) These costs not only include the data collection and handling but also the sample preparation and storage activities.

Third, once the basic cost of measurement has been determined, the observer then has to decide whether the merits of the specific method are worth the expected expenditures. At the present time, it is not clear what a general list of merits would comprise. However, two important items in such a list are the characteristic of resolving power and the expected benefits for the observer.

Step 3 of the system is the application of one or more from a wide variety of shape assessment schemes, depending upon the type of information needed. These schemes are briefly summarized in Section 6.4. The tentative outline of the system of three steps is shown in Fig. 6.24.

The Need for Such a System of Shape Analysis

For the analysis of shapes, many methods have been used directly for particle shape determination, others have been proposed, and some have been speculatively considered to be usable in particle shape analysis.

Is such analysis important or necessary? The answer is "yes," but the degree of preciseness required probably depends upon the point of view of the observer. One can distinguish perhaps three basically different concerns. These might be conveniently labeled the hypothetical point of view, the practical point of view, and the objective point of view. Some years ago an opinion poll was conducted in the ceramics industry in order to ascertain, among other things, what individuals in that field consider to be the most important characteristics of a particle. The results listed the following three items as being the most important particle characteristics: size, surface area, and shape.

Based upon what is generally understood about particulate science and technology, it seems reasonable to assume that these three characteristics would also be listed in many other areas in the powders field. Because the data was obtained from an opinion poll, which is an armchair type of ac-

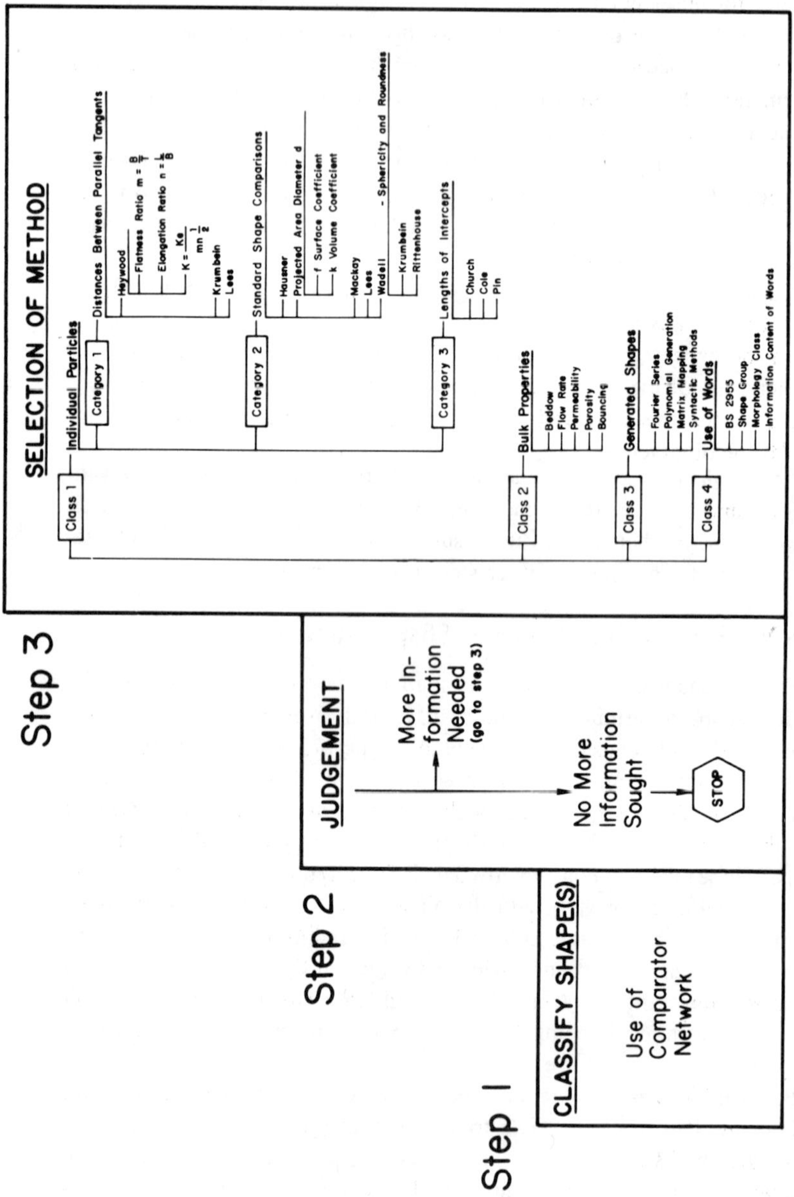
Step 1
CLASSIFY SHAPE(S)
Use of Comparator Network
Step 2
JUDGEMENT
More Information Needed (go to step 3)
No More Information Sought
STOP
Step 3
SELECTION OF METHOD
Class 1
Individual Particles
Category 1
Distances Between Parallel Tangents
Heywood
Flatness Ratio m = B/T
Elongation Ratio n = L/B
K = Ke/mn^(1/2)
Krumbein
Lees
Category 2
Standard Shape Comparisons
Hausner
Projected Area Diameter d
f Surface Coefficient
k Volume Coefficient
Mackay
Lees
Wadell - Sphericity and Roundness
Krumbein
Rittenhouse
Category 3
Lengths of Intercepts
Church
Cole
Pin
Class 2
Bulk Properties
Beddow
Flow Rate
Permeability
Porosity
Bouncing
Class 3
Generated Shapes
Fourier Series
Polynomial Generation
Matrix Mapping
Syntactic Methods
Class 4
Use of Words
BS 2955
Shape Group
Morphology Class
Information Content of Words

Fig. 6.24[126]

tivity, it should be quite safe to label this point of view hypothetical.

The practical answer based upon what people do is more difficult to ascertain because evidence is sparse. Certainly, it is known that particle shape measurements are made in the case of carbon.[46] The process used appears to be quite sophisticated. It is also known that particle shape is most important in paint manufacture,[74] in the varied ramifications of geology, in powder metallurgy,[69] and also in the case of ceramics, already mentioned. Inspection of the literature indicates that there is an increase of interest in particle shape in recent time,[69,75] but this is almost exclusively concerned with research and development rather than with manufacturing production.

It is often the case that when particle shape is stated in a specification, the requirement is stated verbally, with no numerals. Usually one word or phrase is used as a proxy to describe the shape characteristics of the powder. It is clear that the use of a system of particle shape analysis that would afford a much higher precision of particle shape description would in turn make it possible to attain a more precise degree of control over the powder producing process and consequently a higher degree of control over the subsequent powder processing steps would be possible. We are proposing the very general argument that once the two basic parameters of particle size and particle shape are described with some exactness, the resultant increase in preciseness affords corresponding improvements in industrial processing, and this throughout the whole field of powders. The same argument can be applied in the case of research and development problems in powder science and technology.

Therefore it is concluded that not only is the development of a system of particle shape characterization a problem of intellectual worth and interest, but also it is a problem of the greatest practical and theoretical significance for all facets of the powders field.

One generally applicable method used in obtaining successively more precise descriptions of particle shapes follows.

Step One

There are three distinguishable fundamental methodological concepts which may be utilized for the whole or for a part of Step 1 in the shape analysis system under consideration. These are:[76]

1. the membership-roster concept,
2. the common property (or feature) concept, and

3. the clustering concept.

The membership-roster concept works in the following way: patterns belonging to a known pattern class are compared with the unknown pattern of the particle being examined. When the pattern of the unknown sample is found to correspond more closely with the pattern of one of the known class patterns, than with others, the sample may then be classified as being a member of that class. The common property concept assumes that samples with patterns belonging to the same class possess similar features. The sample pattern is examined, its features are noted, compared with the stored patterns, and classified into the class set with which the best match can be made. In the case of the clustering concept, if the pattern vectors are real numbers, a pattern class can be characterized by the clustering properties of these samples in the pattern space.

The implementation of the above concepts may be carried out by using one of three methods (or a combination of them). These methods are heuristic methods, mathematical methods, and linguistic or syntactic methods. Briefly, the heuristic methods are based on "intuition" and experience and usually make use of the membership-roster, the common property concepts, or both. The mathematical methods formulate the rules of classification and measurement in a mathematical framework and make use of the common property concept and the clustering concept. The syntactic methods make use of the fact that a pattern can be described by a hierarchial structure of subpatterns. This is analogous to the syntactic structure of languages* and permits the use of formal language theory in pattern recognition problems.

In the design of our proposed system of particle shape analysis, what concept(s) should be involved and which methods should be used?

Step 1 in our particle shape analysis system should be basically a heuristic one using the membership-roster concept, because, as there is a dearth of information concerning the shapes to be analyzed, a very flexible subsystem is needed.

One major criticism of using the membership-roster concept is that it is an ad hoc approach to the problem and has little or no underlying theory. Although this is true it should be clear that the fundamental purpose of any particle shape method is to relate shape characteristics to the basic physical laws which control the formation of the shapes and to the adaptation of the particle to various environmental forces to which it may be

*A language is generated from its basic components by a grammar; similarly a pattern is generated from pattern primitives by specific rules (i.e., by a grammar).

subjected. Consequently, the criticism that heuristic methods are relatively "simple-minded" in concept is tolerable as long as it provides a means of obtaining useful information about fundamental relationships between shape, particle formation, and adaptation. These relationships form our underlying theory. It is, therefore, the purpose of this present effort to develop Step 1 so that it is consistent with the aforementioned fundamental objectives. Once this is accomplished, more refined methods of shape analysis may then be selected in Step 2 and utilized in Step 3 in order to clarify the nature of these fundamental factors and the ways in which they operate in determining particle shape. It naturally flows from this that the whole system of analysis can then be further improved.

Choice of Methodological Concept

A heuristic Step 1 was chosen on the grounds of flexibility. In addition, the user of Step 1 should need to do as little work as possible in using it, a minimum of equipment should be involved, the method should not be inconsistent with further theoretical development, and finally Step 1 should not require a high degree of special skill for successful operation.

It has already been noted that heuristic methods can be used to permit the observer to allot a specific particle to a particular shape category. The samples of the particles taken from the set are in the first instance, allotted individually to their appropriate categories of pattern. Suitable generalizations concerning the particle set which the sample represents can then be made.

The following examination of how an observer who does not use this system analyzes samples confirms that the comparator subsystem is an appropriate design because it takes advantage of available skills.

It has been established that the shape of a particle is the recognized pattern of relationships among all of the points chosen to represent the external surface. Now, clearly, when an observer examines the shape of an object, the rapid eye scanning that occurs in no way attempts to account for all the points of surface, but instead performs some sort of sampling operation. If the observer is very familiar with the shape being observed, the scan is quickly completed and the classifying decision almost instantaneously made. The interesting point here is that the first thing an individual does in deciding on the shape of a particular object is to use the "built-in" *comparator based system of classification*, with the categories structured by the individual's experiences and presumably based upon some logical process.

It is interesting and important to note, that when carrying out the classifying activity, *the various categories that one allots the particular object's shape are not exactly or very clearly defined.* Thus it can be said that the object's shape is "rounded," "spheroidal," "more or less spherical," "a rough sphere," "a rounded appearance," and so on. All of these descriptions convey the idea of sphericity without stating that the object is a sphere. And, again, what may be rounded to one may be described by another as spheroidal. Both observers may each quite clearly understand the other with little difficulty, or they may not, to an extent which in some part must depend upon the commonality of experience of the observers.

To return to the original argument, the fundamental basis upon which Step 1 is founded should be geometrical for the following reason: a geometrical basis would be a generalization broad enough to cover both examples; it would also possess the capability of being specific to each, as long as a satisfactory procedure for developing shape categories can be provided.

In the case of particles produced from liquids, minimization of surface energy is of importance in the shape formation,[77] and in the case of particles produced by comminution, a minimum energy of fracture path is of importance[78] in the shape formation. The former produces rounded shapes and the latter can produce angular shapes.

An alternative to a geometrical basis could be the use of the letters of the alphabet as the basis for the pattern in the Step 1 shape comparators.* These 26 letters have the merit of being very familiar to us all. One can develop some quite interesting shapes from a letter (or even a number such as 2) and the letters can always be further broken down into more simple geometrical shapes. However, geometry is clearly more fundamental than this proposed use of a familiar set of symbols. Another alternative is the use of verbal descriptors.**

Because the observer is going to perform Step 1 with a minimum of ancilliary apparatus, it can safely be assumed that the unaided eye is the only optical instrument used. Therefore, the observation made earlier that the various categories to which the particular object's shape is allotted are not exactly or clearly defined becomes important as it contains the clue to the method of generating the required shape categories sought.

---

*Note: This is the result of a discussion with Dr. Han C. Wu.

**Note: This suggestion is due to Professor L. A. Zadeh.

Mathematical Treatment

The shape of a particle is an intrinsic characteristic; it represents the quality of a portion of matter.[27] Therefore, to compare the shape of one particle with that of another is to compare their qualities with reference to some specific characteristic(s) (basically a pattern in each case). This is facilitated by applying the recently introduced mathematical concept of the fuzzy set.*

In mathematics, a set is usually conceived as having a precisely defined boundary such that one can distinguish readily between those members that belong and those that do not. In an ordinary set, therefore, there are two levels of belonging represented by 0 for nonmember and 1 for a member. There are *only* two possibilities as regards membership. By way of contrast, and of great practical interest in the real world, all members of sets do not share all characteristics to the same degree. It is often extremely difficult to decide, whether a particular object is a member of a set or not. In response to this problem, the concept of the fuzzy set has been introduced.[79] A fuzzy set is a class in which there is graded membership.[80] Consider a familiar example to illustrate the point: club membership. Using a normal set concept there are only two levels of belonging, members and nonmembers. This can be represented by the group (1,0).

In the case of the fuzzy set we can conceive of various grades of membership, for example: life member, annual member, monthly member, weekly member, honorary member. These can be represented by group shown in Table 6.13.**

Note that the notion of "belonging" has a different role in the case of fuzzy sets. Thus, all members have in common only that their membership function is positive (i.e., lying somewhere between zero and one), but their grade of memberhsip runs the gamut between zero and one. The relevance of the fuzzy set to the problem in hand is shown in the following example: consider a group of shapes consisting of a sphere and a variety of ellipses of gradually increasing eccentricity. Five specific shapes, one sphere and four ellipsoids, are arbitrarily chosen. If the ratio of the minor to the major axis is taken as the membership criterion the grades of membership could be stated as shown in Table 6.14.

---

*Note: The author is grateful to Dr. D. Penrod who first brought the topic of the fuzzy set to his attention.

**This example is a step function. More practical examples are continuous. Some idealized continuous functions as the S and π type are illustrated in Fig. 6.49.[88]

Table 6.13

A FUZZY SET OF CLUB MEMBERSHIP CATEGORIES[126]

Grade	*Category*
1.0–0.9	life member
0.0–0.8	annual member
0.8–0.7	monthly member
0.3–0.1	weekly member
0.1–0.0	honorary member

Table 6.14

FUZZINESS IN PARTICLE SHAPE CATEGORIZATION[126]

Grade	*Category*
1.0–0.9	spherical
0.9–0.8	bulky ellipsoid
0.8–0.4	thin ellipsoid
0.4–0.2	sharp ellipsoid
0.2–0.0	needle

Therefore, if the shape comparators are used the fuzzy sets must be used too. In the shape comparator method, an observer does not have a clear and sharp way of distinguishing between one shape and another. The boundaries between one category or class and another are not sharply defined.

In a fuzzy set, the source of fuzziness may be because the goals are not sharply defined, the constraints are not sharply defined, or both are not sharply defined.[80] Statements illustrating fuzziness in respect to particle shape are e.g., particles are flakey, particles are rounded, etc. The meaning of flakey and of rounded are not sharply defined. When either the goals, the constraints, or both constitute classes of alternatives whose boundaries are not sharply defined the decision process (e.g., the judgment as to the shape of the particle) is an example of decision making in a fuzzy environment. Consider the following example:

x should be in the 'vicinity' of x_o when x_o is a constant,

and its analog:

a particle is 'more or less' spherical.

The source of fuzziness in the mathematical statement is "vicinity" and in the verbal statement it is "more or less."

The fundamental point here is that the source of imprecision can be fuzziness rather than randomness, as is commonly assumed in much of science. Clearly, always assuming that the source is randomness must be guarded against. In actuality, the source may be fuzziness because goals, constraints, or both are only loosely defined. See additional characteristics of fuzzy sets given below.

There is a clear distinction between fuzziness and randomness. Thus, randomness is concerned with the degree of uncertainty of membership or nonmembership of a particular object in an ordinary set.* By contrast, fuzziness is concerned with grades of membership intermediate between full membership and nonmembership. In Fig. 6.25 a Venn diagram for a normal set is contrasted with a Venn diagram for a fuzzy set. Four grades of membership were assumed for purposes of illustration. A fuzzy set, therefore, consists of a class of objects with a continuum of grades of membership. Such a set is characterized by a membership function which assigns to each object a grade of membership ranging from zero to one.[79] The following specifics can be defined:

an empty fuzzy set,
the complement of a fuzzy set,
containment,
union of two fuzzy sets, and
intersection of two fuzzy sets.

Additionally, many identities which obtain for ordinary sets can be extended to fuzzy sets, such as De Morgan's laws, and distributive laws.

Finally, many algebraic operations can be conducted on fuzzy sets, as they can be on ordinary sets:

algebraic product,
algebraic sum, and
absolute difference.

---

*By the term ordinary set a nonfuzzy set is implied.

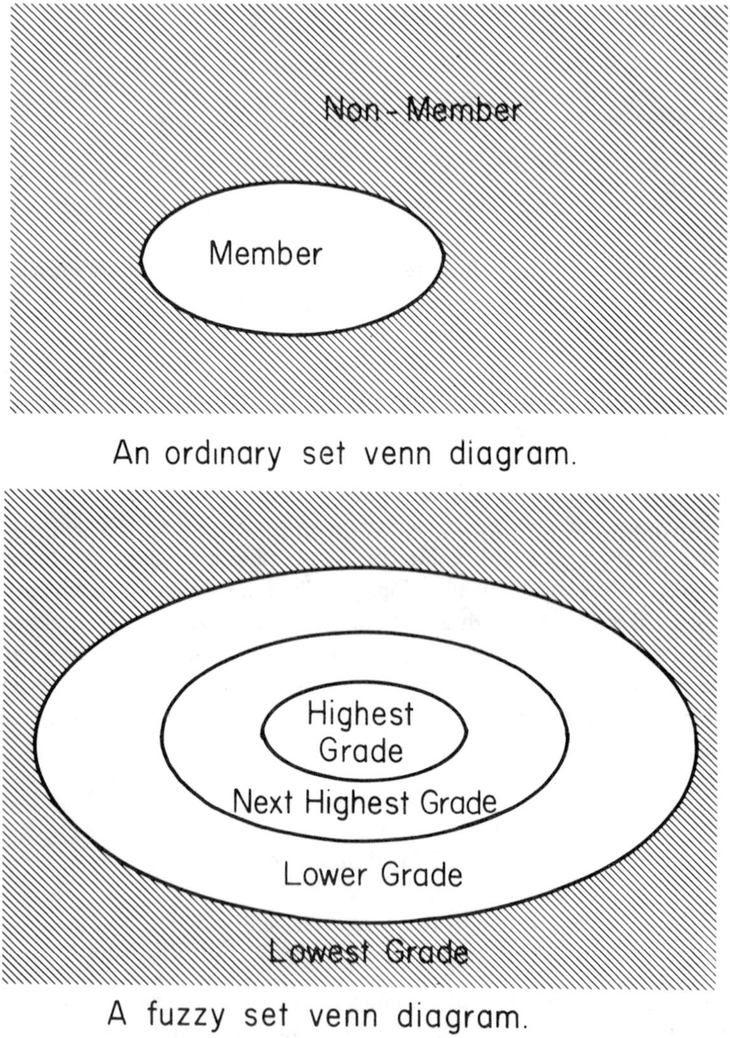

Fig. 6.25[126]

These definitions, identities of operations, are more fully described in Section 9 of this chapter.

The use of fuzzy sets in analyzing shapes of particles is illustrated in the following example. Let X be a space of objects with generic element x. Thus

$$X = \{x\}$$

A fuzzy set A in X is characterized by a membership function $f_A(x)$ which associates with each object in X a real number in the interval $[0, 1]$, with the value $f_A(x)$ at x representing the grade of membership of x in A.

The intersection of two fuzzy sets A and B with membership functions $f_A(x)$ and $f_B(x)$ respectively is a fuzzy set $C = A\ B$.

C has a membership function

$$f_C(x) = \min\ [f_A(x), f_B(x)]\ x \epsilon X$$

or

$$f_C = f_A \wedge f_B \tag{6.15}$$

A graphical representation of the intersection of two fuzzy sets in X is illustrated in Fig. 6.26. The heavy lines denote the intersection. Note that x belongs to fuzzy set A if $f_A(x)$ is positive but that its degree of membership is variable relative to others in the set. (Compare this to an ordinary set in which x either belongs here $f_A(x) = 1$, or x does not belong here, $f_A(x) = 0$.) More importantly, in the case of a fuzzy set if $f_A(x)$ is positive, then the information needed is not just that it belongs but its degree

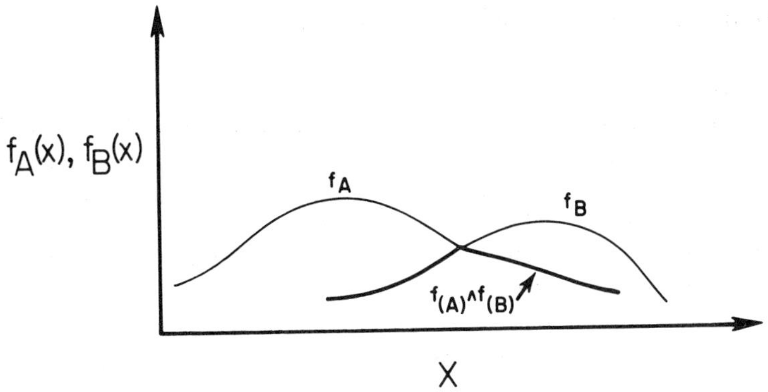

Fig. 6.26[126] Graphic illustration of the intersection of two fuzzy sets A and B.

of membership.

The Macro-Design of a Geometric Comparator and Selection of Primary Shapes

The comparator consists essentially of a graphical representation of a network of shape classes. Familiarity with terniary phase diagrams in ceramics, chemistry, metallurgy, or chemical engineering indicates that the binary mixing on a two-dimensional sheet of paper can be easily represented. With some effort, terniary mixtures can be represented on a sheet of paper, but this, for all practical purposes, will be the limit. The network will, therefore, consist of a series of triangles. The apex of each triangle will be occupied by an easily and unequivocally identifiable shape. The transformation within the triangle from the shape at one apex to the shape at another apex will be designed in discrete steps. Each triangle will, therefore, constitute the intersection of two fuzzy sets.

In practice all scales of measurement are arbitrary and their justification must be based initially upon the logic of their construction followed by their utility. Although measures used are related to fundamental quantities, the majority of these definitions of relationships are expedient, rather than based upon principle and so must be justified with a carefully constructed logic. Any justification of the network comparator must also be firmly based on the usefulness of the device. It is therefore expected that it should be altered as experience dictates how to improve it. It is also probable that powder particles of differing materials or manufacturing methods or both will need to have different comparator scales developed especially for their analysis.

With comparators scales established, the degree of fit can be expressed in a fuzzy set. In the overall design of the shape analysis system the human operator, who makes a preliminary judgment as to the shape classification of the fine particles, will wish to communicate with the binary computer which manipulates the analytical data. Although usually the computer uses ordinary sets, a person more commonly uses fuzzy sets. Consequently, comparator design must provide the link between the human perception of shape and the machine analysis of shape and pattern so that the two can process shape information usefully. Consequently, it is imperative that we incorporate the concept of fuzzy sets in our thinking in the iterative design task of forming a system for particle shape analysis.

Geometrical Comparator: First Version and Its Ramifications

1. The base line of the comparator consists of a set of equi-dimensional, regular figures of 3, 4, 5, 6, 7, 8 and an infinite number of sides—a circle. These are the basic shapes.
2. The basic shapes are modified in one of three ways or in a combination of two ways:
   a) extension in the ratios of height to base of
      2 to 1
      4 to 1
      8 to 1
      10 to 1
   b) contraction in ratios of height to base of
      1/2
      1/4
      1/8
      1/10
   c) skew at an angle of 30° and at an angle of 60°.
3. All shapes are normalized to the same perimeter.
4. An example of one of the comparator charts is shown in Fig. 6.27.

A trial run using 50 or so particle shapes from the Particle Atlas was carried out with the result that the observer* made some recommendations for improvement. The suggested improvements were made and the particle shapes in the Atlas were then analyzed. The results are shown in Table 6.15. Also, profiles only were used as is shown in Fig. 6.28.

A further series of experiments was conducted** in order to improve the speed of use of the comparator and also as a check on its reproducability and general format. In this experiment, published shapes of limestone pieces[78] were used. They are shown in Fig. 6.29. To these 25 shapes, another set of six shapes, from the comparator itself, was added. Two sets of 25 human subjects were utilized for testing purposes.

The first version of the comparator was very difficult to use (as had been found with the shapes from the Atlas). Consequently responses were as low as 44 percent and averaged out at approximately 60 percent—that is

*Ms. Holly Hoffman conducted this survey.
**Mr. Keenan Kelly conducted this survey.

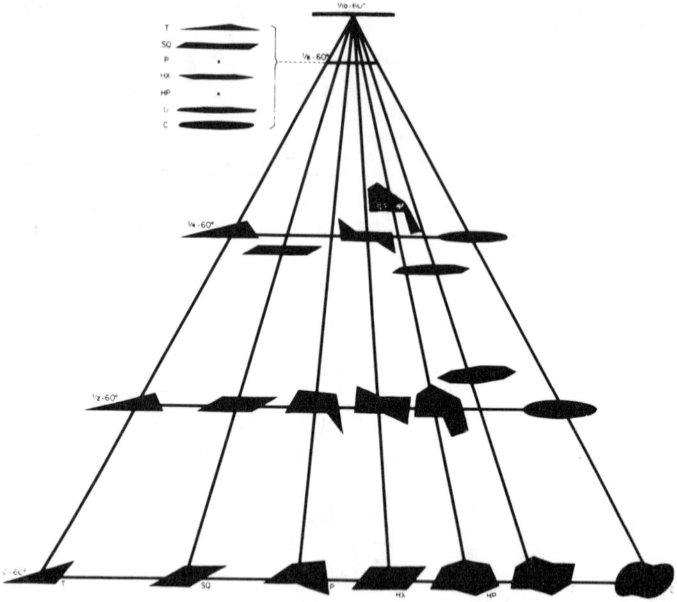

Fig. 6.27

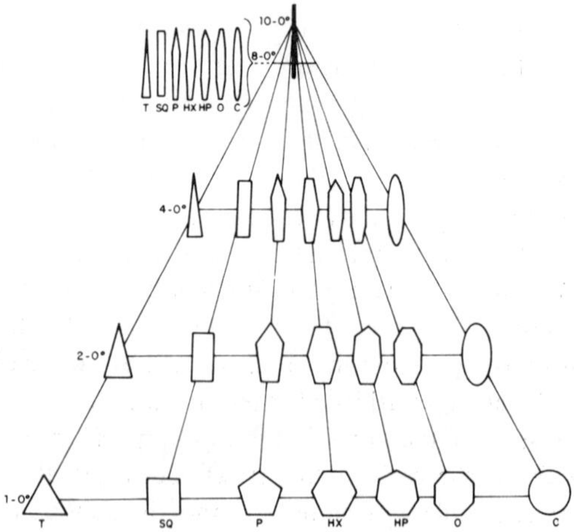

Fig. 6.28

Table 6.15

ANALYSIS OF PARTICLE ATLAS PARTICLE SHAPES USING VISUAL SHAPE COMPARATOR (Modified First Version)

TOTAL NUMBER OF SHAPES	729 (some samples categorized as a result of two different features)
TOTAL CLASSIFIED	607 (83 percent of total)
THOSE NOT CLASSIFIED	122 (17 percent of total)
Too small (powders)	61 (8 percent of total)
Too much texture	43 (5 percent of total)
No close shape	18 (2 percent of total)
OF TOTAL CLASSIFIED	
55 percent were in elongated categories	28 percent in no elongation
28 percent were in squashed categories	10 percent in 2 elongation
19 percent were in no skew categories	5 percent in 4 elongation
43 percent were in 30° categories	.8 percent in 8 elongation
21 percent were in 60° categories	14 percent in ½ elongation
5 percent were in no skew or deformation	6 percent in ¼ elongation
18 percent were fibers	2 percent in 1/8 elongation
	18 percent fibers

to say, only 60 percent of the observers were able to record an opinion as to the classification of the limestone chip. The remainder of the observers (some 40 percent of them) were, on average, unable to classify the chips using this first visual comparator.

The rating increases to 100 percent classified for the modified version of the comparator. In addition, the identification of the shape with respect to the number of sides was much improved. All observers reported improved attitude towards the revised version of the comparator and speed of classification was much increased. The results obtained with a small group of observers in the case of the known shapes is shown in Table 6.16. The final modified version is shown in Fig. 6.30.

At the time of writing, a visual shape comparator has not been interfaced with a computer. It appears that a fundamentally different design concept may be required, including the use of fuzzy descriptors. It could also contain some syntactic elements. Whatever the final outcome, the design of a visual particle shape classifier (comparator) is a challenging task.

Fig. 6.29 Sketch of silhouettes of sample of limestone pieces.[78]

Table 6.16

CLASSIFICATION OF "KNOWN" SHAPES USING MODIFIED VERSION OF THE COMPARATOR

Shape	(1)	(2)	(3)	(4)	(5)	(6)
Percent correct	65	25	33	25	100	60
Percent in correct category of n sides	85	88	75	100	100	75
Percent in correct category of stress	65	25	75	100	100	100

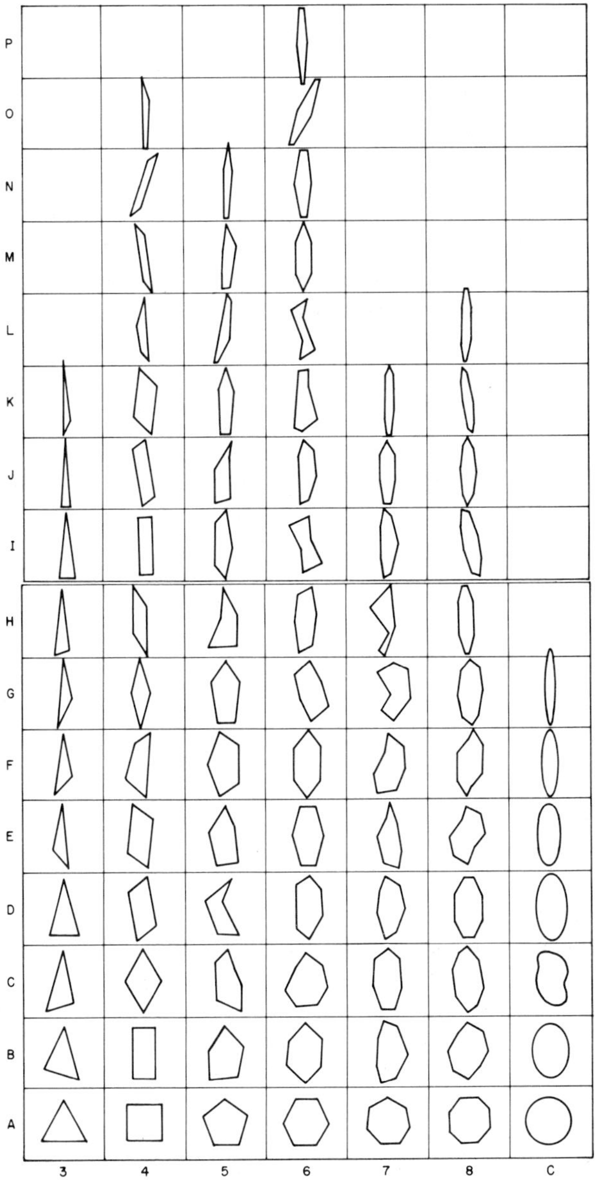

Fig. 6.30 (Reprinted with permission from CRC Press, J. K. Beddow, *Advanced Particulate Morphology*, 1979).[122]

6.7 FEATURE EXTRACTION

The Use of Fourier Descriptors to Identify Particle Shape

It has been shown that a particle profile* can be considered to be a population of points from which a sample set of points in the form of (x, y) coordinates can be abstracted. This set of coordinates may then be transformed into a set of polar coordinates, (R, θ), with the centroid as the origin. The function $R(\theta)$ can then be described in terms of a Fourier series of the general form

$$R(\theta) = A_o + \sum_{n=1}^{\infty} A_n \cos(n\theta - \alpha_n) \qquad (6.16a)$$

in which A_n are the Fourier coefficients and α_n are the phase angles. It may be noted that

$$A_n = (a^2 + b^2)^{1/2}$$

and

$$\alpha_n = \tan^{-1} \frac{b_n}{a_n}$$

in which a and b are coefficients in the more standard form of the Fourier relationship, that is

$$R(\theta) = A_o + \sum_{n=1}^{\infty} (a_n \cos n\theta + b_n \sin n\theta) \qquad (6.16b)$$

The set of coefficients $\{A_n\}$ and $\{\alpha_n\}$ may then be used in order to obtain a regenerate which is compared with the original for faithfulness of reproduction.[81] The significance of obtaining a satisfactory regenerate is that it indicates that the sets $\{A_n\}$ and $\{\alpha_n\}$ contain much of the information of the morphology of the original particle profile. This procedure therefore, effectively condenses a large number of points of the profile via a sample set of (x, y)'s and $R(\theta)$ down to a set of $\{A_n\}$'s and $\{\alpha_n\}$'s. Now, because any one powder contains an unknown diversity of morphic types, it is necessary to take a large number of profiles in any one case. The data

* At the present time profiles are being studied. 3 D image processing of particles is just being started. See for example references 138 and 139.

handling problem is therefore of the greatest importance. As the set of $\{A_n\}$'s may have any number of terms it eventually must be decided at what value of n enough information exits. These two aspects of the problem may be summarized as:

1. At what value of n is the series truncated? and
2. What features should be abstracted from the set of $\{A_n\}$'s and what, if anything, will be their physical significance?

The objective of the experiment reported here was to see if samples obtained from two different powders could be differentiated on a basis of the properties of the sets of $\{A_n\}$ and $\{\alpha_n\}$.

No decision rule is available to decide how to truncate the series. The limit was therefore set at A_{30}. This means that a second objective of the experiment was to determine if sufficient information would be forthcoming from a set of 30 coefficients. It has been observed[82] that there are other types of functions which may be used to represent profiles such as the square functions of the Walsh, Radamacher, and Harr type. These functions and the possible ways of using them are discussed elsewhere.[82]

Before moving on to the details of feature selection, it is convenient at this stage to consider, in graphical form, the meaning of the beginning coefficients of Eq. (6.16). Plots of $A_o + A_1 \cos\theta$ versus θ, $A_o + A_2 \cos 2\theta$ versus θ, $A_o + A_3 \cos 3\theta$ versus θ, and so on, are given in Fig. 6.31. Also, it has been observed that, in general, the A_n's are smaller for larger n. From this it can be seen that a convenient way to look at the Fourier series is that the early terms give the main features and the later terms successively represent the finer and finer scale features.

Feature Selection

In this section an attempt is made to define a set of morphic features of a particle which apparently can be handled by processes of human pattern recognition while being simultaneously abstracted, in some form, from the $\{A_n\}$'s and $\{\alpha_n\}$'s of the corresponding Fourier equation. These features are:

1. The Centroid Aspect Ratio–which is the maximum length through the centroid divided by the perpendicular dimension also through the centroid–abbreviated as CAR.

   There are various methods of determining the aspect ratio of a powder particle. One definition is the maximum length divided by

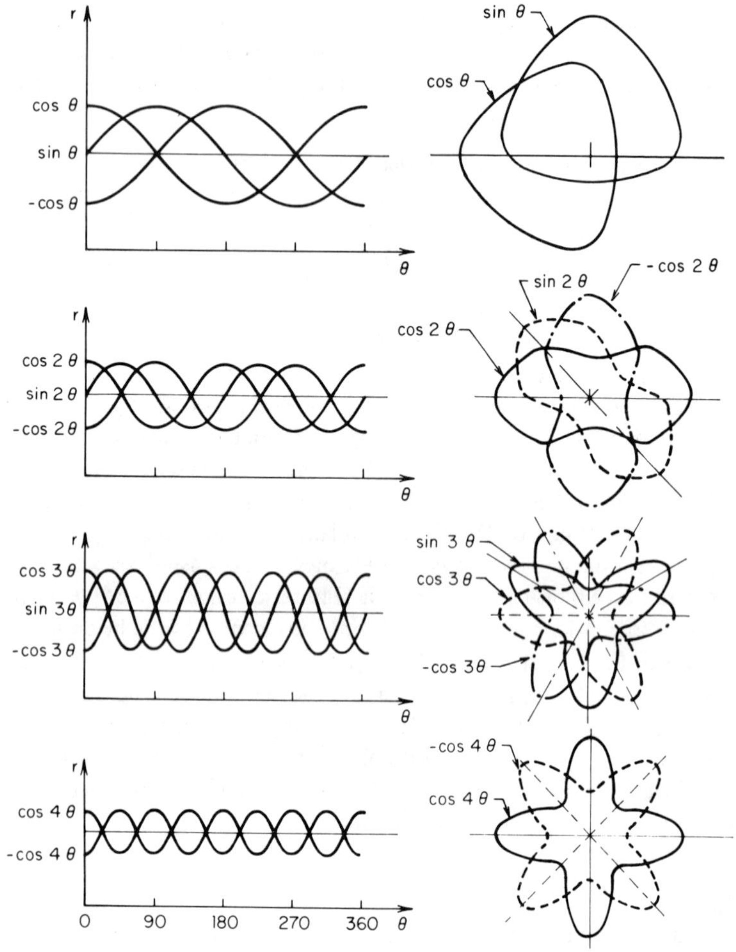

Fig. 6.31 Plots of regenerates[94] for various A_n's.

the shortest dimension at right angles to the maximum length. Because the Fourier method reported here evaluates the position of the particle centroid as an integral procedure, it was considered convenient to obtain the CAR in the manner described. Furthermore, it is possible that the aspect ratio based upon the centroid of a particle may actually be of more fundamental significance than the aspect ratio defined in any other manner. For example, one would hypothesize that CAR should correlate more closely with the effect of a gravitational field on a particle than the alternate definition of the aspect ratio.

2. The Lumpiness, abbreviated as L.
3. The Roughness, abbreviated as R.
4. The Texture, abbreviated as T.

   The definitions of the features of lumpiness, roughness, and texture are summations of arbitrary groupings of coefficients. Clearly, other groupings of coefficients with perhaps other operands may eventually prove to be more effective.

5. The Symmetry, abbreviated as S.

Before going into the details of these features it is necessary to mention two aspects of finding the Fourier coefficients. First, before finding the coefficients the angles θ associated with radii are modified so that the radius with $\theta = 0$ coincides with the longest axis. This helps to reduce the effect of the distribution of particles and facilitates computing the CAR from the Fourier coefficients. Secondly, the Fourier coefficients are divided by A_o to reduce the effect due to size variations.

1. Centroid Aspect Ratio (CAR)

A. Using the following two forms of the Fourier equation:

$$\text{distance from centroid} = A_o + \Sigma\, a_n \cos n\theta + \Sigma\, b_n \sin n\theta, \text{ and}$$

$$\text{distance from centroid} = A_o + \Sigma\, A_n \cos (n\theta - \alpha_n)\,.$$

B. The maximum dimension through the centroid is given by $R(0) + R(\pi)$, in which

$$R(0) = A_o + a_1 \cos 0 + a_2 \cos 2\,0 + a_3 \cos 3\,0 + a_4 \cos 4\,0\,,$$

and

$$R(\pi) = A_o + a_1 \cos \pi + a_2 \cos 2\pi + a_3 \cos 3\pi + a_4 \cos 4\pi\,.$$

Hence $R(0) + R(\pi) = 2A_o + 0 + 2a_2 + 0 + 2a_4 = 2A_o + 2 \sum_{n=1} a_{2n}$, i.e., the longest dimension $= 2A_o + 2 \sum_{n=1} a_{2n}$.

C. By convention, the shortest dimension is given by $R(\pi/2) + R(3\pi/2)$ since it occurs at 90° to the longest diameter, i.e., at $\theta = \pi/2$ and at $\theta = 3\pi/2$. The b terms get cancelled.

$$R(\pi/2) = A_o + a_1 \cos \pi/2 + a_2 \cos 2\pi/2 + a_3 \cos 3\pi/2 + a_4 \cos 4\pi/2 + a_5 \cos 5\pi/2 \,,$$

and

$$R(3\pi/2) = A_o + a_1 \cos 3\pi/2 + a_2 \cos 2\; 3\pi/2 + a_3 \cos 3\; 3\pi/2 + \ldots ;$$

hence

$$R(\pi/2) + R(3\pi/2) = 2A_o - 2a_2 + 2a_4 - \ldots = 2a_o + 2 \sum_{n=1} (-1)^n a_{2n}$$

D. Thus,

$$\text{CAR} = \frac{2A_o + 2 \sum_{n=1}^{n/2} a_{2n}}{2A_o + 2 \sum_{n=1}^{n/2} (-1)^n a_{2n}} \tag{6.17}$$

in which n = the number of coefficients.

This is the ratio of the maximum dimension *through the centroid* divided by the dimension perpendicular to same. It is termed the centroid aspect ratio.

2. Lumpiness, Roughness, Textures

The shape characteristics are measured on a scale of gradually increasing resolving power. The set of Fourier coefficients is divided arbitrarily into three portions:

A. the sum of the first n_1 members are characterized by the term lumpiness,

B. the sum of the intermediate members n_2 are characterized by the roughness, and

C. the sum of the highest members n_3 are characterized by the term texture.

The following computations attempt to develop the general formula for the summation of coefficients. The summations are respectively:

$$L_k = \sum_{n=1}^{n=n_1} A_n \tag{6.18}$$

$$R_k = \sum_{n=n_1}^{n=n_2} A_n \tag{6.19}$$

and

$$T_k = \sum_{n=n_2}^{n=n_3} A_n \tag{6.20}$$

Also the value of T_k is compared with T_r which is the sum of the squares of the differences, between the original radii and the radii of the regenerate with an arbitrary number of coefficients–ten in this case.

The essence of this calculation is to determine the radial lengths at set intervals of θ for the original [these $R_o(\theta)$ values are already available at the start of the program], and to determine the radial lengths at the same intervals of θ for the regenerate with $n_k = 10$. The regenerate radius length is

$$R_r(\theta) = a_o + \sum_{n=1}^{n=10} A_n \cos(n\theta - \alpha_n)$$

This form is selected because the input coefficients are produced in this form:

$$T_r = \sum_{\theta} |R_o(\theta) - R_r(\theta)|$$

3. Symmetry

In this context the term symmetry implies the result of a set of operations which when applied to a specific segment of the profile causes it to come into register (coincidence) with another segment of the same profile.

In this particular application, the first type of symmetry operation considered is that of mirror symmetry across the maximum dimension through the centroid.

The conventional sin/cos representation may be used for convenience for a moment as the profiles in Fig. 6.32 show. The sin terms represent simple shapes with little mirror symmetry and the cos terms represent those simple shapes with mirror symmetry. Therefore, in the second form of the Fourier equation $A_n \cos(n\theta - \alpha_n)$, the α's may be taken to represent the extent of mirror symmetry in any specific profile, as follows.

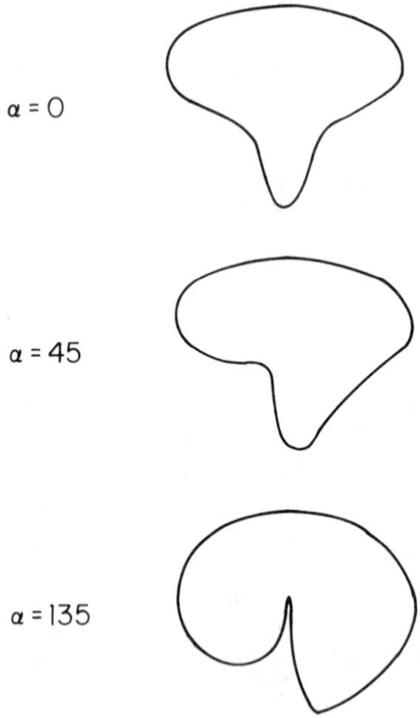

Fig. 6.32

In general, if $\alpha = 0$, the figure has mirror symmetry. If α is greater than zero, the figure is less symmetrical to an extent that depends upon the value of α. In general, the higher α is, the less symmetry the figure possesses.

Note that if A_n's are zero (A_o is finite), then the regenerate is a circle and highly symmetrical. In this case α has no significance.

Similarly, if the A_n's are quite small (for example, less than 1/20th of a_o) then the α's have little or no influence on developing symmetry in a regenerate and therefore they contain no information concerning symmetry.

This method is therefore limited to cases in which the A_n's are at least within an order of magnitude of a_o. For example, consider A_n (1-5) if A_1, A_3, and A_5 are large, their corresponding α's are inspected. If these are small, the figure is considered to have mirror symmetry.

In general, therefore, the asymmetry is measured as the sum of the absolute values of all α_n:

$$\text{Asymmetry} = \sum_{n \in n_4} |\alpha_n| \tag{6.21}$$

in which n_4 is the set of indices such that A_{n_4} is large.

Experimental

Two powders are selected–atomized copper/lead (70/30), and sponge iron. They are size separated by sieving and two samples from each are chosen–the +170 and −400 fractions respectively. After spreading the particles on glass slides, they are photographed to yield 50 or 60 particles in the picture. A slide is prepared and projected onto a flat surface. The profiles are traced by hand such that one profile is obtained per standard size 8½ X 11 page. Clearly, in such a procedure, some detail is lost in obtaining the profiles but the method retained sufficient differentiation (judging by visual results) to permit the experiment to be carried out satisfactorily. Next, the profiles are converted to sets of (x,y) coordinates on punched IBM cards using a point gathering device called "Oscar." This device uses cross wires and is tedious to use (but it works). The data on the punched cards is processed using a program to develop the features already discussed. In all, for the present purpose 15 profiles are processed per fraction, yielding the following data:

Fraction / *Powder*	+170	−400
Fe	15	15
Cu–Pb	15	15

Sample results are given in Table 6.17. The missing columns show where some errors are made in collecting the original (x, y) coordinates. In these cases the program did not run. For purposes of illustration, the features chosen for further processing were CAR, L, R, S.

After certain features are selected and their values obtained for a small number of particles of two specified powders, the information is processed in order to learn if there is any particular pattern of occurrence of these features which can be unequivocally identified with either of the particle sets. There are methods of varying degrees of complexity for carrying out this operation. A very simple approach at this initial stage of the research program is chosen. The following graphs are constructed:

	CAR	L	R	S
CAR			Fig. 6.33	
L			Fig. 6.34	Fig. 6.35
R				Fig. 6.36
S				

It is thus treated as a problem of pattern recognition in two dimensions, although, of course it could be treated as one in four dimensions. The graphs shown in Figs. 6.33–6.36 represent searches for pattern in two-dimensional spaces of four types. The observer can, therefore, use visual observation in order to develop a linear decision function.

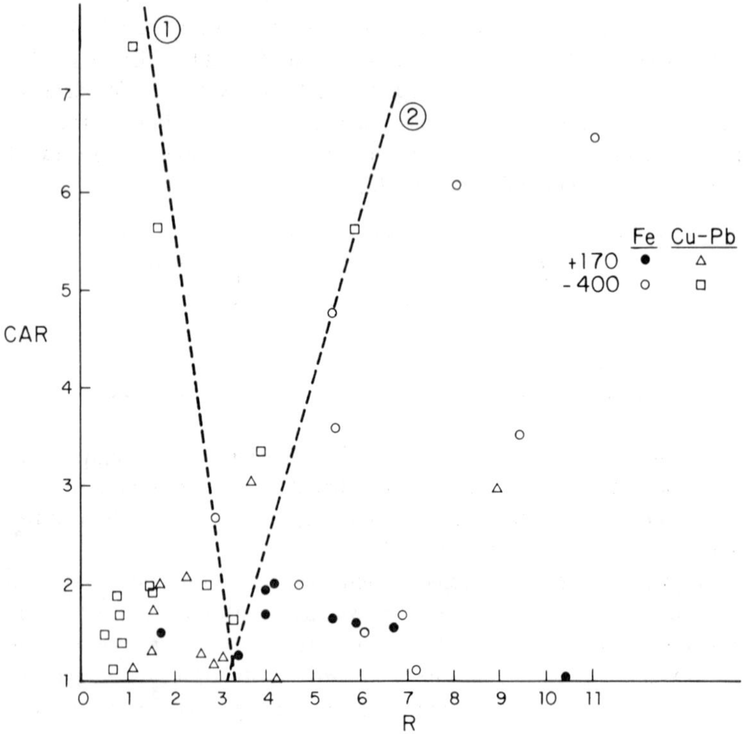

Fig. 6.33 CAR vs R.[94]

Table 6.17

FEATURES FOR Cu Pb +170#[94]

	1	2	3	4	5	6	7	8	9	10	11
CAR	1.24	2.09	1.77	1.31	3.01	1.18	2.0	3.06	1.18	1.3	0.92
L	8.57	11.29	7.67	5.30	20.35	5.95	10.0	23.8	5.3	10.77	12.9
R	3.13	2.32	1.58	1.56	8.92	2.86	1.73	3.7	1.14	2.6	4.2
T_C	.87	1.28	1.01	1.53	3.27	0.8	1.37	31.3	30.5	1.0	38.7
T_R	15.73	21.0	22.94	15.2	46.1	18.0	18.4	40.9	11.0	20.0	88.0
S	109.9	418.0	27.81	179.2	515.0	248.0	210.0	286.0	210.0	498.0	640.0
A_o	36.44	33.3	35.27	36.61	38.25	31.76	31.3	43.0	31.9	37.5	39.9

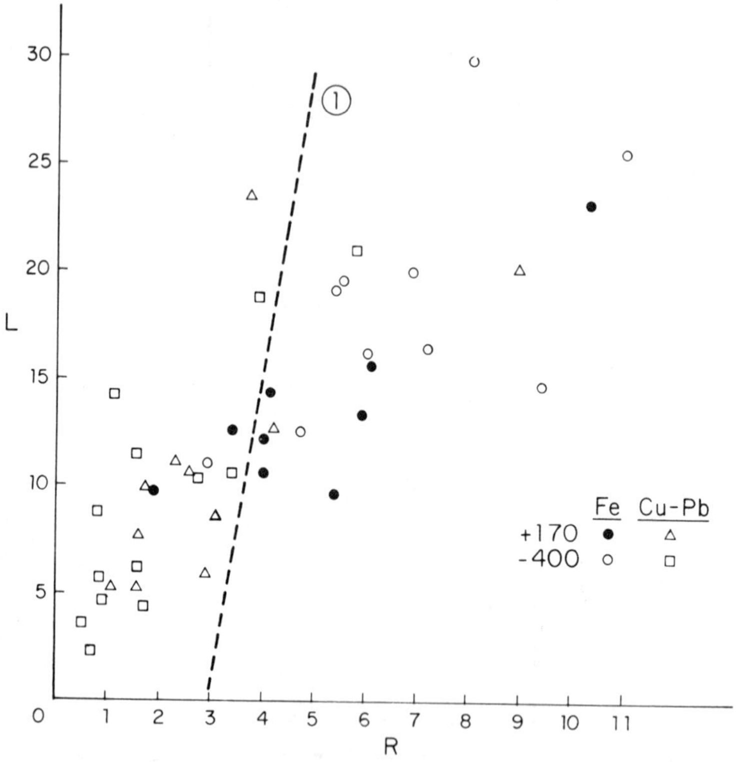

Fig. 6.34 L vs R.[94]

Table 6.18 shows the number of two dimensional items misclassified for a small number of trial and error linear decision functions. The table should be read in conjunction with the appropriate graph, as designated in the table.

The choice of decision function in any one case is illustrated in the case of lines 1 and 2 of Fig. 6.35. If one chooses line 1, then 6 Cu-Pb items are "misclassified" and only one Fe item is misclassified. If line 2 is chosen instead of line 1, then there are only two Cu-Pb items misclassified but there are now two Fe items misclassified. The investigator has to make the judgement as to which is the most advantageous choice. The guide for choice in this example may be stated as: minimize the total number of items misclassified with either an added restriction on the Fe misclassification or no restriction on the Fe misclassification.

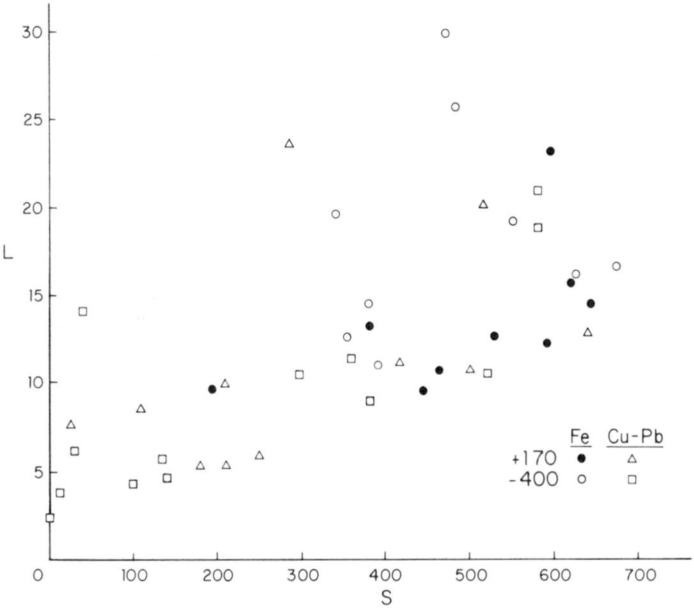

Fig. 6.35 L vs S.[94]

Table 6.18

TRIALS TO FIND LINEAR DECISION FUNCTIONS FROM FIGS. 6.33, 6.34, AND 6.36[94]

Topic	*Items Line 1* Cu-Pb	*Items Line 1* Fe	*Misclassified Line 2* Cu-Pb	*Misclassified Line 2* Fe
CAR vs R See Fig. 6.33	6	1	2	2
L vs R See Fig. 6.34	3	3		
R vs S See Fig. 6.36	3	3	4	2

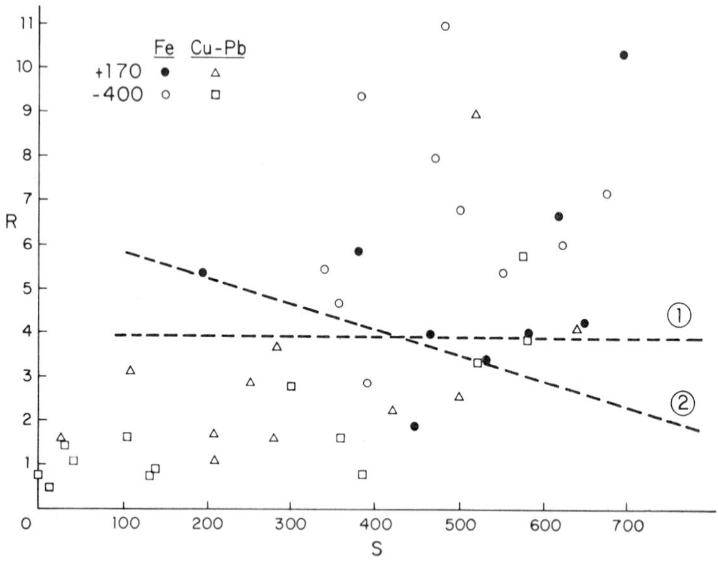

Fig. 6.36 R vs S.[94]

The experiment has demonstrated that it is possible to abstract features from mathematical descriptors of particle profiles which in themselves may have physical significance in terms of human perception of shape. It is one of the eventual aims of our research to develop decision functions defined in multidimensional space that would enable classification of various powders and facilitate the understanding of the relationships of various morphological features to the properties of the powders themselves.

6.8 PARTICLE SIGNATURE AND THE MELOY EQUATIONS

A plot of Log *An* versus *n* in the case of a Fourier series which represents a particle shape has the general form of a line with a negative slope. Fig. 6.37 shows a plot of Log *An* versus *n* for five particles. The profiles of the five particles are shown in Fig. 6.38. This plot has been termed the signature of the particle[82] and the envelope of all of the sample plots for a representative set of particles is termed the signature of the powder (see Fig. 6.39). Therefore, one has to represent the envelope in some conven-

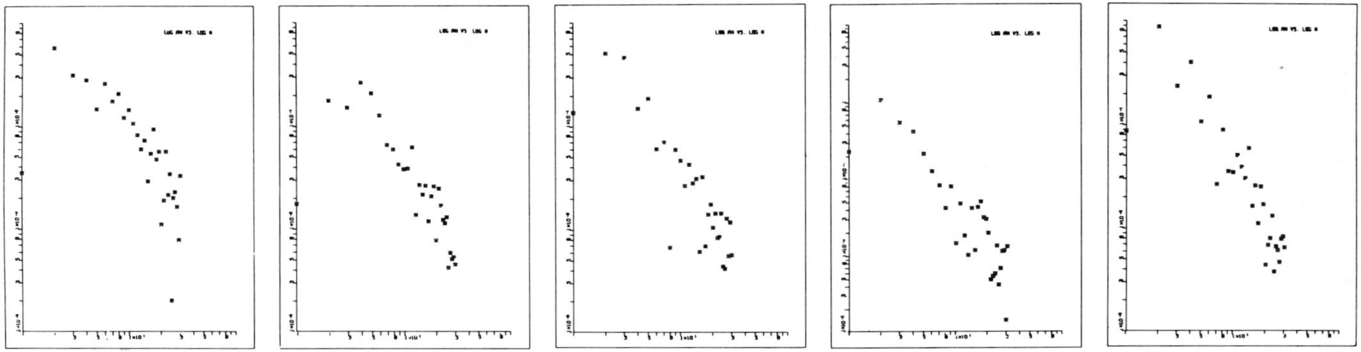

Fig. 6.37 Particle signatures.[122]

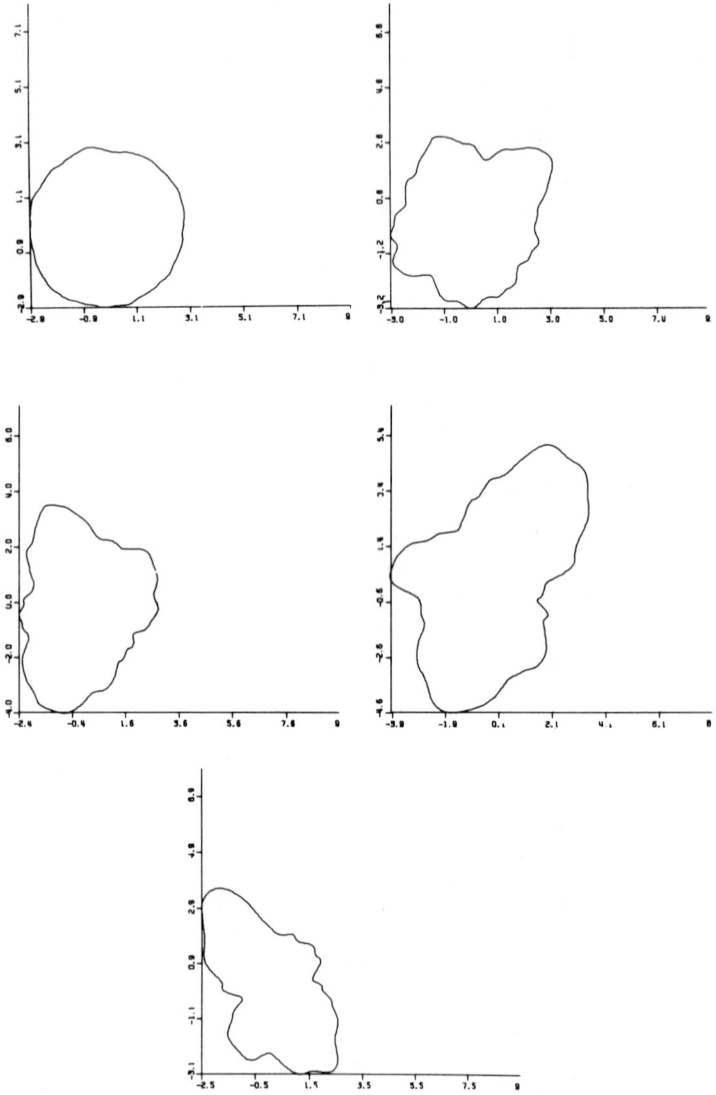

Fig. 6.38 Particle profiles corresponding to particle signatures of Fig. 6.37.[122] (Reprinted with permission of CRC Press, J. K. Beddow, *Advanced Particulate Morphology*, 1979)

ient, succinct, and accurate manner in order that the totality of particle signatures can be represented as a powder signature.

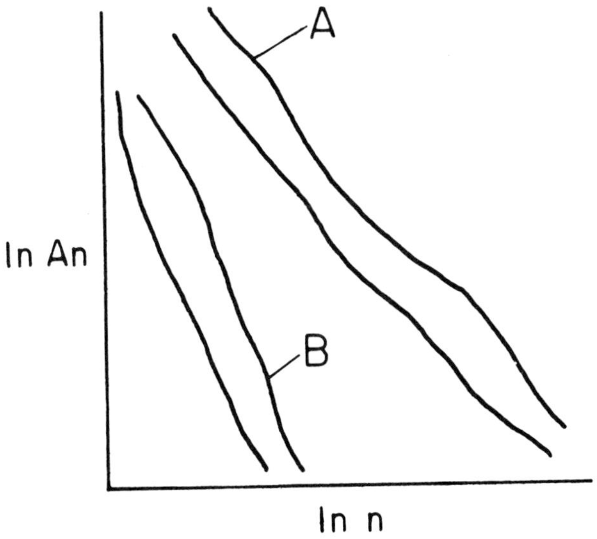

Fig. 6.39 Illustrating the concept of powder signature applied to two powders–powder *A* and powder *B*.[127]

How can this be done? If the mean values of the Fourier coefficients are taken and used to represent the powder signature, the resultant information is insufficient. In fact, it is misleading as is shown in Fig. 6.40 which is a plot of particle profiles for some sponge iron powder particles. The profile in the middle of the figure is the mean regenerate. It is quite obvious that the averaging procedure merely erases differences between particles and produces a bland representation of the particle set. However, the slope of the curve is a useful indicator. But one should not entirely ignore the effect of variation of (α_n). This has been treated in Fig. 6.32 for designated values of α_n.

Another approach which is currently under investigation is to obtain the mean and also the variance properties of the $\{An\}$'s, plotted as a signature curve. It is likely that random generation of regenerate profiles from the statistical properties of a "representative" particle signature would produce a clustering in the pattern space that would closely agree with the clustering properties of actual particles of the real powder. If this be so,

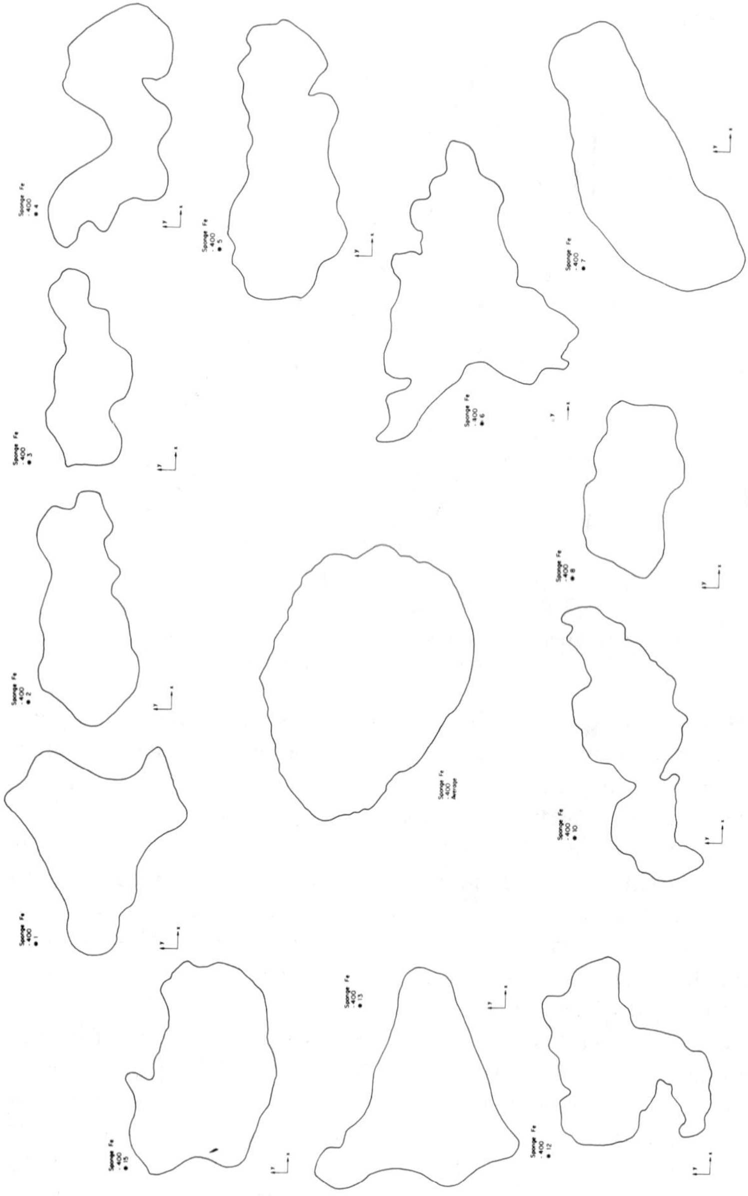

Fig. 6.40[69]

then the statistical properties of the powder signature would be a useful tool in fine particle science.

Particle Morphology Postulates

Three postulates have been proposed[82] as follows:

1. All usable information for broad morphological characterization is contained in the amplitude of the coefficients, not in their phase relationships.
2. Particles from a homogeneous background have similar signatures. That is, particles with a similar genesis, chemical composition, and history should have similar signatures.
3. Homogeneous powders consist of particles whose signature falls on a relatively narrow path (i.e., in a tight envelope).

One might not entirely agree with all of the broad implications of the above postulates, but as of 1977, they represent an imaginative step forward.

The previous section of the chapter showed that the $\{\alpha_n\}$ terms have a significance with regard to the symmetry properties of the particles of the powder. Clearly, postulate 2 should not be interpreted in a simple deterministic way. Postulate 3 has been investigated to a limited extent to date. The search for pattern in the four types of two-dimensional space indicates that the meaning of the term "homogeneous" powder will become more clear as more work is done in the future.

Another very interesting development which has occurred quite recently is the proposal of a series of equations termed the Meloy Equations (after their proposer.[82]

Square Wave Functions[83]

Fourier functions operate over an interval 0 to 2 π, they are orthogonal, normalized, and complete. The term orthogonal means that the product of any two functions integrated over an interval is zero, unless they are identical. The term normalized means that the integral of the function squared over the interval equals 1. The term complete means (among other things) that all of the functions in the set are defined. However, Fourier functions suffer from some serious computational disadvantages. For example, it is possible to state the Fourier form for a line integral (or a surface integral) but, because the integrals contain unspecified terms multiplied to trigonometric functions, they cannot be integrated in closed form and for this reason uses for Fourier representations of surface features are very limited.

Fortunately, there are square wave functions for which the representation of such surface features is much easier. Three of these functions are the Walsh Function, the Rademacher Function and the Haar Function, respectively. These functions are illustrated in Fig. 6.41-6.43, respectively.

The term Walsh Function was named in honor of the inventor of the function. The terms Sal and Cal are composite terms made up from sine/ Walsh and cosine/Walsh combinations respectively.

A single function $\mathrm{Wal}(j,\theta)$ can be defined in order to relate the three functions:

$$\mathrm{Wal}(2i,\theta) = \mathrm{Cal}(i,\theta),$$

$$\mathrm{Wal}(2i-1,2) = \mathrm{Sal}(i,\theta),$$

and

$$i = 1,2,\ldots\ldots\ldots\ldots, \tag{6.22}$$

in which i is the sequency and is one-half of the number of zero crossings per second. Note from examination of the Walsh displays in Fig. 6.41, that the intervals are not equally spaced as are those of a Fourier function.

Note also that $\mathrm{Wal}(0,\theta)$ is the same as R_o; $\mathrm{Sal}(2,\theta)$ is the same as R_2 and so on. The advantage of representing particle profiles with square waved functions instead of with Fourier functions may be understood from Figs. 6.44 and 6.45. Thus the elipticity and triangularity are preserved by the Walsh representation. But in addition, the new representation makes it possible to identify and calculate specific morphological features. For example, in both figures, it is clear that the extra length of line (i.e., of the profile, as represented) is $8A_2$ and $12A_3$ respectively. This represents the excess arc length for the profile that is excess over the base circle radius A_o. Using this approach, Meloy has carried out the following line of reasoning in order to compute functions that are intended to represent the morphological characteristics of the particles. Five such morphological features are discussed: excess arc length, excess surface area, excess edge length, volume of positive blocks, and number of corners.

Arc Length

With the treatment of Walsh functions in the line integral, the arc length

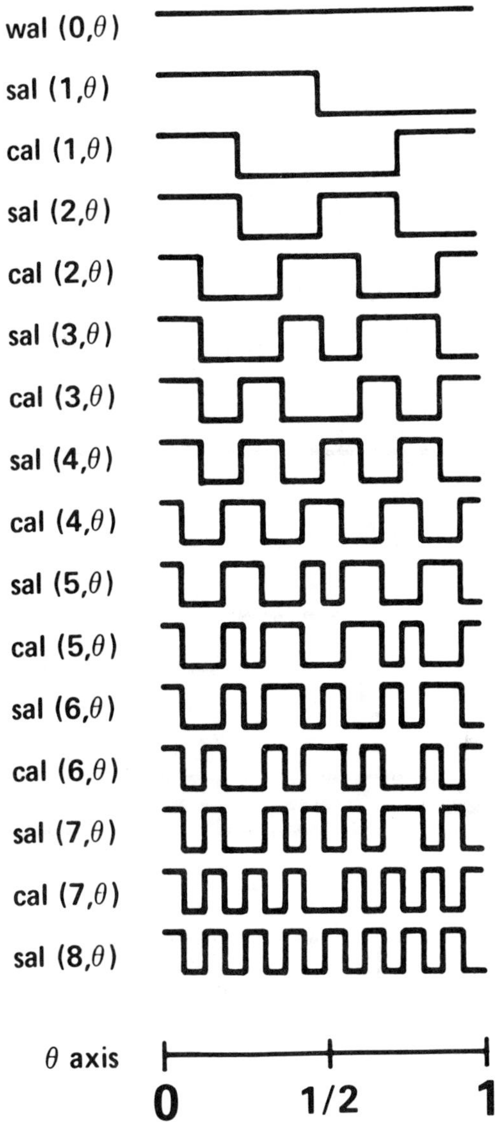

Fig. 6.41 Walsh Functions.[82a,128]

can be described by two terms:

$$L = 2\pi A_o + 8A_2 + 12A_3 + \ldots\ldots = 2\pi A_0 + 4 \sum_{n=2}^{\infty} N \cdot An \qquad (6.23)$$

This equation seems to be reasonable if the superposition of A_2 and A_3 are drawn out, and where the peaks of A_3 were shifted with a 15-degree clockwise rotation in relation to those of A_2. But, if a 30-degree shift is made, the second term of excess edge length is reduced to

$$8A_2 + 8A_3 + \ldots\ldots$$

Similar to the result from the 30-degree shift, it is found that

60°, 90°, 120°, 150°, 180°, 210°, 240°, 270°, 300°, 330°, and 360°

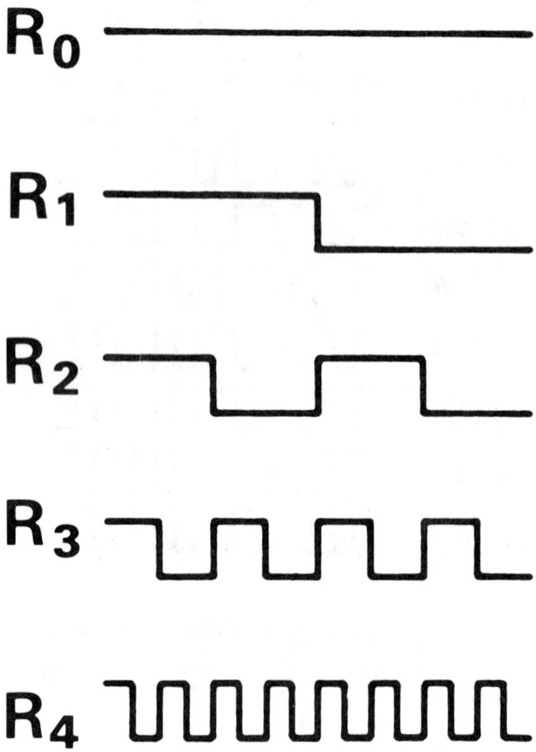

Fig. 6.42 Radamacher Functions.[82a, 128]

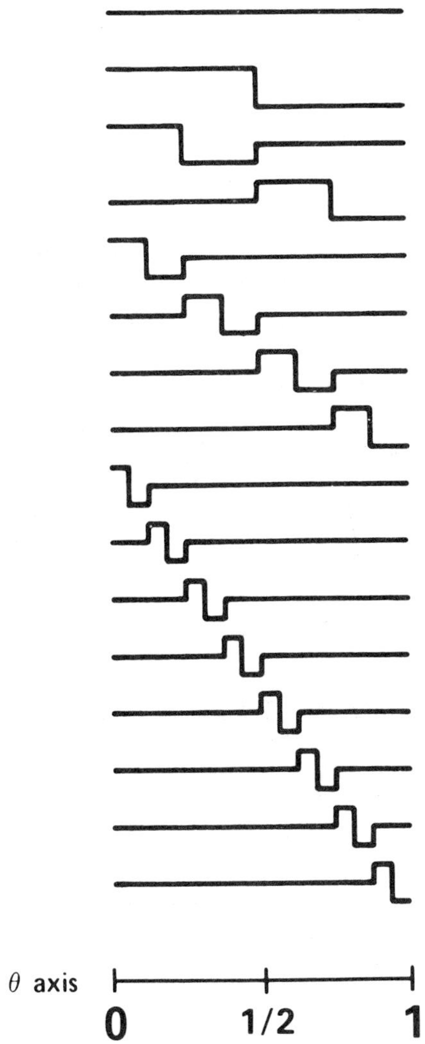

Fig. 6.43 Haar Functions.[82a, 128]

also have the same expression. This means that Eq. (6.23) is a translation variant.

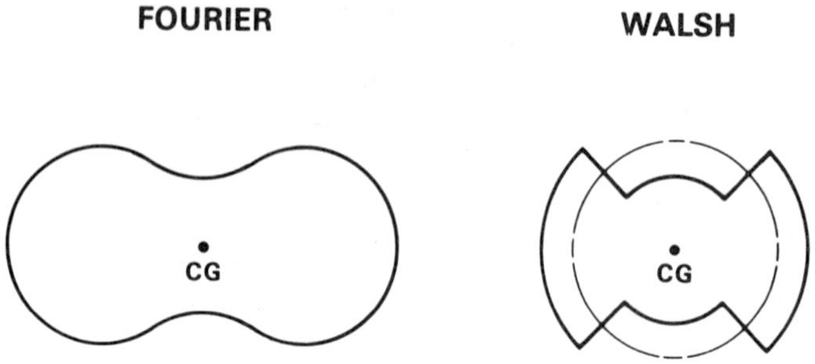

Fig. 6.44 Second coefficient–Aspect ratio.[82a, 128]

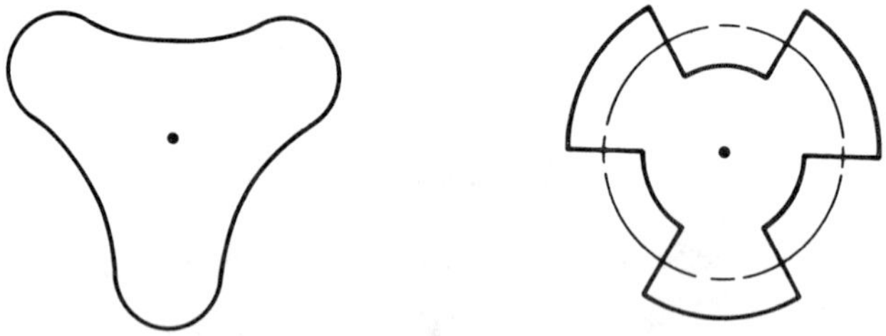

Fig. 6.45 Third coefficient A_3 triangularity.[82a, 128]

Physical Significance of 3-D Walsh Function

1. Cubing the Sphere

It is assumed that the number of square wave peaks are in the ratio of $1/\pi$ between the flat square surface and the sphere surface. If the circumference of the sphere has n positive waves then there are n^2 positive waves on the flat square surface and n^2/π on the sphere surface.

The surface area of a sphere having a unit in radius is

$$A = 4\pi R^2 = 4\pi \qquad (R = 1)$$

By assumption of square waves, the total top area of positive blocks is:

$$A/2 = 2\pi$$

There are n^2/π positive blocks on a sphere, the top area for each block is:

$$2\pi/n^2/\pi = 2\pi^2/n^2$$

We assume that each block has a square base, such that the base length of the positive block is:

$$\sqrt{2\pi^2/n^2} = \sqrt{2}\,\pi/n \tag{6.24}$$

Each positive block on the sphere's surface has a height of An and a square base $\sqrt{2}\,\pi/n$ on a side.

2. Excess Surface Area

There are n^2/π of blocks each with four sides, a height of $2An$, and a base length $\sqrt{2}\pi/n$. Since only half of the side area is attributable to any one block, the side area for each block is

$$\tfrac{1}{2}(2An \times \sqrt{2}\pi/n \times 4) = 4\sqrt{2}\pi An/n$$

in which

$$n = 2, \quad \text{the side area} = 2 \times \frac{4\sqrt{2}\pi A_2}{2} = 4\sqrt{2}\pi A_2$$

$$n = 3, \quad \text{the side area} = 4 \times \frac{4\sqrt{2}\pi A_3}{3} = \frac{16\sqrt{2}}{3}\pi A_3$$

$$n = 4, \quad \text{the side area} = 6 \times \frac{4\sqrt{2}\pi A_4}{4} = 6\sqrt{2}\pi A_4$$

and

$$n = n, \quad \text{the side area} = \frac{n^2}{\pi} \times \frac{4\sqrt{2}\pi An}{n} = 4\sqrt{2}nAn$$

The total side area of all blocks of all frequencies is

$$\text{Side Area} = 4\sqrt{2}\pi A_2 + \frac{16\sqrt{2}\pi}{3} A_3 + 6\sqrt{2}\pi A_4 + 4\sqrt{2}\sum_{n=5}^{\infty} nAn \tag{6.25}$$

3. Edge Length

Each block has four vertical edge lines. Only one-fourth of them can be assigned to a block. The edge length for each block is:

$$\frac{1}{4} \times 4 \times 2An = 2An$$

in which

$$n = 2, \quad \text{edge length} = 2 \times (2A_2) = 4A_2$$

$$n = 3, \quad \text{edge length} = 4 \times (2A_3) = 8A_3$$

$$n = 4, \quad \text{edge length} = 6 \times (2A_4) = 12\,A_4,$$

and

$$n = n, \quad \text{edge length} = \frac{n^2}{\pi} \times (2An) = \frac{2}{\pi} n^2 An\,.$$

The resulting edge length is:

$$\text{Vertical Edge Length} = 4A_2 + 8A_3 + 12A_4 + \frac{2}{\pi} \sum_{n=5}^{\infty} n^2 An \qquad (6.26)$$

The horizontal edge length on the top of each block is:

$$4 \times \frac{\sqrt{2}\pi}{n} = \frac{4\sqrt{2}\pi}{n}$$

and the total horizontal edge length is

$$\text{Horizontal Edge Length} = 4\sqrt{2}\pi + \frac{16\sqrt{2}}{3}\pi + 6\sqrt{2}\pi + 4\sqrt{2} \sum_{n=5}^{\infty} n \qquad (6.27)$$

It has been shown in Section 6.7 that a two-dimensional profile can be adequately represented by a Fourier series. The square wave functions can also be used in a similar manner and making use of this idea, it has been proposed that a physical property of a particle can be represented by:

$$\text{Physical Property} \propto A_o + K \sum_{2}^{\infty} n^{\alpha} A_n^{\beta} \qquad (6.28)$$

in which

A_o = average radius of the particle,
K = constant,
α,β = integers $0 \rightarrow 3$,
n = the number of the coefficient, and
A = the value of the coefficient.

Using this approach the set of equations shown in Table 6.19 has been developed.[82]

Table 6.19[82a, 128]

THE MELOY EQUATIONS FOR MORPHOLOGICAL FEATURES

β \ $\alpha \to$	0	1	2	3
0	A_o	$A_o + k_{21} \sum_2^\infty n$	$A_o + k_{31} \sum_2^\infty n^2$	$A_o + k_{41} \sum_2^\infty n^3$
1	$A_o + k_{12} \sum_2^\infty A_n$	$A_o + k_{22} \sum_2^\infty nA_n$	$A_o + k_{32} \sum_2^\infty n^2 A_n$	$A_o + k_{42} \sum_2^\infty n^3 A_n$
2	$A_o + k_{13} \sum_2^\infty A_n^2$	$A_o + k_{23} \sum_2^\infty nA_n^2$	$A_o + k_{33} \sum_2^\infty n^2 A_n^2$	$A_o + k_{43} \sum_2^\infty n^3 A_n^2$
3	$A_o + k_{14} \sum_2^\infty A_n^2$	$A_o + k_{24} \sum_2^\infty nA_n^3$	$A_o + k_{34} \sum_2^\infty n^2 A_n^3$	$A_o + k_{44} \sum_2^\infty n^3 A_n^3$

The first row of functions in Table 6.19 do not appear to represent a physically meaningful property except for A_o, which is the mean particle diameter is rigorous and it is unequivocal. There is clearly something worth pursuing here in order to develop an improved definition of particle size. Again in Table 6.19, the form has already been calculated, e.g., the edge length of Eq. (6.26) which appears in row 2, column 3, the side area of Eq. (6.25) which is in row 2, column 2. A 3-D simulation of the particle surface is shown in Fig. 6.46. This helps the visualization of the volume of the positive blocks and the top surface area.

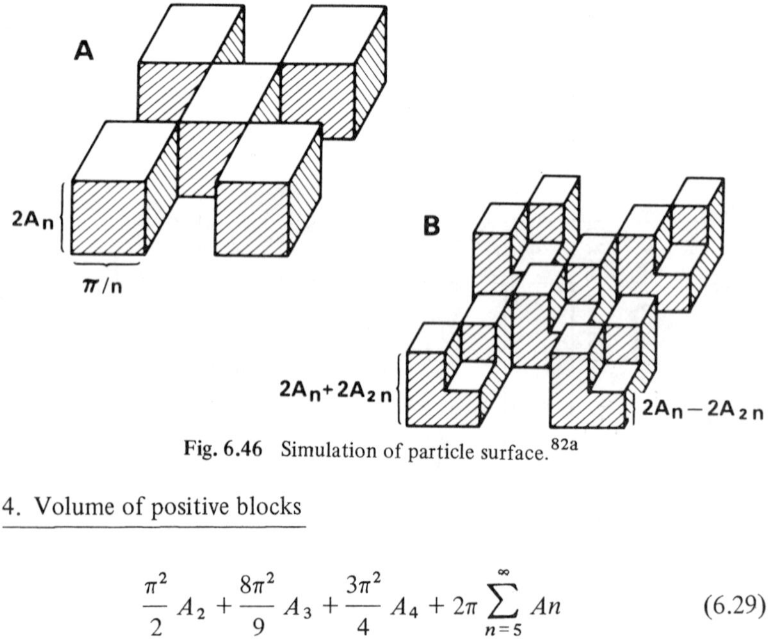

Fig. 6.46 Simulation of particle surface.[82a]

4. Volume of positive blocks

$$\frac{\pi^2}{2} A_2 + \frac{8\pi^2}{9} A_3 + \frac{3\pi^2}{4} A_4 + 2\pi \sum_{n=5}^{\infty} An \tag{6.29}$$

This term is found in the second row of column 1.

Top Surface Area. In row 3, column 1, the terms may be predicted to represent excess top surface area. As is known, the total top surface area of all positive blocks is:

$$\frac{1}{2} \times (4\pi A_o^2) = 2\pi A_o^2$$

in which $\beta = 2$. The equation is similar to that in row 3, column 1.

5. Activity corners

In the application of the square waves to the physical properties of particles, $\Sigma\, n^{\alpha} An^{\beta}$, one must consider the relationship between n and An, and the determination of n when it approaches infinity. Here is a linear relation between log An and n, and the value of An closes to zero as n nears infinity (the signature). One can choose an upper limit of n from a meaningful physical parameter such as the unit crystal size–10A, or choose a limit of n to truncate the formula without the loss of significance of the equation.

To calculate the number of corners on a particle, it is postulated that a pyramid with one apical corner is used for simulating a corner on a particle, and the sharper the apex the greater the reactivity of the corner. As n increases, the top of each block is so small as to be regarded as a single point. So, one can transform the vertical blocks into pyramids with one apical corner.

The base angle of a pyramid, ψ, is measured as

$$\psi = \tan^{-1} \frac{An}{\sqrt{2\pi}/2n} = \tan^{-1} \frac{2An \cdot n}{\sqrt{2\pi}}$$

and the equation of corners becomes

$$\Sigma\, N^2 g = k\, \Sigma N^2 \left(\tan^{-1} \frac{2An \cdot n}{\sqrt{2\pi}} \right)^2 = \frac{2k}{\pi^2} \Sigma N^4 An^2 \qquad (6.30)$$

The potential use of the Meloy Equations is therefore to represent various features of the particle surface such as edges, ledges, and corners. It may thereafter be possible to compute these measured features of particulate materials and relate them to the various chemical and physical responses which they exhibit. Furthermore, in conjunction with other approaches already mentioned, these equations permit the scanning of the range of morphological characteristics–microscopic, mesoscopic, and macroscopic. This is something that was not possible to do before.

Some interesting work has been reported[84] using the semimicroprobe. One of the reported results is shown in Fig. 2.41. Many of the morphological features and the microchemistry are analyzed and reported in a comprehensive way. For example, the figures quoted include the total area and the area of type 2 material (chalcopyrite) which by difference gives the remaining area of quartz, and so on. This approach has been described as the shape-texture characterization of a particle by Julius Tou of the Center for Information Research, University of Florida.

6.9 PROPERTY REPRESENTATION

In this section a graphical way of representing powder morphology and, in some detail, the potential of fuzzy sets in powder characterization are considered.

Feature-Space Representation of Particle Shape

In order to be able to represent the complex patterns of relationships to be observed in particle morphology analysis, it is necessary to develop new ways of representing these relationships. A recent paper discusses a prototype system of feature-space representation.[85] The new method is a three-dimensional plot of form and roundness. A triaxial graph in the plane of the paper is used by the observer to plot the "form" characteristics of the particles under study. These three characteristics are thickness, breadth, and length respectively. The axial lengths are scaled from zero to one such that the following relationship holds:

$$\left\{ L + B + T = 1 \right\}$$

This method of graphical representation has been used to analyze form characteristics of samples of sand and SiC. The results are sets of contours as shown in Fig. 6.47. The roundness of the particles is represented in the vertical direction as shown in Fig. 6.48. Those familiar with terniary phase diagrams will recognize the inherent difficulties which one meets when using three-dimensional plots to represent complex systems. However, it must be remembered that complicated and cumbersome though the 3-D plot may be, it constitutes a distinctly simpler generalization than the plethora of data that it represents. This new method is therefore a significant advance for particle morphology analysis; it points a way that representational methods might go and it also highlights the need for sophisticated computer interfaced imaging systems in order that more data can be more easily amassed.

Properties of Fuzzy Sets*

Fuzzy sets have already been referred to in Section 6 of this chapter

*Some of the illustrative examples here are taken from the excellent summary, "Fuzzy Sets–Notation, Terminology and Basic Properties."[88]

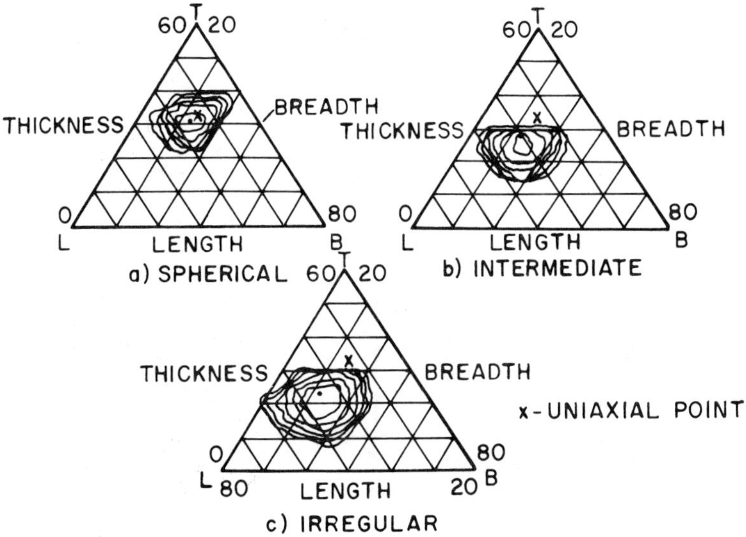

Fig. 6.47[85] Triaxial contour diagrams of shape-sorted sands.

in connection with the problem of visual comparator design. Fuzziness refers to definitions necessarily suffering from vagueness, ambiguity, and ambivalence. It appears that the main fuzzy property of "inexactness" has three distinct aspects:[86]

generality:	the set applies to a variety of situations,
ambiguity:	the set describes more than one distinguishable subconcept, and
vagueness:	the set has no precise degree of membership.

It has been claimed that the theory of fuzzy sets is rich enough in its operations and properties to be of genuine use in constructing models for a wide variety of situations.[87] In this section, some of the properties of fuzzy sets are examined in order to investigate their potential use in powder science.

The symbols U, V, etc. are used to denote a specific universe of discourse, such as a set of persons who are living in a town; the set of all of the nuclear weapons that we have; the set of all real, even numbers, etc.

If A is a nonfuzzy subset of U and it consists of members $u_1, u_2, \ldots,$ then

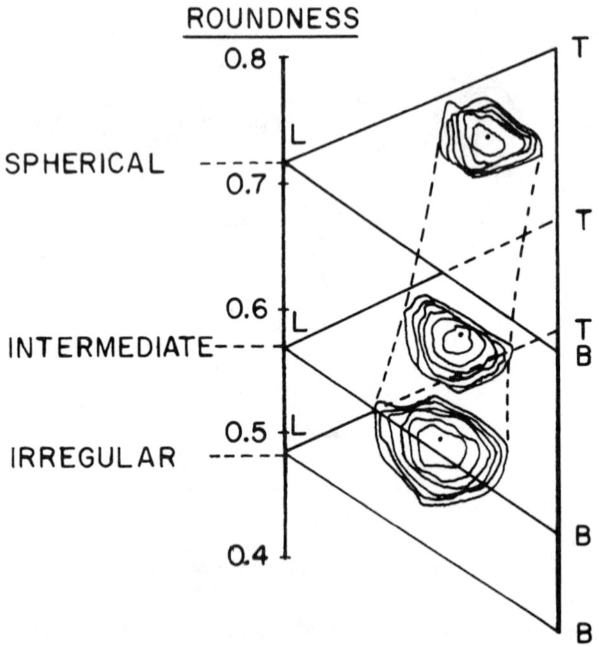

Fig. 6.48[85] Form-roundness feature-space representation of shape-sorted sands.

$$A = u_1 + u_2 + \ldots + u_n \tag{6.31}$$

in which the plus sign denotes "and."

If we let F be a finite fuzzy subset of U, then it can be said that

$$F = \frac{\mu_1}{u_1} + \ldots + \frac{\mu_n}{u_n} \tag{6.32}$$

in which μ_i/u_i is the grade of membership of u_i in F, and

$$\mu_i = 1, 2, \ldots,$$

$$0 \leqslant \mu_i \leqslant 1,$$

$$\mu_i = 0 \text{ denotes no membership, and}$$

$$\mu_i = 1 \text{ denotes full membership.}$$

For example, if

$$u = a + b + c + d \tag{6.33}$$

then

$$A = a + b + d$$

and

$$F = 0.3a + 0.9b + d$$

And, if

$$u = 0 + 0.1 + 0.2$$

then, for example:

$$F = \frac{0.3}{0.5} + \frac{0.6}{0.7} + \frac{0.8}{0.9} + \frac{1}{1}$$

i.e., the grade of membership of 0.5 in u is 0.3 and so on.

Two types of membership functions have been distinguished:

$$S(u, \alpha, \beta, \gamma) \tag{6.34}$$

and

$$\pi(u, \beta, \gamma) \tag{6.35}$$

These are shown in Fig. 6.49.

If F and G are fuzzy subsets of U, their union, $F \cup G$ *intersection,* $F \cap G$ *bounded-sum,* $F \oplus G$, and *bounded-difference,* $F \ominus G$, are fuzzy subsets of U defined by:

$$F \cup G \triangleq \int_U \mu_F(u) \vee \mu_G(u)/u \quad \text{Union,} \tag{6.36}$$

$$F \cap G \triangleq \int_U \mu_F(u) \wedge \mu_G(u)/u \quad \text{Intersection,} \tag{6.37}$$

$$F \oplus G \triangleq \int_U 1 \wedge [\mu_F(u) + \mu_G(u)]/u \quad \text{Bounded-sum, and} \tag{6.38}$$

$$F \ominus G \triangleq \int_U 0 \vee [\mu_F(u) - \mu_G(u)]/u \quad \text{Bounded-difference,} \tag{6.39}$$

in which $\vee$ and $\wedge$ denote max and min, respectively. "$\triangleq$" signifies "is defined as"; "$\int_U$" signifies the union of fuzzy singletons μ_i/u_i. For example, with u defined as in Eq. (6.33),

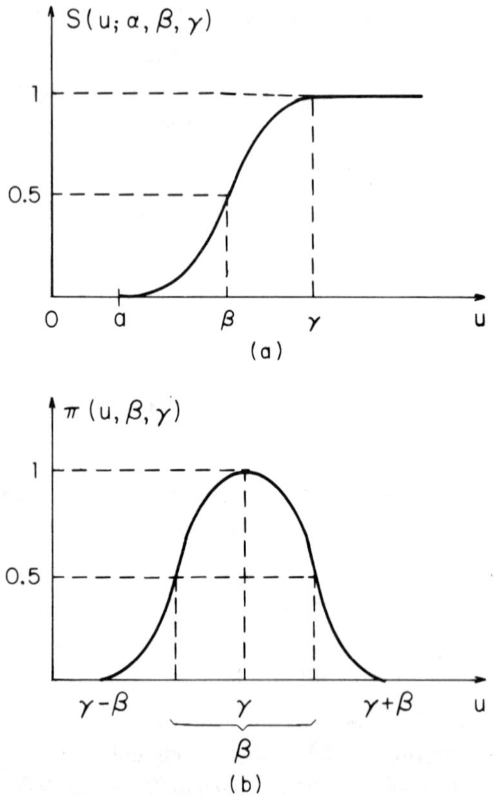

Fig. 6.49[88] Plots of S and π functions.

Let

$$F = \frac{0.4}{a} + \frac{0.9}{b} + \frac{1}{d} \tag{6.40}$$

and

$$G = \frac{0.6}{a} + \frac{0.5}{b} \tag{6.41}$$

Then

$$F \cup G = \frac{0.6}{a} + \frac{0.9}{b} + \frac{1}{d} \tag{6.42}$$

$$F \cap G = \frac{0.4}{a} + \frac{0.5}{b} \tag{6.43}$$

$$F \oplus G = \frac{1}{a} + \frac{1}{b} + \frac{1}{d} \tag{6.44}$$

and

$$F \ominus G = \frac{0.4}{b} + \frac{1}{d} \tag{6.45}$$

Also, the compliment of F, designated F', is defined as

$$F' = \int_u \frac{1 - \mu_F(u)}{u} \tag{6.46}$$

From Eq. (6.44) therefore, F' is in that case

$$F' = 0.6a + 0.16 + c \tag{6.47}$$

Eqs. (6.32)–(6.45) and Eq. (6.47) are plotted in Fig. 6.50.

If α is a real number, then F^{α} is defined as:

$$F^{\alpha} \triangleq \int_u \frac{[\mu_F(n)]^{\alpha}}{u} \tag{6.48}$$

In the case of fuzzy set F of Eq. (6.40),

$$F^2 = \frac{0.16}{a} + \frac{0.81}{b} + \frac{1}{d} \tag{6.49}$$

and

$$F^{½} = \frac{0.2}{a} + \frac{0.3}{b} + \frac{1}{d}$$

F^2 and $F^{½}$ are illustrated in Fig. 6.51.

Linguistic Labels

The operations which have just been demonstrated can be associated with appropriate linguistic labels.

For example, the operation of intersection of two fuzzy sets F, G may be identified with the label "and":

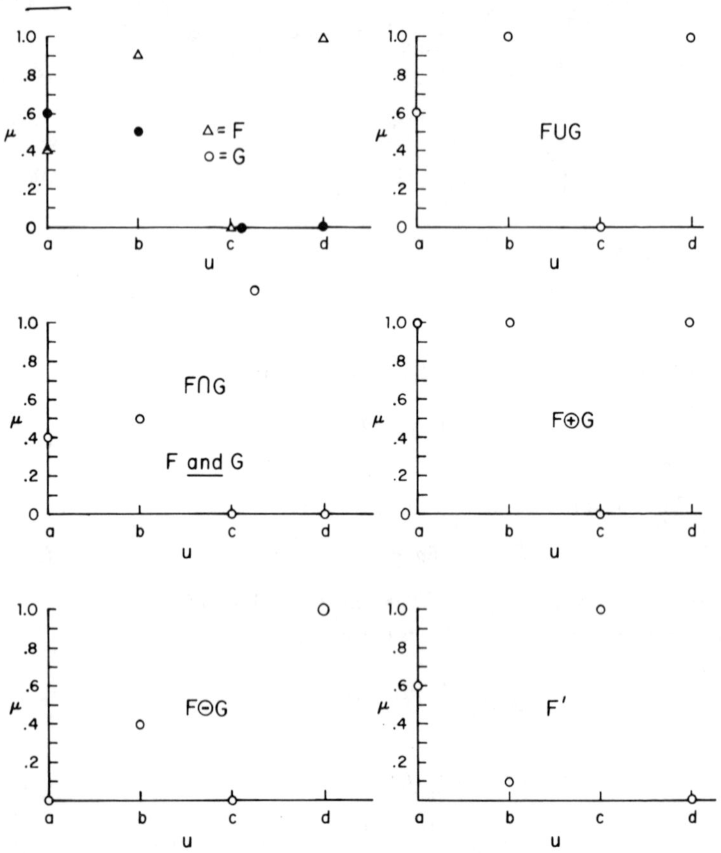

Fig. 6.50 (Reprinted with permission of CRC Press, J. K. Beddow, *Advanced Particulate Morphology,* 1979).[122]

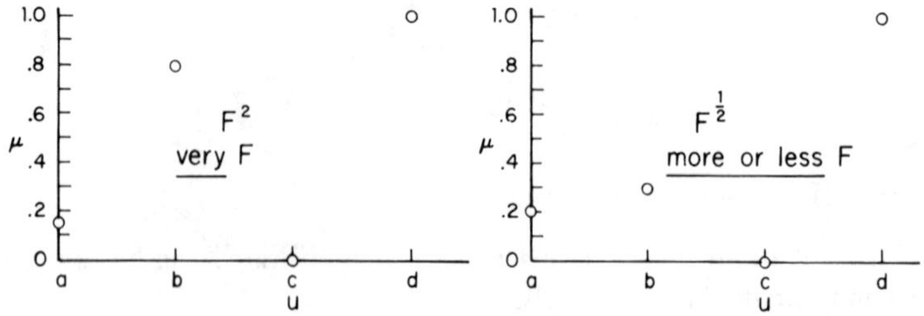

Fig. 6.51 (Reprinted with permission of CRC Press, J. K. Beddow, *Advanced Particulate Morphology*, 1979).[122]

$$F\ \underline{\text{and}}\ G \triangleq F \quad G \tag{6.51}$$

Similarly,

$$F\ \underline{\text{or}}\ G \triangleq F + G = \int_u \frac{\mu_F + \mu_G - \mu_F \mu_G}{u} \tag{6.52}$$

and

$$\underline{\text{very}}\ F = F^2 \tag{6.53}$$

Also,

$$\underline{\text{more or less}}\ F = F^{1/2}\ . \tag{6.54}$$

F $\underline{\text{or}}$ G for the fuzzy sets defined in Eqs. (6.40) and (6.41) is shown in Fig. 6.52.

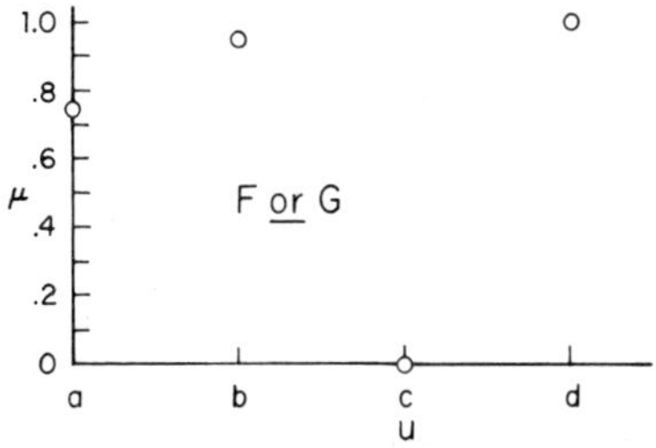

Fig. 6.52 (Reprinted with permission of CRC Press, J. K. Beddow, *Advanced Particulate Morphology*, 1979).[122]

Cartesian Product, Fuzzy Relations, and Composition

1. If $F_1, \ldots, F_n$ are fuzzy subsets of $U_1, \ldots, U_n$, then the *Cartesian product* of $F_1, \ldots, F_n$ is a fuzzy subset of $U_1 \times \ldots \times U_n$ defined by:

$$F_1 \times \ldots \times F_n = \int_{U_1 \times \ldots \times U_n} \frac{\mu_{F_1}(u_1) \wedge \ldots \wedge \mu_{F_n}(u_n)}{u_1 \ldots u_n}\ . \tag{6.55}$$

For example, $F \times G$ of Eqs. (6.40) and (6.41) is a collection of ordered pairs thus:

$$F \times G = \left(\frac{0.4}{a} + \frac{0.9}{b} + \frac{1}{d}\right) \times \left(\frac{0.6}{a} + \frac{0.5}{b}\right)$$

$$= \frac{0.4}{a,a} + \frac{0.4}{a,b} + \frac{0.6}{b,a} + \frac{0.5}{b,b} + \frac{0.6}{d,a} + \frac{0.5}{d,b}$$

2. A *fuzzy relation* R from a set X to a set Y is a fuzzy subset of a Cartesian product $X \times Y$.
3. If R is a relation from X to Y, and if S is a relation from Y to Z, then *the composition* of R and S is a fuzzy relation $R \circ S$ defined as follows:

$$R \circ S \triangleq \int_{X \cdot Z} \frac{\bigvee_{y} [\mu_R(X,Y) \wedge \mu_S(Y,Z)]}{(X,Z)} \tag{6.56}$$

For example, if $R \circ S = M$, what is M if R and S have the following characteristics:

$$R \circ S = M$$

$$[0.1 \;\; 1 \;\; 0.2] \circ \begin{bmatrix} 0.7 & 0.9 & 0.3 \\ 0.6 & 1 & 0.4 \\ 0.5 & 0.7 & 1 \end{bmatrix} \quad ?$$

For an answer, first take the min Cartesian product:

$$= \begin{bmatrix} 0.1 & 0.1 & 0.1 \\ 0.6 & 1 & 0.4 \\ 0.2 & 0.2 & 0.2 \end{bmatrix}$$

Therefore,

$$M = [0.6 \quad 1 \quad 0.4] \tag{6.57}$$

Linguistic Variables

A linguistic variable v is one whose values are words or sentences in natural or artificial verbal languages. For example, if *size* is interpreted as a linguistic variable, then its term's set, $T(v)$, is the set of its linguistic values. This could be

$$T(\text{size}) = \text{little} + \text{big} + \text{very little} + \text{not little} + \text{very big} + \text{very very little} + \text{more or less little} + \ldots \qquad (6.58)$$

Each of the terms in T(size) is a label of a fuzzy set in the universe of discourse. The terms *little* and *big* are designated as *primary terms.*

Now, a language has a *syntactic rule* which deals with the structure of the language generated. That is to say, the rule defines the well-formed sentences in $T(v)$.

A language also has a *semantic rule* which defines the meaning of the terms in $T(v)$. If the compatability functions (the membership functions) of little and big are designated by μ little and μ big, respectively, then from Eqs. (6.53) and (6.54) the compatability functions of the nonprimary terms can be derived as follows:

$$\mu \textit{ very little} = (\mu \text{ little})^2 \qquad (6.59)$$

$$\mu \textit{ more or less big} = (\mu \text{ big})^{1/2} \qquad (6.60)$$

$$\mu \textit{ not very little} = 1 - (\mu \text{ little})^2 \qquad (6.61)$$

Plots of the compatability (membership) functions of these terms are given in Fig. 6.53, for purposes of illustration. In this connection, it should be pointed out that the problem of measuring the fuzzy set is not dealt with, to any major extent, in the literature. It has been observed, however, that "Probably, we should not expect particular numerical values of shortness to be meaningful (except 0 and 1), but rather that their ordering. Observers may agree on whether one man is shorter than another. But their use of the word *short* will never stabilize."[87] It is now easier to understand the three attributes of fuzzy sets mentioned earlier–generality, ambiguity, and vagueness, as shown by the illustration in Fig. 6.54.

Illustrative Potential Applications of Fuzzy Sets

Now, three examples are considered in which fuzzy sets are applied to problems which are important in powder science, and specifically in powder characterization. These are:

1. taxonomy,
2. graphical representation of data, and
3. inexact inferences.

1. Taxonomy. Consider the following verbal descriptors: round, lumpy, and not lumpy. From our earlier discussion (Section 7 of Chapter

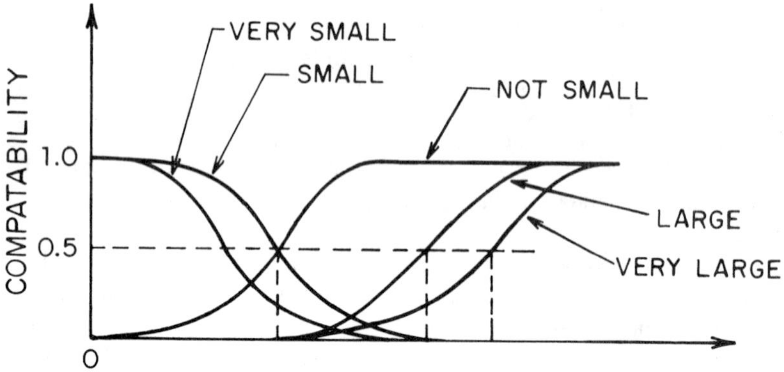

Fig. 6.53

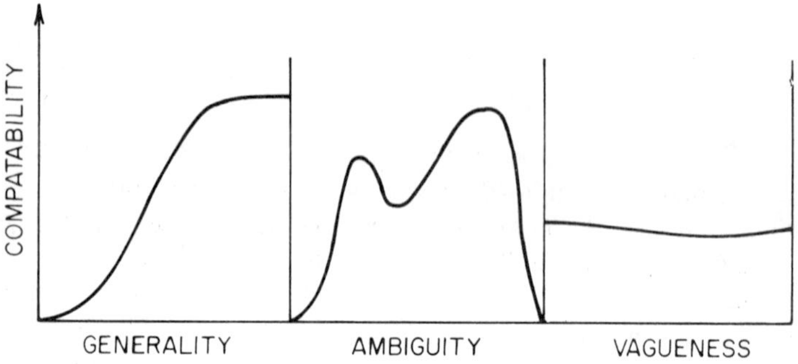

Fig. 6.54

4) it was seen how the arbitrary term "lumpy" was defined. A lumpy particle is therefore a nonround particle. The question is: would the descriptor *not lumpy* have the same meaning as the verbal descriptor *round* when applied to a particle? Experts can argue for many hours over this, but what must be determined is if fuzzy sets can assist in manking the decision.

If it is assumed that the meaning of lumpy may be defined as

$$\mu \text{ lumpy} \triangleq \int_0^1 \frac{\mu \text{ lumpy}(\nu)}{\nu} \tag{6.62}$$

then the descriptor *not lumpy* has the meaning defined as

$$\mu \text{ not lumpy} \triangleq \int_0^1 \frac{1 - \mu \text{ lumpy}(\nu)}{\nu} \tag{6.63}$$

In this case, the meaning of the term round might be defined as

$$\mu \text{ round} \quad \triangleq \int_0^1 \frac{\mu \text{ lumpy}(1-\nu)}{\nu} \tag{6.64}$$

(Of course, the meaning of round could be defined in another way.)

If we now assume certain values for the compatability functions of the universe of discourse of "lumpy" as follows:

$$\text{Lumpy} = \frac{0}{a} + \frac{0}{b} + \frac{0}{c} + \frac{0}{d} + \frac{0}{e} + \frac{0}{f} + \frac{0.3}{g} + \frac{0.5}{h} + \frac{0.7}{i} + \frac{0.9}{j} + \frac{1}{k} \tag{6.65}$$

then

$$\text{Not Lumpy} = \frac{1}{a} + \frac{1}{b} + \frac{1}{c} + \frac{1}{d} + \frac{1}{e} + \frac{1}{f} + \frac{0.7}{g} + \frac{0.5}{h} + \frac{0.3}{i} + \frac{0.1}{j} + \frac{0}{k} \tag{6.66}$$

and

$$\text{Round} = \frac{1}{a} + \frac{0.9}{b} + \frac{0.7}{c} + \frac{0.5}{d} + \frac{0.3}{e} + \frac{0}{f} + \frac{0}{g} + \frac{0}{h} + \frac{0}{i} + \frac{0}{j} + \frac{0}{k}. \tag{6.67}$$

Eqs. (6.65), (6.66), and (6.67) are plotted in Fig. 6.55.

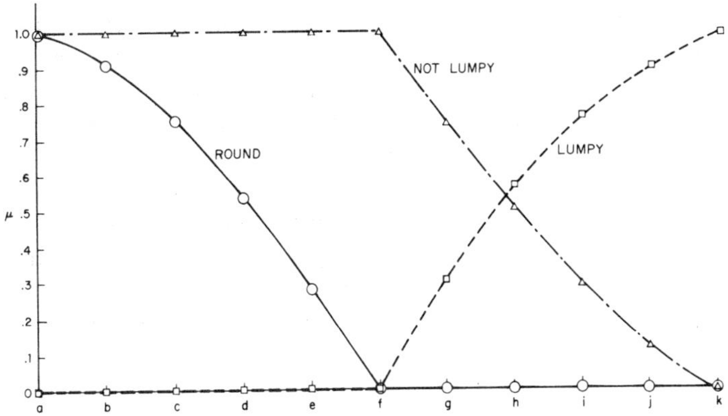

Fig. 6.55 Lumpy, Not Lumpy, and Round–Compatability Functions versus Universe of Discourse.[128]

This shows that on this basis the term *round* does not have the same meaning as *not lumpy*. However, it also shows that it would be quite possible to define the membership function curve for *round* in such a way that it would be practically indistinguishable from *not lumpy*, if so desired. This could be done if the curve for *round* in Fig. 6.55 were shifted to the right (indicated by dashed line). This example indicates the potential utility of using fuzzy sets as a tool in attempts to precisely define the meaning of the terms used in fine particle science and technology.

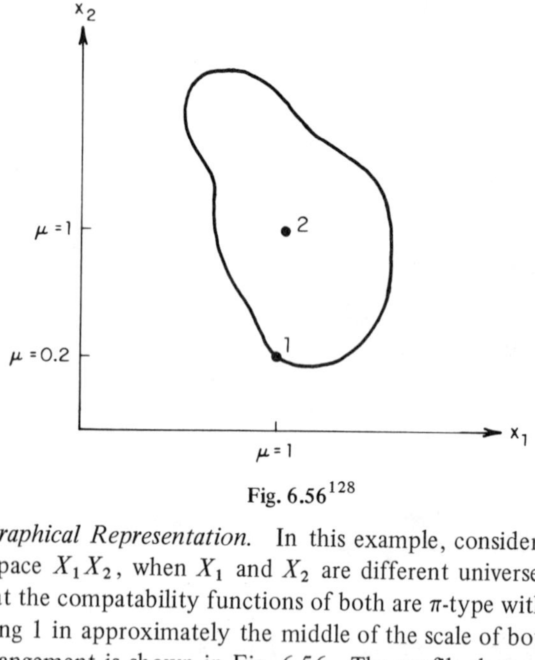

Fig. 6.56[128]

2. *Graphical Representation.* In this example, consider a two-dimensional space X_1X_2, when X_1 and X_2 are different universes of discourse such that the compatability functions of both are π-type with the membership being 1 in approximately the middle of the scale of both dimensions. This arrangement is shown in Fig. 6.56. The profile shown is intended to represent an envelope around a cluster of data points plotted on the graph. The question which is asked is: what is the grade of membership of point 1 and what is the grade of membership of point 2? If it is assumed that X_1 might be a scale of particle roughness and X_2 might be a scale of coloration, the compatability functions corresponding to 1 and 2 are as follows:

$$\mu_{(X_1,X_2)_1} = \frac{1}{X_1}\ \frac{0.2}{X_2}$$

$$= \frac{0.2}{X_1X_2} \qquad (6.68)$$

and

$$\mu_{(x_1 x_2)_2} = \frac{1}{X_1} \frac{1}{X_2}$$

$$= \frac{1}{X_1 X_2} \tag{6.69}$$

Which is to say that the membership or compatability function of point 1 is 0.2 and of point 2 is 1.

3. Inexact Inferences. In powder science and technology, information must be summarized to extract from the collections of masses of data concerning particle and powder characteristics, those and only those subcollections which are relevant to the performance of the task at hand.[89] *By its nature, a summary is an approximation to what it summarizes.* The potential use of fuzzy sets is illustrated with an example of an inferential statement. This is termed a *fuzzy inference.*

Using the terminology already defined, consider the statement:

"If L is $\underline{\text{small}}$, then A.D. is $\underline{\text{high}}$, else A.D. is $\underline{\text{not high}}$", (6.70)

in which L is lumpiness, A.D. is the apparent density of the powder, and "else" may also be read as "otherwise."

The following fuzzy inference is now posed:

"If L is $\underline{\text{not very small}}$, then A.D. is $\underline{(\ ?\ )}$." (6.71)

The problem is to complete the statement with the correct inference. Assume the following universes: L, lumpiness, and A.D., apparent density. The problem is solvable by stating it in terms of the composition, already mentioned. Thus, the fuzzy subset of apparent density is induced by the fuzzy subset of lumpiness and is given by the composition of *R* thus:

$$\text{A.D.} = \text{L} \circ \text{R} \tag{6.72}$$

Assume that

$$\text{L is } \underline{\text{small}} = \frac{1}{1} + \frac{0.4}{3} + \frac{0.1}{5} \tag{6.73}$$

and

$$\text{A.D. is } \underline{\text{high}} = \frac{0}{5} + \frac{0.5}{7} + \frac{1}{10} \tag{6.74}$$

Using the already stated semantic rules:

$$\text{L is } \underline{\text{not small}} \quad = \frac{0}{1} + \frac{0.6}{3} + \frac{0.9}{5} \tag{6.75}$$

$$\text{L is } \underline{\text{not very small}} \quad = \frac{0}{1} + \frac{0.84}{3} + \frac{0.99}{5} \tag{6.76}$$

$$\text{A.D. is } \underline{\text{not high}} \quad = \frac{1}{5} + \frac{0.5}{7} + \frac{0}{10} \tag{6.77}$$

$$\text{A.D. is } \underline{\text{very high}} \quad = \frac{0}{5} + \frac{0.25}{7} + \frac{1}{10} \tag{6.78}$$

$$\text{A.D. is } \underline{\text{not very high}} = \frac{1}{5} + \frac{.75}{7} + \frac{0}{10} \tag{6.79}$$

The Cartesian product of (L is small) X (A.D. is high) is R_1 as follows:

R_1 =

A.D. High / L Small	5	7	10
1	0	0.5	1
3	0	0.4	0.4
5	0	0.1	0.1

The Cartesian product of (L is not small) X (A.D. is not high) is R_2 as follows:

R_2 =

A.D. Not High / L Not Small	5	7	10
1	0	0	0
3	0.6	0.5	0
5	0.9	0.5	0

It is necessary to find the composition, which is the max-min product of $\mu_{(\text{L } \underline{\text{not very small}})}$ with $R_1 \cup R_2$. That is

$$\underline{(\,?\,)} = \mu_{(\text{L } \underline{\text{not very small}})} \circ R_1 \cup R_2 \tag{6.80}$$

Using the previously derived values of R_1 and R_2 their union is computed and the composition is taken as follows

$R_1 \cup R_2 =$

	5	7	10
1	0	0.5	1
3	0.6	0.5	0.4
5	0.9	0.5	0.1

$$(\,?\,) = \left[\frac{0}{1} + \frac{0.84}{3} + \frac{0.99}{5}\right] \circ \left[\begin{array}{c|c|c|c} & 5 & 7 & 10 \\ \hline 1 & 0 & 0.5 & 1 \\ 3 & 0.6 & 0.5 & 0.4 \\ 5 & 0.9 & 0.5 & 0.1 \end{array}\right]$$

$$= \begin{bmatrix} 0 & 0 & 0 \\ 0.6 & 0.5 & 0.4 \\ 0.9 & 0.5 & 0.1 \end{bmatrix}$$

i.e.,

$$(\,?\,) = \left[\frac{0.9}{5} + \frac{0.5}{7} + \frac{0.4}{10}\right]. \tag{6.81}$$

Inspection of Eqs. (6.72) to (6.79) indicates that Eq. (6.81) is nearest to Eq. (6.77).

That is to say, the correct inference, or at least the most correct inference that can be calculated, indicates that:

"If L is not very small, then A.D. is not high." (6.82)

It would be less correct to say that:

"If L is not very small, then A.D. is not very high." (6.83)

As an illustration, the three equations are plotted in Fig. 6.57.

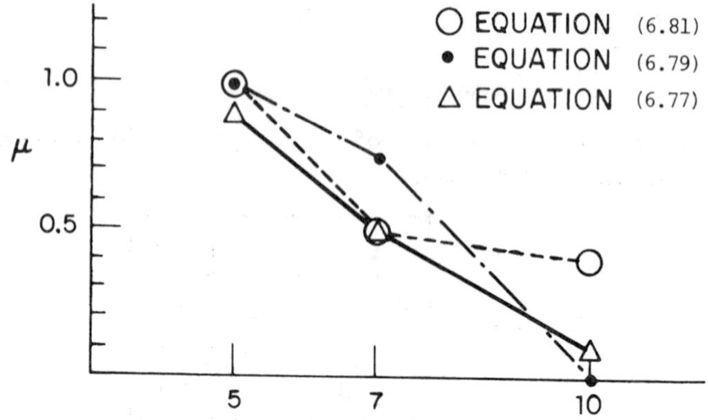

Fig. 6.57[128]

6.10 PRINCIPLES OF STEREOLOGY[90]

Quantitative stereology is a generalized body of methods relating the quantitative statistical properties of 3-D structures to those of their 2-D sections or projections. It extrapolates from two to three dimensional space. For example, the surface area or volume of a particle can be determined by observing and measuring its projected image.

Measurements are usually obtained by penetrating the sample space with test probes, random, or otherwise. The test probes can be slices, planes, lines, or points.

In order to characterize a material completely, the metric relationships as well as the geometrical and topological properties of a microstructure must be considered. On this subject, a variety of topics are included. They are Geometrical Probabilities, Particle Characteristics, Size Distribution, Curvature, Topological Relationships, and Projected Images.

Basic Symbols and Equations

Table 6.20 lists some of the basic symbols: V = volumes, S = curved surfaces, A = flat surfaces, L = lines, and P = points. The S-term is a microstructural quantity and has a fixed value, whereas A is a variable quantity depending on the random cut of the section plane. The combined terms are ratios of a microstructural quantity to a test quantity, wherein V_T, A_T, and L_T represent the test volume, test area, and test length, respectively. Thus, S_V, for example, is the surface area per unit volume of the microstructure. These combined symbols are merely a convenient way of writing the fraction. Each symbol has a definite geometrical meaning and an associated dimension. Thus, the dimensionality of the combined terms is readily apparent, as well as the dimensional consistency of the equations.

Table 6.20[90]

BASIC SYMBOLS

V	
	$V_V \equiv V/V_T$
S	
	$S_V \equiv S/V_T$
A	
	$L_A \equiv L/A_T$
L	
	$P_L \equiv P/L_T$
P	

Basic measurements are shown graphically in Fig. 6.58. The six combined symbols in the top line all represent simple counting measurements. Examples of each are given in the microstructures below. P_P represents points that fall within a selected phase divided by the total number of test points, P_L is the number of points of intersection with linear elements in the microstructure per unit length of test line, and P_A means the number of point features in the microstructure per unit test area.

Particle Characteristics

All the above equations are for systems of particles. Now, individual particles of convex shape and some general relationships that apply to them as well as to systems of convex particles must be considered.

Cauchy in 1832 derived an equation relating the surface area S to the mean projected area $\bar{A}'$. For a convex body, $S = 4\bar{A}'$. Note that the average has been taken over all orientation 4π.

Another important parameter for particles is the Mean Intercept Length $\bar{L}_3$. The subscript 3 refers to the dimensionality of the particle. If the particle is penetrated with straight lines from all possible positions at all possible angles, the Mean Intercept Length is then simply

$$\bar{L}_3 = \frac{L_L}{N_L} \tag{6.84}$$

in which L_L is the total intercept length and N_L is the number of penetrations.

The equation can be reduced to

$$\bar{L}_3 = 4V/S \tag{6.85}$$

and

$$V = \bar{L}_3\bar{A}' \tag{6.86}$$

Another parameter is the Mean Intercept Area $\bar{A}$. It is found that

$$V = \bar{A}\bar{H}' \tag{6.87}$$

in which $\bar{H}'$ is the mean projected height (average height of $\bar{A}'$).

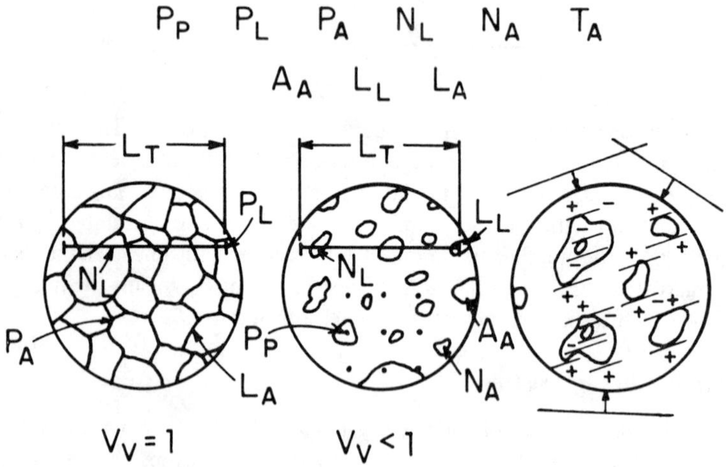

Fig. 6.58 Basic measurements in stereology.[90]

Topological Relationships

Some characteristics of microstructure features, not dependent on size or shape, are called topological properties. They are unchanged by deformation such as stretching or bending as long as they do not involve cutting or joining. These are the number of corners, edges, faces, and grains in microstructure.

For Polyhedron (3-D)

$$C - E + F = 2\,; \tag{6.88}$$

Convention: 4 cells meet at Corner
3 cells meet at Edge, and
2 cells meet at Face

For Polygon

$$C - E = 0 \tag{6.89}$$

Convention: 3 cells meet at Corner, and
2 cells meet at Edge.

It is interesting to note that for most particles, five is the predominant number of edges per polygonal face, but the range can be extended from three-sided to eight-sided faces. This appears rather uniform for simple cells.

A Potential Scheme for Powder Characterization[91]

It is possible to establish six features for characterization of particles of a powder. These include three features pertaining to the "size" characteristics of the particles and three features concerning the "shape" characteristics of the powder particles. The six features are specified below:

Size

1. Mean Radius:

$$S = 4\pi\gamma^2 = 4\bar{A}$$

and

$$\gamma = \sqrt{\bar{A}/\pi} \tag{6.90}$$

in which

$$\bar{A} = \frac{1}{n}\sum_{i=1}^{n} A_i$$

2. Mean Feret Horizontal Diameter, $\bar{F}_h$

$$\bar{F}_h = \frac{1}{n} \sum_{i=1}^{n} F_{hi} \tag{6.91}$$

3. Mean Intercept Length, $\bar{L}_3$

$$\bar{L}_3 = \frac{1}{n} \sum_{i=1}^{n} \left(\frac{A_i}{I_{hi}}\right) \tag{6.92}$$

and

$$I_h = \text{horizontal intercept.}$$

Shape

1. Shape Factor One, SF_1:

$$SF_1 = \frac{\bar{A}^3}{V} \qquad \text{for } V = \bar{L}_3\bar{A}\,,$$

$$= \frac{\bar{A}}{\bar{L}_3^2} \tag{6.93}$$

2. SF_2:

$$SF_2 = \frac{I_h}{F_h} \tag{6.94}$$

3. SF_3:

$$SF_3 = \frac{\bar{A}'}{\bar{L}^2} \tag{6.95}$$

$$L_2 = \text{perimeter}$$

$$\bar{L}_2 = \frac{1}{n} \sum_{i=1}^{n} L_{2i}$$

It should be possible, to measure these six features for powders of the same material but made by different methods, for example, in order to develop pattern classes of the features for each method of powder production. This would involve determination of the appropriate decision functions.

One practical problem with measuring a given sample is to be able to examine the features of a desired set of particles. An interesting example of this problem being solved in the case of two types of alpha phase in a metallographic section is shown in Fig. 6.59.[92] In the photomicrographs

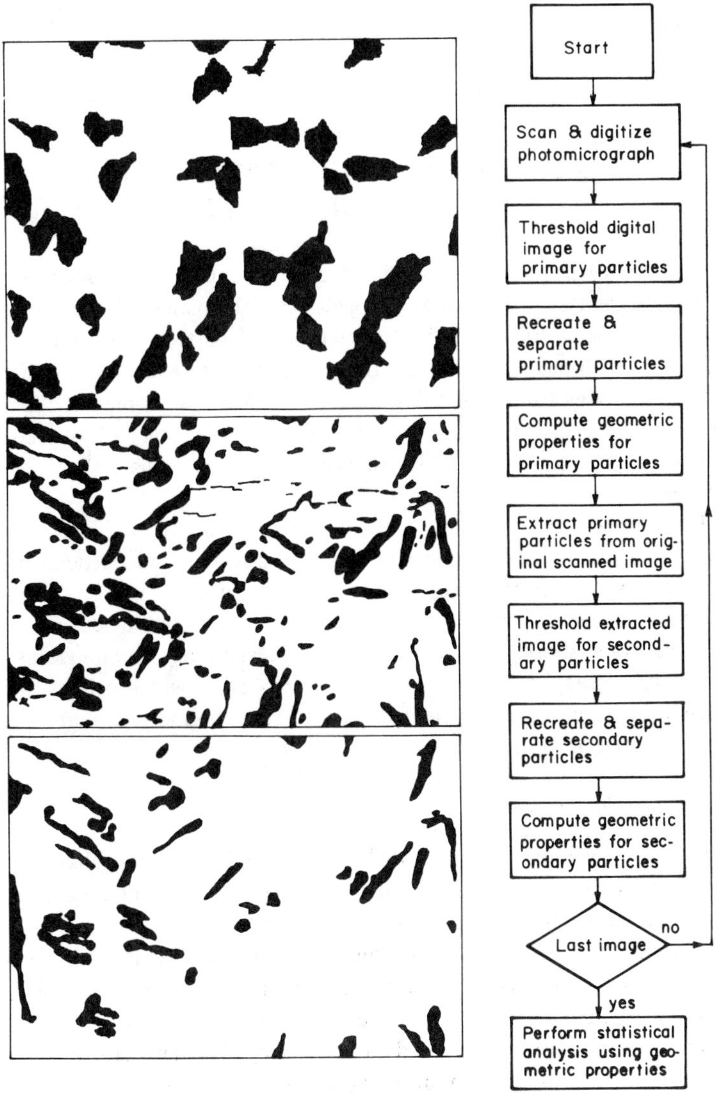

Fig. 6.59 General Flowchart of Mead Technology Laboratories Particle Analysis System.[92]

shown, the primary alpha particles were separated from the structure (see their absence in the center view) so that their features could be determined. The secondary particles were treated in the same way and for the same reason. For additional details on specimen preparation, the reader should consult the Particle Atlas.[93]

6.11 DETERMINISTIC, STATISTICAL AND FUZZY CLASSIFIERS

In order to pursue the desired objective of relating the properties of particulate matter to the morphological characteristics of the particles therein, many of the sophisticated methods of pattern recognition that have been developed during the last two decades must be understood. Once understood, they can be used as a tool. Some of the work reported here is part of that ongoing process. A useful starting point for pattern recognition is given in Reference 108. Reference 109 gives a list of articles on fuzzy sets.

The ultimate test of this approach lies in the ability to predict both particle behavior and macroscale properties of powders and particulate agglomerates in engineering applications. It is hoped and expected that such an ability follows from the understanding of morphological analysis.

In closing it should be observed that Fourier analysis of sand grains has been very successfully used in geological research in which origins, transportative mechanisms and other factors pertaining to sand grains are related to spectra of Fourier coefficients.[110]

Types of Classifiers Compared

The following discussion combined with the results show that a comparatively simple deterministic model is capable of classifying a training set of particle features. Not all errors are acceptable, as Table 6.21 shows. But considering the simplicity of the approach, the results must be described as very encouraging and should stimulate work on morphic features, although some statistical properties have been studied.[96]

What has not been studied is the use of fuzzy classifiers, although some of the properties and potential applications of fuzzy sets have been reported.[97] Two interesting models have been proposed as follows:

The first model leads to a nonfuzzy[98] or a fuzzy[99–104] ISODATA algorithm depending upon the value of m:

$$J_m(U,\underline{v}) = \sum_{K=1}^{n} \sum_{i=1}^{c} (U_{ik})^m \| \underline{X}_k - \underline{V}_i \|^2 \qquad 1 \leqslant m < \infty \tag{6.96}$$

in which J_m is the generalized total squared error incurred by representing each fuzzy subset by a fuzzy cluster center $\underline{v}$. One seeks to minimize this error. The number of fuzzy subsets is designated as c, n is the number of features, U is the fuzzy subset, x is the feature vector. If m is unity, then J_1 is the well-known within-group sum of squares. The case in which m equals 2 has been investigated using the classic data of Anderson,[105] which is the set of 150 four-dimensional vectors in one of three subspecies of Iris.[106] The analysis was highly satisfactory and this indicates a similar prognosis in morphological analysis of particulates.*

Table 6.21[90]

BASIC EQUATIONS

$$V_V = A_A = L_L = P_F$$

$$S_V = \frac{4}{\pi} L_A = 2P_L$$

$$L_V = 2P_A$$

$$\bar{K}_m = \frac{\pi}{2} \frac{\Delta T_A}{P_L} = \frac{\pi}{2} \frac{N_A}{N_L}$$

The second model suggests the way in which the problem of formulating a total presentation of physical properties of a particulate might be approached. Thus,[107]

$$f_A(x) = \sum_{i=1}^{n} c_i^* \phi_i(x) \tag{6.97}$$

in which f_A is the membership function of a fuzzy set A in Σ^n, c_i^*'s are undetermined constants, and $\phi_i(x)$ are known, linearly independent, bounded functions of the vector x in Σ^n .

*In a personal communication, May 1979, Professor Bedzek reports the development of a classifier that incorporates both fuzzy and statistical measures.

It appears to be appropriate to use a deterministic classifier in the initial design/training stage of the development of an analytical capability for morphological analysis. This has been done in the following two cases, both of which represent feasibility studies. The results encouraged further study of the methods reported on.

In the case of four metal powders using a form of Eq. (6.30) in which α is zero and β is unity, the summation was arbitrarily split into three groups.[94] This approach was discussed in Section 7 of this chapter. The case of four metal powders with stereological-like descriptors using a Quantimet 720 System[117] is now discussed.

1. Deterministic Trainable Pattern Classifier

This is a set of linear mathematical equations. Each equation is termed a decision function.

The objective of the classification process is to yield decisions concerning the pattern class* into which the specific pattern** can be classified. There are two main steps:

a. collect measurements from the patterns to be recognized. Arrange these measurements in pattern vectors, and
b. classify (i.e., group into appropriate sets) the pattern vectors.

In the work reported here the collected measurements were obtained from images of particles projected onto the screen of a Quantimet 720 Image Analyzing System.

The classification of the pattern vectors was carried out using a deterministic trainable pattern classifier. This classifier is a mathematical scheme which does not use the statistical properties of the pattern classes. The word trainable in the title means that the classifier is established using known pattern classes, by means of a learning process that takes place only during the design phase of the pattern recognition system. Once acceptable results have been obtained with the training set of patterns, the system is applied to the task of actually performing recognition on samples drawn from the environment in which it is expected to operate.

Samples of four powders were selected for this study as follows: (see Fig. 6.60)

Gas Atomized Cu-Pb,

*Pattern Class: a category determined by some common attributes.
**Pattern: the description of any member of a category representing a pattern class.

Sponge Fe,
Electrolytic Fe, and
Electrolytic Cu.

The study is based upon the following two assumptions:

a. powders of a given material produced by different methods possess different shapes, and
b. powders of different materials produced by a given method possess different shapes.

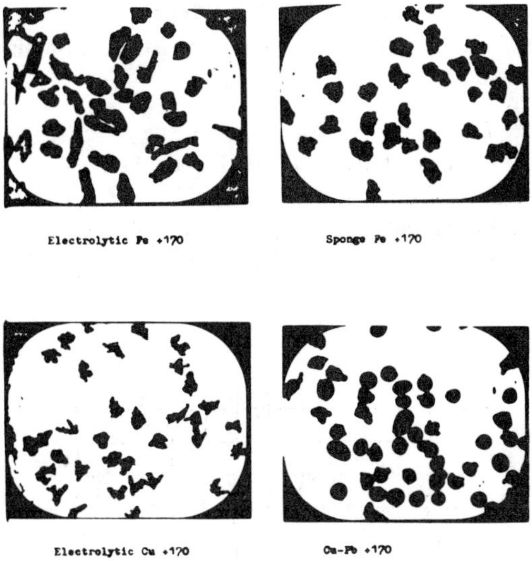

Fig. 6.60 Photomicrographs of particle profiles (magnification 60 X).[117]

2. Feature* Selection

Any pattern which can be recognized and classified possesses a number of discriminatory properties or features. The first step in any recognition process is to consider the problem of what discriminatory features to select and how to measure these features. The number of features needed to successfully perform a given recognition task depends on the discriminatory

*Features: individual parameters that make up the pattern.

qualities of the chosen features.

Feature selection and extraction plays a central role in pattern recognition. It is extremely difficult, if not impossible, to find the complete set of discriminatory features for a pattern class.

In particle characterization research, two types of features are being looked into, morphological and stereological. The former maps the outline of the projected image of a particle into an equation, whereas the latter characterizes a particle by its geometrical structures. Experimentally it is very difficult to take three-dimensional measurements of a particle, so its two-dimensional projected image is usually used.

Projections of a hundred particles of each type of powder were analyzed by a Quantimet 720. The following measurements were obtained:

a. area A,
b. perimeter, L,
c. number of vertical intercept, I_v (see Fig. 6.61)
d. horizontal feret diameter, F_L (see Fig. 6.61)
e. number of horizontal intercept, I_n (see Fig. 6.61), and
f. vertical feret diameter, F_v (see Fig. 6.61).

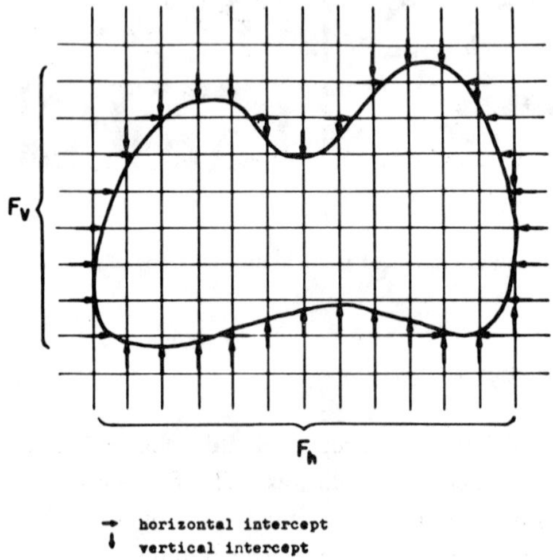

Fig. 6.61 Illustrating four measurements of a particle profile.[117]

From these, five dimensionless features, which are variants of shape and size independent of particle size, were derived. They are either average values of a group of particles or values derived from the use of quantitative stereological methods.[90] These features are:

$$\begin{aligned} &\text{AR: } \bar{F}_v/\bar{F}_h\,, \\ &\text{SF1: } \bar{A}/\bar{L}_3^2\,, \\ &\text{SF2: } \bar{A}/\bar{L}^2\,, \\ &\text{SF3: } \bar{I}_v/\bar{F}\,, \text{ and} \\ &\text{SF4: } \bar{I}_h/\bar{F}\,. \end{aligned} \tag{6.98}$$

Two size dependent features, $\bar{r}$, mean radius and L_3 mean intercept length, were also included.

$$\bar{r} = \sqrt{\bar{A}/\pi}\,, \tag{6.99}$$

and

$$L_3 = ½\left(\frac{\bar{A}}{\bar{I}_h} + \frac{\bar{A}}{\bar{I}_v}\right) \tag{6.100}$$

3. Decision Function

In Fig. 6.62, two hypothetical pattern classes are shown. They are conveniently separated by a line. Let

$$d(\vec{x}) + w_1x_1 + w_2x_2 + w_3 = 0 \tag{6.101}$$

be the equation of the separating line in which the w's are parameters and x_1, x_2 are the general coordinate variables. From the figure a pattern x that belongs to class ψ_1 yields a positive quantity when substituted into $d(\vec{x})$. Similarly, $d(\vec{x})$ becomes negative upon substitution of any pattern from ψ_2. Therefore, $d(x)$ can be used as a decision function since, given a pattern of $\vec{x}$ of unknown classification, $\vec{x}$ can be said to belong to ψ_1 if $d(\vec{x}) > 0$, or ψ_2 if $d(\vec{x}) < 0$. If the pattern lies on the separating boundary (i.e., the line defined by $d(\vec{x})$, the indeterminate condition $d(\vec{x}) = 0$ is obtained.

These concepts need not be restricted to two classes. Once a specific decision function (or functions if more than two classes are involved) has been selected, the problem becomes the determination of the coefficients

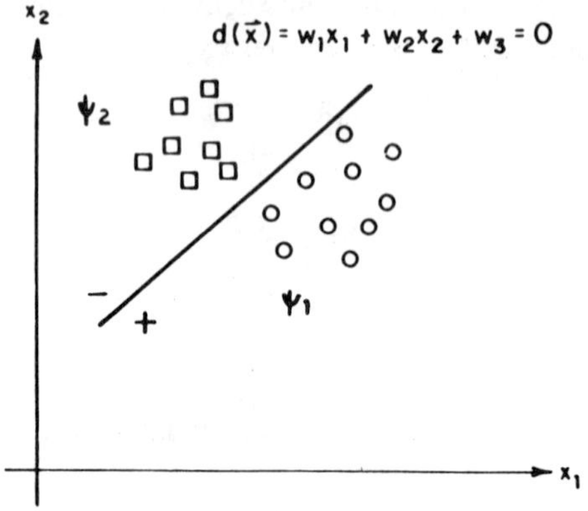

Fig. 6.62 A simple decision function for two pattern classes.[117]

in the equation.

For the n-dimensional case, the general linear decision function is of the form

$$d(\vec{x}) = w_1x_1 + w_2x_2 + \ldots + w_nx_n + w_{n+1} \tag{6.102}$$

which can be expressed in vector form as

$$d(\vec{x}) = \vec{w}^t\vec{x} \tag{6.103}$$

in which

$$x = (x_1, x_2, \ldots, x_n, 1)^t$$

and

$$w = (w_1, w_2, \ldots, w_n, w_{n+1})$$

are called the augmented pattern and weight vector respectively, and t indicates the transpose.

For n particles, there are n equations. If m features are used, the equations are the following:

$$\left.\begin{array}{l} w_1x_{11} + w_2x_{12} + \ldots + w_mx_{1m} + w_{m+1} > 0, \\ w_1x_{21} + w_2x_{22} + \ldots + w_mx_{2m} + w_{m+1} > 0, \text{ and} \\ \quad\bullet \qquad\qquad\qquad\qquad\qquad \bullet \\ \quad\bullet \qquad\qquad\qquad\qquad\qquad \bullet \\ \quad\bullet \qquad\qquad\qquad\qquad\qquad \bullet \\ w_1x_{n1} + w_2x_{n_2} + \ldots + w_mx_{nm} + w_{m+1} > 0, \end{array}\right\} \tag{6.104}$$

in which w's are coefficients to be determined, and x's are features.

In this study, each pattern class is considered apart from all other classes by a simple decision surface. There are M decision functions with the property

$$d_i(\vec{x}) = \vec{w}_1^t\vec{x} = \begin{cases} > 0 \text{ if } x \\ \qquad\qquad i, i = 1, 2, \ldots, M, \\ < 0 \text{ otherwise} \end{cases} \tag{6.105}$$

in which $\vec{w}_1 = (w_{i1}, w_{i2}, \ldots, w_{in}, w_{in+1})^t$ is the weight vector associated with the ith decision function.

To determine the decision function for the ith pattern class, the scheme is considered as a two-class problem ψ_i and $\bar{\psi}_i$ in which $\bar{\psi}_i$ denotes all classes except ψ_i. If the patterns of $\bar{\psi}_i$ are multiplied by -1, the equivalent condition $\vec{w}^t\vec{x} > 0$ for all patterns is obtained.

The problem reduces to that of finding a vector $\vec{w}$ such that the system of inequalities

$$\vec{X}\vec{w} > 0$$

is satisfied, when

$$\vec{X} = \begin{array}{c} \vec{x}_1^t \\ \vec{x}_2^t \\ \bullet \\ \bullet \\ \bullet \\ \vec{x}_n^t \end{array} \tag{6.106}$$

The learning algorithm used here to determine $\vec{w}$ belongs to the gradient technique. Basically, gradient schemes provide a tool for finding the minimum of a function. Upon it, many iterative schemes have been devised.

If a certain criterion function* is chosen so that it achieves its minimum value whenever $\vec{w}^t\vec{x}_i$ 0, then finding the minimum of the function for all i, $i = 1, 2, \ldots, n$, is equivalent to solving the given system of linear inequalities. Different algorithms use different criterion functions.

The general gradient descent algorithm may be written as

$$\vec{w}(k+1) = \vec{w}(k) - c \left. \frac{\partial J(\vec{w},\vec{x})}{\partial \vec{w}} \right|_{\vec{w}=\vec{w}(k)} \tag{6.107}$$

in which

$\vec{w}(k+1)$ is the new value of $\vec{w}$,

$c > 0$ dictates the magnitude of the correction,

$J(\vec{w},\vec{x})$ is the criterion function which is chosen with the situation, and

$\dfrac{-\partial J}{\partial \vec{w}}$ is the negative of the gradient points in the direction of the maximum rate of decrease of J.

The specific criterion function used in this study is:

$$J(\vec{w},\vec{x},\vec{b}) = \frac{1}{2}\sum_{j=1}^{N} (\vec{w}^t\vec{x}_j - b_j)^2 = \frac{1}{2}\left|\left|\vec{X}\vec{w} - \vec{b}\right|\right|^2 \tag{6.108}$$

in which $\vec{b} = (b_1, b_2, \ldots, b_N)$ are all positive.

4. Experimental Results

The design of the experiment is shown in the block diagram in Fig. 6.63.

Block 1 Data Treatment. Raw data from the Quantimet 720 measurements was treated in the following manner:

a. It was assumed that any ten projections taken from a sample of particles of the same powder can be treated as representing one three-dimensional particle.
b. The mean values of these measurements were substituted into Eqs. (6.95) through (6.97) to obtain the five shape features and the two size features for ten representative three-dimensional particles. These results are used as the training sets of patterns.

*A criterion function may have the form $J(\vec{w},\vec{x}) = (|\vec{w}^t\vec{x}| - \vec{w}^t\vec{x})$, for example; it is a minimum when the condition of Eq. (6.104) is satisfied.

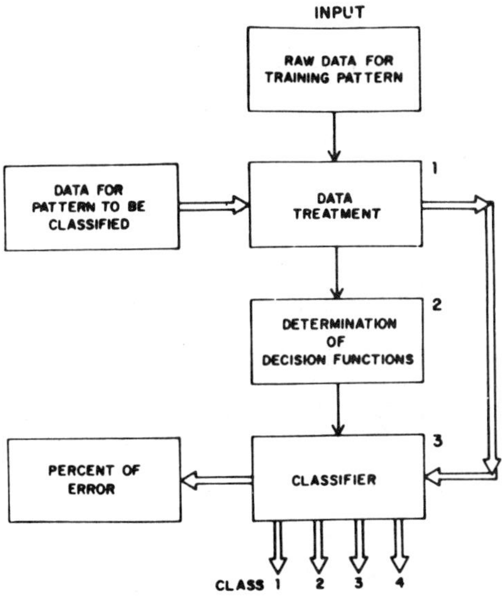

Fig. 6.63 Design of experiment.[117]

Block 2 Determination of Decision Function. The training patterns from Step 1 were put into an iterative scheme using the algorithm derived from Eq. (6.95). Four decision functions were obtained in the case of the four powders as shown in Fig. 6.60. The four decision functions were composed of different sets of features according to the following scheme:

set 1 – AR, SF1, SF2, SF3, SF4, r, L3,
set 2 – AR, SF1, SF2, SF3, SF4,
set 3 – AR, SF2, SF3, SF4, and
set 4 – AR, SF1, SF3, SF4,

All but set 1 are particle size independent.

The number of features in the training patterns were varied in order to obtain different sets of decision functions so that the discriminatory quality of these features can be compared. The decision coefficients (the w's) were obtained for the four powders.

Block 3 The Use of the Classifier. Data obtained from powders to be classified by the Quantimet 720 was fed into the data treatment process with the outcome of pattern vectors. Patterns to be tested should have the same features as the training set in that particular set of decision functions. They therefore should be classified by the decision functions according to these features.

The number of misclassifications for each set of decision functions and for each of five test samples is given in Table 6.22.

Table 6.22[117]

NUMBER OF MISCLASSIFICATION/TOTAL NUMBER OF TEST PARTICLES FOR THE FOUR SETS OF DECISION FUNCTIONS

Class	*Decision Functions*			
	Set 1	Set 2	Set 3	Set 4
Gas-Atomized Cu-Pb + 170	0/60	48/49	54/99	30/99
Electrolytic Fe + 170	0/60	7/99	9/99	7/99
Electrolytic Cu + 170	0/60	0/100	0/100	8/100
Sponge Fe + 170	0/60	11/100	28/100	20/100
Gas-Atomized Cu-Pb 270/325	–	11/102	0/102	83/102
Σ (errors)		77	91	148

Summary and Update

This chapter describes a new research approach to the characterization of fine particles. The basic properties of fine particles may be described in terms of four general characteristics: size, shape, chemistry, and physics.

The physics properties include electrical and thermal conductivities, optical behavior, and density. Chemical properties include gross as well as point to point chemical analysis. The properties of size and shape may be incorporated in the collective title of particle morphology. Particle morphology may be defined as that part of Particulate Science and Technology concerned with the form of finely divided matter and the structures, homologies, and processes of change which govern that form.

Importance of Particulate Systems

Particulate morphology is basic to powder technology. Until recently it has not been possible to define the size of the particle unequivocally and also get the complete morphological information characterizing a particle or powder (a set of particles).

Civilization is built on particulate systems: cement, soil, brick, mortar, paint, glass, and metals are a few examples. Our food is grown on soils that are complex particulate systems; oil and water lie underground in semiconsolidated particulate systems; but for all the importance of particulate systems little coordinated research is done in the field.

When most people think of industry they think of liquids like petroleum, yet the overwhelming material handled by industry is in particle form. Tens of billions of tons of particulate material are handled by industry: mining–iron, coal, limestone, sand, gravel, phosphate; chemicals–cement, plastics, fertilizer, paint, metals, foods, glass, ceramics, brick, cinder block; construction–concrete, soils, sediment; farming–grains, produce, soils; waste products–garbage, refuse, sewage, sludge, fly ash, mine tailings, stream sediment. Some 70 percent of the pollution problems in the U.S. are due to particulates.

Eleven billion tons of waste particulates discarded by U.S. industries such as mining, chemical, and coal in 1974 caused problems, enough to cover the state of Delaware to a height of 2 m. On this material one cannot grow grass, cannot build, nor even predict its behavior. Should the U.S. ever utilize its oil or uranium shales then these figures will quintuple.

The best method in present use for flue gas desulfurization results in one-third ton of "soupy sludge" for every ton of coal burned. During the next ten years, if EPA standards are to be met for new, stationary coal-burning plants, the annual production of sludge will be 300 million tons.

Sixty percent of water pollution in the U.S. is caused by particulate matter such as phosphate slimes, sewage sludge, and acid mine drains. These toxic materials do not settle out in a reasonable time. Currently there is no satisfactory treatment of these particulate and sludge wastes–nothing will grow on them, they foul our water, choke our land, and pollute our air. The handling of particulates is a growing national and international problem. Though tens of billions of dollars are being invested yearly in obsolete, energy-intensive technologies to cope with this problem, chemical and equipment companies have no profit incentive to support the basic research in particulate technology. In the United States, federal agencies such as EPA, the Bureau of Mines and ERDA have an interest in the particulate problem, and, although they do support applied research and demonstration projects, none has the mission to support the *fundamental engineering research* needed for cost and energy effective solutions.

Morphological characterization of particulates is the most pressing problem. Now theoretical advances in the field, coupled with the use of recently developed instruments and mathematical tools, will allow us to predict the morphological characteristics and relate these characteristics theoretically and experimentally to the bearing strength, permeability, resistance to compaction, flow, and solubility of particulate systems. This particulate technology will be immediately applicable to such diverse fields as farming, water resources, chemical processing, and, above all, pollution abatement. From a small sample, engineers will be able to take mine tailings, morphologically characterize the particles, determine their leach rate, their compaction rate, and their long-term bearing strength.

Despite the obvious importance of particulates, it is only comparatively recently (within the last 15 years) that the industrial nations have begun to realize that powder technology–particulates–is a field of considerable scientific and technological concern. Indeed, it is only within the last few years that the United States Federal Government has instituted a basic research program in this area.[111] In the light of this situation, it is not surprising that there are more major technical problems than in older companion areas of solids, liquids and gases, respectively. Furthermore, many of these technical particulate problems are effectively intractible for the moment because they are dependent upon the successful solution of the more basic scientific problems in particulates. At the present time, the most pressing of these basic problems is particle characterization, specifically, particle morphology.[112]

With such a diversity of industrial interest it should be clear that the problems of particulate science and technology are interdisciplinary in nature, which is to say that no specialty has or can have the entire complement of technical skill and philosophical enterprise to solve many of these problems. This is especially true of the problems of particulate morphology.

With few exceptions, physicists and chemists are deeply concerned with the molecule, the atom, and subatomic particles. By contrast, the building blocks of particle technology are the particles themselves which cover a wide range of types–from polymolecular aggregates to large sized gravel. Much of the spectrum of particulates is in the province of the engineers and therefore it is most appropriate that we in engineering accept the responsibility for advancing the technology of particle morphology analysis.

Interest in particle size analysis has been intense for the past 30 years[113] but due to a lack of theoretical formulation, interest in particulate morphology, although present, has been less intense.[69] In the last decade, however, there has occurred a quickening of scientific and engineering involvement in the area resulting in a profound theoretical breakthrough.

Particle Size

If we consider a particle of any arbitrary shape and desire to measure its size, two general categories of measurement are available to us:

1. Essentially static methods of measurement in which some dimension or diameter is measured using light microscopy, electron microscopy or x-ray methods, etc. A basic step in this approach is to measure off the length of an intercept. By repeating this for many particles, the data is manipulated using statistical methods in order to yield mean diameters, standard deviations and so on.

   A moment's thought concerning this type of measure shows that in the case of the particle of arbitrary shape, the intercept is a proxy of the actual size of the particle and the size is not a simple thing to either conceive of or to measure. In fact the intercept is simply the length of the line x_1y_1 to x_2y_2 where x_1y_1 is a point on the profile and x_2y_2 is another point on the profile but at the other end of the intercept. Clearly, the intercept is completely arbitrary and any one particle can yield a myriad of intercepts of different lengths. Even if the concepts of the Martins diameter or the Feret diameter are introduced, it is impossible to get away from the arbitrariness of the choice of intercept and, more importantly, its utter inadequacy for describing the size of the particle. One must add to this the belief that if a property is not described and measured adequately on one particle, no amount of statistical juggling is going to improve the quality of the information so obtained on many particles.

2. The other main group of methods is essentially dynamic in character. In these methods, the set of particles is placed in an environment which is subsequently disturbed. The members of the particle set react differently to these imposed environmental impulses and these different reactions (and interactions) evoke a spread or spectrum of responses from the test sample. The observer then quite arbitrarily assumes that this spread of responses is due to size and imposes on the data the interpretation that the data indicates the size distribution of the particle set under test. Examples of methods such as these include sieving, sedimentation, elutriation, centrifuging, and viscometry. All these methods are eminently suitable for quality control but they are severely lacking when it comes to scientific rigor.

Particle Shape

The problems associated with the measurement of particle size pale to insignificance when compared with those of the classical methods of particle shape measurement. Quite apart from the extreme tediousness of some of the methods of shape measurement (some of them were so laborious that virtually only their progenitors ever used them, or claimed to use them) the main criticism that can be levelled at them is once again their lack of scientific rigor. Consider the following two examples:

1. One often quoted shape factor is the aspect ratio. This is variously defined, however, a brief definition is the ratio of the maximum length divided by the minimum length at right angles to the maximum length. This method attempts to reduce the shape of the particle to a number. It ignores therefore the question of the dimensionality of particle shape just as it ignores the other shape characteristics. But, and perhaps this is more serious, the method takes very little information about the particle and yet tries to say something definite and important about the shape.
2. There is a number of apparently sophisticated measuring instruments on the market that carry out a number of measurements on particle images. These measures include intercept, area, and perimeter of the profile. Using these starting data, various shape factors may be calculated. Again, one must criticize these measures on the

grounds that they do not yield unequivocal measures of particle shape. For example some function of perimeter divided by area is not unique and furthermore it fails to describe the shape of the particle. At the very minimum one would hope to be able to recognize the significant shape features of the particle profile by inspection of the shape factors, but this is not obtained. The information is useful, however, in quality and process control work. It is misleading in a rigorous scientific sense.

Morphological Analysis

There are two stages in the historical development of morphological analysis. In the first stage, advantage was taken of the incredible human ability to observe, recognize, and codify particles of differing shape. This finds its penultimate expression in the *Particle Atlas*[114] which is a unique and valuable guide to the identification of fine particles using the powers of human observation in conjunction with some very fine science. It is difficult to say when the second and most recent development started. However, the three main proponents of this most recent development are Beddow,[81] Ehrlich,[110] and Meloy.[82] The basis of their approach is described here.

Before one rushes off to start measuring particle shape it is a mandatory prerequisite that one defines it: "Particle shape is the pattern of all the points of the surface of the particle." From this definition a number of corrollaries may be deduced:

1. Shape is an intrinsic property of a portion of matter. Whereas size defines the extent of the particle, shape describes the form that the matter takes.
2. Shape can be the property of an abstraction just as much as that of a real particle.
3. Shape is concerned exclusively with the outline of the external surface of the particle. Internal pores, for example, are excluded from this consideration except in so far as they intercept the surface of the particle.
4. The shape characteristics go all the way from the gross outline down to the very minutest detail that can be observed.
5. To differentiate from the internal detail of the particle, the latter will be termed *texture.*[115] Therefore the microchemical composition from point to point within the particle is a textural property,

for example.

6. Differences between one shape and another are perceived by noting differences between the respective patterns of the points of the surfaces concerned.

In order to measure the shape of a particle profile a sample set of coordinate points of the particle profile that adequately represents all of the points of the surface according to the above definition must be obtained. Once obtained, this data set has to be condensed to a minimum while retaining the original pattern of the data that is the shape information. Note that the modus operandi here is to obtain as much information as is convenient and then condense this information to an acceptable minimum.

The method of obtaining the sample set of profile points is termed digitization. This set of $\{x,y\}$ coordinates is then converted to polar coordinates $\{R,\theta\}$ * which gives a graph as shown in Fig. 6.64. The curve may then be converted into an equation by means of an orthogonal transform. In the example shown in Fig. 6.64 a Fourier series has been used. The general form of this equation is the well-known Fourier equation given in Eq. (6.16), et. seq.

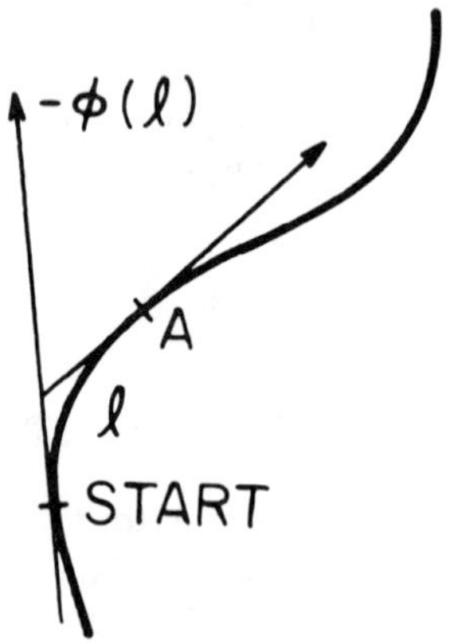

Fig. 6.64 Parametric representation of a profile.[132]

This series has many interesting properties but the most useful one for our purpose is that one can take the number of terms (the order of n) to as high as one pleases in order to reduce the error below a given level. This is illustrated in Fig. 6.65 which shows the increasing exactness of reproduc-

*in the $\{R,\theta\}$ method under discussion.

tion of an original profile (the shaded one) due to the increasing number of harmonics used in this analysis. Thus, if only three harmonics are used the general outline is there but not the details. Much closer reproduction is obvious when 20 harmonics are used. The proof that the terms in the equation (i.e., the set of coefficients $\{A_n, \alpha_n\}$) really represent the information originally present in the profile is given by the ability to regenerate that same profile from the set of coefficients.

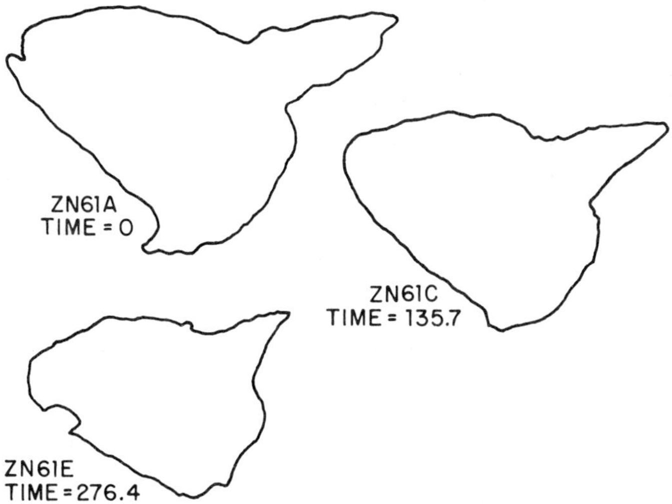

Fig. 6.65 Profiles of Zn61 at successive stages of attack by HCl.[95]

The individual coefficients themselves represent different types of geometrical features. For example, A_2 indicates the aspect ratio; A_3 the triangularity and so on. This is shown in Fig. 6.66. There are other orthogonal functions including the square wave Walsh transform and also the pule wave Haar transform. Each of these has its own advantages and disadvantages as compared with the Fourier.

Feature Selection

With the geometric information of the profile transformed into mathematical information in the form of sets of coefficients, the latter now represents the shape data. In order to analyze this new form of the data we can call upon the many and varied techniques of pattern recognition.

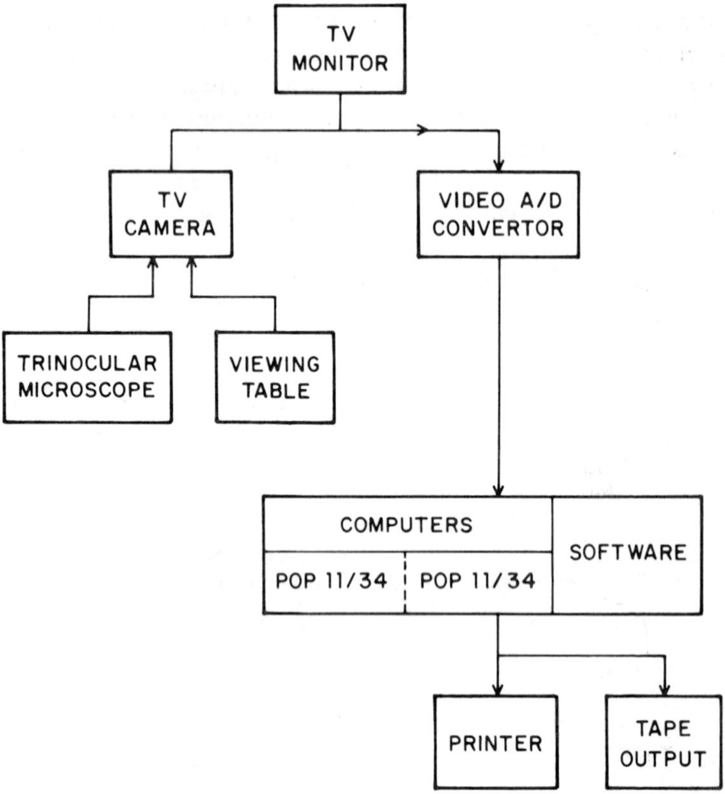

Fig. 6.66 Automated Image Analyzing System[95]

These have been reviewed in this connection.[60] The proponents of morphological analysis have developed analytical techniques in three broad categories corresponding to three different approaches. These are Signature Analysis (Meloy), Harmonic Analysis (Ehrlich), and Pattern Classifiers (Beddow).

Signature Analysis.[81] A plot of log A_n versus n gives a decreasing curve to which a straight line first approximation may be applied (see Fig. 6.37). This line is termed the signature of the particle (actually Meloy has called this the fingerprint of the particle and the assembled fingerprints of many particles he has called the signature) (see Fig. 6.39). The line may be described by the intercept on the y-ordinate and the slope. This signature therefore represents a considerable compression of the information originally in the profile but the statistical properties of the coefficients and phase angles have to be concomittently considered for this approach to be successful. Basic research work is currently proceeding in this area.

Harmonic Analysis.[116] The principle underlying this approach is to assemble the information about the individual coefficients in the form of distribution curves of the same. This is plotted as a histogram. The subsequent analysis has three stages and the investigator can stop at any particular stage depending on the type of information required of the analysis.

Stage 1 involves χ^2 analysis and the end result of this is that the investigator is able to determine the list of significant coefficients. As an example say that two particle sets are being compared and contrasted and it is found that the following coefficients are significantly different:

$$A_2, A_3, A_8, A_{11}, A_{15}, A_{16}, A_{18}$$

This is analogous to the well-known spot test in chemistry. In this example we can say that we have found that the two particle sets differ on the basis of the above coefficients and therefore that we have every expectation that they can be differentiated on the basis of these same coefficients also.

Stage 2 carries out an R-mode factor analysis on the histograms in order to deduce the "basic factors" inherent in the set of coefficients listed above. These factors are linearly related to the coefficients. As an example there might be three such factors X, Y, and Z which are linearly related to the earlier list of coefficients. This technique is analogous to chemical qualitative analysis.

Stage 3 conducts a Q-mode factor analysis in order to determine the proportions of basic shapes present in given samples of particulate materials. This technique in this application is therefore analogous to chemical quantitative analysis. These techniques of harmonic analysis developed by Ehrlich represent a powerful tool with which he and his coworkers have answered some very interesting questions of concern in Geology.

Classifiers. There are three general categories of classifiers—deterministic, statistical, and fuzzy. A deterministic classifier capable of separating input data (the coefficients in one form or another) into shape classes has been developed.[117] This has been applied to four particle sets initially with modest success. Since that time, successive improvements in technique have decreased the misclassifications scores substantially. Whereas the deterministic classifier ignores the statistical properties of the input data, statistical classifiers use these properties to discriminate the data. We have investigated three types of classifier (Linear Discriminate Function; Quadratic Discriminate Function, and the Fix-Hodges Rule) using as input data a small number of Walsh coefficients (three) of four particle sets. The misclassification scores have been gratifyingly low.[118]

The Fuzzy classifier of importance is that due to Bezdek.[99 et seq] This classifier depends on the application of fuzzy set theory to the data. An ordinary set has two grades of membership associated with it–one for membership and zero for nonmembership. The object under consideration is either a member of the set or it is not. Alternatively, a fuzzy set has grades of membership going all the way from zero to one. This concept is most useful when applied to complex multidimensional properties such as morphology of fine particles.

Applications

In such a new discipline, applications are necessarily few and far between, however, three are given as follows:

Natural Tracers. Ehrlich has shown that the sand issuing from rivers along the Alaskan coast does not move off the continental shelf into the deep ocean but instead it moves along the beach dominated shallows and this intermingles with the sand effluent from the issuing rivers.[119] Ehrlich has also shown that the sand in Charleston Harbor in South Carolina comes from the sea. The sand is not river sand.[120] All of these studies and more have been predicated on the morphological properties of the sand grains.

Modeling Morphological Features.[82] Meloy proposed the use of square waves (Walsh functions) to represent the particle profile instead of the Fourier waves. Using this approach he was able to develop functions said to represent certain features of the particle surface including (see Fig. 6.46): excess surface area of the particle–that is the surface area of the particle over and above that of the sphere of average radius; edge length of the particle surface–this is the total length of the edges observed when the surface is represented by a series of square waves. Some of these characteristics have been reported in the case of metal powders.

6.12 INTERPRETATION OF COEFFICIENTS

Although in special circumstances, individual coefficients have been interpreted as indicating specific geometric features. For example, A_2 = elongation; A_3 = triangularity and so on. However, in general, this is not true. For example a scalene triangle has coefficients shown in Table 6.23.

Table 6.23

COEFFICIENTS OF SCALENE TRIANGLE

Harmonic	A_n/R_0
1	1.10771
2	0.43515
3	0.10824
4	0.17576
5	0.13549
6	0.09469
7	0.10198
8	0.03492
9	0.08158
10	0.02372

in which $A_n^2 = (a_n^2 + b_n^2)$

πR_0^2 = area of particle profile

An inspection of the set of coefficients shows clearly that A_3 is small. This illustrates the danger of giving an interpretation to only one coefficient. It is believed that the proper way to interpret the coefficients is as a set and not a coefficient-by-coefficient basis.

Up until recently[137] all of the methods outlined above have been utilizing the Fourier coefficient A_n and ignoring the phase angle α_n. What this means is that the input data $\{x, y\}$ has been converted into another data form $\{A_n, \alpha_n\}$, part of which is subsequently ignored by the analysis. The reasons for this are not hard to find; the coefficient A_n is invariant with respect to rotation, the phase angle α_n is not. The analysts rightly chose an invariant descriptor and in the process they tended to ignore the other portion of their data. This state of affairs should no longer be tolerated. The formulation of a solution to this problem–the provision of invariant descriptors–is discussed here.

Desirable Properties of the Descriptors[130]

What should be the properties of the invariant descriptors? How many are required? How can the researcher conduct further analysis on the descriptors? Is dependence on empirical descriptors necessary or can descriptors be developed as a natural consequence of the fundamental properties of the particle? Is it not more desirable to be able to calculate the descriptors from the first principles rather than develop an expedient solution to the problem? Will the descriptors possess simple physical interpretations? Will the descriptors relate in a simple way to basic statistical properties of the radial distribution?

A consideration of the above questions leads to the formulation of a potential list of desirable descriptor properties as outlined below. Note that this list is not exhaustive:

1. The descriptors should be derivable from first principles and they should relate to the inherent properties of the Fourier series. They should not be empirical formulas.
2. The descriptors should be easily calculated from the input data.
3. They should be demonstrated to be invariant.
4. They should be capable of treatment with the usual statistical methods, preferably standard, widely accepted methods so that a broad spectrum of workers can follow and use the analysis techniques.
5. Any physical interpretations of the meanings of the descriptors should be unequivocal and preferably capable of rigorous definition, if at all possible.
6. The descriptors should be capable of relating to the fundamental physical-chemical properties of particles and their surfaces and ultimately to the mode of formation of the particle (i.e., relating to the mechanisms occurring during particle formation) and also ultimately to the chemical and mechanical behavior of the particle in specific environments.
7. One should be able to describe sets of particles in terms of the invariant descriptors without losing precision inherent in the data, using if possible, standard statistical techniques.
8. It should be possible in principle at least, to recalculate the particle profile from the set of invariant descriptors.
9. The descriptors should be invariant to a change of image size. This is especially important with regards to the use of modern image

analyzing instruments.

Definition of the Descriptors

In the case of Eq. (6.16b), the amplitudes, a_n and b_n, are not independent of the initial starting point (i.e., rotationally variant) but they are related to the shape of the profile in that its boundary can be reproduced by Eq. (6.16b).

The following size and shape terms are defined. These are *rotationally invariant* and related to the amplitudes, a_n and b_n, as follows.

The size term is defined as

$$R_0 = \sqrt{a_0^2 + \tfrac{1}{2} \sum_{n=1}^{\infty} (a_n^2 + b_n^2)} \tag{6.109}$$

in which

R_0 is termed "equivalent radius" and is the radius of a circle having the same area as that of the particle profile, and

n is the order of the coefficient.

The shape terms are defined as

$$L_0 = a_0/R_0 \tag{6.110}$$

$$L_{1,n} = 0 \qquad \text{for all } n \tag{6.111}$$

$$L_{2,n} = \frac{1}{2R_0^2} (a_n^2 + b_n^2) \tag{6.112}$$

$$L_{3,n} = \frac{3}{4R_0^3} (a_n^2 a_{2n} - b_n^2 a_{2n} + 2a_n b_n b_{2n}) \tag{6.113}$$

Relationships Between Invariant Terms and Radial Distribution

The size and shape terms are related to the mean and the moments about the mean of the profile radial distribution. This is discussed and demonstrated as follows:

The Size Term R_0. The area of the particle profile may be stated in terms of the Fourier coefficients:

$$\text{Area} = \int_0^{2\pi} \int_0^{R(\theta)} r\,dr\,d\theta = \pi \left[a_0^2 + \tfrac{1}{2} \sum_{n=1}^{\infty} (a_n^2 + b_n^2)\right] \tag{6.114}$$

Keeping in mind that a circle of radius R_0 has an area of πR_0^2 the terms are equated by

$$\pi R_0^2 = \pi[a_0^2 + \tfrac{1}{2} \sum_{n=1}^{\infty} (a_n^2 + b_n^2)] \tag{6.115}$$

This leads directly to Eq. (6.109).

The Shape Term a_0. From the calculus, it is known that the average value of a function, $y = f(x)$ for $a \leqslant y \leqslant b$, is given by

$$\overline{Y_{ave}} = \frac{1}{b-a} \int_a^b f(x)\,dx$$

By analogy:

$$\overline{R(\theta)} = \frac{1}{2\pi} \int_0^{2\pi} rR(\theta)\,d\theta = a_0 \tag{6.116}$$

Therefore, a_0 is the average or mean radius.

The Other Shape Terms. Similarly,

$$\overline{R(\theta)^2} = \frac{1}{2\pi} \int_0^{2\pi} R(\theta)^2\,d\theta = R_0^2 \tag{6.117}$$

So,

$$R_0^2 - a_0^2 = \overline{R(\theta)^2} - \overline{R(\theta)}^2 = \overline{[R(\theta) - a_0]^2}$$

Therefore, from Eq. (6.109):

$$\overline{[R(\theta) - a_0]^2} = \tfrac{1}{2} \sum_{n=1}^{\infty} (a_n^2 + b_n^2) \tag{6.118}$$

Eq. (6.118) is the second moment about the mean of the radial distribution of the particle profile. Making use of the general form of the moment equation,

$$\mu_n = \frac{1}{2\pi} \int_0^{2\pi} [R(\theta) - a_0]^n d\theta = \overline{[R(\theta) - a_0]^n}$$

the higher moments can also be expressed as functions of the Fourier coefficients, a_n and b_n .

The third moment about the mean is derived in a similar fashion and is:

$$\mu_3 = \frac{1}{2\pi} \int_0^{2\pi} [R(\theta) - a_0]^3 \, d\theta$$

$$= \tfrac{3}{4} \sum_{n=1}^{\infty} (a_n^2 a_{2n} - b_n^2 a_{2n} + 2a_n b_n b_{2n}) \qquad (6.119)$$

This leads directly to Eq. (6.113).

To summarize, the size and shape terms are directly related to the mean and moments about the mean of the particle profile radial distribution as follows:

$$\mu_0 = L_0 R_0 \quad \text{(mean radius)} \qquad (6.120)$$

$$\mu_1 = R_0 \sum_{n=1}^{\infty} L_{1,n} = 0 \qquad L_{1,n} = 0 \text{ for all } n \qquad (6.121)$$

$$\mu_2 = R_0^2 \sum_{n=1}^{\infty} L_{2,n} \qquad (6.122)$$

$$\mu_3 = R_0^3 \sum_{n=1}^{\infty} L_{3,n} \qquad (6.123)$$

in which μ_0 is the mean of the radial distribution and μ_1, μ_2, and μ_3 are the first, second, and third moments about the mean of the radial distribution, respectively.

Demonstration of Rotational Invariance

Consider a particle with coefficients a_0, a_n, b_n. Let the particle be rotated an angle θ' such that the new coefficients are a_0', a_n', b_n'. In this case, the original particle representation $R(\theta)$ is related to the new, rotated particle representation by the relationship

$$R'(\theta) = R(\theta + \theta') \qquad (6.124)$$

The required rotational transformation equations are obtained from Eq. (6.124) simply by expanding the trigonometric terms on the right-hand side of Eq. (6.124). The coefficients for the nth $\cos n\theta$ term and the coefficients for the nth $\sin n\theta$ term are collected together. The coefficients for these terms on the left- and right-hand side of Eq. (6.124) are then equated. This procedure leads to the following set of relationships:

$$a_0{}' = a_0 \tag{6.125}$$

$$a_n{}' = a_n \cos n\theta' + b_n \sin n\theta' \tag{6.126}$$

$$b_n{}' = b_n \cos n\theta' - a_n \sin n\theta' \tag{6.127}$$

These are the rotational transformation equations for an (R, θ) particle. (Note that each $\sin n\theta$ term and each $\cos n\theta$ term are independent on the interval 0-2π.)

When Eqs. (6.125), (6.126), and (6.127) are substituted into Eqs. (6.110), (6.111), (6.112), and (6.113), it is found that

$$a_0' = a_0 \tag{6.128}$$

$$(a_n')^2 + (b_n')^2 = a_n^2 + b_n^2 \tag{6.129}$$

$$(a_n')^2 a_{2n}' - (b_n')^2 a_{2n}' + 2a_n' b_n' b_{2n}' = a_n^2 a_{2n} - b_n^2 a_{2n} + 2a_n b_n b_{2n} \tag{6.130}$$

and

$$R_0' = R_0$$

That is to say, the size and shape terms are rotationally invariant. It follows then that $\mu_0, \mu_1, \mu_2, \mu_3$ are also rotationally invariant.

Size Change Invariance

When an image is either magnified or reduced in size during the morphological analysis process it is especially important that the morphic descriptors remain invariant. For example, let both the x-axis and the y-axis be scaled by a factor α. It will be shown that the magnitude of the radii change but that the angular relationships do not change:

Definition:

$$r \equiv \sqrt{x^2 + y^2}$$
$$\theta \equiv \tan^{-1} \frac{y}{x} \tag{6.131}$$

Also by definition:

$$x' = \alpha x$$

and (6.132)

$$y' = \alpha y$$

in which " ' " indicates the new scale.

Then,

$$r_s = \sqrt{(\alpha x)^2 + (\alpha y)^2} = \alpha\sqrt{x^2 + y^2} = \alpha r \quad (6.133)$$

and

$$\theta_s = \tan^{-1} \frac{\alpha y}{\alpha x} = \tan^{-1} \frac{y}{x} = \theta \quad (6.134)$$

in which the subscript, s, means scaled. It follows that

$$R_s(\theta) = \alpha R(\theta) \quad (6.135)$$

In Eq. (6.135) $R_s(\theta)$ is the scaled Fourier series. Using this equation, a scaled equivalent radius is determined:

$$R_{0s} = \alpha R_0 \quad (6.136)$$

And, from Eq. (6.109)

$$R_{0s}^2 = \alpha^2 R_0^2 = (\alpha a_0)^2 + ½ \sum_{n=1}^{\infty} (\alpha a_n)^2 + (\alpha b_n)^2 \quad (6.137)$$

Dividing through by $(\alpha R_0)^2$ eliminates the scaling factor, α, completely. This procedure ensures that the descriptors are normalized with respect to the equivalent radius R_0. This is shown in Eqs. (6.110) - (6.113).

Conclusion

This discussion has demonstrated that invariant morphological Fourier descriptors can be used to characterize the morphic features of fine particles. The descriptors presented involve both size and shape terms with the following properties:

1. They are derived from first principles and relate to the properties of the Fourier series.
2. They are easily calculated from the input data.
3. They are invariant.
4. They indeed should be treatable with normal statistical methods.
5. They are rigorously defined.
6. Recalculation of the original coefficients and hence regeneration of

the original profile should be possible.

7. The descriptors are invariant to magnification or diminution of the particle under examination (this excludes consideration of resolving power of the system).

The relationships between the descriptors and fundamental physical-chemical properties of fine particles, their mode of formation, and their behavior have yet to be determined.

More Generalized Methods of Analysis

Many particles possess relatively simple (R,θ) shapes but many others have much more complex outlines which contain reentrants. A reentrant produces multiple values of R at certain values of the angle θ, and the Fourier transformation cannot handle multivalues. Therefore a more generalized method of analysis is required.

The general approach currently being developed[131] is called the (ψ, S) method, in which the profile is parametrized as a function of: (ψ) the angle between the tangent and the radius, *and* the distance traversed round the profile (S). This method gives a relationship for the area of the particle and other parameters which, however, do not have the same physical meaning as those derived via the (R,θ) method. Virtually all types of profiles can be analyzed using the (ψ, S) method.

Another method called the (ϕ,l) method, which can also analyze complex shapes has been reported in the literature.[132] It has a disadvantage because it does not yield an expression for the area. However, it does have the capability for analyzing complex shapes.

The (ϕ, l) Method[132]

The profile is parametrized by its arc length l and change of slope $\phi(l)$ from its starting point (Fig. 6.64), i.e., $(x,y) \rightarrow (l, \phi(l))$. Total arc length (or perimeters) are normalized by defining:

$$t = \left(\frac{l}{L}\right) 2\pi$$

$$\phi^*(t) = \phi\left(\frac{l}{L}\, 2\pi\right) + t$$

The interval of interest is $t = [0, 2\pi]$ for $l = [0,L]$.

The function $\phi^*(t)$ is expanded into a Fourier series:

$$\theta^*(t) = \mu_0 + \sum_{k=1}^{\infty} (A_k \cos kt + B_k \sin kt) \tag{6.138}$$

$$\mu_0 = \frac{1}{2\pi} \int_0^{2\pi} \phi^*(t)\, dt \tag{6.139}$$

$$A_k = \frac{1}{\pi} \int_0^{2\pi} \phi^*(t) \cos kt\, dt \tag{6.140}$$

$$B_k = \frac{1}{\pi} \int_0^{2\pi} \phi^*(t) \sin kt\, dt \tag{6.141}$$

The morphological information is contained in the coefficients (A_k, B_k). As the ordered pair, $(l, \phi(l))$ is unique for every point on the profile, multi-valuedness no longer appears. The profile is fully represented by (A_k, B_k).

One interesting application of the (ϕ, l) method is in the study of chemical reactions. For example, the reaction of Zn particles in HCl has been reported on.[133] Fig. 6.65 shows three representative profiles from the reaction sequence of particle Zn6l. The corresponding Fourier coefficient data from the (ϕ, l) method are given in Table 6.24. Note that the sum of the coefficients was used as a simple indicator of the progress of the acid-zinc reaction. The data in Table 6.24 and Fig. 6.65 shows that as the reaction proceeds, there is a gradual smoothing out of the profile accompanied, in the early stages, by the elimination of the two protruberances on the left side. This elimination of surface detail is indicated in the reduction of the value of $\sum_{n=1}^{\infty} A_n$. However, once the secondary attack on the profile occurs, the value of $\sum_{n=1}^{\infty} A_n$ sharply increases.

Towards Efficient Data Gathering

At the present time methods used to digitize the particle profiles are totally or partially manual in nature. An automated system is currently being developed in at least two laboratories.[134] The diagram shown in Fig. 6.66 gives the outline of the digitizing system under development.

Table 6.24

VALUE OF SUM OF COEFFICIENTS Zn61
AT VARIOUS TIMES OF REACTION

Time (s)	$\sum_{n=1}^{\infty} A_n$
0	4.95
60	3.85
136	3.70
200	3.75
275	4.55

The Image Analyzing System[135] is based upon two PDP 11/34 minicomputers linked together by a high speed communications link. The master PDP 11/34 has 64K words of memory, a 15M byte disk, a 9 track magnetic tape and a line printer. The slave PDP 11/34 has 32K words of solid state memory and dual floppy disks. A real-time multitasking operating system (RSX-11M) is utilized on the master CPU while the slave runs a simpler operating system (RSX-11S) dedicated to a single user. Network software (DecNet) supports program and data transfer between the two computers.

A Hamamatsu vidicon camera system provides image digitizations completely under computer control. This system provides eight bit A/D conversion at a computer selected resolution (256, 512, or 1024 lines). Density slicing at an adjustable threshold can be performed by the camera interface. Using a Bausch & Lomb light table, both photographs and transparencies may be digitized. Image display is accomplished using a Ramtek 9050 display system together with a high resolution Mitsubishi color monitor. This image display device is microprocessor-based and permits fairly sophisticated interactive image manipulation. Graphic and alphanumeric information may be superimposed on an image, particle boundaries may be traced using a cursor and intensity pseudo-color transformations can be performed without extensive computation.

Hard copy display is done using a Polaroid camera as a line printer. Software support for our image processing equipment has been and is being developed in house.

The image may be presented to the camera in a number of alternate ways:

1. as a photograph using a carefully illuminated viewing table,
2. as a tracing using a microscope with a drawing head,
3. using a microscope, the images of the particles themselves may be digitized.

At the time of writing, the system is now operational digitizing two particles per minute. Ehrlich has a similar system digitizing one particle per minute. The expected rate of digitization is 50-60 particles per minute.

Conclusion

This chapter has reviewed recent developments in particulate morphology analysis. The interested reader may refer to the literature already cited and especially to references 122 and 128.

REFERENCES

1. Jelinek, Z. *Particle Size Analysis.* New York: Halsted Press, 1974 (copyright is held by Ellis Horwood, Ltd. Publishers, England).
2. Allen, T. *Particle Size Measurement.* London: Chapman and Hall, 1968.
3. Johnson, J., paper entitled "Particle Characterization Using the Photoscan," presented at Fine Particle Society meeting, Philadelphia, August, 1975.
4. Silverman, L., Billings, C., First M. *Particle Size Analysis in Industrial Hygiene.* New York: Academic Press, 1977.
5. Lawrence, L. R., and Beddow, J. K. *Powder Technology* 2 (1968): 253–259.
6. Williams, J. C., and Shields, G. *Powder Technology* 1 (1967): 134.
7. Ahmed, K., Smalley, I. J. *Powder Technology* 8 (1973): 69–76.
8. Lawrence, L. R., and Beddow, J. K. *Powder Technology* 2 (1968): 125–130.
9. Drinker, P., and Hatch, T. *Industrial Dust.* New York: McGraw Hill, 2nd ed., 1954.
10. Brown, J. J., et al. *American Journal of Public Health* 40 (1950): 450–459.
11. Lippmann, M., Harris, W. B. *Health Physics* 8 (1950): 155–163.
12. ICRP Task Group on Lung Dynamics; Health Physics 12 (1966): 173–207.
13. Breuer, H., *Staub* 24, (1964): 324–329.
14. Mercer, J. *Aerosol Technology in Hazard Evaluation.* New York: Academic Press, 1973.
15. Watson, H. H. *American Industrial Hygiene Association Quart.* 15 (1954): 21–25.
16. Fairs, G. L. *Journal of the Society of Chemical Industry* 62 (1943): 374.
17. Cartwright, J. *British Journal of Applied Physics* 5, Supplement 3, S109 (1954).
18. Benoit, H., Holtzer, A. M., Doty, P. *Journal of Physical Chemistry* 58, 635 (1954).
19. Hlousek, M. *Chemical Prumysl* 9, 32 (1959).
20. Lange, H. *Kolloid – Z* 223, 24 (1968).
21. Rayleigh, Lord. *Philosophical Magazine* 41, 107, 274, 447 (1871).

22. Van de Hulst. *Light Scattering by Small Particles.* New York: Wiley, 1957.
23. Whitby, K. T. "Symposium Particle Size Measurement," *ASTM* (1958): 3-23.
24. Daeschner, H. W. *Powder Technology* 2 (1968/69): 335-348.
25. Roberts, T. A., and Beddow, J. K. *Powder Technology* 2 (1968): 121-124.
26. Rose, H. E. *Journal of Japanese Society of Powder and Powder Metallurgy* 22, 2 (1975): 55-70.
27. Beddow, J. K. "On Schemes for Shape Analysis of Individual Particles." Report A390 ChMe, 1974-007, University of Iowa.
28. Katz, Y. H., and Doyle, W. L., R.M. 3412 NASA, Rand Corporation, December 1964.
29. Shaw, A. C. *Journal of ACM,* Vol. 17, No. 3 (July, 1970): 453-481.
30. Ledley, R. S., and Ruddle, F. H. *Scientific American* (April, 1966): 40-46.
31. Anderson, R. H. *An Introduction to Linguistic Pattern Recognition.* P-4669 Rand Corporation, July, 1971, also American Bankers Association E-13B font character set is used on all of our checks.
32. Stallings, W. W. *Computer Graphics and Image Process* 1 (1972): 47-65 (paper actually deals with Chinese characters).
33. Rosenberg, R. *International Journal Man-Machine Studies* 6 (1974): 1-12.
34. Guzman, A. *Machine Intelligence* 6, ed. by Meltzer, B., and Michie, D., Edinburgh: Edinburgh University Press, (1971): 325-376.
35. Fu, K. S. *Sequential Methods in Pattern Recognition and Machine Learning.* New York: Academic Press, 1968.
36. Fu, K. S. *Syntactic Methods in Pattern Recognition.* New York: Academic Press, 1974.
37. Heywood, H. "Symposium on Particle Size Analysis." *Trans. Inst. Chem. Engr.* 22 (1947): 14-24.
38. Krumbein, W. C. *Journal of Sedimentary Petrology* 11, 2 (1941): 64-72.
39. Lees, G. *Sedimentology* 3 (1964): 2-21.
40. Lees, G. *Journal of British Granite and Whinstone Federation* 4, 2 (1964).
41. Allen, T. *Particle Size Measurement.* pp. 18-21, London: Chapman & Hall (1968).
42. Wadel, H. *Journal of Geology* 40 (1932): 443-457.
43. Wadel, H. *Journal of Geology* 43 (1935): 250-280.

44. Heywood, H. H. *Proceedings of Second Lunar Science Conference.* vol. 13, pp. 1989–2001, 1971.
45. Hausner, H. H. *Planseeberichte fur Pulvermetallurgie* 14, 2 (1966): 75–84.
46. Medalia, Avrom I. *Powder Technology* 4 (1970–71): 117–138.
47. Crayton, P. H. *Powder Metallurgy International.* vol. 6, No. 4 (1974): 198–200.
48. Church, T. *Powder Technology* 2 (1968–1969): pp. 27–31.
49. Cole, M. Reprinted from *American Laboratory,* Vol. 3, No. 6 (1971). Copyright 1971 by International Scientific Communications, Inc., pp. 19–29. See also Fischmeister, H. F. "Automatic Measuring and Scanning Devices in Stereology." in *Stereology.* ed. by Elias, H. New York: Springer-Verlag OHG, 1967, and ed. by Elias, H. "Proceedings of the Second International Congress for Stereology," Chicago, 1967. New York: Springer-Verlag OHG, 1967.
50. Chalkley, Cornfield, and Park. Science 110 (1949): 295–297.
51. Kostelnick, M., and Beddow, J. K. *Advances in Powder Metallurgy 4 Processes,* pp. 29–48, edited by Hausner, H. H. New York: Plenum, 1971.
52. Roberts, T., and Beddow, J. K. *Powder Technology* 2 (1968–69): 121–124.
53. Grey, R., and Beddow, J. K. *Powder Technology* 2 (1968-69): 323–326.
54. Kostelnick, M., Kludt, E. H., and Beddow, J. K. *International Journal of Powder Metallurgy* 4 (4) (1968): 19–28.
55. Mackey, R. D. *Civil Engineering and Public Works Review.* February, 1965.
56. Shergold, F. A. *Magazine of Concrete Research* 5, 13 (1953): 3–10.
57. Chaplin, T. K., personal communication.
58. Riley, G. S. *Powder Technology* 2 1968/69: 315–319.
59. Beddow, J. K., Vetter, A. F., Philip, G., Paper 25b AIChE Meeting, Dec. 1–5, 1974, Washington, D.C.
60. Beddow, J. K., Philip, G., and Nasta, M. D. *Planseeberichte für Pulvermetallurgie* 24 (1976): 167–174.
61. Tou, J. T., and Gonzales, R. C. *Pattern Recognition Principles.* Reading: Addison Wesley, 1974.
62. Zingg, T. *Schweizerische Mineralogische und Petrogrphische Mitteilunger* 15 (1935): 39–140.
63. Rosslein, D. *Quarry Managers Journal* 30, 4 (1946).
64. Heywood, H. *Journal of Imperial College Engineering Society* 8 (1954).

65. Bird, K. E. "The Assessment of Particle Shape and Its Influence on Sizing Analysis." Thesis, University of London, UK, 1966.
66. British Standard 2955, 1958, Glossary of terms relating to powders.
67. Riley, R. E., and Magnuson, J. E. *Characterization of Commercial Refactory Metal and Metal Carbide Powders.* U.S. Atomic Energy Commission, Los Alamos, N.M., Report LA-3894 NC25, 1968.
68. Rittenhouse, G. *Journal of Sedimentary Petrology* 13, 2 (1943): 79–81.
69. Beddow, J. K., Vetter, A. F., Sisson, K. *Powder Metallurgy International* 8, 2 (1976): 69–76. and 8, 3 (1976).
70. Brill, E. L. "Character Recognition Via Fourier Descriptions." *WESCON Tech.* Paper Session 26, 1968.
71. Morton, R. R. A., *ASTM STP* 504 (1972): 81–94. More recently see: Johnson, J. E., and Rosen, L. J., "Particle Characterization Using the Photoscan." paper presented at Fine Particle Society Meeting, Philadelphia, August, 1975.
72. Chomsky, Noam. *PGIT.* vol. 2, pp. 113–124, 1956.
73. Stover, E. R. *Technical Report AFML-TR-67-56.* Wright Patterson AFB, Ohio, 1967.
74. Cadle, R. D. *Particle Size.* New York: Reinhold, 1965.
75. Naylar, A. G., Rubin, G. A., and Kaye, B. H. "Fine Particles, Second International Conference." edited by Kuhn, W. H., pp. 28–42, *Electrochemical Society,* Princeton, 1974.
76. Tou, J. T., and Gonzales, R. C. *Pattern Recognition Principles.* Reading: Addison Wesley, 1974.
77. Rao, P. "Shape and Other Properties of Gas Atomized Metal Powders." Ph.D. thesis, Drexel University, 1973.
78. Lowrison, G. C. *Crushing and Grinding.* Cleveland: CRC Press, 1974.
79. Zadeh, L. A. *Information and Control* 8 (1965): 338–353.
80. Bellman, R. E., and Zadeh, L. A. *Management Science* 17, No. 4 (December, 1970): B141–B164.
81. Beddow, J. K., and Philip, G. C. *Planseeberichte für Pulvermetallurgie* 23 (1975): 3–14.
82a. Meloy, T. P. *Powder Technology* 16, 233 (1977).
82b. Meloy, T. P. *Powder Technology* 17, 27 (1977). These two papers represent a significant advance in morphological analysis at this time.
83. Henning, F. Harmuth. *Transmission of Information by Orthogonal Functions.* second edition, New York: Springer-Verlag 1972.

84. Grant, G., Hall, J. S., Reid, A. F., and Zuiderwyk, M. "Multicompositional Particle Characterizations Using SEM-Microscope," *Scanning Electron Microscopy* (Part III), IITRI, Chicago, (1976): 401–408.
85. Davies, R. *Powder Technology* 12 (1975): 111–124.
86. Black, M. *Philosophy of Science* 4 (1937): 427–455.
87. Goguin, J. A. *Synthese* 19 (1968/69): 325–373.
88. Zadeh, L. A. "Fuzzy Sets and Applications." pp. 27–39, New York: Academic Press, 1975.
89. Zadeh, L. A. *IEEE Transactions on Systems, Man and Cybernetics,* vol. SMC-3, 1 (Jan 1973): 28–44.
90. Underwood, E. E. *"Basic Stereology."* Praktische Metallographie (Sonder band der) issues, Band 5, 5 8–14, 1975.
91. Lew, Garland, MS Thesis, Univ. of Iowa, Iowa City, Iowa, Materials Engineering Div., 1978.
92. DeKaney, A. *Industrial Research* (Jan 1974): 44–47.
93. McCrone, W. C., Delly, J. G. *Particle Atlas.* second edition, Michigan: Ann Arbor Science, June, 1976.
94. Beddow, J. K., Philip, G. C., Vetter, A. F., and Nasta, M. D. *Powder Technology* 18 (1977): 19–25.
95. Beddow, J. K., Vetter, A. F., Luerkens, D. W., Collins, S., "Morphological Analysis of Fine Particles" to be presented at the Partec Conference, Nuremberg, W. Germany, 1979.
96. Beddow, J. K., Nasta, M. D., and Philip, G. C. "Characteristics of Particle Signatures." presented at *International Meeting of Powders and Bulk Solids Handling* published in proceedings, Industrial and Scientific Conference Management Inc., pp. 185–191, Chicago, May, 1977.
97. Beddow, J. K., Philip, G. C., and Nasta, M. D. "Fuzzy Sets in Property Representation." ibid., pp. 391–398.
98. Hall, G. H., and Hall, D. J. "ISODATA, a Novel Method of Data Analysis and Pattern Classification." Stanford Research Institute, Menlo Park, Ca., April, 1965.
99. Bezdek, J. "Fuzzy Mathematics in Pattern Classification." Ph.D. thesis, Cornell University, Ithaca, 1973. (See also more recent work by Bezdek reported in reference 122.)
100. Bezdek, J. "Feature Selection for Binary Data; Medical Diagnosis with Fuzzy Sets." proc. National Computer Conference, edited by Winkler, S., pp. 1057–1068, Montvale: AFIPS Press, 1976.
101. Bezdek, J. "Cluster Validity with Fuzzy Sets." *Journal of Cybernetics* 3, 3 (1974): 58–71.
102. Bezdek, J. "Mathematical Models for Systematics and Taxo-

nomy." proc. Eighth International Conference on Numerical Taxonomy, edited by Estabrook, G., pp. 143–164, San Francisco: Freeman, 1975.

103. Bezdek, J., and Dunn, J. "Optimal Fuzzy Partitions: A Heuristic for Estimating the Parameters in a Mixture of Normal Distributions." *IEEE Trans. Comp.* C-24 (1975): 835–838.

104. Bezdek, J. "A Physical Interpretation of Fuzzy ISODATA." *IEEE Trans., SMC,* SMC-6 (1976): 387–390.

105. Anderson, E. "The Irises of the Gasce Peninsula." *Bull. Am. Iris Soc.* 59 (1935): 2–5.

106. Bezdek, J. "Numerical Taxonomy with Fuzzy Sets." *Journal of Mathematical Biology* 1, 1 (1973): 57–71.

107. Chang, C. V. "Fuzzy Sets and Pattern Recognition." Ph.D. thesis, University of California, Berkley, 1967.

108. Tou, J., and Gonzales, R. *Pattern Recognition Principles.* Reading: Addison Wesley, 1975.

109. Zadeh, L, Fu, K., Tanaka, K., and Shimura, M. *Fuzzy Sets and Their Application to Cognitive and Decision Processes.* New York: Academic Press, 1975.

110. Ehrlich, R., and Weinberg, B. *J. of Sedimentary Petrology* 40, 1 (1970): 205–212.

111. NSF Program on Solids and Particulates Processing, Director Dr. M. Ojalvo.

112. Ad Hoc Committee Meeting on Particulate Morphology of the Fine Particle Society, Chicago, August, 1976.

113. See for example books by Allen (2), Cadle (reference 11, Chapter 5) (1) in *Particle Size Analysis.*

114. McCrone, W., et al., *Particle Atlas.* Ann Arbor Science Publishers, Ann Arbor, Michigan, 1973, 1979.

115. Tou, J., Oral Presentation, NSF Residential Research Workshop on Advanced Particle Morphology, University of Iowa, Iowa City, August, 1977, see also Tou, J., "Automatic Pattern Processing Approach to Particle Morphology Study," *Proceedings of the International Powder and Bulk Solids Handling and Processing Conference,* Rosemont, Ill., May 16, 17, 18, 1978, Industrial and Scientific Conference Management, Inc., Chicago, Ill., (1978): 152–157.

116. Ehrlich, R., Markovich, J. V., and Weinberg, B. "New Techniques for Stratographic Analysis and Correlation—Fourier Grain Shape Analysis Louisiana Offshore Pliocene." *J. Sed. Pet.* 46, 1 (March, 1976).

117. Beddow, J. K. "Physical Characterization of Particulates." Powtech 77 Proceedings of 4th International Powder Technol-

ogy and Bulk Solids Conference, Harrogate, England, London: Heyden and Son, 1977, pp. 99–105.

118. Wang, C., Beddow, J. K., and Vetter, A. F. "Statistical Classifiers in Particle Morphology Analysis." Proceedings of International Powder and Bulk Solids and Handling and Processing Conference, Chicago, 1978, Industrial and Scientific Conference Management, Chicago, 1978, pp. 158–164.
119. Yarus, J. M., Ehrlich, R., and Molnia, B. F. "Bottom Sediment Circulation Patterns in the Gulf of Alaska–Fourier Grain Shape Analysis." Geological Society of America Meeting, Seattle, Abstract, July 1978.
120. Van Nieuwenhuise, D., Yarus, J. M., Przygocki, R. S., and Ehrlich, R. "Sources of Shoaling in Charleston Harbor." *J. Sed. Pet.*, 48, 2, 373–383, 1978.
121. Lew, G., Beddow, J. K., Vetter, A. F. "Morphological Coefficients of Particle Profiles derived from Walsh Coefficients," *Powder Metallurgy International,* in press.
122. Beddow, J. K., and Meloy, T. P. (eds), *Advanced Particulate Morphology*. CRC Press, 1979, Boca Raton, Florida.
123. Rittenhouse, G. *J. Sed. Pet.* 11, 2 (1941): 64–72.
124. Stairmand, C. J. *Trans. Institute Chem. Engr.*, London, 29, 15, 1951.
125. Broersma, G. *Behavior of Granular Materials.* Culemborg, The Netherlands: Stam Technische Boeken, 1972.
126. Beddow, J. K. *J. Res. Assoc. Powd. Tech.*, Japan, 14, 11, 1977, 618–626.
127. Beddow, J. K. and Vetter, A. F. *J. Powd. and Bulk Solids Tech.* 1, 1 (1977): 42–44.
128. Beddow, J. K., Meloy, T. P., eds. *Testing and Characterization of Powders and Fine Particles.* London: Heyden & Son Ltd., 1979.
129. Hurst, W. and Spence, G. *J. Brit. Ceramic Soc.* 6 (1969): 60.
130. Luerkens, D. W., Beddow, J. K., and Vetter, A. F. "Invariant Fourier Morphological Descriptors," to be submitted to *Powder Technology*.
131. Luerkens, D. W., Beddow, J. K., and Vetter, A. F. "A General Method of Morphological Analysis" to be submitted to *Trans AIME.*
132. Fong, S-T., Beddow, J. K., and Vetter, A. F. "Refined Method of Particle Shape Representation." *Powder Technology.* 22, 1, 1979. See also: Zahn, C. T. and Roskies, R. Z. "Fourier Descriptors for Plain Closed Curves." *IEEE Trans. Comput.* vol. C-21, (March 1972): 269–281. Raudseps, J. G. "Some

Aspects of the Tangent-Angle vs. Arc Length Representation of Contours." Ohio State University Research Foundation, Columbus, Rep. 1801–6, Astia Ad 462877, March 1965.

133. Fong, S-T., Beddow, J. K., and Vetter, A. F. "A Technique to Study the Effect of a Chemical Reaction on Particle Morphology." *Proceedings of the International Powder and Bulk Solids Handling and Processing Conference.* Rosemont, Ill., May 16, 17, 18, 1978, Industrial and Scientific Conference Management, Inc., Chicago, Ill. (1978): 214–218. See also: Fong, S-T., Beddow, J. K., and Vetter, A. F. "Acid-Zinc Reaction Predictably Affects Particle Morphology." *Powder Technology,* 23, 2 (1979): 187–190.
134. Ehrlich, R., Dept. Geology, Univ. South Carolina, Columbia, SC, and Beddow, J. K., Univ. of Iowa, Iowa City, IA.
135. This description is due to S. Collins, Image Analyzing Laboratory, Information Engineering, Univ. of Iowa, Iowa City, IA.
136. Ehrlich, R., personal communication, May 1979.
137. This problem is now being studied in some laboratories. For example, see Davies, K. W. and Hawkins, A. E., "Harmonic Description of Particle Shape" Partikel Technologie Conference, Nürnberg, W. Germany, 24–26 Sept., 1979.
138. Baker, R., Walker, W., "Computation of Double Fourier Coefficients in Shape Analysis," Dept. of Mathematics and Anatomy, Univ. Auckland, Private Bag, Auckland, New Zealand.
139. Weichert, R., and Huller, D., "Measurement of Volume and Shape of Particles by Three Dimensional Image Analysis," Partikel Technologie Conference, Nürnberg, W. Germany, 24–26 September, 1979.

Chapter 7

PHYSICAL–CHEMICAL PROPERTIES

7.1 AN ELEMENTARY STARTING POINT

Just as little is known about the internal structure-particle morphology continuum at the present time, there is also insufficient knowledge about the effects of these particle characteristics upon the particle interactions within a force field(s). It is clear therefore, that this chapter about the physical-chemical properties of fine particles is an elementary beginning to a story that remains to be written.

Fine particles are particularly influenced by force fields and a great deal of scientific effort has been employed in order to elucidate the responses of particles in various types of fluxes. When the particles are large enough to be quite easily discerned and handled, the effects of the force fields are either not so important, relative to gravity, or not so observable or both. When the particles are extremely small, the effects are more difficult to measure and more subtle to disentangle. Interest in this area is prompted by the need to clean up our environment (the removal of particulate matter from effluents) and also the interest in separation technology, both wet and dry, for minerals including our recently rediscovered friend and ally—coal.

This chapter deals with a variety of properties and processes under the general heading of physical-chemical properties including: dielectric and optical properties and effects, photophoresis, electrical properties, Brownian motion, diffusiophoresis, thermophoresis, interparticle forces, deposition, magnetic, and thermal properties.

7.2 VISUAL APPEARANCE

As inspection of the Particle Atlas shows it is possible to observe and identify extremely small particles in the optical microscope down to 1 μ in size weighing approximately 10^{-12} g or 1 picogram. Using the electron microscope, methods have been developed for examination and identification of much smaller single particles down to 10^{-18} g or 1 milliemich, where an emich is 10^{-15} g. These particles may be as small as 100 Å or 0.01 μ in size. At this small size, if the material substance of which the particle is composed is not amorphous, either the optical or electron microscope may be used to evaluate the likely crystal structure by careful analysis of the appearance of the particle. The specifications of the seven crystal systems are given in Table 7.1. The drawings of the 14 Bravais lattices are given in Fig. 7.1. It must be borne in mind that the actual appearance of such a simple figure as a cube may be difficult to recognize unless one has carefully considered the possible alternate appearances. For example, a cube of salt is shown on the left hand side of Fig. 7.2. On the right hand side of the figure one can see the possible distortions of one of the three common crystal forms of the cubic system—the octahedral form. A particle may be colored or it may be colorless. It may be isotropic or anisotropic (pechroic). It may be transparent or opaque. It may show a different color by transmission than by reflection. Some examples of these categories of particles, with their Particle Atlas code designation are given in Table 7.2.

Table 7.1

GEOMETRY OF CRYSTAL SYSTEMS[1]

System	*Axes*	*Axial angles*
Cubic	$a_1 = a_2 = a_3$	All angles = 90°
Tetragonal	$a_1 = a_2 \neq c$	All angles = 90°
Orthorhombic	$a \neq b \neq c$	All angles = 90°
Monoclinic	$a \neq b \neq c$	2 angles = 90°; 1 angle ≠ 90°
Triclinic	$a \neq b \neq c$	All angles different; none equal 90°
Hexagonal	$a_1 = a_2\ (= a_3) \neq c$	Angles = 90° and 120° (or 60°)
Rhombohedral	$a_1 = a_2 = a_3$	All angles equal, but not 90°

Dielectric Material Properties

When an electric field is maintained between two parallel electrodes

Table 7.2

A SAMPLE OF PARTICLES OF VARIOUS APPEARANCES[2]

Visual Appearance	*Code*	*Example*
Transparent, Colorless, Low Refractive Index, Isotropic	0.000000	Spores, Pollens, Gelatin, Glass
Transparent, Colorless, High Refractive Index, Isotropic	4:000100	Periclase, Diamond, Slag, Thoria
Transparent, Colorless, Low Refractive Index, Anisotropic	9:001001	Talc, Aspirin, Straw, Wool, Nylon
Transparent, Colorless, High Refractive Index, Anisotropic	13:001101	Diacron, Caffeine, Kyanite
Transparent, Colored, Low Refractive Index, Isotropic	17:010001	Oil Soot, Leather Dust, Fungal Conidiophores
Transparent, Colored, High Refractive Index, Isotropic	20:010100	Bit. Coal, Slag, Sphalerite
Transparent, Colored, Low Refractive Index, Anisotropic	25:011001	Insect Parts, Human Hair, etc.
Transparent, Colored, High Refractive Index, Anisotropic	29:011101	Malachite (green), Rhodonite (pink)
Opaque, Colorless	33:100001	All Sawdust, Incinerator & Cigarette Ash
Opaque, Colored	48:110000	Rubber, Oxide Metal Shot, Copper

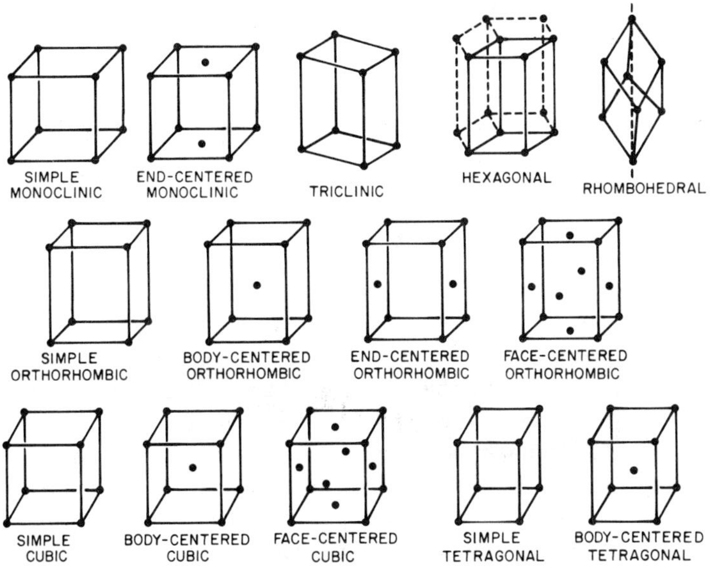

Fig. 7.1 Space lattices. These 14 Bravais lattices represent the basic three-dimensional repetition patterns which fill space. When extended into space, each Bravais point of a lattice has equivalent surroundings.[1]

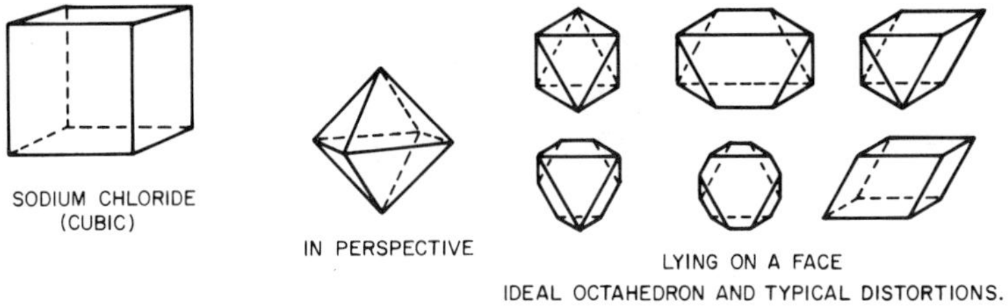

Fig. 7.2[2]

between which a material substance is placed, the charge density contributed by the material is given by

$$P = (k-1)\in_o \epsilon \tag{7.1}$$

in which

ϵ is the electric field applied
k is the dielectric constant and is $\in/\in_o$,
$\in$ is the permittivity of the material in farads/m, and

$\in_o$ is the permittivity of a vacuum in farads/m.

The polarization may be composed of four distinct contributions as:

$$P = P_e + P_i + P_o + P_s\,. \tag{7.2}$$

When P_e is the electronic polarizability, this condition is due to the shifting of the atomic nucleus and electrons in opposite directions in an applied electric field. A dipole moment develops and specifically

$$P_e = N\alpha_e \epsilon \tag{7.3}$$

in which

N is the number of atoms per unit volume, and

α_e is the electronic polarizability $4\pi\in_o R^3$, R being the atom radius.

When P_i is the ionic polarizability, this condition is due to the shifting of ions of opposite charge to the electrodes of opposite signs.

Both P_e and P_i are removed when the applied electric field is removed. These dipoles are therefore not permanent, they are said to be induced. Both the electronic and ionic polarizabilities are insensitive to temperature.

When P_o is termed the orientation polarization, this condition is due to the presence of permanent dipoles which although randomly oriented in the absence of an electric field, become preferentially oriented in an electric field. The orientation polarization is a function of temperature as well as of applied electric field, thus:

$$P_o = N\left(\frac{p^2}{3k}\right)\frac{\epsilon}{T} \tag{7.4}$$

in which

p is the permanent dipole moment, and

k is the Boltzmann constant, and

T is the temperature in °K.

When P_s is the space charge polarization, if small particles of a conducting (or semiconducting) phase are dispersed in an insulating matrix, an applied electric field displaces the electrons in a positive direction.

Dielectric Constant

In the case of a solid material there can be large numbers of dipoles per unit volume, as compared with a liquid. This relatively greater dipole proximity means that the size of the so-called local field must be taken into account. The strength of the local field at a point depends upon the distance of the point from the dipole and also upon the orientational relationship between the two. Assuming eight-fold coordination (i.e., assuming a bcc structure) the strength of the local field is

$$\epsilon_{\text{loc}} = \epsilon + \frac{1}{3}(P/\in_o) \tag{7.5}$$

This is termed the Lorentz field.

If the polarization is written as

$$P = N\alpha\in_{\text{loc}} \tag{7.6}$$

then Eqs. (7.5) and (7.6) in combination give the form for the dielectric constant of a solid as follows:

$$k = 1 + \frac{N\alpha}{(\in_o - N\alpha/3)} \tag{7.7}$$

In general it can be said that the dielectric constant does not change with increase of temperature. There are exceptions to this, however. A special case can be made for amorphous solids (plastics) in which between the glass transition temperature, T_g, and the melting point, T_m, there is a gradual rise in the value of k due to movement of the polar sites in the polymer molecules.

In the case of alternating current, the higher the alternating field frequency, the less tendency for orientation polarization to occur until finally only electronic polarization can be observed. Above 10^{16} Hz the polarization drops to zero. The natural frequency of electronic vibrations about an atom is 10^{16} Hz.

Dielectric Losses

The effect of polarization is caused by an applied electric field. The effect is not instantaneous. It takes time and so also does the reverse behavior (polarization is reversible). The charge density on the parallel electrodes is a function of the dielectric constant of the material between the electrodes:

$$D = \epsilon_o k \epsilon \tag{7.8}$$

If the field is varied in a cyclic manner, in general there will be a lag between the expected and actual charge density as follows:

$$D = D_m \sin(wt - \delta) \tag{7.9}$$

when

m indicates the maximum value,

δ is the loss angle.

The power loss due to the hysterises effect is expressed as

$$PL = \frac{w\epsilon^2 \in_o}{2} k \tan \delta \tag{7.10}$$

in which

$\tan \delta$ is termed the loss tangent and is equal to k''/k'. The tag$'$ indicates the in phase component (real) and the tag$''$ indicates the out of phase component (imaginary); and

$k \tan \delta$ is termed the loss factor.

The way in which the dielectric constant and the loss factor vary with frequency are shown in Fig. 7.3.[1] Some "typical" values of dielectric properties are given in Table 7.3.

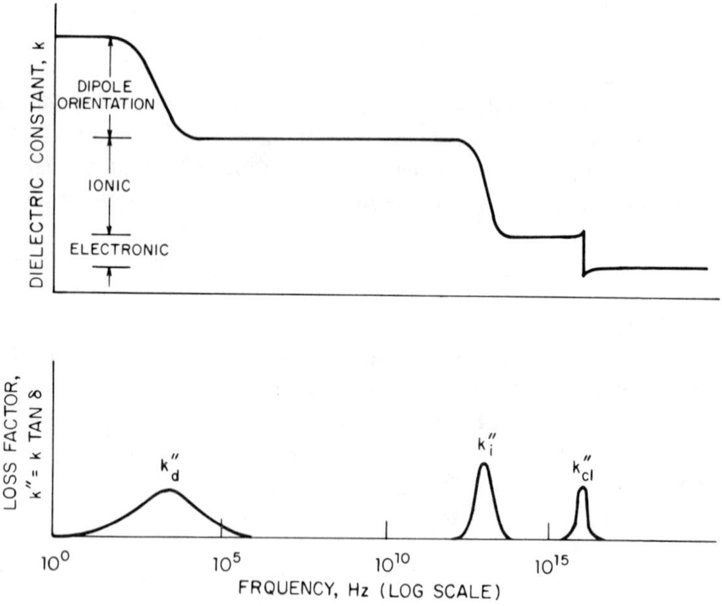

Fig. 7.3 Dielectric constant and loss factor versus frequency. The dielectric constant, K, approximates K' in any usable dielectric insulator.[1]

Table 7.3

SOME DIELECTRIC PROPERTIES OF MATERIALS[1]

Material	*Property*	*Units*	*Numerical Value*
Argon	Polarizability ($\alpha_e + \delta i$)	10^{-40} farad.m^2	1.4
NaCl			8.9
KCl			15.3
CO	Dipole moment (p)		0.3
H_2O			6.2
CH_3Cl			5.5
Porcelain	Dielectric constant (k) 60 Hz		6
	Loss tangent (tan δ) 60 Hz		0.010
Soda-lime glass	Dielectric constant, 60 Hz		7
	Loss tangent, 60 Hz		0.1 [high due to Na^+]
Fused silica	Dielectric constant, 60 Hz		4
	Loss tangent, 60 Hz		0.001
Nylon (6,6)	Dielectric constant, 60 Hz		4
	Loss tangent, 60 Hz		0.02
PVC	Dielectric constant, 60 Hz		7
	Loss tangent, 60 Hz		0.1

Index of Refraction

The index of refraction is related to the electronic polarization such that

$$n = \sqrt{k_e} \tag{7.11}$$

In terms of the velocity of light, the refractive index is simply

$$n = c_o/c \tag{7.12}$$

in which

n is the refractive index of the substance,
c is the velocity of light in the substance, and
c_o is the velocity of light in a vacuum.

The effect is to slow down the velocity of the light as it passes through the substance. The higher the atomic weight of the element, the slower the light passes through. For example, NaF has n of 1.336 while NaBr has a value of 1.775 for n. The most common method of measuring the refractive index is by immersing the particle in a liquid of known refractive index such that the particle disappears (there may be a problem, however, if the particle tends to dissolve in the test liquid).[2]

In the case of noncubic crystals, however, the electronic polarizability varies with the direction in which the light vibrates due to the nonsymmetrical local fields in these substances. This variation means that the index of refraction is also anisotropic. Therefore, to determine the index of refraction of a noncubic crystalline material, it is necessary to utilize polarized light. For example, consider a tetragonal crystal. This material is essentially a crystal with a longitudinal axis for which there is a corresponding refractive index designated epsilon and two lateral axes for which the refractive index is designated omega. Polarized light vibrates in a direction perpendicular to the direction of its light beam. Therefore, if the tetragonal crystal is mounted with its c axis parallel to the light beam and then immersed in a liquid of refractive index equal to the omega value, the crystal disappears twice per revolution of the mounted crystal. By comparison, isotropic crystals "stay disappeared" all the time. Some examples of refractive indices for anisotropic substances are given in Table 7.4.[2]

7.3 ABSORPTION

For many covalent and ionic substances "k" is effectively zero for both

Table 7.4

REFRACTIVE INDICES OF SELECTED SUBSTANCES[2]

	Refractive Index					
Substance	n	n_w	n_e	n_α	n_β	n_γ
Kf	1.352	–	–	–	–	–
Quartz (SiO_2)	–	1.544	1.553	–	–	–
Diamond	2.42	–	–	–	–	–
Corundum (Al_2O_3)	–	1.768	1.760	–	–	–
Carborundum (SiC)	–	2.654	2.697	–	–	–
Borax ($Na_2B_4O_7$,$1OH_2O$)	–	–	–	1.447	1.469	1.472
Talc (3MgO, $4SiO_2$, H_2O	–	–	–	1.539	1.589	1.589
Malachite [$CuCO_3$,$Cu(OH)_2$]	–	–	--	1.655	1.875	1.909

From McCrone, W. C. and Delly, J. G., *Particle Atlas,* Tables XX, XXI, XXII, p. 77-78, Ann Arbor Science Publishers, 1973, 1979.

ionic and electronic polarizability. As shown in Fig. 7.3(b), this condition indicates that these substances are essentially transparent to radiation of 5×10^{14} Hz, which is the visible part of the spectrum. Opacity and color of substances exist because of the phenomenon of absorption, which involves the partial absorption of incoming white light and a partial transmission of the rest of the spectrum. These substances may be contrasted to metals which contain effectively free electrons that absorb the incoming light energy and preclude transmission (except, that is, in the case of very thin foil which can transmit light).

There are numerous ways in which a material may come to reflect color. For example, the introduction of a chemically different atom into a matrix may make the crystal seem colored. Cr^{3+} ions reflect color in ruby as follows: when the incoming radiation reaches the Cr^{3+} ion, the latter absorbs a quantum of blue light which excites an electron. The electron then makes a nonradiative transition down to a lower energy level from which it later makes a radiative transition down to the original ground state. The last transition reflects red light. Therefore, in effect the process appears to convert incoming blue light into an emitted red light.

Yet another way of developing color in a normally transparent substance is by means of an F center. An F center is formed in the following way: Consider the case of KBr in which there is an anion vacancy. This vacancy is surrounded by K^+ ions on their appropriate lattice sites and consequently possesses a net positive charge. An electron attached to this positively charged vacancy is able to jump around there from one K^+ ion to another. In fact its resonant frequency is such that it just about corresponds to an optical wavelength of 6200 Å. This frequency transmits the blue-green color and damps out the orange-red wavelengths. The light intensity is related to the intensity of the light entering the particle as follows:

$$I = I_o e^{-\beta x} \tag{7.13}$$

in which

I is the transmitted intensity of the light,
I_o is the initial light intensity,
x is the thickness of the crystal, and
β is termed the absorption coefficient.

Absorption and Scattering

There are two main consequences of the interaction between light rays and small or colloidal particles (in this sense the term colloidal particles is used to describe the size range from below 1 μ down to 30/100 Å or so). The absorption of light heats up the particle which in turn interacts with the enveloping fluid. This effect is termed photophoresis.[3] The other major effect is due to the radiation pressure on the particle which results from the scattering of light by the particle (but there is some effect due to absorption, too). The theoretical description of scattering of light by particles may be conveniently divided into three regimes:

the particle diameter is less than the wavelength of the light,
the particle diameter is about the same as the wavelength of the light, and
the particle diameter is much more than the wavelength of the light.

In the case of the first regime, the diameter is less than 0.05 μ. It is termed Rayleigh scattering.[4]

In the case of the second regime, the diameter is in the range of 0.05 through 2 μ. It is termed Mie scattering.[5]

In the case of the third regime, the diameter is more than 2 μ.

Rayleigh Scattering

The basic idea behind this derivation is that the oscillating field of the incident radiation induces a corresponding electrical moment within the particle. This induction then acts as a linear electrical oscillator and radiates light, except in the direction corresponding to that of the vibrations themselves. A schematic description of the angular distribution of the intensity of the scattered light is shown in Fig. 7.4.[6]

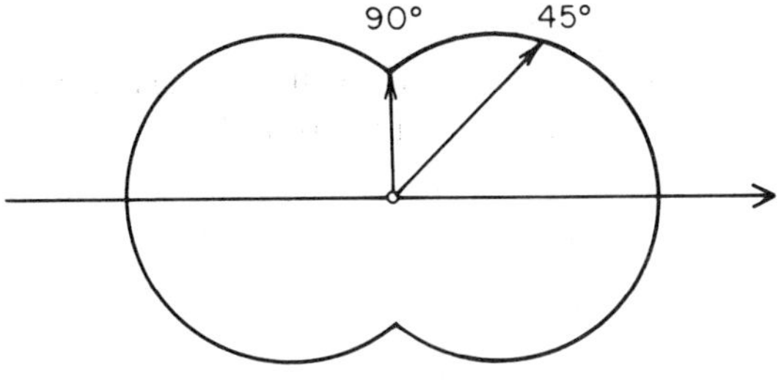

Fig. 7.4 Radiation pattern for small dielectric spheres (Rayleigh scattering). The radius indicates the intensity of light in that direction. For example, the radii at 45° and 90° indicate the respective intensities (based on ref. 6).

For optically isotropic spheres illuminated with unpolarized light, the intensity of the scattered light is expressed by:

$$I(\theta) = \frac{\alpha^4 D^2}{8R^2}\left(\frac{k^2-1}{k^2+2}\right) \cdot (1+\cos^2\theta) \cdot I_o \qquad (7.14)$$

in which

$I(\theta)$ is the intensity of the scattered light at angle,

D is the diameter of the particle,

R is the distance from the particle to the point at which the scattered light is viewed,

α is $\Pi D/\lambda$,

I_o is the initial intensity of the illumination, and

λ is the wavelength of the light.

The ratio of the total radiant energy scattered by the particle to the incident radiant energy is termed the scattering cross section and is expressed as

$$C_{\text{sca}} = \frac{8}{3}\,\alpha^4\left(\frac{k^2-1}{k^2+1}\right)\frac{\Pi}{4}\,D^2 \equiv E\,\frac{\Pi}{4}\,D^2 \qquad (7.15)$$

in which

C_{sca} is the scattering cross section, and

E is the extinction coefficient or area efficiency factor:

$$= \frac{\text{(total flux scattered and absorbed by the particle)}}{\text{(flux geometrically incident on the particle)}}$$

The scattered light intensity I is composed of two components:

$$I(\theta) = i_1 + i_2 \tag{7.16}$$

in which i_1 is the intensity associated with vibrations normal to the plane of observation and is equal to the first part of Eq. (7.14). Thus

$$i_1 \text{ is } \frac{\alpha^4 D^2}{8R^2}\left(\frac{k^2-1}{k^2+2}\right) I_o$$

and

$$i_2 \text{ is } i_1 \cos^2\theta$$

Fig. 7.5 illustrates Eq. (7.16) for an isotropic particle.

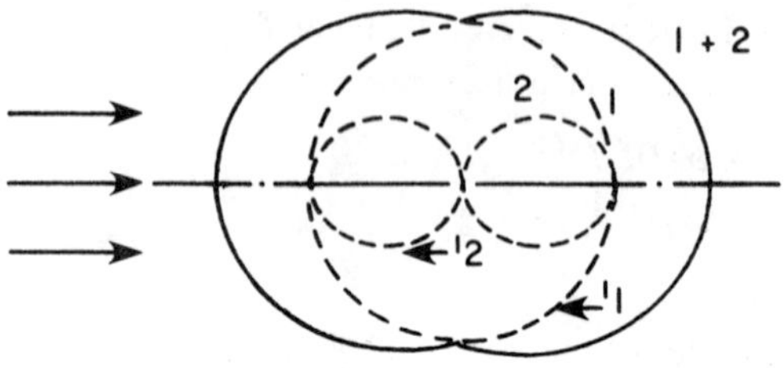

Fig. 7.5 Illustrating Eq. (7.16).[7] Note: forward and backward scatter are equal.

Mie Scattering

The scattering relationship becomes very complicated above a particle size of 2 μ. In the intermediate range of 0.05–2 μ, a more tractable expression has been developed and is known as Rayleigh-Gans scattering. This expression is derived by assuming that each volume element of the particle scatters light independently of every other volume element. The form of the Rayleigh equation is modified accordingly:

$$I\theta = \frac{\alpha^4 D^2}{8R^2}\left(\frac{k^2-1}{k^2+2}\right)(1+\cos^2\theta)\, I_o \cdot \frac{1}{9}\,(k^2+2)^2 G^2(\theta,\alpha) \tag{7.17}$$

in which $G(\theta,\alpha)$ is a function and the values have been tabulated.[7]

When the particle is large as compared with the wavelength of the incident radiation, the scattering is a resultant of the light waves originating from all of the various portions of the particle. The computations are necessarily complicated and tedious. A typical form of the Mie equation is

$$S = \frac{\lambda^2}{2\Pi} \sum_{\nu=1}^{\infty} \frac{a_\nu{}^2 + p_\nu{}^2}{2\nu + 1} \tag{7.18}$$

in which

S is the total scattering of one spherical particle per unit intensity, and

a_ν, p_ν are functions of x (i.e., $2\Pi r/\lambda$) and β (i.e., $2\Pi rm/\lambda$).

A polar plot for Mie scattering is drawn in Fig. 7.6.[8] It shows that the intensity of scattering becomes more marked in the forward direction as the particle size increases.

Rayleigh, Rayleigh-Gans, and Mie scattering may serve as a basis for measuring the size and also the concentration of aerosols. These possibilities are summarized in Table 7.5.[9]

The illumination of a monodispersed aerosol by a beam of white light produces a series of spectra as the angle between the incident beam and the scattered direction is increased. These spectra are known as higher order Tyndall spectra. This is a basis for one method of measurement which is reported in Table 7.5 under the abbreviated title of HOTS.

Light Transmission or Extinction

The passage of light into a suspension of particles always results in some attenuation of the incident beam. The total removal of the incident flux from the beam is termed extinction. This removal is due both to scattering and to absorption (where this occurs). The relationship between the initial light intensity and the level of intensity after the light has travelled a distance into the suspension is an exponential one and is termed the Bouguer Law.[10]

$$I = I_o \exp(-\tau l) \tag{7.19}$$

in which

τ is the extinction coefficient of the suspension as a whole, and

l is the thickness of the suspension.

The product τl is called the optical density (turbidity) and the quotient

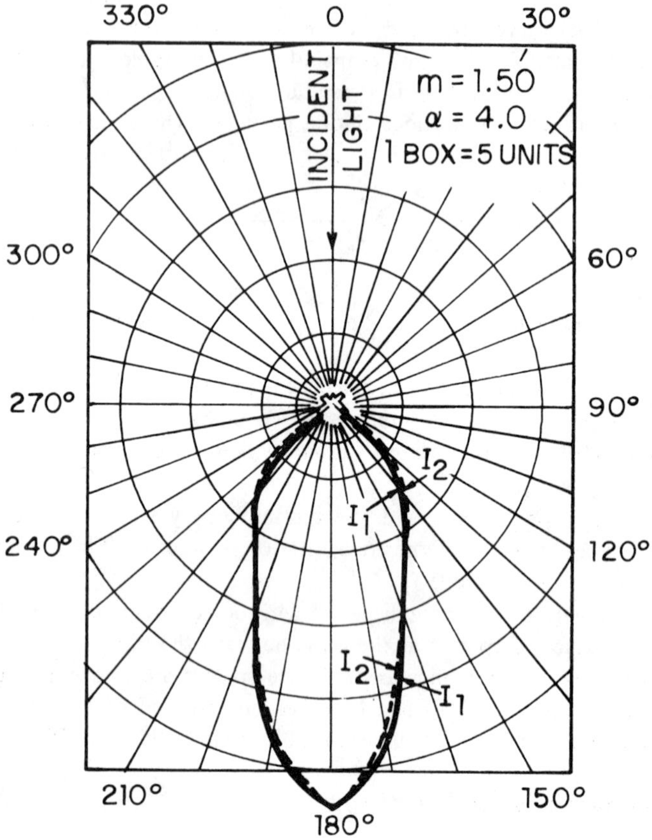

Fig. 7.6 Mie scattering.[8]

I/I_o is called the transmittance.

Eq. (7.19) may be rewritten in terms of the individual particle scattering coefficient E as follows:

$$I = I_o \exp(-naEl) \tag{7.20}$$

in which

n is the number of particles per unit volume, and

a is the cross section of the particles which is Πr^2 in terms of the particle radius.

A diagram of the essential features of an apparatus is shown in Fig. 7.7.[9] An important feature of this equipment is that none of the scattered light

Table 7.5

MAIN POSSIBILITIES FOR MEASURING SIZE AND CONCENTRATION OF AEROSOLS BY SCATTERING

Particle-Size and Light-Scattering Regime	$d \ll \lambda$ Dipole (Rayleigh) $\alpha < 0.3, d < 0.05\ \mu$	$d \approx \lambda$ Intermediate (Mie, Rayleigh-Gans) $0.3 < \alpha < 3/(m-1)$, $0.05\ \mu < d < 2\ \mu$	$d \gg \lambda$ Large: Diffraction, Refraction, Reflection, Absorption (Bricard) $3/(m-1) < \alpha, 2\ \mu < d$
		LIGHT SCATTERING	
Scattering Phenomena (single particles)	7. Relative angular distribution same for all particle sizes, shapes, refractive indices, transparent or absorbing. Symmetrical forwards and backwards. $I_1(\theta) = \alpha^4(m^2-1)^2/2\pi(m^2+2)^2, I_2(\theta) = I_1(\theta)\cos^2\theta$, unless metallic particles.	8. For all shapes, scattering becomes increasingly forwards-directed and main forward lobe develops within ± 45°. Subsidiary lobes in pattern also develop, about 2α in number. These lobes are less conspicuous for polydisperse transparent particles and for absorbing particles, mono- or polydisperse.	9. Forward lobe approximates increasingly to Fraunhofer diffraction, forward intensity coefficient $I_0 = \alpha^2/4\pi$ (total forward intensity *per particle* $\propto d^4$) angular width $220°/\alpha$, largely independent of particle composition or of shape of particles randomly oriented. Range in d or λ smooths out subsidiary lobes giving patterns approximated, for transparent particles, up to angle $\theta = 2\cos^{-1} 1/m$, by refraction and external reflection, and for absorbing particles up to nearly 180° by external reflection.
Polarization Measurements $I_2(\theta)/I_1(\theta)$ and *Disymmetry Measurements* $I_{1,2}(90° - \phi)/I_{1,2}(90° + \phi)$ Monochromatic light	10. Finding $I_2(\theta)/I_1(\theta) = \cos^2\theta$, or $I_1, I_2, I_{1+2}(90° - \phi) = I_1, I_2, I_{1+2}(90° + \phi)$ shows $d < 0.05\ \mu$ regardless of shape, size, m_0, k.	11. If m_0, k are known, particles are spheres and fairly monodisperse, θ suitably chosen (90° not necessarily best for polarization, or 45° and 135° for disymmetry), polarization or disymmetry ratios vary monotonically up to $\alpha < 5$ ($d < 1\ \mu$) and can be used for sizing.	12. Polarization and disymmetry ratios oscillate with increasing α and give no unique measure of size.
High-order Tyndall Spectra: (HOTS), white light	13. No spectra, but 90° scattered light noticeably blue.	14. Angular spacing of reds $\approx 180°/\alpha$. Use tables or reported measured values if m_0 known. Not practicable to observe when particles absorbing, or $d > 2\lambda$, or range in $d >$ few %, but background of numerous particles in Rayleigh region tolerable.	
Total flux at defined angle (or range of angles) outside, forward lobe	15. Transparent particles any shape, flux *per particle* $\propto d^6\lambda^{-4}(m^2-1)^2(m^2+2)^{-2}$. Absorbing particles, any shape, flux per particle $\propto d^2\lambda^{-1}$.	16. Flux per particle $\propto d^n, 2<n<6$, where n and factor of proportionality depend on size, shape, m_0, k.	17. Flux per particle $\propto d^2$ when $d > 2\ \mu, \theta = 30°$ and $d > 10\ \mu$ $\theta = 90°$ (to ±20%). Dependence on m_0 least at 40° for transparent particles (most shapes), greatest about 90°. Difference between absorbing and transparent particles less at 90° than at 40°.
Total flux integrated over forward lobe		18. Varies similarly to E (entries 2 and 3), approximately equal to geometrically-incident flux when $d > 1$ or $2\ \mu$, regardless of m_0, k, shape size.	
Flux ratio at two angles well within forward lobe		19. For $\alpha > 1$ ($d > 0.3\ \mu$) gives unique and reasonably accurate estimate of size regardless of particle composition and shape. Impracticable when α large, angles too small.	

be allowed to reach the photocell. The light that is permitted to reach here is the undeviated light only.

By direct substitution in Eq. (7.20), in the case of very small particles that obey the Rayleigh relationship, the transmittance is written:

$$I/I_o = \exp \frac{128\ {}^5 r^6}{3\lambda^4}\left(\frac{k^2-1}{k^2+2}\right) nl \qquad (7.21)$$

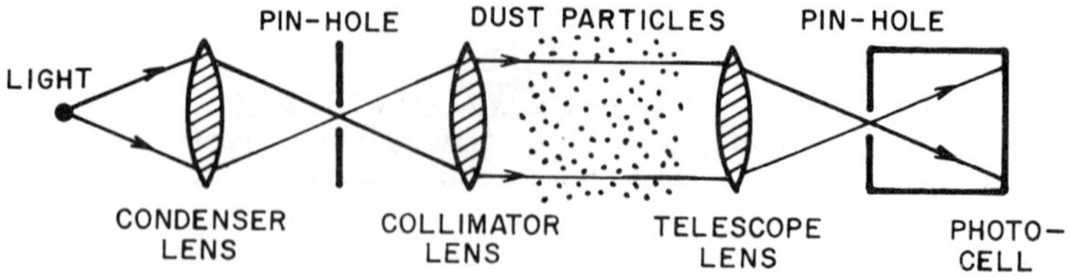

Fig. 7.7 Essential of extinction apparatus.[9]

In the case of suspensions containing larger particles, some complications arise due to the forward scattering of the light, the proportion of which increases with the increase of particle size. Fortunately, there are numerous reports, in the literature which deal with this problem. They are reported in some detail elsewhere.[9] A table showing the possibilities for measurement of size and concentration of aerosols by extinction is given in Table 7.6.[9]

In general fairly simple equipment may be used to measure size and concentration characteristics of suspension of particles. The phenomenon of scattering is more complicated and requires more finesse both experimentally and theoretically. However, it has been observed that more details may be obtained about the aerosol from the latter phenomenon.[9] The reader is referred to Tables 7.5 and 7.6 and also to reference 9 for further details.

The greater the turbidity, the shorter is the range of visibility. Smoke particles suspended in the air are especially effective in cutting down visibility as shown in Table 7.7.[6] It is particularly interesting to note that the quantity of material in tobacco smoke is about two orders of magnitude less than the quantity of air that it so dramatically affects.[6]

Radiation Pressure

The Maxwell theory of electromagnetic radiation relates the energy and velocity of the radiation to the momentum in the following way:

$$m = u/c \tag{7.22}$$

The effect of radiation pressure therefore depends upon the rate of removal of momentum from the forward direction of the incident beam by the particle. This can be illustrated in the following way: if the cross section for scattering is C_{sca} such that this represents the radiant energy removed from the incident beam by scattering alone, if the cross section for extinc-

Table 7.6

MAIN POSSIBILITIES FOR MEASURING SIZE AND CONCENTRATION OF AEROSOLS BY LIGHT EXTINCTION

The choice of method is strongly influenced by the nature of the particulate suspension, by what is known about its nature beforehand, and by the circumstances and purpose of the measurement. This chart is intended mainly as a guide for measuring particle size and concentration in unknown aerosols to an accuracy of ±20 percent.

Particle Extinction Coefficient or Area Efficiency Factor: $E \equiv$ (Total flux scattered and absorbed) ÷ (Geometrically incident flux).

The Bouguer Law: Transmittance $f/f_0 = \exp(-naEx)$ where f_0 and f are incident and emergent undeviated fluxes for path x through suspension containing in unit volume n particles of mean projected area a and mean extinction coefficient E.

Refractive index of particle: $m = m_0(1 - ik)$; absorption coefficient $k = 0$ for transparent particles.

Particle size parameter: $\alpha = \pi d/\lambda$. For wavelength $\lambda = 0.5\ \mu$, $\alpha = 6$ means diameter $d = 1\ \mu$.

Particle Scattered Intensity Coefficient: $I(\theta) \equiv$ (Intensity in direction θ from forward direction) ÷ (Geometrically incident flux), resolvable into polarized components I_1 and I_2 in, and normal to, observing plane.

Note: All references to irregular particles are for the average over random orientations and for complete lack of symmetry.

Particle-Size and Light-Scattering Regime	$d \ll \lambda$ Dipole (Rayleigh) $\alpha < 0.3, d < 0.05\ \mu$	$d \approx \lambda$ Intermediate (Mie, Rayleigh-Gans) $0.3 < \alpha < 3/(m-1)$, $0.05\ \mu < d < 2\ \mu$	$d \gg \lambda$ Large: Diffraction, Refraction, Reflection, Absorption (Bricard), $3/(m-1) < \alpha, 2\ \mu < d$
	LIGHT EXTINCTION		
Extinction Phenomena (single particles)	1. Transparent: $E = 8\alpha^{4/3}(m^2-1)^2(m^2+2)^{-2}$ Absorbing: $E \propto \alpha$ Particle shape enters only as constant factor, roughly same for equal-volume particles.	2. Transparent: $E \propto \alpha^{3,2,1,0}$ reaches max. for sphere when $\alpha(m-1) \approx 2$, reaches plateau for irregular particle randomly oriented when $\alpha(m-1) \approx 3$, $(d \approx 2/3\ \mu)$. Absorbing (carbon): E reaches plateau for sphere when $\alpha \approx 2$, $d \approx 0.3\ \mu$; for irregular when $\alpha \approx 3, d \approx 0.5\ \mu$.	3. Transparent sphere: E settles to 2 after some oscillation. Oscillations partly or wholly smoothed if size-range wide enough. Transparent irregular, absorbing spherical, irregular, $E \approx 2$ independent of α.
Transmission Measurements: several λ, small acceptance angle, 10 percent $< f/f_0 <$ 90 percent	4. Finding $E \propto \lambda^{-4}$ shows $d < 0.05\ \mu$ for transparent particles regardless of shape or size-distribution. Finding $E \propto \lambda^{-1}$ shows $d < 0.05\ \mu$ only if particles known to be absorbing; otherwise means $d > 0.05\ \mu$. Can estimate mean particle volume: surface diameter only if also mass concentration and particle density.	5. Finding $E \propto \lambda^{-n}, 0 < n < 4$, shown $0.05\ \mu < d < 1\ \mu$.† Can estimate particle area-concentration and area-diameter by fitting log-log plot of $\log(f_0/f)\lambda : \pi/\lambda$ to log-log plot of theoretical $E : \alpha$ curve, the more accurately, the better m_0, k, shape and relative spread in size are known. († Unless absorbing.)	6. Finding E independent of λ shows $d < 1\ \mu$ (can be transparent, polydisperse spheres or irregular particles, any absorbing aerosol), so deduce area concentration. To obtain mean size, must know mass or number concentration‡ also. Finding E oscillates with λ shows monodisperse spheres, get conc'n. and size by fitting to theoretical curve. (‡ Gives mean volume/surface or area diameter.)

Table 7.7

COMPARISON OF TURBIDITY OF AIR, SMOKE, AND RAIN[6]

	Number per cc	*Turbidity* (cm^{-1})
Air	3×10^{19}	6×10^{-12}
Smoke	1×10^{7}	2×10^{-1}
Rain	1×10^{-3}	2×10^{-5}

tion is C_{ex} such that this is the radiant energy removed from the incident beam per unit time by scattering plus absorption, and if C_{pr} is the cross section for radiant pressure, then, the quotient C_{pr}/c is the net time rate of removal of momentum from the forward direction. This latter has been related to the scattering and extinction in the following expression:[10]

$$C_{pr}/c = \frac{1}{c}(c_{ext} - \langle\cos\theta_{sca}\rangle \cdot C_{sca}) \tag{7.23}$$

in which the quantity $\langle\cos\theta_{sca}\rangle$ is an asymmetry factor and represents the fraction of the scattered radiation whose component is in the forward direction. Therefore $\langle\cos\theta_{sca}\rangle \cdot C_{sca}$ describes the component of the scattered radiant energy that is returned in the forward direction. Dividing each factor by the cross section of the particle produces the equation in more or less standard form:

$$Q_{pr}/c = \frac{1}{c}(Q_{ext} - \langle\cos\theta_{sca}\rangle\, Q_{sca}) \tag{7.24}$$

The quotient Q_{pr}/c is termed the radiation pressure.

Photophoretic Force—Photophoresis[3]

The occurrence of this phenomenon moves in three steps:

1. Electromagnetic energy is absorbed by the particle and reappears as thermal energy, which may or may not be evenly distributed around the particle.
2. The interior of the particle is heated up. This heat flows to the particle surface by conduction. This event is illustrated in Fig. 7.8.[6]
3. The thermal energy and its associated momentum is radiated away from the surface (and also conducted and convected if a fluid surrounds the particle). The forces resulting from momentum transfer of this sort in a fluid are called radiometric forces.[6]

The calculation of the local distribution of thermal energy—termed the source function—is the key to being able to calculate the photophoretic force. A study of this problem has been alluded to.[6]

Radiation Pressure with Absorption

Just in the same way that some of the scattered radiation is returned in the forward direction, so, too, when some of the incident radiation is absorbed, a portion of it is returned in the forward direction. A general state-

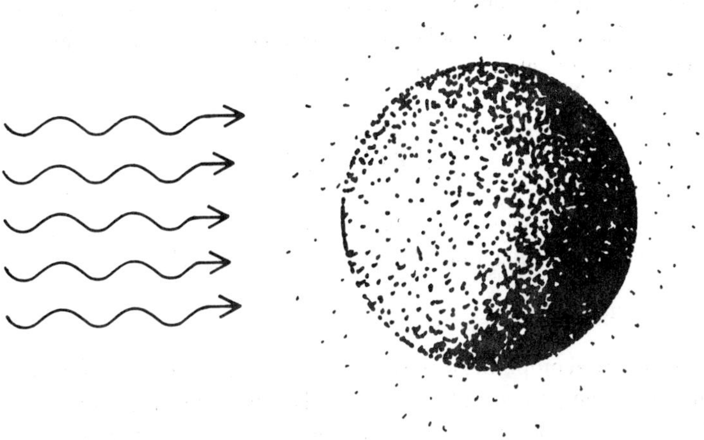

Fig. 7.8 Temperature distribution within an illuminated, partially absorbing sphere. Darker shading indicates higher temperature; dots outside the sphere represent gas molecules.[6]

ment for this state of affairs has the following form:

$$\frac{C_{pr}}{c} = \frac{1}{c}\left[(C_{sca} - \langle\cos\theta_{sca}\rangle \cdot C_{sca}) + (C_{abs} - \langle\cos\theta_{abs}\rangle \cdot C_{abs})\right] \tag{7.25}$$

The absorbed energy that is returned in the forward direction (of the incident beam) is in the form of radiant energy. The cos θ terms are the assymmetry terms, as before.

The simplest application of the type of relationship presented in Eq. (7.25) is the case of a particle in a vacuum. A more complex but perhaps more useful case is that of a particle in a fluid. In this latter case, convection and conduction of the fluid must also be taken into account.

In the solar system the effect of radiation pressure is important, especially in connection with dust settling on the earth which originally comes from the outer space of the solar system. It is estimated that 1000 tons per day of dust fall on the earth in this way. These dust particles are affected in two ways: the first is that their centrifugal force is supplemented by the radiation pressure and they may therefore fly off into space out of the solar system altogether; the second is that they are subject to a drag force and as this reduces the particle angular momentum, they tend to return towards the sun. The path of the earth crosses this never ending stream of particles going this way and that, collecting 1000 tons per day from it. The effects of radiation pressure are real enough as have been shown using laser light to levitate 20-μ glass spheres.[11]

7.4 ELECTRICAL PROPERTIES

The electrical charging of small particles may occur in one of three ways: contact charging, field charging, and diffusion charging.

Contact Charging

The electrical properties of powders are much influenced by the internal structure of the particles of the set and also by the previous history of the particles. These factors influence the charging of the particles, as do the properties of the material substance of which the particles are composed, i.e., metal, semiconductor, or insulator.[12]

Internal pores or fissures reduce the continuity of the material and this reduces the electrical conductivity of the particles of the set.

The point of contact between particles depends upon the size and shape of the particles. Only in extreme conditions can applied pressure promote hot spots to occur at these points of contact.[12]

Of practical importance is the effect of wetting of a powder. When a particle is wetted by a liquid electrolyte, ions of a particular sign are preferentially adsorbed on the particle surface. This layer of ions attracts in its turn a layer of compensating ions from the liquid to form a space charge layer around the particle. This condition is illustrated in Fig. 7.9.[13]

The layer of ions around the particle is called the electrical double layer. The tightly bound layer immediately surrounding the particle is termed the Stern Layer[14] and the more diffuse layer around this is termed the Gouy Layer.[15] An applied electric field gives rise to a mechanical force on the surface. This effect is the basic cause of the phenomenon of electrophoresis in which particles, due to the effect of the double layer, move under the action of an applied electric field. Another phenomenon, the streaming potential, is caused by a flowing liquid which removes the more loosely bound ions and yet leaves in place those bound to the surface.

Both the zeta potential and the streaming potential can be brought about with quite low ionic concentrations.

Of course, just because a particle is no longer wet does not mean that it no longer shows any effect of the previous liquid—quite the contrary. For

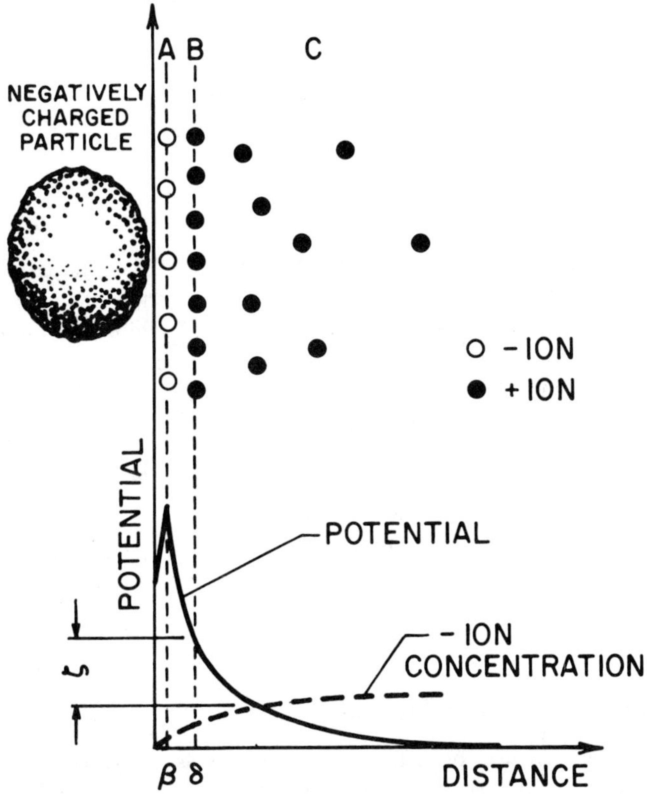

Fig. 7.9 Concept of the zeta potential.

example: the closely bound ions may be retained and so ions of opposite charge will tend to become rather loosely attached, sooner or later. This process may be viewed as one in which the surface stabilizes.

Assuming that the Hertz formula[16] can be applied to the contacts between two (ideally) spherical particles and also assuming that only elastic deformation occurs, the radius of contact is expressed by:

$$a = 1.1\left(\frac{FR}{2E}\right)^{1/3} \tag{7.26}$$

in which

R is the particle radii,

E is the elastic modulus of the particle substance,

$\frac{1}{3}$ is Poisson's ratio, and

F is the force pressing the two spheres together.

This relationship may be simplified by inserting the value of the pressure in the powder (P)

$$a = 1.1R(\pi P/2E)^{1/3} \tag{7.27}$$

Therefore in a mass of particles in a static assembly, the interparticle contact area severely constricts the flow of electric current. The resistance of a circular electrode of radius "a" feeding into an infinite medium of resistivity ρ is $\rho/4a$ and therefore the resistance of each contact area is $\rho/2a$. By simple comparison of the resistance of a string of beads to that of a cylinder of the same diameter, one can show that[12]

$$\rho_a/\rho = \pi R/4a \tag{7.28}$$

The above relationship is only an approximation but it does show that the apparent resistivity of a powder may be orders of magnitude of the resistivity of the solid substance of which it is composed. By substitution

$$\rho_a/\rho = 0.61(E/P)^{1/3} \tag{7.29}$$

Theoretical Maximum Charging

When two particles of different material (metals in this case) are placed in contact, a double layer is established which causes a levelling up of the Fermi levels in the two metals. When the two particles are separated, some of the charge is neutralized by the electrons which cross the gap by a quantum-mechanical tunnel effect. However, there is a residual of charge left which constitutes the observed electrostatic charge.[16] Using this theoretical approach, the separation charge for two dissimilar metals is calculated to be:

$$\frac{(\phi_1 - \phi_2)R}{600} (1.1511 \log_{10} R/2 + \xi) \text{ e.s.u} \tag{7.30}$$

in which

$\phi_1 - \phi_2$ is the contact potential,

ξ is the complicated function of (a certain mean of ϕ_1, ϕ_2), and $n\sqrt{a/M}$ a parameter.

In practice, contact potentials are less than what are calculated for clean surfaces in a vacuum because of the presence of oxide layers as well as adsorbed gas molecules, etc. The calculation has been carried out and the maximum specific charge for contact charging of metallic particles has been shown to be:

$$1.33 \times 10^{-13} (\phi_1 - \phi_2)(1.151 \log_{10} R + 8.5)/\rho R^2 \text{ coulombs/g} \quad (7.31)$$

This equation is plotted in Fig. 7.10[12] for a contact potential of 1 volt for a range of values of R. It should be noted that the potential is never more than a few times the contact potential,[12] and so one would not normally expect a discharge into a gaseous medium due to contact charging of metallic powder particles.[12]

On the face of it, there is no very obvious reason why contact charges should be able to develop in the case of homogeneous sets of particles. However, such charges do develop and the inference is therefore that homogeneity in the strict sense is an ideal that is not attained in practice. An added point is that if the small particles are crystalline material, the different faces must represent different crystallographic planes which will have inherent differences for charging relative to each other.

Because semiconductors have few charge carriers, the double layer is effectively thicker than in the case of a metal. The charging of semiconductors has been shown to be less than for metals[16] and as the conductivity of a semiconductor is decreased, the contact charging is reduced; eventually no contact charging occurs. The shaded area in Fig. 7.10 concerns semiconductors. In the case of insulators, there are two quite distinct types: electrophilic insulators and electrophobic insulators. Electrophobic insulators, when clean, exhibit negligible charging. Electrophilic insulators become very strongly charged when clean. In fact, they may become charged up to a level at which they can discharge into the surrounding gas.[12]

Diffusion Charging

A particle can receive a charge due to collisions with ions that are in thermally activated random motion. A relationship which describes this process is expressed as follows:[17]

$$n_p = \frac{D_p kT}{2e^2} \ln \left(1 + \frac{D_p \bar{c} \pi e^2 N t}{2kt} \right) \quad (7.32)$$

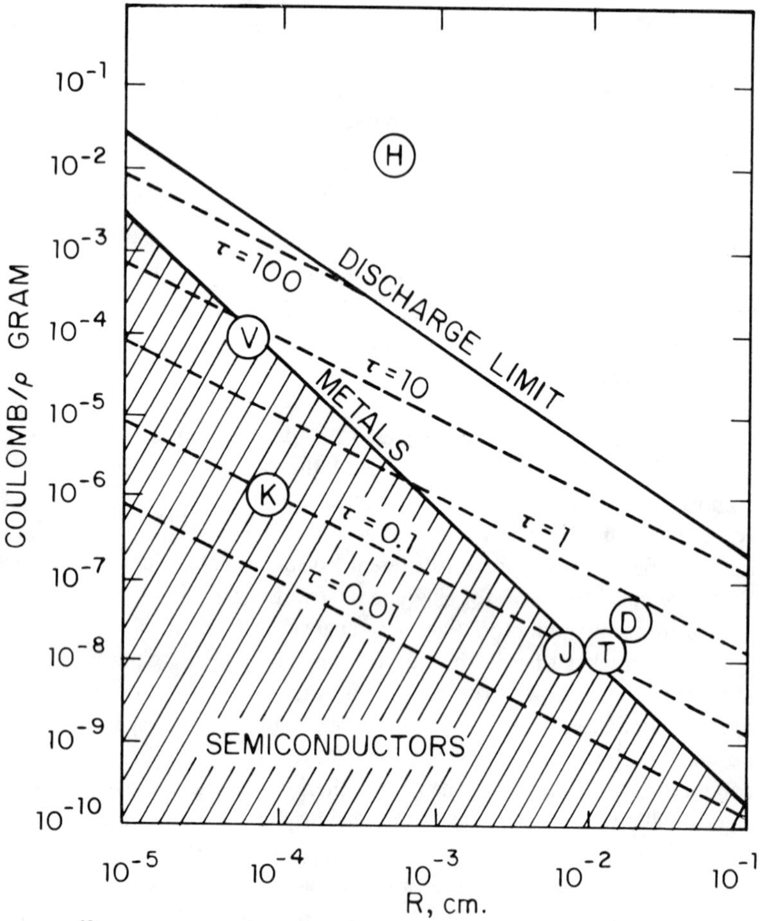

Fig. 7.10[12] Maximum specific charge for contact charging of typical powders. Experimental values for observed charging are:

D Debeau	H Harper (large spheres)
J Jewell-Thomas (sphere/plane)	K Kunkel
T Thomas	V Vollrath

τ = charge density

in which

n_p is the charge on an initially neutral aerosol,
N is the concentration of unipolar ions/cc,
t is the time,
$\bar{c}$ is the average speed of the ions,

e is the elementary unit of charge,
D_p is the particle diameter, and
k is Boltzmann's constant.

Field Charging

In the case of an electric field of sufficient intensity being applied, the resultant ordered motion of the ions may in effect drive the ions onto the particle and thereby charge them. It can be assumed that the ions travel along the lines of force and are unaffected by the diffusion charging. The relationship for this process or mechanism is expressed by:[17]

$$n_p = \left(1 + 2\frac{k-1}{k+2}\right)\frac{ED_p{}^2}{4e}\left(\frac{\pi NeZit}{\pi NeZit + 1}\right) \tag{7.33}$$

in which Zi is the charge on the species of ion.

Charge Limit

There is a natural electrical charge that can be carried by a particle. This level is dependent upon the physical properties of the material substance of the particle. Three such limits have been identified: the electron limit, the ion limit and the Rayleigh limit.[18]

When the surface field intensity becomes sufficiently great, a negatively charged particle emits electrons and positively charged particles emit ions. For a spherical particle the level of the limiting charge is:

$$n_p = \frac{D_p^2 E_s}{4e} \tag{7.34}$$

in which

E_s is the level of surface field intensity at which emission of electrons or ions will occur.

In the case of a fluid, specifically a liquid, when the pressure exerted by the electric field exceeds the surface tension of the drop, which is tending to keep the liquid in one piece, then the drop disintegrates. The limiting charge for this process (the Rayleigh limit) is

$$n_p = \frac{(2\pi\sigma)^{1/2}}{e} D_p^{1.5} \tag{7.35}$$

in which σ is surface tension. Particle charge limits for these processes are shown in Fig. 7.11.[19]

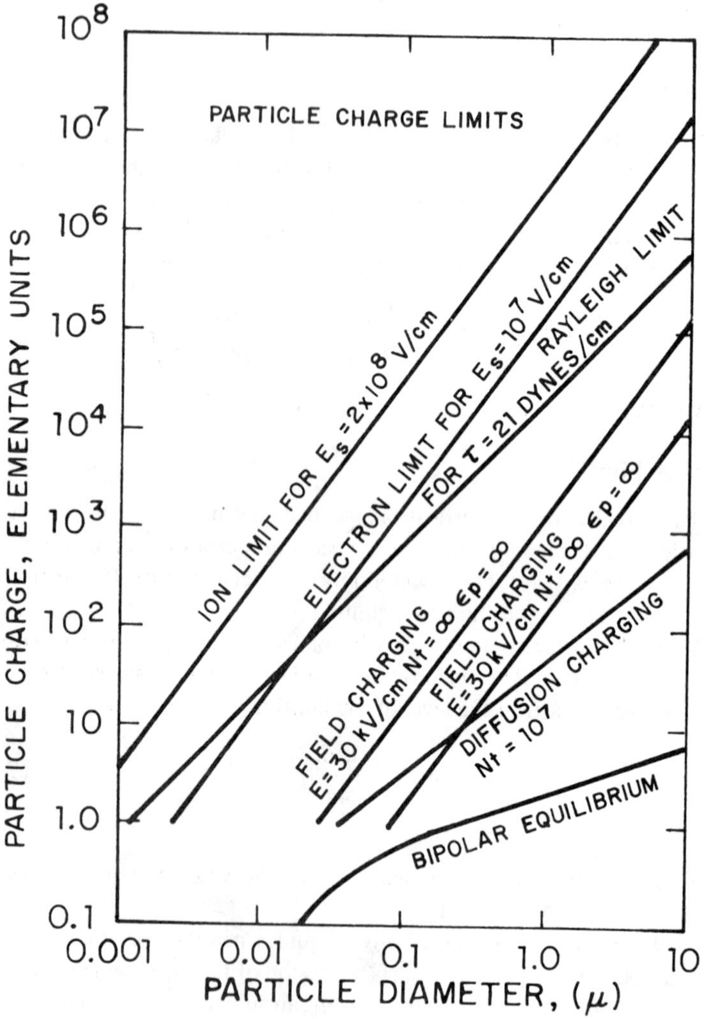

Fig. 7.11 Particle charge limits.[19]

7.5 BROWNIAN MOTION

It is a familiar story of the Botanist Robert Brown who inadvertently made the major discovery that free microscopic particles move around in an enveloping fluid in more or less random manner due to the imbalance of molecular impacts.[20] If one stops to think about this Brownian motion, even today it is still exciting to realize that such an important connection between molecular motion and the behavior of microscopic/macroscopic forms can be so easily observed.

The relationship between the particle displacement and time is expressed by:[21]

$$\overline{\Delta}_x^2 = 2Dt \qquad (7.36)$$

in which

$\overline{\Delta}_x^2$ is the displacement squared,
D is the diffusion coefficient, and
t is the time.

Eq. (7.36) may be rewritten for two conditions: for large and for small particles. In the case of the large particles, Stokes Law applies [see Eq. 7.37)] and in the case of small particles, in which the particle is small relative to the mean free path of the gas, the Cunningham correction is applied [see Eq. (7.38)]:[22]

$$\overline{\Delta}_x^2 = \frac{2RTt}{N'3\pi Nd} \qquad (7.37)$$

and

$$\overline{\Delta}_x^2 = \frac{2RT(1 + 2A\lambda/d)t}{N'3\pi\eta d} \qquad (7.38)$$

in which

N' is Avogadro's number,
d is the mean free path,
η is the gas viscosity, and
A is a constant.

A Brownian rotational motion is also observed and has been accounted for by:[23]

$$\overline{\Delta}_r^2 = \frac{2RT}{N'\pi d^3 \eta} \qquad (7.39)$$

The mean free path λ is not as simple to conceive as some diagrams in the literature would have one believe. The path is not jagged but consists of a series of curves. A curve of such type is illustrated in Fig. 7.12.[24] The mean free path is defined as the average distance a particle travels before its velocity has turned through 90°.[24]

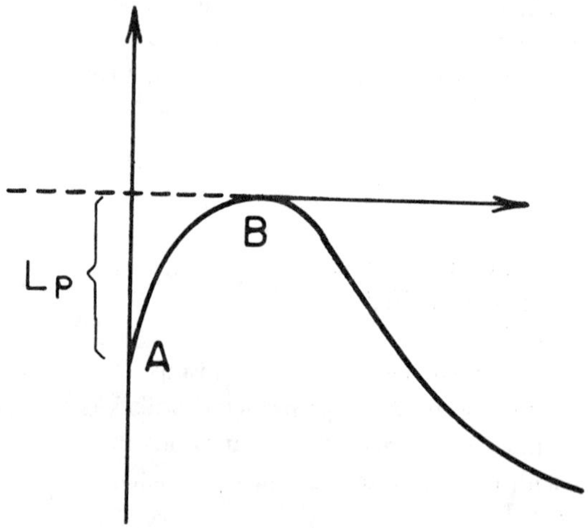

Fig. 7.12 Particle mean free path.[24]

To illustrate the effect of particle size upon the mean free path length, the data are presented in Table 7.8.[24] For the calculation, unit density of particles and room temperature were assumed.

The deposition of small (less than 0.5 μ diameter) particles in the lower reaches of the lungs is due to Brownian motion.[26]

Table 7.8

MEAN FREE PATH OF BROWNIAN MOTION VS PARTICLE DIAMETER CALCULATED AT ROOM TEMPERATURE[24]

Particle diameter micron	cm $\times 10^{-6}$	D cm^2/s
1	1.58	2.75×10^{-7}
0.1	1.19	6.81×10^{-6}
0.01	3.01	5.23×10^{-4}

Diffusiophoresis and Thermophoresis

In an isothermal gas mixture a suspended particle moves under the influence of existing concentration gradients. In a binary mixture with a zero (vanishing) average molecular velocity, the particle tends to move along with the flux of the heavier gas molecules. This phenomenon is termed diffusiophoresis. It has been observed experimentally.[27] A special case involving the denudation of aerosol particles around an evaporating water drop was also reported.[28]

Thermophoresis has been known for many years now[29] and is observed as the motion of suspended particles towards the region of lower temperature in a fluid containing a thermal gradient. Under suitable illumination, a dust free zone can be observed around a hot body in a dust laden atmosphere. The phenomenon is used in thermal precipitators.[24]

In practice, there are many examples of natural and industrial processes in which water is either being evaporated or condensed. In these cases there are both water vapor and thermal gradients existing which may well be additive in their effects on particle velocities.[30]

A discussion of the potential role of diffusiophoresis in scavenging fine particles from the atmosphere has been presented in the literature.[30] It has been observed that the total "wash out" of radioactive dust* due to rain drops is due to four possible contributors: Brownian diffusion, direct condensation on particles, impaction, and diffusiophoresis. The assessment of the contribution to the wash out ratio W (defined as the specific activity of the rain in ΓΓC/Kg divided by the specific activity of the air in ΓΓC/Kg) indicated that if $W = 300$ (say) then W_{BD} due to Brownian diffusion could be as high as 300 in the case of rain drops of 10 μ radius and particles of 0.05 μ after ten hours. By comparison, W for diffusiophoresis was estimated at about one. It was observed that the contribution of diffusiophoresis might be increased if a "cycle" process occurs. This cycling would consist of evaporation-condensation-evaporation, etc., on the same nucleus.[30] An experimental demonstration using a simple condenser system was reported in which the efficiency of a combination of diffusiophoresis and thermophoresis was found to be quite high. The results are shown in Fig. 7.13.[30] In a separate experiment, the thermophoretic and diffusiophoretic forces were shown to be additive in their effects. This is illustrated in Fig. 7.14.[30]

The theory of the two phenomena has been developed and the velocities attained by the particles depends upon their size. There are thus small par-

*Present in the atmosphere due to nuclear explosion.

ticle and large particle theories. The respective equations are given in Table 7.9 with appropriate symbols explained therein.[31]

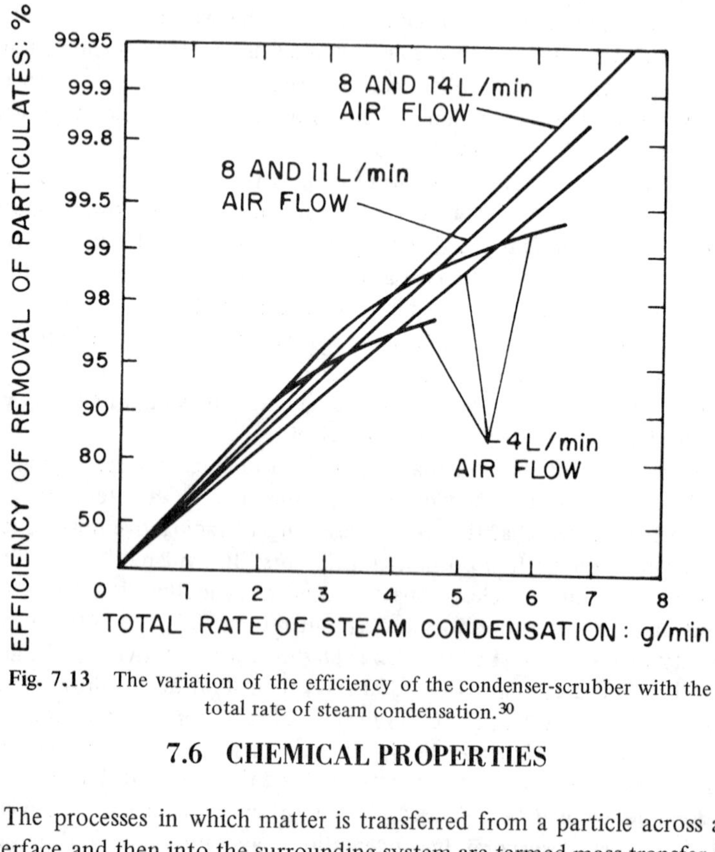

Fig. 7.13 The variation of the efficiency of the condenser-scrubber with the total rate of steam condensation.[30]

7.6 CHEMICAL PROPERTIES

The processes in which matter is transferred from a particle across an interface and then into the surrounding system are termed mass transfer.[33] The particle may be either solid or liquid but in this discussion only solid particles are treated. It is convenient to divide the processes into different types: dissolution of a solid in a liquid, sublimation, and burning-reaction of a solid with a gas. It would be overloading the discussion to include remarks about the order of a chemical reaction, etc. The interested reader may refer to standard texts on this subject matter. It should be observed that with the implementation of energy policies world-wide, there is room for increasing funding in this area at both the fundamental and applied levels.

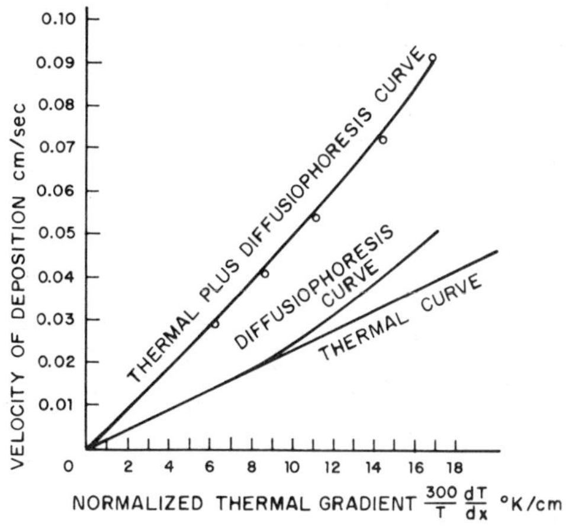

Fig. 7.14 Velocity of particles in superimposed thermal and water-vapor pressure gradients.[30]

Reaction Kinetics of Powders

Many models for the reactions of powder particles proposed over the years[32–41] can be conveniently divided into two main categories: The first is linear kinetics in which the reaction rate is proportional to the area of the reaction interface; the latter moves at a constant rate. The second is the parabolic kinetics, in which the rate of reaction is controlled by transportation of reacting species through a (more or less) coherent coating of the reaction product.

The terms linear and parabolic are intended to convey the relationship between the reaction rate and the fraction reacted. A listing of some of the equations resulting from the models previously referred to has been examined and it has been shown empirically that an equation of the general form of Eq. (7.45) satisfies most of the equations.[42]

$$\frac{dF}{dt} = K'zt^{-z}(1 - \beta F)^m \tag{7.45}$$

in which

F is the fraction reacted,
t is the time,
z depends upon the kinetic regime,

Table 7.9

THEORETICAL PARTICLE VELOCITIES FOR THERMOPHORESIS AND DIFFUSIOPHORESIS[31]

	Thermophoresis	*Diffusiophoresis*
Small Particles	$V = -\frac{1}{5(1+\pi a/8)} \frac{\lambda_{trans}}{p} \nabla T$ (7.40) where a = coefficient of thermal reflection, λ_{trans} = translational part of gas heat conductivity, ∇T = temperature gradient, and p = pressure in gas.	$V = -\frac{m_1 - m_2}{m + \sqrt{m_1 \cdot m_2}} D \nabla \gamma_1$ (7.41) where m_1, m_2 are molecular masses of a binary mixture, and $\nabla \gamma_1$ is gradient of relative concentration
Large Particles	$V = -\frac{2\lambda}{2\lambda + \lambda_i} \sigma \frac{\lambda}{p} \nabla T_\infty$ (7.42) where λ = heat conductivity in gas, λ_i = heat conductivity of particle, ∇T_∞ = uniform temperature gradient, and $\sigma \cong 1/5$, a constant For particles with low conductivity $[(\lambda_i/\lambda) \approx 10]$ this Eq. (7.42) is satisfactory.	$V = -\sigma_{12} D \nabla \gamma_{1\infty}$ (7.43) $\sigma_{12} = A \frac{m_1 - m_2}{m_1 + m_2} + B \frac{d_1 - d_2}{d_1 + d_2}$ (7.44) σ_{12} = empirical expression for the diffusion slip factor, A, B = constants, $\nabla \gamma_{1\infty}$ = relative concentration of gradient of species 1 at $r = \infty$ from particle, d_1, d_2 are molecular diameters (gas)

K' is a constant,
m is an index of reaction that depends upon the kinetic regime and the particle shape, and
β is a dimensionless factor.

The models that Eq. (7.45) appears to represent satisfactorily are listed in Table 7.10.[34,42]

Table 7.10[34]

$f(\alpha) = Kt$

Function		*Equation*
$D_1(\alpha)$	parabolic $z = 1/2$	$\alpha^2 = kt$
$D_2(\alpha)$		$(1-\alpha)\ln(1-\alpha) \quad \alpha = kt$
$D_3(\alpha)$		$[1-(1-\alpha)^{1/3}]^2 = kt$
$D_4(\alpha)$		$1 - 2/3 - (1-\alpha)^{2/3} = kt$
$F_1(\alpha)$	linear $z = 1$	$-\ln(1-\alpha) = kt$
$R_2(\alpha)$		$1-(1-\alpha)^{1/3} = kt$
$R_3(\alpha)$		$1-(1-\alpha)^{1/3} = kt$
zero order		$\alpha = kt$

Refer to Eq. (7.45)

Dissolution of a Solid in a Liquid

As the particle size decreases, the solubility of the particle increases. This has been known for a long time now.[43,44] A number of expressions has been developed to relate particle solubility to particle characteristics. These refer to specific substances: NaCl[45] and KCl.[46] An equation from the latter reference is:

$$\log \frac{S}{S_o} = \frac{2\sigma_m M}{2.303\rho RT} \frac{1}{r} \tag{7.46}$$

in which

S is the solubility of the particle,
S_o is the normal solubility,

σ_m is the mean interfacial tension,
M is the molecular weight of the material,
ρ is the density of the material, and
r is the radius of the particle.

Taking into consideration the variabilities within the system in which the particle is dissolving, the expression for the dissolution process may be stated in terms of the particle diameter and the diffusion coefficient utilizing the dimensionless Sherwood, Reynolds, and Schmidt numbers as follows:

$$Sh = a(Re)^q (Sc)^p \tag{7.47}$$

in which

Sh is the Sherwood number and is $k'd_v/D_L$; this is a measure of the local loss of substance;
Re is the Reynolds number;
Sc is the Schmidt number and is $\mu_L/\rho_L D_L$;
a is a coefficient;
p and q are exponents;
d_v is the volume mean diameter;
k' is a dissolution rate coefficient;
D_L is the diffusivity;
μ_L is the liquid viscosity; and
ρ_L is the liquid density.

Some typical values for D_L for particles are listed in Table 7.11.[13]

Table 7.11

D_L VALUES FOR SOME PARTICLE FLUID SYSTEMS

Particle in Liquid	D_L cm^2/s $\times 10^5$, *diffusivity*
Benzoic acid/water 25°C[47]	1.9
Cinnamic acid in water at 30°C[47]	1.4
NaOH in water 25°C[48]	1.1

The simple processes described in Eq. (7.47) are as follows:

1. the atom/molecule leaves the surface of the particle, entering the boundary layer;

2. motion through the boundary layer, in which there is no flow, or laminar flow, is by diffusion along a concentration gradient; and
3. once outside the boundary layer, the species are whisked off by the motion of the surrounding liquid.

In practice, it is recommended that a pilot study be conducted in order to establish the better system to use and its optimum operating conditions.[13] Choices as to the system to be used include: should the process be a batch process or should it be a continous one? should agitation be provided? (it usually is) and if so, what stirrer, velocity, and so on should be employed? what size of vessel must be used? In general, Eq. (7.47) may be used in conjunction with the experimentally acquired data in order to evaluate the system performance. Some data obtained by using the particle diameter as the characteristic length and the particle motion as the velocity is shown in Table 7.12:[49]

Table 7.12

DISSOLUTION OF SOLID PARTICLES IN MOVING FLUID[49]

System	*Re*	*a*	*q*	*p*
Benzoic acid/H_2O[49]	100–1000	0.95	0.50	0.33
Benzoic acid/H_2O[49]	500–8000	0.58	0.50	0.33
Benzoic acid/H_2O[49]	1000–100,000	0.94	0.50	0.31

Agitation of the vessel can also be dealt with using Eq. (7.47) if one uses the velocity of the stirring device as being the velocity term in *Re* and the diameter of the stirrer is used as the characteristic length term.[52] A satisfactory treatment of this topic in a theoretical manner is still awaited.[13]

Fixed beds introduce an added complication into an already cloudy theoretical picture. They are, however, used but their use is best retained for the dissolution of larger particles simply because smaller ones create too much of a pressure drop. Fluidized beds suffer from one disadvantage that as the particles dissolve and decrease in size they can become part of the effluent from the bed. Equations for a fixed bed and for a fluidized bed are given as[3]

Fixed Bed[53]

$$(Sc)^{0.58} \cdot \frac{k'}{\rho_L u_{OL}} = \begin{vmatrix} 2.40\,(Re'_M)^{-0.66} & 0.08 < Re < 125 \\ 0.442\,(Re'_M)^{-0.31} & 125 < Re < 5000 \end{vmatrix} \tag{7.48}$$

Fluidized Bed[54]

$$Sh = 2.0 + 1.031\left(\frac{du_{OL}\rho_L(1-\epsilon)}{\mu_L}\right)^{0.50} Sc^{0.33} \tag{7.49}$$

where

$Re'_M = du_{OL}\rho_L/\mu_L\epsilon$,

u_{OL} = empty column velocity (superfixed),

ϵ = bed porosity, and

ρ_L = liquid density.

Sublimation

When a solid is heated and transforms directly into a vapor, sublimation is said to have occurred. A well-known example of such a substance is solid CO_2 which simply vaporizes when heated at atmospheric pressure. The important factor that determines under what conditions sublimation might occur is the triple point and its location as shown in Fig. 7.15.

Sublimation occurs when the temperature and pressure are below the triple point. In practice most substances do not sublimate when heated at atmospheric pressure. However, sublimation may be achieved if the pressure is lowered. Simple sublimation involves the passage of the sublimated vapor from one evacuated chamber to another (a condenser chamber). Entrainer sublimation is used when the vapor pressure of the vapor produced is low. In this method, a carrier inert gas moves the desired vapor because the vapor does not move unaided with a sufficient rate.

Theoretically, sublimation is treated in a manner analogous to evaporation and the same equation can be used to describe both processes with the proviso that sublimation may effectively reduce the temperature of the particle well below that which is estimated because the process is endothermic. If care is taken to establish the most likely temperature, the equation for rate of sublimation is:[55]

$$-\frac{dM}{dt} = \frac{4\pi MD}{RT}(\delta P_T - P)\frac{r}{D/rv\alpha + 1} \tag{7.50}$$

in which

- D is the diffusion coefficient for vapor in the gas phase,
- P is the particle vapor pressure at an active temperature,
- P_T is the particle vapor pressure at an ambient gas temperature,
- δ is the relative saturation of the particle vapor in gas,
- r is the particle radius,
- v is $(RT/2\pi M)^{1/2}$,
- α is a sublimation coefficient,
- M is the mass, and
- t, is the time.

If there is some turbulence in the gas, a modified form of Eq. (7.47) may be used. In general, data for D is quite limited. It has been noted that in the case of NaCl in Argon at 1272°C, the value of D at 1.3 atmospheres is 0.85 cm^2/s while at 0.221 atmospheres, D is 3.5 cm^2/s.[56]

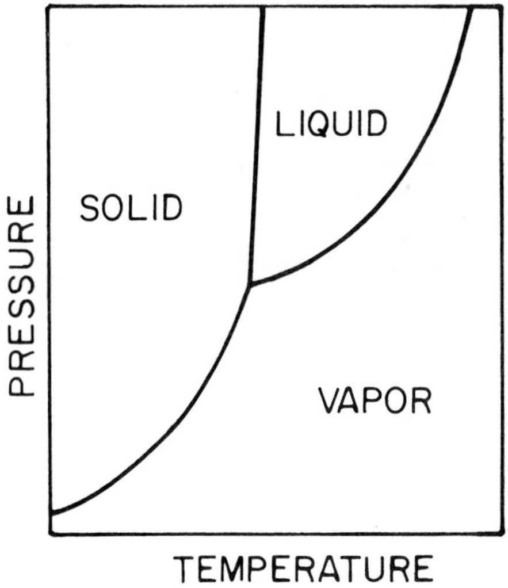

Fig. 7.15 Phase diagram for sublimation.

7.7 ADHESION AND DEPOSITION OF PARTICLES

The term adhesion describes the more or less indiscriminate attachment of particles to surfaces (including each other's). The term deposition describes the processes by which the particles reach these surfaces.

Although all molecular and atomic forces have the same basic origins, it is conventional to treat them as different from and independent of each other.[57] Four items are discussed: Van der Waals forces, Ionoelectrostatic forces, surface forces, and other factors.

Van der Waals Forces

Although the atom or molecule may be considered to be neutral in an average sense, the fluctuation of positions of the electrons at any particular moment gives rise to the formation of a dipole which, in turn, is attracted to other dipoles nearby. This so-called London-Van der Waals force is expressed by:[58]

$$F = \lambda r^{-7} \tag{7.51}$$

in which

- λ is the Van der Waals constant for attraction, and
- r is the distance between the two molecules that are attracted to each other.

This equation is not strictly applicable at very short intermolecular distances nor is it applicable at longer range. The approach treats intermolecular forces as being electrostatic in origin. The constant λ includes the polarizability of molecules, a measure of the distortion of a molecule caused by an electric field. These measures are not known. The quantum electrodynamic approach produces an equation of the following form:[59]

$$F = Kr^{-8} \tag{7.52}$$

in which

- F is the attractive force between two molecules,
- K is a constant (which also, alas, contains the polarizability property), and
- r is the intermolecular distance apart.

Both Eqs. (7.51) and (7.52) have the disadvantage that they only apply to isolated molecular species. However, it is important to be able to apply a theory to the problem of the attractive force between particles each of which consists of many molecular species. This theoretical problem has been reported elsewhere[60] and is briefly reviewed here.

The interaction of particles may be regarded as occurring through the medium of the fluctuating electromagnetic field within the particle. This field extends beyond the bounding surface of the particle. In part it exists as standing waves that are damped exponentially as one moves away from the particle. Even at absolute zero, these waves do not vanish as they are associated with the zero point vibrations of the radiation field. The calculation of the fluctuating electromagnetic field between the two particles is based upon the introduction into the Maxwell equations of a "random" field. The starting field is therefore:

$$\text{curl}\, E = i\frac{\omega}{c}H \tag{7.53}$$

and

$$\text{curl}\, H = -i\frac{\omega}{c}\epsilon E - i\frac{\omega}{c}K \tag{7.54}$$

in which

$e^{-i\omega t}$ is the time factor, and

K is the random field.

From this starting point, the calculation of the force of attraction is made. The resultant is further applied to the case of small interparticle distances and large interparticle distances, respectively. The formulas obtained are complicated but they contain no quantities that cannot be measured.

For the case of small separations

$$F = \frac{\hbar}{8\pi^2\rho^3}\int_0^\infty \left(\frac{\epsilon(i\xi)+1}{\epsilon(i\xi)-1}\right)^2 d\xi \tag{7.55}$$

in which

F = attractive force,

ρ = distance of separation, and

ω = $i\xi$ for imaginary values of ω.

In the case of large interparticle separations, the force is inversely proportional to 1^4. Calculations of the attractive forces in the case of large objects have been compared with experimental data and the results are shown in Figs. 7.16 and 7.17.[61] The theory certainly seems to fit the data obtained, even though the calculations are complex and tedious. There is an added advantage because in the case of two molecules a short distance apart, the force is found to be proportional to r^{-7} and at slightly larger distances ($\approx$ 200 Å) it is proportional to r^{-8} . This relationship means that

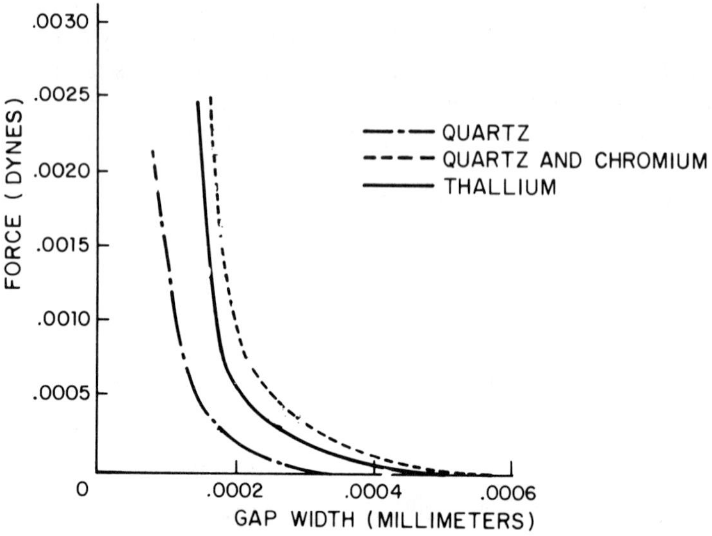

Fig. 7.16 Attractive force versus gap width, theory versus experiment (a) Flat plate and spherical lens (10 cm radius) (based on reference 61).

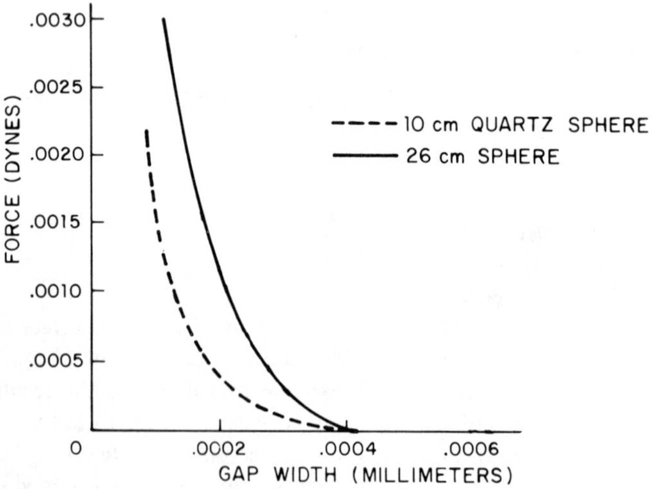

Fig. 7.17 Attractive force versus experiment (b) flat plate with quartz spheres (10 cm radius) (based on reference 61).

the previous results drop out as special cases. An alternative method is that assuming that the energy of interaction of the dispersive forces is additive, use the following integral:[62]

$$E = -\int_{v_1}\int_{v_2} \frac{q_o^2 dv_1 dv_2}{r^6} \tag{7.56}$$

in which

E is the total energy of interaction in ergs,
v_1, v_2 are the total volumes of particles 1 and 2 respectively,
dv_1, dv_2 are differential elements of the two particles,
q_o is the number of atoms per cc, and
r is the distance apart.

In the case of two ideally smooth spheres of differing radius, the force is represented by:

$$F = \frac{\pi^2 q_o^2 \lambda}{12r^2}\left(\frac{d_1 d_2}{d_1 + d_2}\right) \tag{7.57}$$

And if it is assumed that the diameter of one of the spheres is infinite this relationship reduces to that for a particle attracted to an infinite flat plate

$$F = \frac{\pi^2 q_o^2 \lambda d_1}{12r^2} \tag{7.58}$$

In terms of A, the Hamaker constant

$$F = \frac{A d_1}{12r^2} \tag{7.59}$$

in which

r is the distance between the particles, and
A is Hamaker's constant and is $\pi^2 q_o^2 \lambda$. "A" has been determined in various laboratories and ranges from 0.05–10.[57]

When the two particles come into direct contact, Born repulsion is exhibited.

Iono-Electrostatic Forces

If the particles are submersed in a liquid, the situation is more complicated for two reasons: First, there may exist concentration gradients of dissolved species within the liquid. This condition is important in the case of ionic species being present because of their proclivity for the formation of layers. These layers cause repulsive forces, the magnitude of which is dependent upon the composition in the liquid and also the surface potential of the particles. Second, the liquid structure adjacent to the particle may differ from the bulk liquid structure and this condition may produce results that vary for "unexplained" reasons.

It is for this reason that there are two minima on the force-distance diagram instead of one. These forces have been termed iono-electrostatic forces. The conventional force distance diagram in Fig. 5.3 is therefore changed to one with two minima as shown in Fig. 5.4.

In the limiting case in which two spherical particles of equal size contact at one point, as the moisture content in the space between the particles is reduced, the force acting at the point of contact tends towards the upper limiting value of[64]

$$f = 2\pi\sigma R \tag{7.60}$$

in which

σ is the surface energy at the water-air interface, and
R is the radius of the spherical particles.

Alternatively, where the distance between the two particles is of the order of the intermolecular distance, the attractive force N_m can be approximated by[64]

$$N_m = 2\pi\sigma_{1-2} R \tag{7.61}$$

in which σ_{1-2} is the surface tension at the interface between the particle and the liquid.

The iono-electro repulsion has been shown to be[64]

$$N_i = \pi R\beta/\delta \tag{7.62}$$

in which

N_i is the iono-electronic (repulsion) force,
β is the $aD(kTze)^2$,
δ is the intermolecular distance,
a is a coefficient,
D is the dielectric permeability,
e is the electron charge, and
z is the ion valence.

Remembering that the convention has repulsion positive and attraction negative, the force barrier that in effect represents the tensile strength of the bond between the two particles in the liquid due to the intermolecular and iono-electrostatic force sum is expressed by

$$P = N_m + N_i = (-2\pi\sigma_{1-2}R + \pi R\beta/\delta)\,\frac{1}{\pi R^2}$$

and therefore

$$P = \frac{1}{R}\left[\left(\beta/\delta\right) - 2\sigma_{1-2}\right] \tag{7.63}$$

It should be noted that the strength depends upon the inverse of particle size.[64] Consideration of the possible effects of iono-electrostatic forces has been reported for two basic classes of particulate matter classified as:

1. primary minerals such as quartz, feldspar, and mica,[65] and
2. secondary minerals such as clay type insoluble in water.

The mechanical behavior of the primary mineral type is said to be mainly due to particle shape as a variable. The clay minerals exhibit properties that have been ascribed to the iono-electrostatic forces heretofore described. For example, at one extreme kaolinite has a crystal structure that consists of assymetrical layers that are about 15 Å apart. The other extreme is represented by montmorillonite which consists of symmetrical layers. In kaolinite, hydroxyls and oxygens face each other from within adjacent layers. In montmorillonite the boundaries of adjacent layers are occupied by oxygens of the same sign. The within-layer bonding is relatively strong in both minerals. These differences in crystalline layered structures can account for the very different behaviors of these two minerals in the presence of water. Kaolin particles swell to a certain extent when wetted and montmorillonite particles just keep on swelling until they can decompose into their elementary layers.[64]

The origin of these effects is described in Fig. 7.18.[64]

A number of cases have been considered for the iono-electrostatic interaction of flat parallel surfaces:[66]

1. If the charges are like and the potentials about equal, repulsion effects are observed at any distance.
2. If the potentials are different, but of like sign, the repulsion first increases as the surfaces approach each other, reaches a maximum, and thereafter decreases until a mutual attraction is experienced, ultimately leading to a Born repulsion.

3. If the potentials are of unlike signs, the surfaces attract each other as they approach from infinity until the Born repulsion occurs.

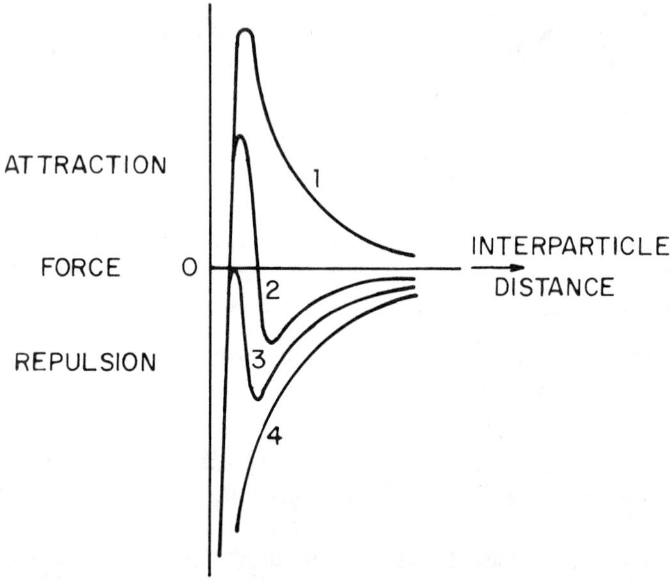

Fig. 7.18 Force-distance diagram. Curve 2–typical of kaolinite. Cuves 3 and 4–typical of montmorillonite.[64]

Curve 1 in Fig. 7.18 may be considered typical of a set of particles in an aggregated state. Introduction of a new and different ionic medium changes this condition. Curve 2 in Fig. 7.18 which may be considered to be representative of kaolinite shows that as the particles approach each other the mutual repulsion increases to a maximum level followed thereafter by a growing attraction, eventually leading to the Born repulsion. Curves 3 and 4 may be considered representative of Montmorillonite. In this case, the particle interactions are usually mutually repulsive except for curve 4 which represents a case in which both the intermolecular and iono-electrostatic forces in the fluid are repulsive. In this same paper, it was demonstrated that in a particulate in which the shape is plate-like, the plastic forces of adhesion are larger than they are in the case of spherical particles. This therefore ascribes the plasticity of kaolinite, etc., at least as partly due to the particle shape.[64]

A very interesting report points out that the ionic strength of a suspension of particles can be controlled to permit particle adsorption and desorption on a collector surface.[67] At low ionic strengths, a large maximum in the energy-distance profile effectively prevents particle adsorption. They are therefore carried through the system and effluxed. At high ionic strengths, there is no large maximum and so the particles can adsorb rapidly on the surface of the collector. However, now the residence time within the system is expected to be rather long as compared with the fluid residence time.

The interesting possibility emerges when the ionic strengths are intermediate such that the particles can adsorb and desorb and the residence time can be selectively controlled. The possibility of separation of particles, based upon their respective bahavior in such a system, is raised. It is to be expected that particle characteristics of significance in this method would include particle size and particle charge density.[67]

Electrostatic Forces

A calculation of the magnitude of electrostatic forces on particles indicates that they may be quite small.[57] For example, if two particles of 1 μ diameter separated by about 10 Å are considered, then the approximate level of the elctrostatic, gravitational, and Van der Waals forces on this single 1-μ diameter particle are shown in Table 7.13.

Table 7.13

FORCES ON SMALL PARTICLES[57]

Electrostatic force	=	5.2×10^{-3}	dynes
Gravitational force	=	5.1×10^{-19}	dynes
Van der Waals force	=	0.4	dyne

when

the charge is 15 e.s.u ,
the particle diameter is 1 μ,
the separation is 10 Å,
the density is 1 g/cc, and
the Hamaker constant is 5×10^{-10} ergs.

The electrostatic force is therefore 10^{-3} times the Van der Waals force.

On the other hand, it appears to be a very well-known "fact" that electrostatic charges on particulates play a large role in the adhesion of particles. Some statements which attest to this fact follow:

"In our first experiments in 1951 we measured forces that were about 5,000 times as large as predicted by any theory."[61]

"... high values for A ... were recently attributed to the presence of electrostatic charges and lower figures were subsequently obtained."[57]

A number of cases is reported in which the electrostatic charge is said to be industrially important: it increases the cohesion of a layer of dust formed in an electrostatic precipitator, and it is well known to be a nuisance in the sieving of dielectrics.

Although contact charging is due to mechanism(s) about which there is some doubt,[68] it is a fact that particles in the cloud are charged and opposite charges on different particles coexist in the cloud.[69]

It was observed that, in the case of a particle of 10-μ diameter with a charge of 600 e.s.u., the charge would have to be concentrated within 0.01 μ of the point of contact so that the electrostatic attraction would equal the molecular attraction.[70] This contribution underlines the observation that much more needs to be known about the relationship between the internal structure of particles and their charge distribution.[57] Of course, this also serves to underline the fact that little is known about the internal structures of matter in finely divided form.

Other Factors Influencing Surface Forces

Of concern in laboratory investigations, surface cleanliness can usually be obtained so that, as far as is humanly possible, the experimental conditions under which measurements are made matches the assumptions of the model being used. In a practical situation it is highly unlikely that such ideal conditions exist and in wet processing it is expected that such correlations might be transient and recurring. However, there is little information in the literature to support this type of conjecture.

Of interest here and relevant to the problem of contact charging is the area of contact between particles of differing shapes and the flatness of the surface.[71] But also of concern is the number of corners possessed by each different shape. The contact areas are listed in Table 7.14.[71]

7.8 PARTICLE CHARACTERISTICS IMPORTANT IN DEPOSITION[72]

The particle characteristics and the system characteristics can be de-

Table 7.14

CONTACT AREA OF PARTICLE DEPENDING UPON SHAPE[71]

Shape	*Area of face* $cm^2 \times 10^{-8}$
Tetrahedron	0.633
Cube	0.650
Octahedron	0.464
Dodecahedron	0.278
Icosahedron	0.166
Sphere	0

scribed in a simple and effective way in order to establish a qualitative understanding of what is going on in deposition processes. A number of mechanisms will be considered: sedimentation in laminar flow, settling from a turbulent environment, electrical deposition, diffusive deposition, inertial deposition, and combined inertial/Brownian deposition. The Reynolds number of a particle is discussed in Chapter 5. Also of interest is the definition of the particle's dimensionless parameter P:

$$P = \frac{\tau v_o}{L} \tag{7.64}$$

in which

τ is termed the relaxation time of the particle,
v_o is termed the characteristic velocity of the particle, and
L is termed the characteristic length of the system.

Essentially the relaxation time indicates how long it takes for the particle to adjust to an applied force. For example, if a particle is at rest and then is subjected to an air stream, in τ seconds this particle reaches $1/e$ of the velocity of the applied air stream. The characteristic velocity is related to the relaxation time in the following way:

$$d_s = \tau v_o \tag{7.65}$$

in which d_s is the stopping or stop distance.

The modeling behind this equation is that of the particle being projected into a still mass of air with an initial velocity v_o and coming to rest at a distance d_s from the point of entry (or projection).

In the case of particles that are heavy enough to settle, the terminal velocity may be stated as

$$v_s \approx \tau g \tag{7.66}$$

in which

v_s is the terminal velocity, and

g is the acceleration due to gravity.

Where the velocity of the flow is low, Stokes law applies and the relaxation time can be stated as being

$$\tau = \frac{2a^2 \rho}{9\eta} \tag{7.67}$$

in which

a is the particle radius,

ρ is the particle density, and

η is the fluid viscosity.

Some exact values of relaxation time for spheres of unit density in air at 20°C and atmospheric pressure have been calculated exactly and are shown in Table 7.15.[72]

If two particles possess identical values of $\dot{P}$ and Re, then in a laminar flow system they have dynamically similar trajectories, although this is complicated if some gravitational settling is involved and cannot be ignored. In a turbulent flow regime, one can consider that small eddies, i.e., small compared with d_s which is the stopping distance, have a negligible effect upon the passage of the particles. When the eddies are large as compared with d_s, the particles can become entrapped in the vortices.

Settling of Particles

This is by far the most common process by which finely divided matter is deposited on surfaces. It can occur in still atmosphere, in an atmosphere moving with laminar flow, and in an atmosphere moving with turbulent flow.

In the case of a still atmosphere, when the particles have travelled downwards to a sufficient extent, they will have reached their terminal velocity. In the Stokes Law region, the gravitational and viscous forces can be equated thus:

$$(m_1 - m_2)g = 6\pi\eta r v \tag{7.68}$$

in which

Table 7.15

RELAXATION TIMES AND TERMINAL VELOCITIES OF SPHERES OF UNIT DENSITY IN AIR AT ONE ATMOSPHERE AND 20°C[72]

Diameter (μ)	*Relaxation time* τ (s)	*Velocity of fall* v_s (cm/s)	
0.1	8.67×10^{-8}	0.000085	
0.2	2.295×10^{-7}	0.000225	
0.3	4.28×10^{-7}	0.00042	
0.5	1.02×10^{-6}	0.0010	Approximately viscous flow region (0.5–50)
1	3.57×10^{-6}	0.0035	
2	1.305×10^{-5}	0.0128	
3	2.802×10^{-5}	0.0275	
5	7.95×10^{-5}	0.078	
10	3.057×10^{-4}	0.30	
15	6.93×10^{-4}	0.68	
20	1.223×10^{-3}	1.2	
25	1.938×10^{-3}	1.9	
30	2.752×10^{-3}	2.7	
50	7.34×10^{-3}	7.2	
100	2.548×10^{-2}	25	
150	4.690×10^{-2}	46	
200	7.135×10^{-2}	70	
300	1.172×10^{-1}	115	
500	2.04×10^{-1}	200	
1000	3.925×10^{-1}	385	

For other cases see references 72, 73.

m_1 is the particle mass,
m_2 is the volume of particle $\times$ medium density,
η is the medium viscosity,
r is the particle radius, and
v is the terminal velocity.

During the period of acceleration before terminal velocity is reached Eq. (7.68) is not correct and should be rewritten as

$$(m_1 - m_2)g = (6\eta\pi r)v + m_1 \frac{dv}{dt}$$

in which $m_1 \dfrac{dv}{dt}$ is the force to produce acceleration or

$$\left(1 - \frac{m_2}{m_1}\right)g = \left(\frac{6\eta\pi r}{m_1}\right)v + \frac{dv}{dt}$$

Letting

$$\left(1 - \frac{m_2}{m_1}\right)g = G$$

and

$$\frac{6\eta\pi r}{m_1} = K$$

then

$$G = Kv + \frac{dv}{dt} \tag{7.69}$$

The solution for this equation is

$$v = \frac{G}{K}(1 - e^{-Kt}) \tag{7.70}$$

so

$$v = \frac{dx}{dt}$$

in which x is the displacement. Therefore

$$\frac{dx}{dt} = \frac{G}{K}(1 - e^{-Kt})$$

and

$$\int dx = \int \frac{G}{K}(1 - e^{-Kt})dt$$

which has the solution

$$x = \frac{G}{K}\left(t + \frac{1}{K}\, e^{-Kt}\right) + C$$

When $x = 0$, $t = 0$, and C can be solved for:

$$x = \frac{G}{K}\left(t + \frac{1}{K}\, e^{-Kt}\right) - \frac{G}{K^2} \qquad (7.71)$$

This equation is difficult to solve for t when x is given, but it can be written as

$$\frac{G}{K}\left(t + \frac{1}{K^2}\, e^{-Kt}\right) - \frac{G}{K^2} - x = 0 \qquad (7.72)$$

Newton's method which lends itself to computer use results in a close approximation of

$$t = f(x)$$

since

$$t_2 = t_1 - \frac{f(t_1)}{f'(t_1)}$$

or

$$t_2 = t_1 - \frac{G/K(t_1 + e^{-Kt}/K) - G/K^2 - x}{G/K(1 - e^{-Kt})}$$

or

$$t_2 = t_1 - \frac{t_1 + e^{-Kt}/K) - 1/K - (K/G)x}{(1 - e^{-Kt})} \qquad (7.73)$$

Substitute a logical estimate of t for t_1, and then compute t_1. Let t_1 equal the computed value of t_2; then compute a new t_2. Repeat this process until $\left|\frac{(t_2 - t_1)}{t_2}(100)\right|$ equals or is less than the maximum percentage of error.

In the case of laminar flow of the medium or atmosphere, provided that there is no shadowing effect, the gas motion has no effect upon the deposition per unit area because the air flow merely produces a lateral movement.

The rate of deposition is expressed by:

$$\text{Rate} = cv_s \tag{7.74}$$

in which

c is the particle concentration close to the surface, and

v_s is the terminal velocity, as before.

When there is a shadowing effect, methods have been worked out that can be used in calculations.[74] The actual extent of deposition on an upward facing surface depends upon the rate of flow of the aerosol: the more inclined the surface, the less the rate of deposition (it is proportional to the cosine of the angle with the horizontal); the higher the air stream speed, the less deposition.

In the case of deposition in conditions of turbulence, two situations can be considered. In the first situation, the air is turbulent and the particulate must be considered to be thoroughly mixed so that no concentration gradient exists. This type of situation exists in cases of deposition from turbulent air in enclosures. Whatever the initial size distribution, the suspension tends towards a specific size distribution in which the frequency of occurrence of particles having velocity v_s out of a total of N particles is:[75]

$$\frac{1}{N}\frac{dN}{dv_s} = \frac{\alpha^{n+1}}{\Gamma(n+1)} v_s{}^n \exp(-\alpha v_s) \tag{7.75}$$

If n is ½, Eq. (7.75) becomes a special case of the Rosin-Rammler distribution. If n is −½, Eq. (7.75) describes a Gaussian distribution of diameter for Stoke's law particles.[73]

Deposition in a Roadway

A good example of deposition in a real situation is the deposition of coal dust in a roadway of a coal mine.[76,77,78]

The concentration of airborne particles decreases as the air moves along the duct or roadway because of deposition which increases exponentially.

The initial concentration c_o is given as

$$c_o = \frac{n}{whd} \tag{7.76}$$

in which

n is the number of particles introduced into the roadway per second ("emitted"),

w is the wind velocity,

h is the height of the roadway, and
d is the width of the roadway.

At a distance x along the roadway from the source, the concentration is expressed as

$$c = c_o \exp \frac{(-v_s x)}{wh} \tag{7.77}$$

The rate of deposition is cv_s [Eq. (7.74)] and so, in this case,

$$\text{Rate} = cv_s = c_o v_s \exp\left(\frac{-v_s}{w} \cdot \frac{x}{h}\right) \tag{7.78}$$

Eq. (7.78) is used to calculate the loss by settling from the air stream of coal dust in a typical underground roadway of a coal mine. Assuming spheres of unit density, the calculated results are given in Table 7.16.[72]

It has been observed from the experience in coal mines that the reason variously sized coal particles seem to deposit at the same time is that the difference in size is really not great throughout most of the distance. The coarser particles drop out of the air stream early, and also they occur less frequently in the detriti generated by the mining operations.[72]

Deposition from the Atmosphere

The second situation in which air turbulence is a factor is that of deposition from the atmosphere. Because the atmosphere is so very extensive and specifically because the extent of turbulence in the troposphere is limited, the conditions are somewhere between complete turbulence and settling in still air. The effect of wind speed and particle size upon the deposition of an aerosol is illustrated in Table 7.17.[79] See also reference 72.

A higher proportion of the larger particles is carried downstream by the wind as compared with the flow along a duct (see Table 7.16 for comparison). Also, quite a high proportion of the particles are dumped quite near the source. Of course, meteorological conditions markedly affect the pattern of deposition. For example, less turbulence promotes a more uniform settlement of the particles downwind. The deposition from smoke clouds can reach areas which are distant from the source. One can quote from the well-known folklore on this aspect: the plume from Canadian smelters may impinge on New York air space if the wind is (un)favorable,[80] the chimneys in England can deposit their particulate product on the tracts of the Soviet Union, the steel plants of Gary, Indiana, exude smoke that

Table 7.16[72]

THE PROPORTION OF PARTICLES EMITTED FROM A SOURCE AT ONE END OF A HORIZONTAL RECTANGULAR DUCT WHICH IS DEPOSITED UP TO DIFFERENT DISTANCES DOWNWIND

Distances down-wind x/h	*Wind velocity* 5 miles/h *Particle diameter* (μ)					*Wind velocity* 10 miles/h *Particle diameter* (μ)				
	5	10	20	50	100	5	10	20	50	100
10	0.003	0.013	0.052	0.274	0.671	0.002	0.007	0.026	0.148	0.427
20	0.007	0.026	0.101	0.473	0.892	0.003	0.013	0.052	0.274	0.671
50	0.017	0.065	0.234	0.798	0.996	0.009	0.033	0.125	0.551	0.938
100	0.034	0.125	0.414	0.959	1.000	0.017	0.065	0.234	0.798	0.006
200	0.067	0.234	0.656	0.998	1.000	0.034	0.125	0.414	0.959	1.000
500	0.159	0.487	0.931	1.000	1.000	0.083	0.284	0.738	1.000	1.000
1000	0.292	0.737	0.995	1.000	1.000	0.159	0.487	0.931	1.000	1.000

Spheres of unit density, zero cross wind concentration gradient, distance downwind measured in units of the duct height, *h*.[72,79]

Table 7.17[72]

FRACTION OF HOMOGENEOUS AEROSOL CLOUDS DEPOSITED UP TO VARIOUS DISTANCES FROM A CONTINUOUS LINE SOURCE ON THE GROUND

Distances down-wind (yds)	*Wind velocity* 2 miles/h *Particle diameter* (μ)					*Wind velocity* 5 miles/h *Particle diameter* (μ)				
	1	5	10	20	50	1	5	10	20	50
10	0.005	0.129	0.37	0.70	0.90	0.000	0.06	0.18	0.49	0.85
50	0.007	0.154	0.42	0.74	0.93	0.002	0.07	0.22	0.54	0.88
100	0.009	0.163	0.44	0.76	0.95	0.002	0.07	0.24	0.57	0.88
500	0.010	0.195	0.49	0.79	0.96	0.004	0.09	0.28	0.61	0.90
1000	0.012	0.208	0.51	0.81	0.96	0.004	0.10	0.30	0.63	0.91
5000	0.014	0.245	0.56	0.84	0.97	0.005	0.11	0.34	0.68	0.93
10,000	0.015	0.263	0.58	0.85	0.97	0.006	0.12	0.36	0.70	0.93

Vertical eddy diffusion coefficient = 0.12, $m = 1.75$

helps nucleate snowfall for areas further southeast.

Different models have been developed by a number of workers. One early model assumed that the smoke cloud expands by eddy diffusion and that sedimentation affects the vertical concentration gradient.[81] This produced an equation for the ground level concentration of particles at the location of the point of maximum deposition:

$$c_o = \frac{M(1.78/p)^{v_s/pw}}{wx^{(1+v_s/pw)}} \exp\left(-\frac{h}{px}\right) \tag{7.79}$$

in which

M = strength of the source (line source),
h = height,
p = 0.05,
w = wind speed, and
x = distance from source.

Plumes

The deposition of an aerosol depends upon a number of factors and one important aspect of this problem is the plume that rises from a smokestack. The orientation of smokestack plumes and their internal temperature and vertical velocity variation are dependent upon a number of factors, including:

height composition mass flux excess temperature	plume properties
wind(s) temperature(s)	atmospheric properties

The theoretical treatment of plumes may be divided into three distinct regimes:

vertical plume theory–in this case there is no ambient wind;
bent plume theory–in this case the wind deforms the effluent; and
multiple plume theory–nearby plumes interact in a nonsimple way.

The approach taken in the case of the vertical plume theory is a straightforward application of the principles for the conservation of mass, momentum, and heat.[82-84] For the bent plume, it is assumed that the effluent immediately acquires the horizontal wind speed. As of 1975, little progress is reported in the studies of the multiple plume problem.[85] It has been observed that experimental observations and theoretical models agreed fairly well for single stacks with little or no wind and for windy conditions but that neither the vertical nor bent over plume theories fitted observed data very well in a certain range of light winds.[85]

Atmospheric Dispersion

The plume effluent spreads and disperses to an extent and in ways that depend upon plume and system conditions. In order to estimate or calculate the space/time distributions of the airborne particulates, a great deal of effort has been expended over the last 50 years. The earlier work was based upon the view that Fick's laws of diffusion were directly applicable.[86] Since that time the approach has become more sophisticated.

The assumed form of the dispersion may vary to an extent that depends upon the type of source and the existing conditions as shown in Fig. 7.19.[87] A fruitful approach appears to be the numerical solution of diffusion equations applied at the following phases[88] (see Fig. 7.20 for illustration):

- Phase 1: the horizontal spread is controlled by horizontal diffusion;
- Phase 2: horizontal dispersion is controlled by wind shear;
- Phase 3: transition;
- Phase 4: the upper boundary ("lid") is fully effective and the dispersion is similar to Fickian diffusion but with a much higher horizontal diffusion coefficient affected.

The complications due to the presence of large size building structures have been discussed in the literature and the use of the Finite Element method is promising. For example, the effect of an adjacent large building has been studied by modeling the building/stack ratio at 2/1 and also at 1/1. The results indicate that the higher the building relative to the stack the more the tendency for backflow of particulate matter towards the base of the building.[89]

Models incorporating plume rise and dispersion have been applied to the task of predicting ground level concentration results for CEGB (UK) smokestacks with favorable results.[90]

Electrical Deposition

The electric field may be uniform (essentially) or it may be nonuniform (deliberately). The charging may be by ionic collision or by ionic diffu-

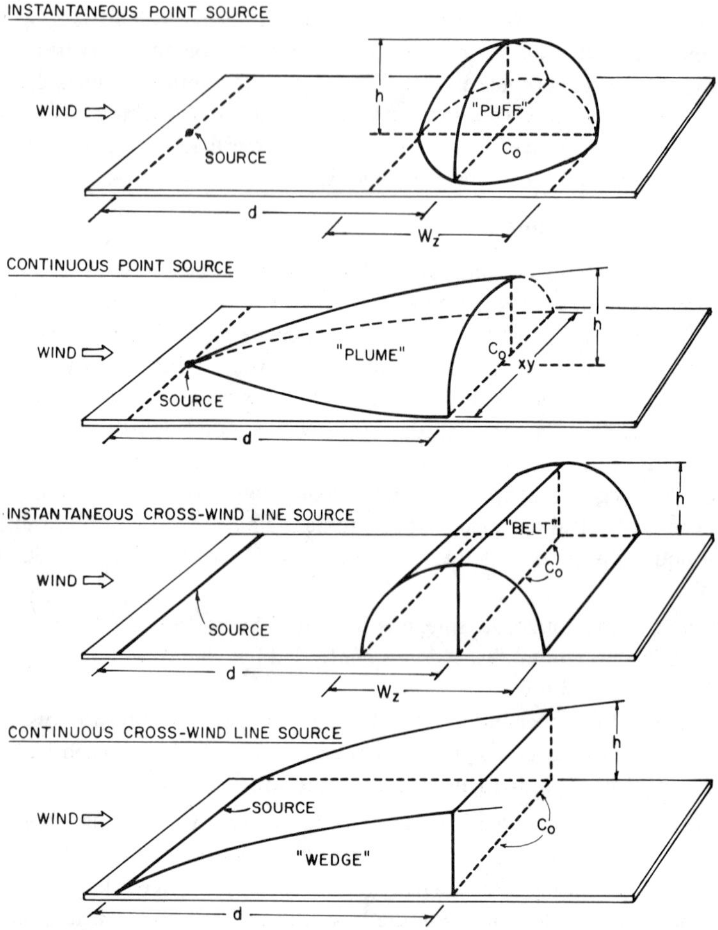

Fig. 7.19 Assumed forms of diffusing clouds from four types of ground level sources. c_o denotes peak concentration.[87]

sion. In the open air, the effect of particle charging on a particle of, say, 0.5 μ is probably not going to be very significant.[72] However, in an appropriately designed system, the effect of charge can be substantial. The electrostatic precipitator is a well established technology. A central electrode is surrounded by an outer electrode so that when the field strength is increased sufficiently, electrical breakdown of the gas occurs near the wire. The mechanisms occurring within the system are complicated and

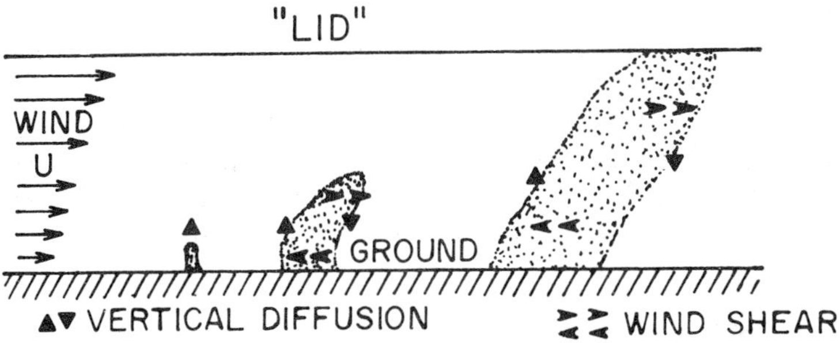

Fig. 7.20 Vertical diffusion acting in a wind shear produces horizontal dispersion.[87]

should be referred to elsewhere.[91] However, the net result is the production of an atmosphere rich in unipolar ions (usually in the range 10^7-10^8 ions per cc). Relationships expressing the influence of system variables upon collision and diffusion charging have been reported [see Eqs. (7.33) and (7.32), respectively].

Whereas the electrical force on an aerosol particle in a uniform electric field is simply proportional to the number of charges carried by the particle and is expressed by:

$$F = Ene \tag{7.80}$$

the force on a particle due to a nonuniform electric field is proportional to the (field)^2 and the (particle radius)^3, termed the dielectrophoretic force and is[92] expressed by:

$$F \alpha r^3 E^2 \tag{7.81}$$

Increasing attention is being given to the use of other force fields in conjunction with electrical force. Some of these include: electric and acoustic fields,[93] fabric filtration with electrostatic augmentation,[94] and fluidized bed with electrostatic augmentation.[95]

The passage of a planar sound wave train through an aerosol tends to increase the collision frequency of the particles and results in coagulation to aid separation of the particulate from the suspension.[96] The superimposition of a sonic field and an a.c. electric field produces a resulting motion of the particles in the aerosol that is a function of the two field frequencies, the magnitude and direction of the two fields and the particle characteristics which include charge, size, and mass. This problem is under investigation.[93]

The effect of electrostatic augmentation of fabric filters upon the filter efficiency has been known for many years.[97] The magnitude of the effect can be observed in Fig. 7.21.[98] The increased collection efficiency obtained through a field being applied or through the particles themselves being charged is said to be due to the effective increase of the collision cross section between the fiber and the approaching particle.[94]

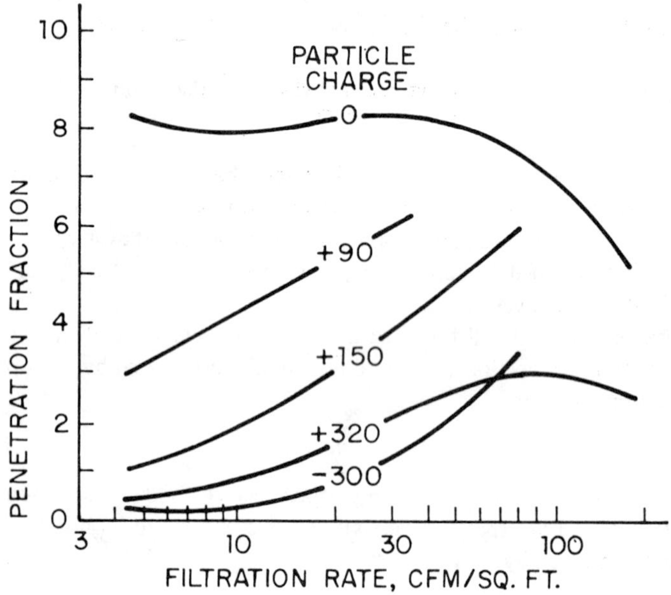

Fig. 7.21 Electrostatic augmentation of fabric filters.[98]

The principle behind the use of electrofluidized beds for cleaning polluted gases is illustrated in Fig. 7.22. The dirty gas fluidizes the bed after it has passed through a charging device. The potential between the electrodes causes the bed particles to act as collection sites. Care must be taken to avoid bubbling and reentrainment to any extent, otherwise the model being used is not applicable.[95]

Diffusive and Inertial Deposition

In a moving aerosol, the cross-stream transportation of particles that must occur for deposition to result is due either to diffusion, inertia, or both. In the case of diffusive transfer, the velocity of deposition has been

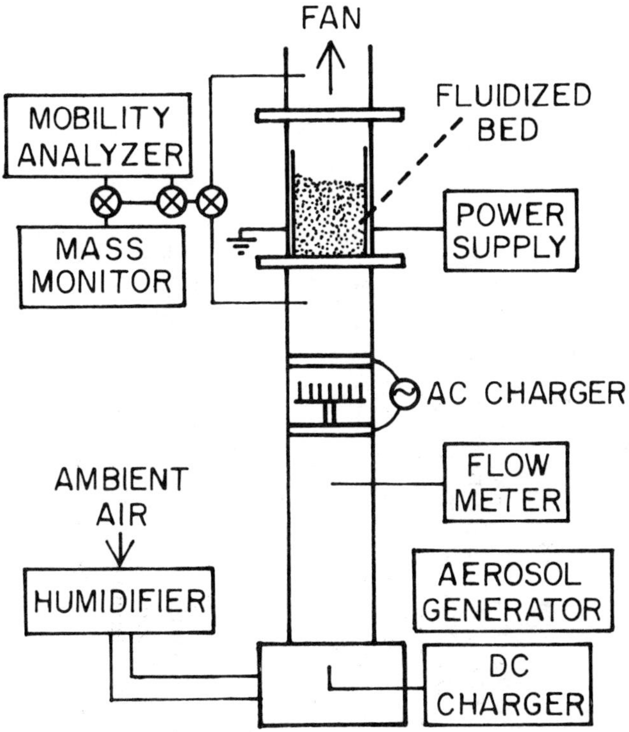

Fig. 7.22 Electrofluidized bed.[95]

calculated for a range of particle sizes and boundary layer thicknesses. The data are shown in Table 7.18.[72] The velocity of deposition is defined as:

$$V = \frac{\text{the mass deposited/cm}^2\,\text{s}}{\text{the mass per cc}} \tag{7.82}$$

in which V is the velocity of deposition in cm/s.

With the model of a boundary layer adjacent to the surface of deposition, the rate of deposition per unit area is

$$R_o = \frac{Dc_o}{\delta} \tag{7.83}$$

in which

Table 7.18

THE VELOCITY OF DEPOSITION (RATE PER UNIT AREA FOR UNIT AIRBORNE CONCENTRATION, R/c): THROUGH DIFFUSION BOUNDARY LAYERS OF VARIOUS THICKNESSES[72]

Particle radius μ	*Velocity of deposition* (cm/s)					
	By gravity.	*By molecular diffusion through boundary layers of thickness* (μ)				
		1	10	100	1000	10,000
10^{-4}	Negligible	7000	700	70	7	0.7
10^{-3}	Negligible	130	13	1.3	0.13	0.013
10^{-2}	Negligible	1.4	0.14	0.014	0.0014	0.00014
10^{-1}	0.000225	0.022	0.0022	0.00022	0.000022	Negligible
1	0.0128	0.0013	0.00013	0.000013	Negligible	Negligible
10	1.2	0.00012	0.000012	Negligible	Negligible	Negligible
100	70	0.000012	Negligible	Negligible	Negligible	Negligible

D is the diffusion coefficient of the particles;
c_o is the concentration outside the boundary layer–i.e., in the aerosol proper, and
δ is the thickness of the diffusion boundary layer adjacent to the surface through which linear deposition occurs.

From Eq. (7.82) the velocity of deposition is

$$V = \frac{R_o}{c_o} \quad \text{or} \quad \frac{Dc_o}{\delta} \cdot \frac{1}{c_o} = \frac{D}{\delta} \tag{7.84}$$

The rate of deposition by gravity is compared with V in Table 7.18.

The indication is that in the case of particles larger than a 1-micron radius, deposition by diffusion through a boundary layer in laminar flow is very small. It is something more than negligible only when the boundary layer is extremely thin.

A special and interesting case occurs when the air stream is interrupted by the presence of a cylinder as shown in Fig. 7.23, a schematic of the process of air filtration. The deposition in this situation occurs by one of two processes (in the absence of an electrostatic effect):

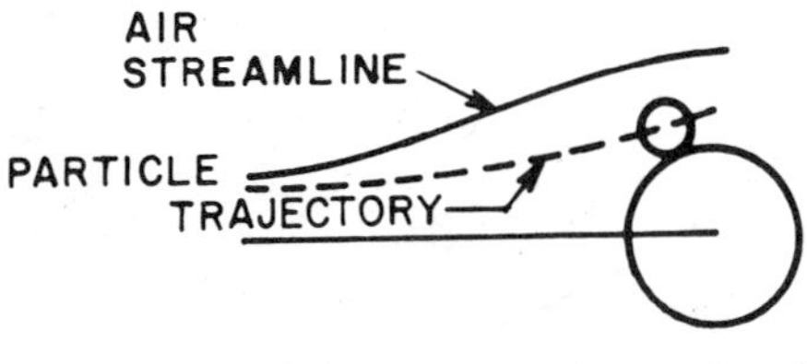

Fig. 7.23 Particle trajectory near a cylinder.[24]

Interception and Impaction

The latter is merely direct contact between the particle and the cylinder. The former occurs when the particle trajectory just permits the particle to touch the cylinder. In either case, deposition may or may not occur and thereafter, reentrainment may or may not occur. An overview in the case of one five-micron diameter fiber is given in Fig. 7.24.[99]

In the case of a laminar boundary layer, the layer thickness increases as the point of measurement moves downstream. This increase affects the velocity of deposition as is shown by the relationship for velocity of de-

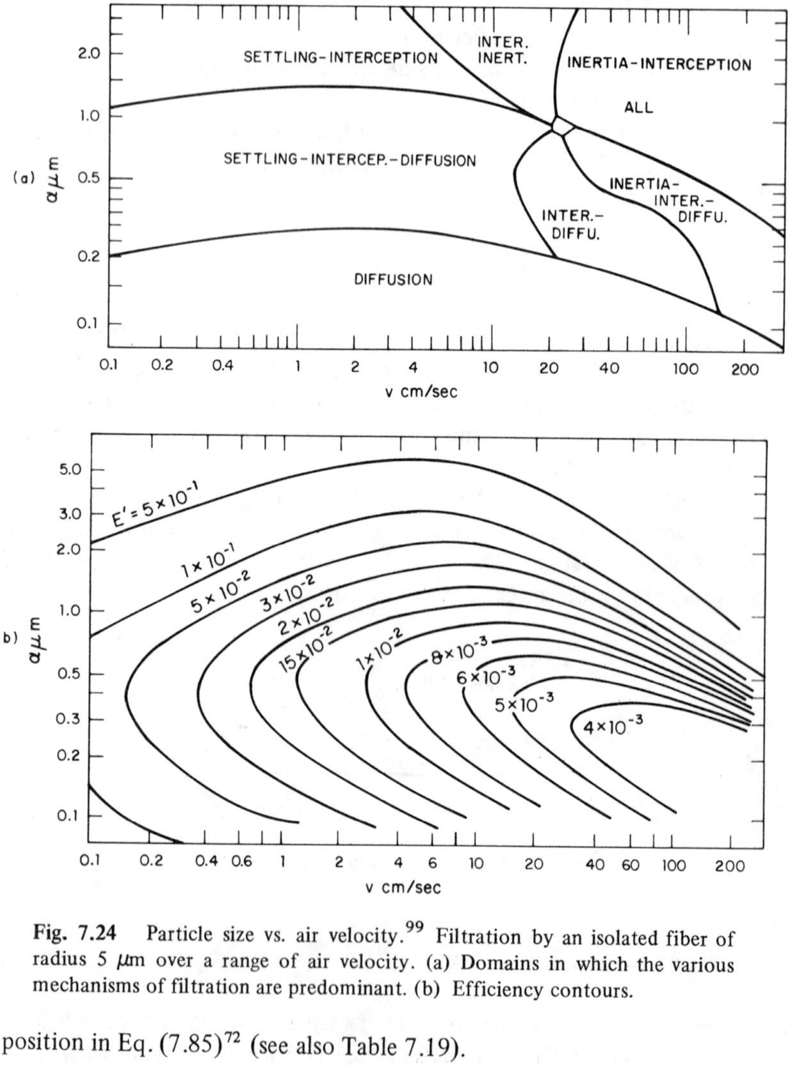

Fig. 7.24 Particle size vs. air velocity.[99] Filtration by an isolated fiber of radius 5 μm over a range of air velocity. (a) Domains in which the various mechanisms of filtration are predominant. (b) Efficiency contours.

position in Eq. (7.85)[72] (see also Table 7.19).

$$V = \frac{0.585 D^{2/3} U_o^{\ 1/2}}{x^{1/2} \upsilon^{1/6}} \tag{7.85}$$

in which

U_o is stream velocity,

x is distance downstream, and

Table 7.19

VELOCITY OF DEPOSITION VS. DOWNSTREAM POSITION THROUGH LAMINAR FLOW LAYER[72]

$D = 0.1\ cm^2/s,\ U_o = 500\ cm/s$

	Downstream →		
x cm	1	10	100
v cm/s	2.2	0.7	0.22

Constant is .332 instead of .585 Eq. (7.85).

υ is kinematic viscosity of air.

The practical consequence of this variation is that, if the design projects a stream towards a surface, the layer thickness will be smaller there and this will permit increased deposition in that region.

The rate of diffusion through a turbulent layer has been shown to be small to the extent that the interposal of even a very thin layer, if it is turbulent, will act as an obstacle to diffusion.[72]

A particle entrained and moving with a gas stream possesses momentum. If the particle inertia is sufficient, it may be able to travel some distance independent of the enveloping gas stream. If the conditions are appropriate, deposition may occur. Deposition may occur in the case of turbulent flow and it may also occur during laminar flow. The latter is more simple to deal with. If the flow is laminar or if the scale of the turbulence is small compared with the stop distance of the particle [see Eq. (7.64)], deposition occurs by sedimentation upon surfaces parallel to the moving aerosol stream.

The case of sets of fibers, as in a fiber filter, has been studied for many years and in consequence there is an extensive literature on the subject (see reference 97, for an example). There has been introduced the concept of flow cells, as illustrated in Fig. 7.25.[97] The problem was developed as one described by the solution of the Navier-Stokes equations for flow transverse to the set of fibers, which were assumed to be parallel. One assumption was that the assembly could be considered as a set of cells of fluid within which the fiber was located such that, at the cell boundary, the shear stress was zero.[100] For mathematical convenience the cell was considered to be cylindrical. A further development of this approach was made when, instead of assuming the absence of shearing stress, a zero vorticity was assumed at the cell boundary. This is the origin of the term Kuwabara cell.[101] This is more fully described in Fig. 7.25.

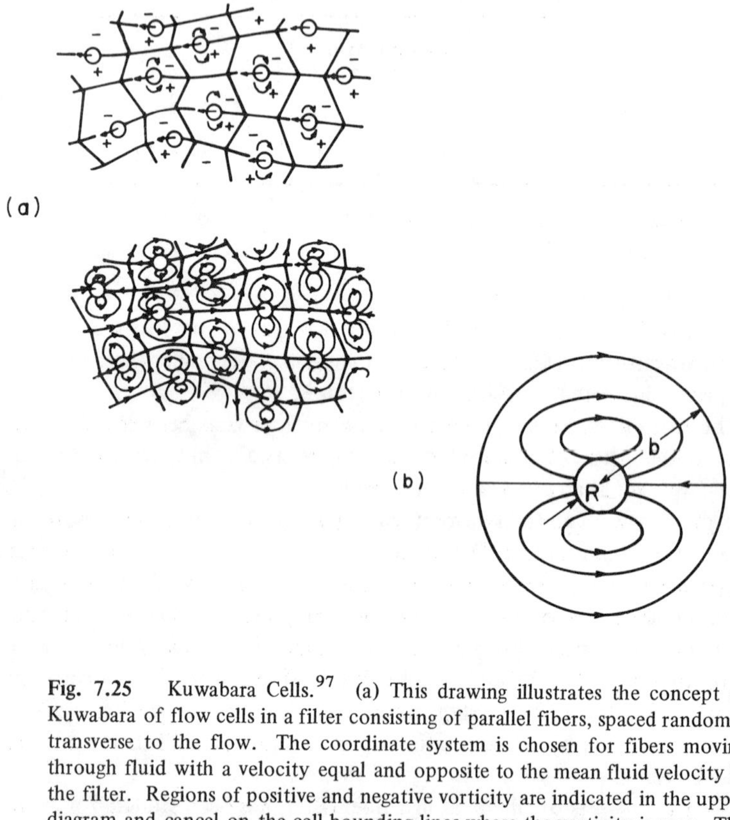

Fig. 7.25 Kuwabara Cells.[97] (a) This drawing illustrates the concept of Kuwabara of flow cells in a filter consisting of parallel fibers, spaced randomly transverse to the flow. The coordinate system is chosen for fibers moving through fluid with a velocity equal and opposite to the mean fluid velocity in the filter. Regions of positive and negative vorticity are indicated in the upper diagram and cancel on the cell bounding lines where the vorticity is zero. The lower diagram shows the circulating flow in the cells. (b) An idealized cell for mathematical treatment. The polygonal cell is reduced to a cylinder of radius b which is coaxial with the fiber of radius R.

Coagulation

Assuming that particles stay together when they make mutual contact and also that they are large as compared with the mean free path of the enveloping gas molecules, it has been shown that the equation for coagulation is[102,103]

$$\frac{dn}{dt} = -\frac{K_o}{2} N^2 \qquad (7.86)$$

in which

n is the total number of particles;

t is time;

K_o is the coagulation constant, and has been shown to have values ranging from 3.0×10^8 cc/min for NH_4Cl to 28.0×10^8 cc/min for SiO_2;[104,105] and

N is the particle concentration.

Integration of Eq. (7.86) yields:

$$(1/N) - (1/N_o) = \tfrac{1}{2} K_o t \qquad (7.87)$$

in which N_o is the initial particle concentration.

If the particle size is small, then the original assumption as to the size being large relative to the molecular mean free path is no longer valid and allowance has to be made by insertion of the Cunningham correction factor, thus

$$-\frac{dn}{dt} = \underbrace{\frac{4}{3}\frac{kT}{\eta}}_{H} \left(1 + \frac{A\lambda}{r}\right) N^2 \qquad (7.88)$$

in which

A is a constant with a value of approximately 0.9,

λ is a mean free path of the gas molecules at temperature,

η is the gas viscosity, and

r is the particle size.

For additional information, including references, the reader should consult reference 106.

7.9 MAGNETISM

Magnetism is the manifestation of the electric charge in motion.[107] There is a number of different types of magnetic material as follows: paramagnetic material, diamagnetic material, ferromagnetic material, and ferrimagnetic material. In a vacuum, a magnetic field of strength H produces a magnetic flux density B according to the relationship:

$$\mu_o = \frac{B}{H} \tag{7.89}$$

in which

μ_o is termed the magnetic permeability of a vacuum. Its units are henries/m,
the units of B are volt · s/m^2 or webers/m^2,
the units of H are amp/m,
a henry is an ohm·s, and
the value of μ_o is $4\pi \times 10^{-7}$ henry/m.

The introduction of a material into the magnetic field alters the magnetic flux density as follows:

$$B = \mu_o \mu_r H \tag{7.90}$$

in which μ_r is the relative permeability.

The magnetic induction B is the total number of lines of induction, which includes the lines of force due to the field H and also the lines of magnetization due to the material M such that

$$B = H + 4\pi M \tag{7.91}$$

The ratio of the magnetization to the field strength is the magnetic susceptibility expressed thus:

$$k = \frac{M}{H} \tag{7.92}$$

in which k is equal to $(\mu_r - 1)$.

The various classes of magnetic behavior are indicated by permeabilities and susceptibilities as shown in Table 7.20.

A graphical description of the weaker paramagnetism and diamagnetism is shown in Fig. 7.26.[1]

Atoms in ferromagnetic materials such as iron, nickel and cobalt, for example, develop a coupling such that they spontaneously align their magnetic moments with each other to form regions of uniformly directional magnetization termed magnetic domains. A beginning condition can be visualized as in Fig. 7.27[108] and the transition region of direction of magnetization across the domain wall is also illustrated there. The magnetization curve in Fig. 7.28[108] shows that in the initial part of the magnetization

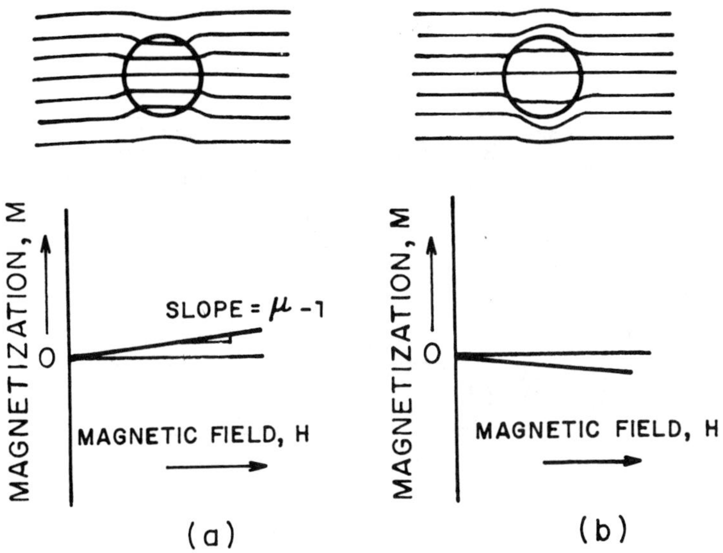

Fig. 7.26 Flux density and magnetization. (a) Paramagnetism: Since a paramagnetic material has positive susceptibility and magnetization, magnetic flux is concentrated through it. (b) Diamagnetism: The susceptibility and magnetization are negative.[1]

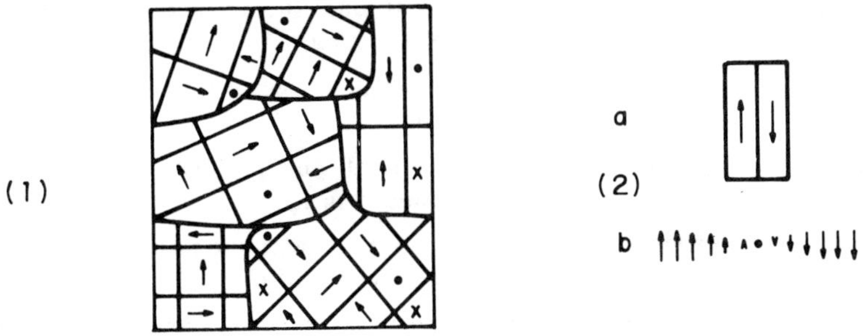

Fig. 7.27 1. Domains in a ferromagnetic material. 2. (a) A domain wall; (b) the transition region corresponding to a wall (based on reference 108).

process, the domains grow by wall motion (the portion of the curve marked ORP) and later by rotation (beyond point *P*).

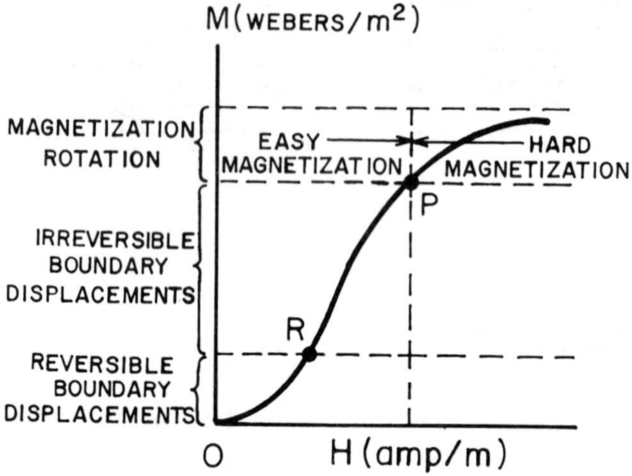

Fig. 7.28 Initial part of magnetization curve.[108]

Table 7.20

TYPES OF MAGNETIC BEHAVIOR AND CHARACTERISTICS[107]

Type	*Characteristics*
Empty space	$k = 0$, since there is no matter to magnetize, and $\mu = 1$.
Diamagnetic	k is small and negative, and μ slightly less than 1.
Para- and antiferromagnetic	k is small and positive, and μ slightly greater than 1.
Ferro- and ferrimagnetic	k and μ are large and positive, and both are functions of H.

The full magnetization curve for a ferromagnetic material has the form shown in Fig. 7.29.[1] It should be noted that when the magnetic field H is reversed, not all of the induction disappears. In fact, as shown, a remnant induction B_s remains. Similarly, it is necessary to reverse the magnetic field to a value of $-H_c$ in order that the induction be completely removed.

H_c is termed the coercivity. A scheme of domain boundary movements is also shown in Fig. 7.29.

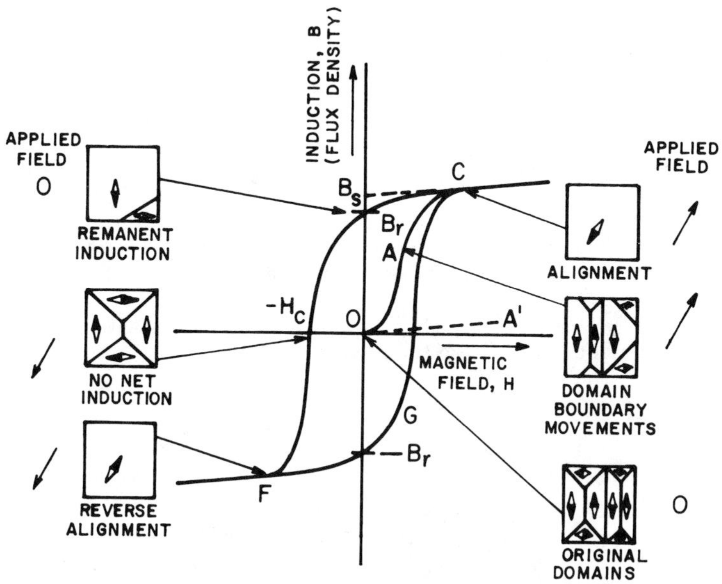

Fig. 7.29 Ferromagnetic hysteresis (hard magnet). The spontaneous polarization requires energy for reversal.[1]

The area bounded within the hysterisis curve is the energy consumed per cycle to magnetize and demagnetize.

Ferrimagnetic materials are similar to ferromagnetic materials in many respects. They possess domains, hysterises, magnetic saturation, and spontaneous magnetization at room temperature. They also possess or exhibit the disappearance of the magnetic property above T_c, the Curie temperature. Lodestone, which is Fe_3O_4 is a naturally occurring ferrimagnetic material. In fact, the most important class of ferrimagnetic materials is that of certain double oxides of iron. A visual, schematic comparison of the ferromagnetic, antiferromagnetic and ferrimagnetic materials is shown in Fig. 7.30.[108] Ferromagnetic and ferrimagnetic effects are the stronger magnetic effects.

Effect of Particle Size on Coercivity

The coercivity of small particles shows a very strong dependence upon

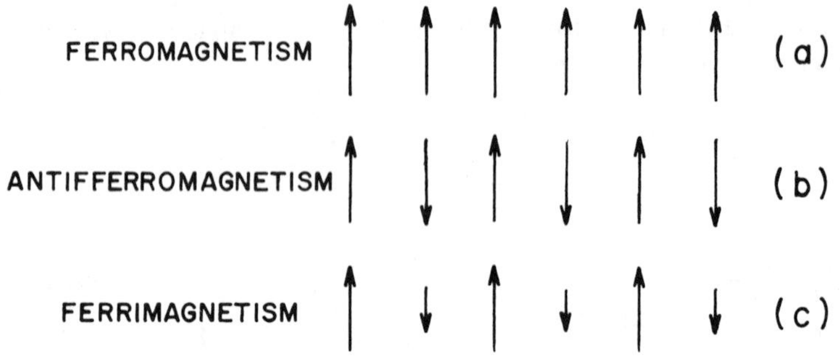

Fig. 7.30 Comparison of ferromagnetic, antiferromagnetic, and ferrimagnetic materials (based on reference 108).

their size as is shown in Fig. 7.31[110] for a number of different systems in the size range from 100 μ down to just above 10 Å.

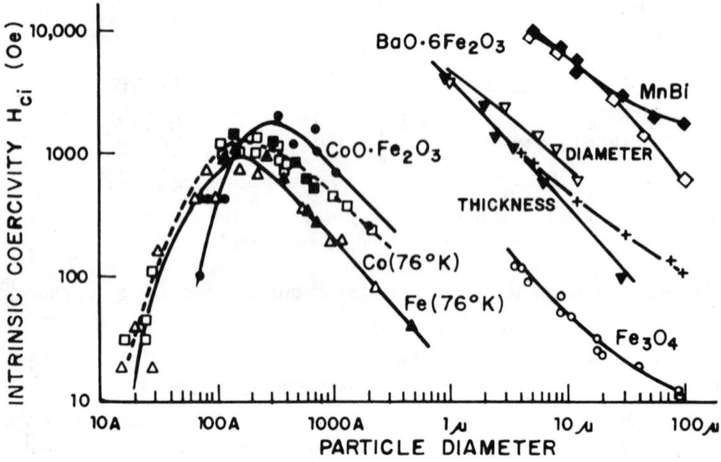

Fig. 7.31 Variation of coercivity with particle size for particles deriving their coercivity principally from crystal anisotropy.[110]

For simplicity, Fig. 7.32[107] which is a schematic representation of Fig. 7.31 is referred to. The curve can be divided into three regions:

Region 1: as the particle size decreases from relatively large values, the coercivity increases up to a maximum value;

Region 2: as the particle size is decreased still further, the coercivity

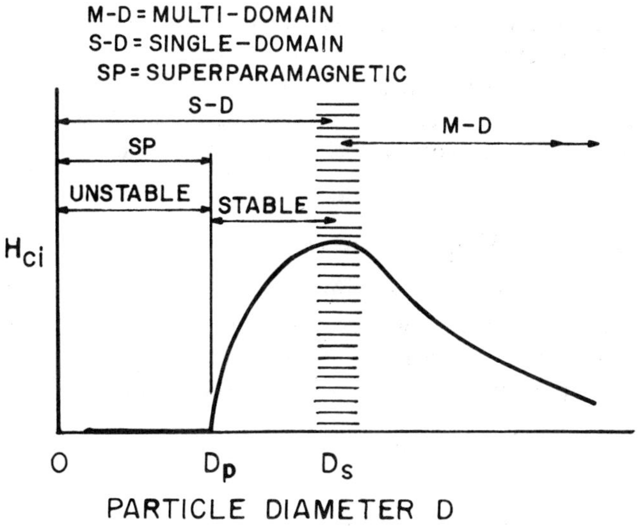

Fig. 7.32 Variation of intrinsic coercivity, H_{ci} with particle diameter D (schematic).[107]

decreases towards zero;

Region 3: at very small sizes, the coercivity is effectively zero.

Region 1. In this region, changes in magnetization occur by domain wall motion. The relationship between the intrinsic coercivity and particle size is expressed by:

$$H_{ci} = a + \frac{b}{D} \tag{7.93}$$

in which

H_{ci} is the intrinsic coercivity,
a,b are constants, and
D is the particle size.

Region 2. Below the critical particle diameter D_s, the particles become single domain particles. It is in this size range (D_s is not well defined[107]) that the coercivity reaches a maximum value. Below a particle size of D_s, the coercivity decreases according to the relationship

$$H_{ci} = g - \frac{h}{D^{3/2}} \tag{7.94}$$

in which g and h are constants.

Region 3. Below D_s changes in magnetization occur by spin rotation, among other mechanisms. The decrease of coercivity is due to thermal effects. Below a critical particle diameter D_p, the thermal effects are now strong enough to spontaneously demagnetize a previously saturated particulate assembly. The coercivity of the particles, termed superparamagnetic, is zero.

Effect of Particle Shape on Coercivity

The magnetic hardness of fine particles is affected by the particle shape. Three modes of magnetization have been considered: coherent rotation, fanning, and curling. In general, the demagnetization field is related to the magnetic field that produced the original magnetic effect as

$$H_d = N_d M \tag{7.95}$$

in which

H_d is the demagnetization field,
N_d is the decay coefficient, and
M is the original magnetization.

The saturation magnetization is related to the particle shape by

$$H = (N_a - N_c)M_s \tag{7.96}$$

in which

M_s is the saturation magnetization, and
a and c indicate the longitudinal and transverse directions respectively.

Calculated coercivities of single domain particles are listed in Table 7.21.

The values of H_{ci} for shape anisotropy listed in Table 7.20 were calculated on the assumption that the spins of all the atoms in the particle kept in step and remained parallel to each other during reversal of magnetization by rotation.[109] Experimental data for iron particles shows a considerable discrepancy with the figures in Table 7.21. For example, H_{ci} for an aspect ratio of more than ten was found to be 1800 oersted.[110] This result is so far out of line with the calculated figure that this discrepancy prompted further investigation and led to the proposal of the modes of fanning and curling. The main conclusion of this theoretical approach is that small particles reverse magnetization by coherent-rotation and that large particles do so by curling. In brief, fanning describes the fan-like

Table 7.21

CALCULATED COERCIVITIES OF SINGLE-DOMAIN PARTICLES (EASY AXES ALIGNED WITH FIELD)* [107]

Shape anisotropy			*Stress anisotropy*		*Crystal anisotropy*
c/a	$N_a - N_c$	H_{ci}(Oe)	σ (lb/in^2)	σ (kg/mm^2)	K_1 (10^5 ergs/cm^3)
1.0	0	0	0	0	0
1.1	0.472	810	340,000	240	7
1.5	1.892	3,240	1,350,000	950	28
5.	5.231	8,950	3,700,000	2,600	77
10.	5.901	10,100	4,200,000	3,000	87
20.	6.156	10,500	4,400,000	3,100	90
∞	$6.283 = 2\pi$	10,800	4,500,000	3,200	93

*$M_s = 1714$ emu/cm^3, $\lambda_{si} = 20 \times 10^{-6}$.

spread of the M_s vectors of successive spheres. It is shown in Fig. 7.33, along with the coherent mode of reversal, with which it may be compared.[107] The curling mode is again illustrated in Fig. 7.34 along with an example of the coherent-rotation mode.[107]

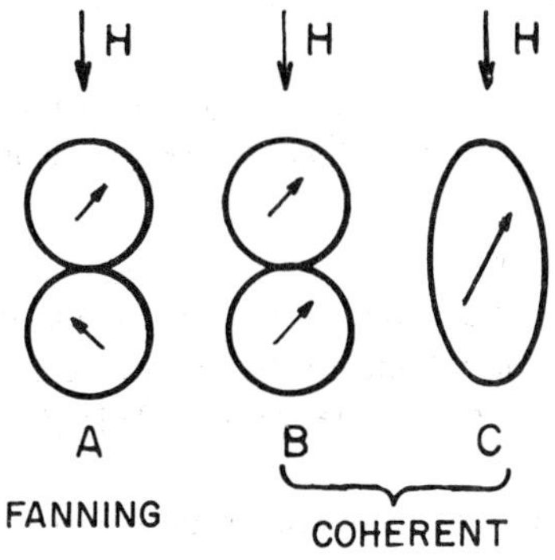

Fig. 7.33 Modes of reversal.[107]

The effects of pits, bumps, and sharp corners is described schematically in Fig. 7.34.[107] The practical effect of these local imperfections is to provide sites at which domains of reverse magnetization may exist even if the

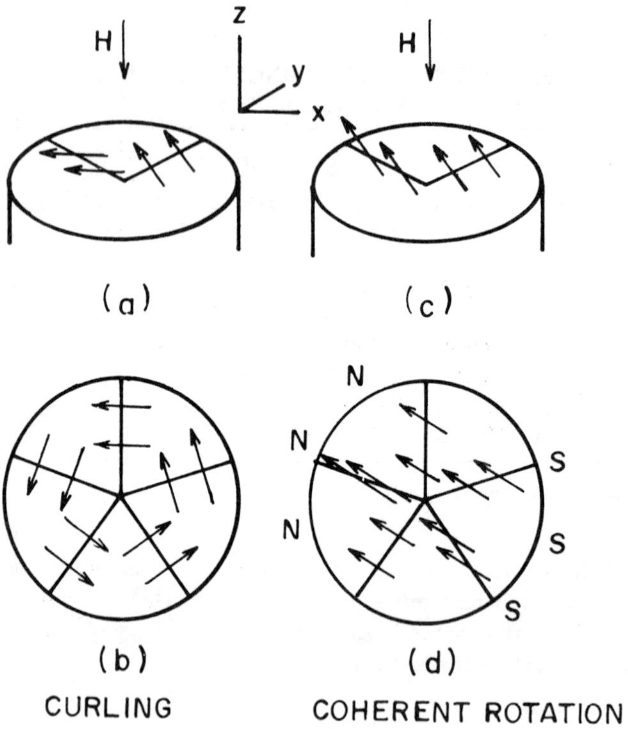

Fig. 7.34 Curling and coherent-rotation modes. (b) and (d) are cross sections normal to the z axis after a 90° rotation from the z direction.[107]

particle as a whole is saturated.[112]

The poles on the walls of a pit act via their field to increase the magnetization at the base of the pit and so this counteracts the applied field H_a which is trying to reverse the magnetization. This effect is shown in Fig. 7.35.[107]

Conversely, the effect of a surface protuberance is to produce local fields which demagnetize the particle and so act in concert with the applied field H_a.[111]

The effect of surface bumps and corners is thought to promote the formation of closure domains as shown in Fig. 7.36.[107]

In the case where particles are contained in a static bed, a relationship has been developed in which the coercivity is related to the bed voidage as

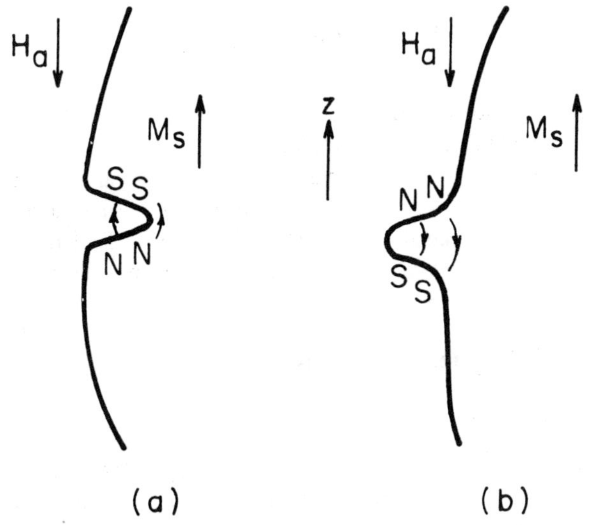

Fig. 7.35 Local fields near pits and bumps.[107]

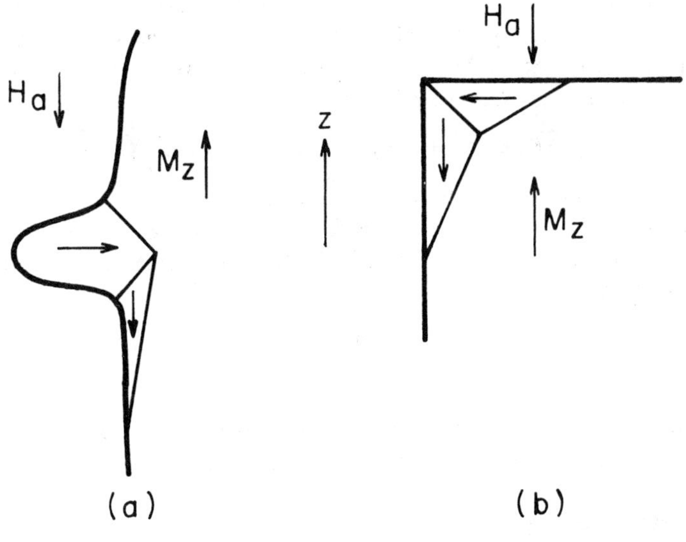

Fig. 7.36 Closure domains near bumps and sharp corners.[107]

follows:

$$H_{ci}p = H_{ci}o(1-p) \tag{7.97}$$

in which

o indicates an isolated particle, and

p is the packing fraction of particles.

Unfortunately, Eq. (7.97) is based on a theoretical approach about which there is some doubt.[113] In fact, there is reasonable agreement of actual data with the theoretical equation for p up to level of about 0.4, but there is a considerable deviation between theory and experiment thereafter.

Internal Particle Effects on Coercivity

1. Crystal Anisotropy

There are several kinds of magnetic anisotropy. Crystal anisotropy is due to the magnetic properties in different directions being different. A useful way of considering this property is to think of the applied field as having to do work against the crystal anisotropy force in order to turn the magnetization vector away from the direction of easy magnetization. Therefore, in order to point M_s in a noneasy direction, it is necessary to store energy in the crystal. This is termed the crystal anisotropy energy and is related to the cosines of the angles which the axes a, b, c make with M_s according to the following relationship:

$$E = K_o + K_1(\alpha_1^2\alpha_2^2 + \alpha_1^2\alpha_3^2 + \alpha_3^2\alpha_1^2) + K_2(\alpha_1^2\alpha_2^2\alpha_3^2). \tag{7.98}$$

Eq. (7.98) applies to the case of a cubic crystal, in which K_o, K_1, and K_2 are constants for a particular material. Some published values of these anisotropy constants are listed in Table 7.22.[107] The relationship between the coercivity and crystal anisotropy is simply

$$H = \frac{2K_1}{M_s} \tag{7.99}$$

The magnitude of this effect may be compared with shape anisotropy in Table 7.21 for a range of aspect ratios up to infinite.

2. Stress Anisotropy

When a material is exposed to a magnetic field its size changes. The

Table 7.22

ANISOTROPY CONSTANTS[107]

Structure	*Substance*	K_1 (10^5 ergs/cm^3)	K_2
Cubic	Fe	4.8	±0.5
	Ni	−0.5	−0.2
	$FeO \cdot Fe_2O_3$	−1.1	
	$MnO \cdot Fe_2O_3$	−0.3	
	$NiO \cdot Fe_2O_3$	−0.62	
	$MgO \cdot Fe_2O_3$	−0.25	
	$CoO \cdot Fe_2O_3$	20.	
Hexagonal	Co	45.	15.
	$BaO \cdot 6Fe_2O_3$	33.	
	YCo_5	550.	
	MnBi	89.	27.

relationship is

$$\lambda_s = \frac{\Delta \ell}{\ell} \tag{7.100}$$

in which subscript s denotes the magnetostriction at magnetic saturation, and $\Delta\ell$, ℓ are the change in length and the length, respectively.

For a cubic crystal, the description of saturation magnetostriction contains quite a complicated appearing relationship:

$$\lambda_{si} = \frac{3}{2}\,\lambda_{100}\left(\alpha_1^2\beta_1^2 + \alpha_2^2\beta_2^2 + \alpha_3^2\beta_3^2 - \frac{1}{3}\right)$$
$$+\ 3\lambda_{111}(\alpha_1\alpha_2\beta_1\beta_2 + \alpha_2\alpha_3\beta_2\beta_3 + \alpha_3\alpha_1\beta_3\beta_1) \tag{7.101}$$

in which

100 and 111 are the crystal directions,
$\beta_1\ \beta_2\ \beta_3$ are the cosines relative to the three crystal axes,
$\alpha_1\ \alpha_2\ \alpha_3$ are the cosines relative to the three crystal axes,
β refers to the direction of magnetostriction, and
α refers to the direction of saturization magnetization.

Applied stress affects the magnetization curve. For example, in the case of nickel, a compressive stress increases the magnetization and a tensile stress decreases the magnetization. Reference to Table 7.23 shows that the magnetostriction of nickel is negative. In the case of a material with a positive magnetostriction the effect is opposite. The influence of stress upon coercivity is indicated in the following relationship:

$$H = \frac{3\lambda_{si}\sigma}{M_s} \tag{7.102}$$

in which σ is the stress level.

Table 7.23[102]

MAGNETOSTRICTION CONSTANTS OF CUBIC SUBSTANCE (UNITS OF 10^{-6})

	λ_{100}	λ_{111}	λ_p^*
Fe	+ 21	− 21	− 7
Ni	− 46	− 24	− 34
$FeO \cdot Fe_2O_3$	− 20	+ 78	+ 40
$Co_{0.8}Fe_{0.2}O \cdot Fe_2O_3$	−590	+120	
$CoO \cdot Fe_2O_3$			−110
$Ni_{0.8}Fe_{0.2}O \cdot Fe_2O_3$	− 36	− 4	
$NiO \cdot Fe_2O_3$			− 26
$MnO \cdot Fe_2O_3$			− 5
$MgO \cdot Fe_2O_3$			− 6

* Experimental values for polycrystalline specimens.

3. Miscellaneous Effects

Inclusions hinder the motion of domain walls and this distraction therefore would be expected to contribute to the magnetic characteristics of larger particles, rather than smaller ones.

Residual stress also hinders the motion of domain walls and would be expected to affect coarse particle magnetic properties. The effects of cold work, in the case of metals, are rather well known. The effect is mainly due to the generation of large numbers of dislocations during the mechanical processing.

Magnetic Separation

Magnetic separation of ferrous material from bulk solids has been practiced for many years now. It is more or less a standard part of any process in which the presence of small (or large) pieces of scrap iron and steel would be an embarrassment, as in food, drugs, etc. High gradient magnetic separation, HGMS for short, are available for benefication processes generally and also in the search for ways to remove certain pollutants from gaseous effluents in the size range 0.3–3 μ.[114,115] The essential design feature of HGMS systems is the insertion of a set of ferromagnetic fibers in a magnetic field of sufficient strength so that the fibers become saturated. The array of line dipoles causes the field in the vicinity of the fibers to change rapidly with respect to displacement. The resultant field gradients produce a magnetic force on particles that are being swept through the fiber set. The idea behind the HGMS design is to retain the magnetized particles on the ferromagnetic fibers and to sweal away the rest of the particles in the fluid stream. Some potential applications of this technology are listed in Table 7.24.

Applications in the case of liquid streams include the removal of weakly magnetic colored particles smaller than 2 μ from clay, cleaning of waste water from steel mills, and its removal of wear particles from hydraulic fluids. It must be observed here that there is a considerable discrepancy between theoretical predictions and observed results in the case of red blood cell removal[119] and in the case of cupric oxide removal of[120] broth from a liquid stream.

7.10 THERMAL CONDUCTIVITY

It is important to differentiate between thermal conductivity and heat transfer. Thermal conductivity is a material property whereas heat transfer is a system property. For example, the thermal conductivity of a particulate material may be known but the heat transmission of a bed of particles of the material depends on (among others) the bed voidage. If the bed is a dynamic assembly, it also transfers heat to an extent that depends on the rate of motion of the bed and its direction of motion. Mathematically, thermal conductivity is a proportionality constant in the heat flow equa-

Table 7.24

POTENTIAL APPLICATIONS OF HGMS
IN PARTICULATE EMISSION CONTROL[115]

Industrial Process	*References*	*Typical Dust Composition, mass basis*
Blast furnace	116	35–50 percent Fe, 12 percent FeO, 0.5–0.9 percent Mn
Basic oxygen furnace	116, 117	90 percent Fe_2O_3, 1.5 percent FeO, 4 percent Mn_3O_4
Open hearth furnace	116, 117	85–90 percent Fe_2O_3, 1–4 percent FeO, 0.5 percent MnO
Electric arc furnace	116, 117	20–55 percent Fe_2O_3, 4–10 percent FeO, 0.5 percent MnO
Silico-manganese furnace	116, 117	4-7 percent FeO, 30–35 percent MnO
Ferro-manganese furnace	116, 118	6 percent FeO, 34 percent MnO
Ferro-chrome furnace	116, 118	7-11 percent FeO, 3 percent MnO, 29 percent total Cr as Cr_2O_3
Coal-fired boiler	117	2-36 percent Fe_2O_3

tion

$$J = H \frac{dT}{dx} \tag{7.103}$$

in which

J is the heat flux,
dT/dx is the temperature gradient, and
H is the thermal conductivity.

In a simplified way, one can appreciate that there can be two modes of heat conduction through a solid, by electron motion and by phonon motion. In a metal, the electrical conductivity is high so the thermal conductivity is due to the electrons. In a semiconductor, there are not very many electrons and so the thermal conductivity is due to phonons. In general:

$$H\ 1/3\ C_v \bar{v} \ell \tag{7.104}$$

in which

C_v is the heat capacity at constant volume per unit volume,
$\bar{v}$ is the average particle velocity, and
ℓ is the particle mean free path.

For a metal, $\bar{v}$ and ℓ are very high (which compensate for the fact that C_v is low). For semiconductors, the free electron density is low, so the phonon mechanism is the only possible alternative mode of thermal conduction. The effect of temperature upon H is quite interesting. At low temperature (say below 20°K, or thereabouts) the mean free path of electrons and phonons is constant and so the change of H with temperature is due to the effect of temperature upon the heat capacity C_v. Thus as the temperature is increased from 2 to ≈20°K, the value of H increases. But above 20°K, the mean free path decreases and so the value of H remains constant or falls off. In the case of noncrystalline solids and concentrated solid solutions, the mean free paths are small at all temperature levels and so any variation in H is due to the effect of temperature upon C_v.

A static particulate bed invariably has a much lower thermal conductivity than the corresponding solid because of the presence of air (small mean free path, for example) and also because of the constriction to heat flow at the points of contact between the particles. Some clear examples of the difference between a solid and its corresponding particulate are shown in Table 7.25.[121]

There are a number of different cases for which it is necessary to be

Table 7.25

COMPARATIVE THERMAL CONDUCTIVITIES OF VARIOUS SOLIDS IN BULK AND POROUS FORM[121]

Solid	λ *at* 300 K (W cm^{-1} K^{-1})		*Porous form*	*Effective porosity* (percent)
	bulk	porous		
Steel	0.58	4.1×10^{-3}	1/8 in. diameter spheres	41.3
Lead	0.33	4.2×10^{-3}	1/16 in. diameter spheres	42
Alumina	0.36	4.0×10^{-3}	166 μm particles	42
Aluminum	2.31	7.2×10^{-4}	wool	98.5
Glass	0.0094	3.6×10^{-4}	silk	93
Carbon	1.0	2.5×10^{-4}	0.01 μm particles	97
Silica	0.014	2.2×10^{-4}	aerogel	95

able to obtain thermal conductivity values. These include: slurries, suspensions, static beds, and static beds with percolating fluids. In all of these cases the variation of conductivity with system parameters is fairly well documented.

Dilute Heterogeneous Media

One approach to explaining thermal conductivity in these media assumes that the particles are spherical, far enough apart to have no effect on each other, and that potential theory can be applied. The derivations were carried out some years ago; they also apply to electrical and magnetic fields.[122] The thermal conductivity of the slurry or suspension is:

$$Hs = H_f \left[\frac{2H_f + H_p - 2x_v(H_f - H_p)}{2H_f + H_p + x_v(H_f - H_p)} \right] \tag{7.105}$$

in which

s indicates slurry, suspension,
f indicates surrounding fluid,
p indicates particle, and
x_v is the volume fraction of p.

This approach can lead to expressions relating thermal conductivity to a number of system variables. Some of these are referenced in Table 7.26.

An alternate approach is to model the suspension after the manner of an electrical circuit. This method leads to expressions for thermal conductivity as a function of fluid and particle conductivities and also of particulate volume fraction.[127]

Table 7.26

EXPRESSIONS RELATING SYSTEM VARIABLES TO THERMAL CONDUCTIVITY

Variable	*Reference*
Hf, Hp, x_v up to x_v of 0.1	Eq. (7.105)
as above with x_v up to 0.3	123
Particle size distribution, x_v	124
Particle shape	125
More than one particulate	126

Static and Dynamic Beds

An overview of the effects of variables on the thermal conductivities of gas-filled (stagnant) particulate beds is illustrated in Fig. 7.37.[128] In the case of beds through which fluid is flowing the radially effective thermal conductivity has been related to system variables as follows:[129]

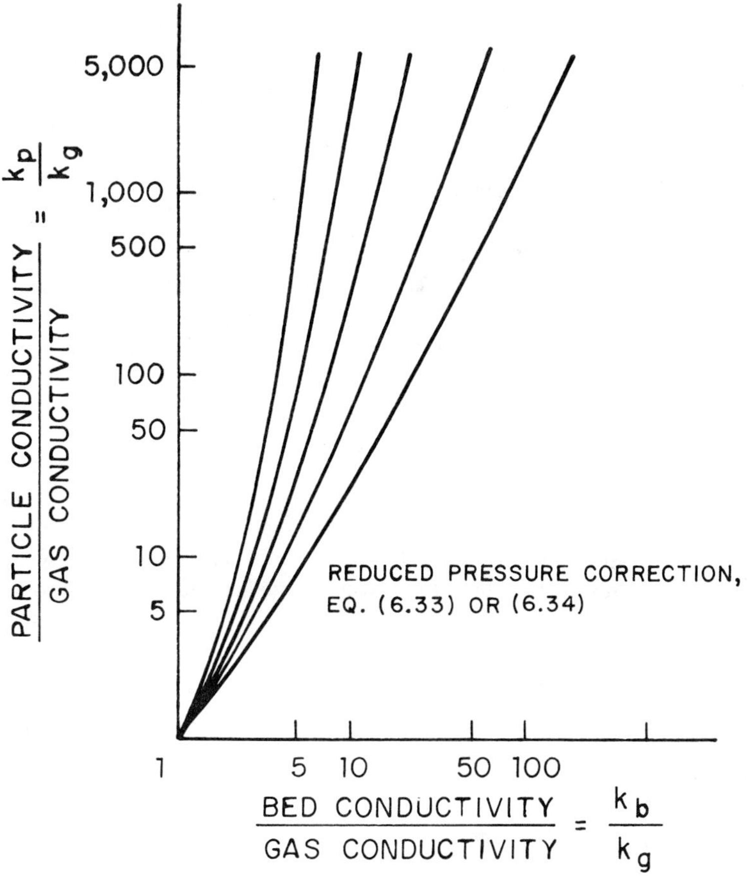

Fig. 7.37 Correlation for the thermal conductivity of packed beds with static gas filling the interstices.[128]

$$\frac{He}{Hg} = \frac{He^o}{Hg} + (\alpha\beta)N_{pr}N_{Rem} \tag{7.106}$$

in which

- e indicates the packed bed,
- g indicates the fluid,
- o indicates motionless fluid,
- α = mass velocity of fluid flowing indication of heat transfer/superficial mass velocity of fluid flowing along empty tube, and
- β = average diameter of solid particle.

$$N_{pi} N_{Rem} = \frac{Cp\mu}{Hg} \frac{DG}{\mu} CpD \frac{G}{Hg}$$

in which

- D is the particle diameter,
- G is the superficial mass fluid velocity, and
- Cp is the specific heat of fluid.

When it is necessary to take into account the effect of radiation, one obtains an expression for the effective thermal conductivity which consists of two parts. The first part deals with the thermal conductivities of the fluid and the particles and the second term takes into account the radiation effect.[13,130]

In the case of a fluidized bed, a correlation for conductivity at right angles to the flow direction has been derived.[131]

Heat Transfer

For a single particle that is stationary within an enveloping fluid, where the fluid is flowing, the limitation placed on the heat transfer is due to the layer of fluid immediately adjacent to the surface of the particle. The heat transfer may be stated in terms of an experimentally derived relationship (which closely follows a theoretical one)

$$Nu = 2 + 0.6\, Re^{1/2}\, Pr^{1/3} \tag{7.107}$$

in which

- Nu is the Nusselt number,
- Re is the Reynolds number, and
- Pr is the Prandtl number.

For a stationary fluid, Nu is 2.

In the case of particulate beds, there can be a number of different considerations, depending upon the system under consideration. For example, the bed can be static and the fluid may be either a liquid or a gas. Fluidized beds may also be employed, in which case they may be fluidizing normally or they may be spouted. A graphical description of the general relationship between fixed, fluidized dense phase beds and dilute phase beds and their heat transfer characteristics is shown in Fig. 7.38.[132]

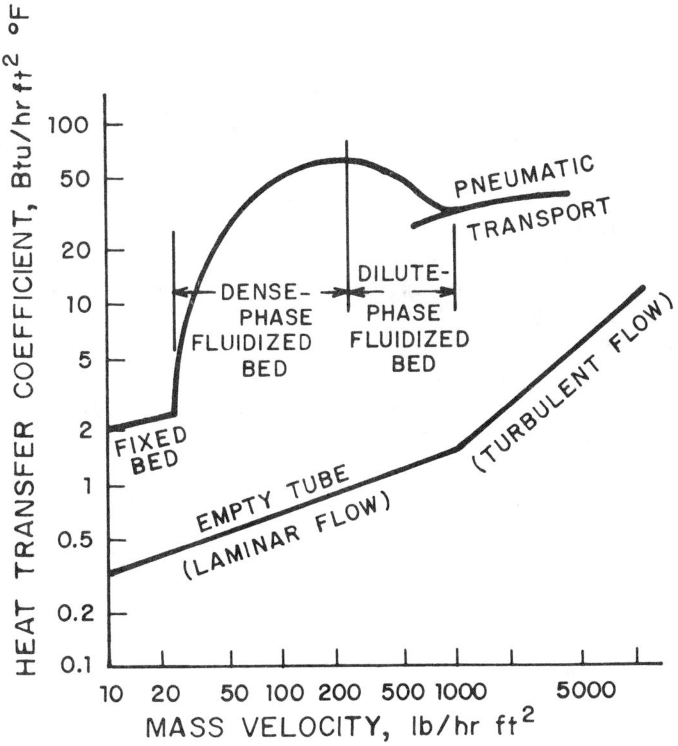

Fig. 7.38 Hypothetical examples of heat transfer to particle beds from the container wall, showing comparison with an empty container.[132]

A simple explanation of these differences in heat transfer is as follows: once the bed becomes fluidized, the increased mixing increases the heat transferred; but when the rate of motion is increased beyond a certain point, the particle-particle contacts are reduced because the bed expands; this results in a lower heat transfer. The experimental and theoretical work reported in the literature should be reviewed by the interested reader as needed. Some appropriate starting references are given in Table 7.27.

In the case of dispersions of fine particles in a gaseous phase, the particles heat up fairly rapidly at first and more slowly thereafter as they reach a pseudo-equilibrium condition.[143] The addition of solid particles to an air stream flowing along a pipe increases the heat transfer from the wall to the air stream.[143]

Table 7.27

EXPRESSIONS RELATING SYSTEM VARIABLES
TO HEAT TRANSFER IN BEDS

Variables	*References*
Static bed-flowing gas (low flow)	133
Static bed-flowing gas (high flow)	134
Static bed-flowing gas (particles generate their own heat)	135
Static bed-liquid flowing	136
Spouted bed-gas flowing	137
Fluidized bed-gas flowing (fluid to particle)	138
Fluidized bed-gas flowing (wall to bed)	139, 140, 141
Dilute phase, gas-solid heat transfer	142

REFERENCES

1. van Vlack, L. *Materials Science for Engineers.* Reading: Addison Wesley, 1970.
2. McCrone, W. C., Draftz, R. G., and Delly, J. G. *The Particle Atlas.* Ann Arbor Science Publishers, Ann Arbor, MI, 1973, 1979, second edition, Vols. I–IV.
3. Ehrenhaft, F. *Physic Zeit* 17 (1917): 352–368.
4. Rayleigh, Lord (J. W. Strutt) *Phil. Mag.* 41 (1871): 107–120 and 274–279.
5. Mie, G. *Ann. Physik* 25 (1908): 377–445.
6. Kerker, M. *Am. Sci.* 62 (Jan-Feb., 1974): 92–98.
7. van de Hulst, H. C. *Light Scattering by Small Particles.* New York: Wiley, 1957.
8. Sinclair, D. *Handbook of Aerosols.* p. 81. USAEC, Washington D.C., 1950.
9. Hodkinson, J. R. *Aerosol Science.* edited by Davies, C. N. pp. 287–357, New York and London: Academic Press, 1966.
10. Debye, P. *Ann. Physik* 30 (1909): 57–136.
11. Ashkin, A., and Dziedzic, J. M. *App. Phys Lett.* 19 (1971): 183–285.
12. Harper, W. R. "Powders in Industry." *SCI Monograph* 14, London (1961): 115–129.
13. Orr Jr., C. *Particulate Technology.* New York: Macmillan, 1966.
14. Stern, O. *Z. Electrochem.* 30 (1924): 508–516.
15. Gouy, G. *J. Phys.* 9 1901: 457–467.
16. Harper, W. R. *Proc. Roy. Soc. A.* 205, 83, 1951.
17. White, H. J. *Trans. Am. Inst. Elec. Engrs.* 70 (1951): 1186–1191.
18. Cohen, E. "Research on Electrostatic Generation and Acceleration of Sub-micron Particles." Report May 1963, Space Tech Labs, Redondo Beach, Calif.
19. Whitby, K. T., and Liu, B. Y. H. in *Aerosol Science.* edited by Davies, C. N. pp. 59–86, New York and London: Academic Press, 1966.
20. Brown, R. *Phil. Mag.* 4 (1828): 161–169. Of course, Brown thought originally that the motion of the particles was in some way self-sustained; however, later he soon observed that the motion was independent of whether the substance of the particles was organic or mineral.

21. Einstein, A. *Ann. Physik* 17 (1905): 549–560.
22. Cunningham, E. *Proc. Roy. Soc.* 83A, 357 (1910).
23. Perrin, J. *Compt. Rendue* 149, 549 (1909).
24. Mercer, T. C. "Aerosol Technology." in *Hazard Evaluation.* New York: Academic Press, 1973.
25. Fuchs, N. A. *Z. Phys. Chem.* A-171, (1934): 199–208.
26. Cadle, R. *Particle Size.* New York: Reinhold, 1965.
27. Schmitt, K. H., and Waldman, L. *Z. Naturf.* 15a, 843 (1960).
28. Facy, L. *Arch. Met. Geophys. Bioklim.* 8, 229 (1955).
29. Tyndall, J. *Proc. R. Inst. GB.* 6, 3 (1870).
30. Goldsmith, P., and May, F. G. in *Aerosol Science.* edited by Davies, C. N., New York and London: Academic Press, 1966.
31. Waldman, L., and Schmitt, K. H. in *Aerosol Science.* edited by Davies, C. N., pp. 137–162, New York and London: Academic Press, 1966.
32. Sharp, J. H., Brindley, G. W., and Arch, B. N. N. *J. Amer. Ceram. Soc.* 49 [7] (1966): 379–82.
33. Hulbert, S. F. *J. Brit. Ceram. Soc* 6 [1] (1969): 11–20.
34. Hancock, J. D., and Sharp, J. H. *J. Amer. Ceram. Soc.* 55 [2] (1972): 74–77.
35. Valensi, Gabriel *C. R. Acad. Sci.* 202 [4] (1936): 309–12.
36. Taplin, J. H. "Proceedings of IVth International Symposium on Chemistry of Cement. pp. 263–66, Washington, D.C., 1960, *National Bureau of Standards Monograph 43,* Vol. I, U.S. Government Printing Office, Washington, D.C., 1962.
37. Taplin, J. H. *J. Chem. Phys.* 59, 1 (1973): 194–99.
38. Ginstling, A. M., and Brounshtein, B. I. *Zh. Prikl. Khim.* (Leningrad), 23, 12 (1950): 1249–59; *J. Appl. Chem.* (USSR) 23 [12] (1950): 1327–38.
39. Avrami, M. *J. Chem. Phys.* 7, 12 (1939): 1103–12, II, ibid, 8, 2 (1940): 212–24, III, ibid, 9, 2 (1941): 177–84.
40. Austin, J. B., and Rickett, R. L. *Trans. AIME* 135 (1939): 396–415.
41. Prout, E. G., and Tempkins, F. C. *Trans. Faraday Soc.* 42 (1946): 463–72.
42. Taplin, J. H. *J. Am. Ceram. Soc.* 57, 3 (1974): 140–143.
43. Ostwald, W. *Z. Physik, Chem.* 34 (1900): 495–503.
44. Freundlich, H. *Colloid and Capilliary Chemistry.* London: Methuen, 1926.
45. Van Zeggeren, F., and Benson, G. C. *Can. J. Chem.* 35 (1957): 1150–1156.
46. Preckshot, G. W., and Brown, G. G. *Ind. Eng. Chem.* 44 (1952): 1314–1321.

47. Steele, L. R., and Geankoplis, C. J. *AIChEJ* 5 (1959): 178–81.
48. Hixson, A. W., and Baum, S. J. *Ind. Eng. Chem.* 36 (1944): 528–531.
49. Treleaven, C. R. *Ind. Chemist* 38 (1962): 231–238.
50. Linton, W. H., and Sherwood, T. K. *Chem. Eng. Prog.* 46 (1950): 258–264.
51. Heath, W. S. "Rate of Dissolution of Crystals." Ph.D. thesis, Syracuse University, 1958.
52. Marangozis, J., and Johnson, A. I. *Can. J. Chem. Eng.* 40 (1962): 231–237.
53. Williamson, J. E., Bazaire, K. E., and Geankoplis, C. J. *Ind. Chem. Eng. Fund* 2 (1963: 126–129.
54. Fan, L. T., Yang, Y. C., and Wen, C. Y. *AIChE J.* 6 (1960): 482–487.
55. Eisner, H. S., Quince, B. W., and Slack, C. *Disc. Farad Soc.* 30 (1960): 86–95.
56. Whitmore, D. H., and Moser, J. B. *J. Chem. Phys.* 33, 918 (1960).
57. Corn, M. in *Aerosol Science.* edited by Davies, C. N. pp. 359–392, New York and London: Academic Press, 1966.
58. London, F. *Z. Phys.* 63 (1930): 245–279.
59. Casimir, H. B. G., and Polder, D. *Phys. Rev.* 73, 360 (1948).
60. Lifshitz, E. M. *Sov. Phys., JETP* 2, 73 (1956).
61. Derjaguin, B. V. *Scientific American* 203, 1 (July, 1960): 47–53.
62. Hamaker, H. C. *Physica* 4 (1937): 1058.
63. Derjaguin, B. V. "Powders in Industry." *SCI Monograph* 14, pp. 102–113, 1961.
64. Nerpin, S. V., and Derjaguin, B. V. "Powders in Industry." *SCI Monograph* 14, 102–113, 1961.
65. Okhotin, V. V. "Physical and Mechanical Properties of Soils Depending Upon Their Mechanical Composition and Degree of Dispersity, 1937, Gushossdor. Quoted in reference 64.
66. Derjaguin, B. V. *Kolloid Zh* 16, B3 (1954).
67. Ruckenstein, E., and Prieve, D. C. *AIChE J.* 22, 2 (1976): 276–283.
68. Loeb, L. B. *Static Electrification.* Berlin: Springer-Verlag, 1958.
69. Kunrl, E. N. *J. App. Phys.* 1950a, 21, 820.
70. Jordan, D. W. *Brit. J. App. Phys.* Suppl 3 (1954): S194–197.
71. Kordecki, M. C., Gladden, J. K., and Orr Jr., C. Final Report B-148, Eng. Expt. Station, Georgia Tech., 1959.
72. Davies, C. N. in *Aerosol Science.* edited by Davies, C. N. pp.

393–446, New York and London: Academic Press, 1966.
73. Davies, C. N. *Trans. Inst. Chem. Eng.* Suppl. 25 (1947): 25–39.
74. Walton, W. H. *Brit. J. App. Phys.* Supp. 3 (1954): 29–37.
75. Lidwell, O. M. *Nature.* 158, 61 (1946).
76. Davies, C. N., Colliery Guard, pp. 442–445, 1954.
77. Dawes, J. G., and Slack, A. Rep. Saf. Mines. Res. No. 105, 1954.
78. Davies, C. N. unpublished report, quoted in reference 73.
79. Johnstone, H. F., Winsche, W. E., and Smith, L. W. *Chem. Rev.* 44, 2 (1949): 353–371. (Also see, C. N. Davies, ref. 72).
80. Kaye, B. H., personal communication.
81. Bosanquent, C. H., Carey, W. F., Halton, E. M. *Proc. Inst. Mech. Eng.* 162, 335 (1950).
82. Morton, B. R. *J. Fluid Mech.* 5, 1 (1959): 151–163.
83. Morton, B. R. *Int. J. Air. Poll.* 1 (1959): 184–197.
84. Morton, B. R., and Middleton, J. *J. Fluid Mech.* 58, 1 (1973): 165–176.
85. Stiener, J.T. Int. Clean Air Conference Rotorua, New Zealand, 1975, vol. 2, pp. 361–380, Clean Air Soc. of Australia and New Zealand.
86. Taylor, G. I. *Phil. Trans Roy. Soc. A* 1, 215, 1915.
87. Wallington, C. E. Int. Clean Air Conference (see ref. 85), vol. 2, pp. 381–402, Clean Air Soc. of Australia and New Zealand.
88. Tyldesly, J. B., and Wallington, C. E. *Quart J. R. Met. Soc.* 91, 158 (1965).
89. Ukeguchi, N., Okamoto, H., and Ohba, R. *Int. Clean Air Conference* (see ref. 85), pp. 330–348.
90. Moore, D. J. *Int. Clean Air Conference* (see ref. 85), vol. 1, pp. 298–319.
91. White, H. J. *Industrial Electrostatic Precipitation.* Reading: Addison Wesley, 1963.
92. Thompson, J. K., Clark, R. C., and Fielding, G. H. "Dielectrophoretic Air Filtration: Progress and Problems." in *Novel Concepts, Methods and Advanced Technology in Particulate-Gas Separation,* Proceedings of NSF Workshop at U. of Notre Dame, ed. T. Ariman, Apr. 20–22, 1977, sponsored by NSF Grant # ENG 77-02016 and EPA R 805148-01.
93. Scholz, P. NSF Workshop "Effects of Electric and Acoustic Field on Particle Collision Rates in Aerocolloidal Suspension." *Novel Concepts, Methods and Advanced Technology in Particulate-Gas Separation,* Proceedings of NSF Workshop at U. of Notre Dame, ed. T. Ariman, Apr. 20–22, 1977, sponsored by NSF Grant # ENG 77-02016 and EPA R 805148-01.

94. Helftricht, D. J. "Electrostatic Filtration and the Addition Design and Field Performance," *Novel Concepts, Methods and Advanced Technology in Particulate-Gas Separation,* Proceedings of NSF Workshop at U. of Notre Dame, ed. T. Ariman, Apr. 20–22, 1977, sponsored by NSF GRant #ENG 77-02016 and EPA R 805148-01.
95. Melcher, J. R. "Electrofluidized Beds for Industrial Space Air Pollution Control," *Novel Concepts, Methods and Advanced Technology in Particualte-Gas Separation,* Proceedings of NSF Workshop at U. of Notre Dame, ed. T. Ariman, Apr. 20–22, 1977, sponsored by NSF Grant # ENG 77-02016 and EPA R 805148-01.
96. Mednikov, E. P. "*Acoustic Coagulation and Precipitation of Aerosols.*" English translation, New York: Consultants Bureau.
97. Davies, C. N. *Air Filtration.* London: Academic Press, 1973.
98. Lundgren, D. A. and Whitby, K. T. *Indus. Engg. Chem.* 4 (1965): 345–349.
99. Emi, H. and Yoshioka, N. First Pacific Chem Eng. Congress, 1972, "Prediction of Collection Efficiencies of Aerosols by Fibrous Filters," New York: AIChE.
100. Happel, J. *Am. Inst. Chem. Eng.* 5 (1959): 174–177.
101. Kuwabara, S. *J. Phys. Soc. Japan* 14, 4 (1959): 527–532.
102. von Smoluchowski, M. *Phys. Z.* 17, 557, 585 (1916).
103. von Smoluchowski, M. *Z. Phys. Chem.* 92, 129 (1918).
104. Whytlaw-Gray, R. and Patterson, H. S. *Smoke.* London: Edward Arnold, 1932.
105. Gillespie, T., and Langstroth, G. O. *Can. J. Chem.* 30, 1003 (1952).
106. Ziebel, G. in *Aerosol Science,* edited by Davies, C. N. pp. 31–58, New York and London: Academic Press, 1966.
107. Cullity, B. D. *Introduction to Magnetic Materials.* Reading: Addison Wesley, 1972.
108. Nussbaum, A. *Electronic and Magnetic Behavior of Materials.* New Jersey: Prentice-Hall, 1967.
109. Stoner, E. C., and Wohlfarth, E. P. *Phil. Trans. Roy. Soc. A.* 240, (1948): 599–642.
110. Luborsky, F. E. *J. App. Phys.* 32 (1961): 171S–183S.
111. Kittel, C., and Galt, J. K. *Solid State Physics* 3 (1956): 437–564.
112. Fowler, C. A., Fryer, E. M., and Treves, D. *J. App. Phys.* 32 (1961): 296S–297S.
113. Wohlfarth, E. P. "Permanent Magnetic Materials." in *Magnetism.* eds. Rado, G. T., and Suhl, H. 3, pp. 351–393, New

York: Academic Press, 1963.
114. Friedlander, F. J., "HGMS: Review of Single Wire Models," *Novel Concepts, Methods and Advanced Technology in Particulate–Gas Separation*, Proceedings of NSF Workshop at U. of Notre Dame, ed. T. Ariman, Apr. 20–22, 1977, sponsored by NSF Grant #ENG 77-02016 and EPA R 805148-01.
115. Drehmel, D. C., and Gooding, C. H. "Magnetic Separation of Particulate Air Pollutants," *Novel Concepts, Methods and Advanced Technology in Particulate–Gas Separation*, Proceedings of NSF Workshop at U. of Notre Dame, ed. T. Ariman, Apr. 20–22, 1977, sponsored by NSF Grant #ENG 77-02016 and EPA R 805148-01.
116. Katari, V., Issacs, G., and Devitt, T. W. EPA-650/2-74-115, October, 1974.
117. Hedley, W. H., Mehta, S. M., and et al., EPA-650/2-74-117, November, 1974.
118. Dealey, J. O., and Killin, A. M. EPA-450/2-74-008, May, 1974.
119. Melville, D., Paul, F., and Roath, S. *IEEE Trans. Magnetics MAG-11* 6 (November 1975): 1701–1704.
120. Oberteuffer, J. A. *MAG-9* 3 (1973): 303–306.
121. Parrott, J. E., and Stuckes, A. D. *Thermal Conductivity of Solids.* London: Pion, 1975.
122. Maxwell, J. C. *A Treatise on Electricity and Magnetism.* second ed., vol. 1, Oxford: Clarendon Press, 1881.
123. Lord Rayleigh. *Phil. Mag.* 34 (1892): 481–502.
124. Bruggeman, D. A. *Ann. Physik* 24 (1935): 636–664.
125. Meredith, R. E. *At Energy Comm.* U.S. UCRL-8667, July, 1959.
126. Johnson, F. A. At En. Res. Est. AERE. R/R 2578, 1958 (UK).
127. Jefferson, J. B., Witzell, O. W., and Sibitt, W. L. *Ind. Eng. Chem.* 50 (1958): 1589–1592.
128. Deissler, R. G., Elan, C. S. National Advisory Committee for Aeronautics, RM, #52C05, "Investigation of Effective Thermal Conductivities of Powders," 1952.
129. Yagi, S., Kunii, D., and Wakao, N. "International Developments in Heat Transfer." ASME, part IV, pp. 742–749, New York, 1961.
130. Schotte, W. *AIChE J.* (1960): 63–67.
131. Gopalarathnam, C. D., Hoelscher, H. E., and Laddha, G. S. *AIChE J.* 7 (1961): 249–253.
132. Wen, C. Y., and Miller, E. N. *Ind. Eng. Chem.* 53, 51 (1961).
133. Cambell, J. M., and Huntington, R. L. *Petrol. Refin.* 31 (1952): 123–131.

134. Yagi, S., and Kunii, D. *AIChE J.* (1960): 97-104.
135. Baumeister, E. P., and Bennet, C.O. *AIChE J.* 4 (1958): 69-74.
136. Sunkoori, N. R., and Kaparthi, R. *Chem. Eng. Sci.* 12 (1960): 166-174.
137. Malek, M. A., and Tendolkar, G. S. *Trans. Ind. Inst. Chem. Engrs.* 6 (1953-54): 90-104.
138. Gupta, A. S., and Thodos, G. *AIChE J.* 8 (1962): 608-610.
139. Wen, C.Y., and Leva, M. *AIChE J.* 2 (1956): 488.
140. Kagi, S., and Kunii, D. "Int. Devs. in Heat Transfer." AIME, pp. 750-759, New York, 1961.
141. Zabrodsky, S. S., ibid. pp. 737-741.
142. Wen, C. Y., and Miller, E. N. *Ind. Eng. Chem.* 53 (1961): 51-53.
143. Sleicher, C. A., and Churchill, S. W. *Ind. Eng. Chem.* 48, 1823, 1956.

Chapter 8

HAZARDS

8.1 THE THREAT TO HUMANKIND

Finely divided matter can be hazardous in a number of ways. One classification of the types of hazards is: dust explosions and fires, health hazards, and desertification. The interesting thing about all of these multifarious hazards is that they are insidious in nature. For example, consider the dust explosion hazard. The plant can contain large quantities of dust on exposed horizontal surfaces for many years. In such a condition most of us tend to discount the ever present danger of a dust explosion. But if there occurs a primary explosion (from whatever cause) the previously quiescent dust layers may be disturbed and the newly formed dust clouds may then act as energy sources for the secondary explosion(s). If one considers that good, solid-looking brick walls may only be able to withstand a pressure of not more than 1–3 psi, it is evident that the secondary dust explosion may well destroy the plant and building at one and the same time. Similarly, we can rather easily become accustomed to breathing in quantities of finely divided materials (including fibers) in our workplaces and also in our homes, gardens, and streets. It may be many years before the effects of such inhalations can become evident as indicated by health problems, mortality expectancy differences or both, as compared with the bulk of the unaffected population. The problem of desertification may be caused by overgrazing, drought, and many other concomitant factors, but the actual advance of the desert is carried out quietly, spasmodically, and insidiously (except for sand storms) by sand particle saltation under the action of the prevailing winds. Just as in the case of the explosion hazard, familiarity breeds indifference to other fine particle hazards, also.

8.2 DUST EXPLOSIONS

An explosion of a portion of matter in finely divided form can be viewed as being a rapid or uncontrolled thermal expansion of the gas initially present and of additional gaseous matter generated by the chemical reactions occurring therein. There occurs a pressure wave ahead of the flame which is assumed not to reach more than ≈ 100 m/s (which is approximately the velocity of sound in air). This phenomenon is termed deflagration and it is the usual way in which dust explosions occur. It should be differentiated from detonation in which the flame and the shock wave move together. There is no well documented case of a dust explosion occurring by detonation.[1] The flame speed varies with the conditions of the system and is a complicated function of the material; its characteristics of size, shape, etc., the plant or apparatus, the time of burning; and many other factors. It is not possible at this time to predict the flame speed. For example, in gases, the maximum flame speed occurs at the stoichiometric ratio. In dust clouds one can have the maximum flame speed at a large excess of the fuel (the dust). It was observed above that it is necessary to provoke the particles to move in order to produce an explosion. This fact is directly related to the existence of two well-defined explosions: the primary explosion and the secondary explosion. The primary explosion may be small simply because there is insufficient material floating around in the vicinity to whip up a really serious explosion. However, what can happen is that the primary explosion, although relatively tame, does agitate the dust into a cloud and this in turn can set the stage for the big secondary explosion. This series of events is common in dust explosions and it is the main reason why it is claimed above that the danger of dust explosions is insidious; operators tend to become accustomed to dust lying around here and there. So long as the layer is undisturbed, this attitude may seem justified. After all, time is money and the dust is in out-of-the-way places on piping, ducts, and the like; the main walkways are normally kept clean. The problem is that even a thin dust layer amounts to billions and even trillions of particles. Because people are familiar with the uses and properties of materials in solid form when it poses no explosive hazard at all, the danger of that same material in dust or fine powder form is discounted although ignition is much easier as the particle size decreases. The sum total of these factors extends the dangers of dust explosions through the following principal industries: agricultural; chemical, including dyestuffs; coal, mining et al; foodstuffs (all); metals; pharmaceutical; plastics; and woodworking. Some examples of grain dust explosions are given in Table 8.1.

Dust Properties

There are some properties of the dusts which can be measured and

Table 8.1

CUMULATIVE TOTALS OF ELEVATOR AND FEED MILL EXPLOSIONS IN THE UNITED STATES DURING THE PERIOD 1958–1978

(data abstracted from a report entitled "Prevention of Dust Explosions in Grain Elevators – An Achievable Goal, USDA, April 1979)

Year	*Cum. Explosions in*		*Cum.*	*Cum.*
	Elevators	*Feed Mills*	*Deaths*	*Injuries*
1958	8	2	2	27
1959	16	4	5	45
1960	24	8	9	63
1961	32	10	9	80
1962	40	11	12	131
1963	48	17	15	161
1964	50	33	18	183
1965	56	36	20	188
1966	65	41	22	210
1967	77	46	23	224
1968	86	53	35	262
1969	87	58	39	275
1970	97	58	40	289
1971	106	59	44	303
1972	108	65	51	326
1973	114	67	53	336
1974	124	72	66	373
1975	130	75	70	392
1976	148	79	92	474
1977	151	87	157	558
1978	158	92	164	605
Av per year	7.9	4.6	8.2	30.25

evaluated. Ignition properties that can be measured include the Minimum Ignition Temperature (MIT), the Minimum Ignition Energy (MIE) and the Minimum Explosion Concentration (MEC). None of these is a "fundamental" property and therefore the apparatus specification within which the tests are made combined with the test procedures are important and should be adhered to strictly. The MIT test is one in which a known amount of dust is introduced into an elevated temperature chamber. The lowest temperature to cause ignition is taken as the MIT. The Godbert-Greenwald Furnace (modified) is a standard apparatus for this test.[2] It is shown in Fig. 8.1. The MIE is measured by noting the least electric spark energy required to ignite. This test is carried out in a vertical tube apparatus. The MEC test is conducted in a Hartmann (vertical tube) apparatus on a known sample. A diagram of this equipment is shown in Fig. 8.2.[2]

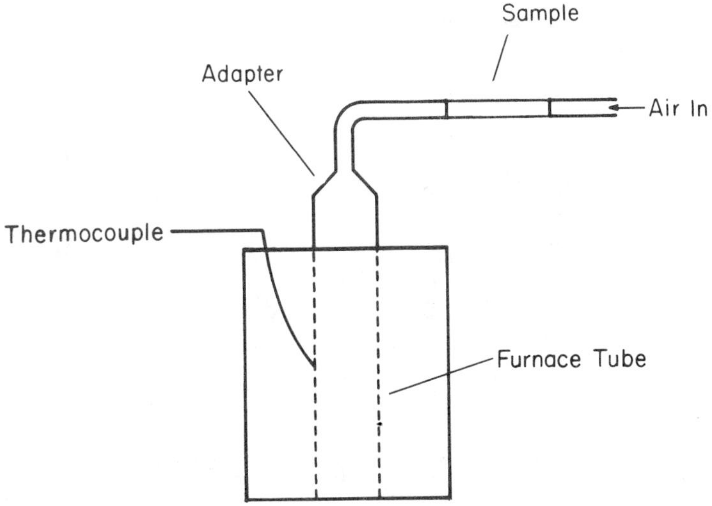

Fig. 8.1 Furnace apparatus for Minimum Ignition Temperature determination.[1,3]

A diagram of the procedure that can be followed in the United States to evaluate dust explosability is shown in Fig. 8.3.[3] Other countries use procedures which differ somewhat. The reader is referred to reference 3 for additional details.

Hazard Evaluation

When appropriate tests are selected from the procedures outlined in Fig. 8.3 the rating of the particulate matter as a hazard can be determined in accordance with the following scheme. First, five properties of the dust are determined:

Minimum Ignition Temperature	MIT,
Minimum Ignition Energy	MIE,
Minimum Explosion Concentration	MEC,

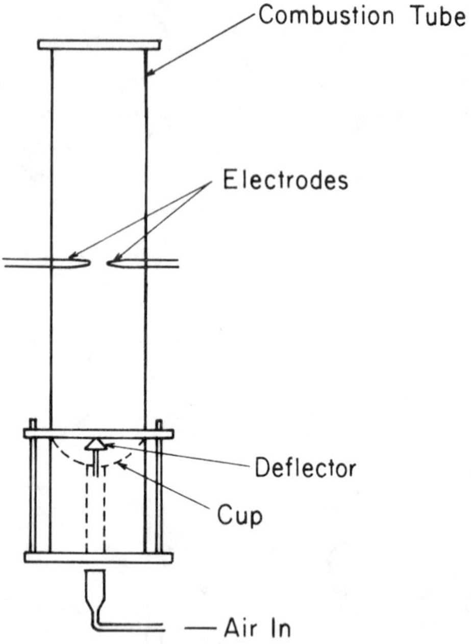

Fig. 8.2 Vertical tube apparatus (Hartmann) for Minimum Explosion Concentration.[1,3]

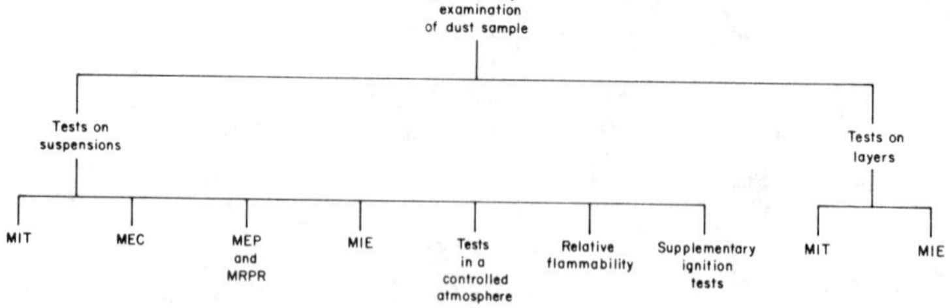

Fig. 8.3 Basic scheme of tests in United States.[1,3]

Maximum Explosion Pressure	MEP, and
Maximum Rate of Pressure Rise	MRPR .

Then, the above set of five properties is also determined for a sample of Pittsburgh coal dust to serve as a standard for comparison.

Two parameters are then calculated from the above test data, the ignition sensitivity and the explosion severity. Both of these parameters are obtained by comparison with the Pittsburgh coal as standard.

$$\text{Ignition Sensitivity} = \frac{\text{MIT} \times \text{MIE} \times \text{MEC (Pittsburgh coal dust)}}{\text{MIT} \times \text{MIE} \times \text{MEC (sample dust)}} \tag{8.1}$$

$$\text{Explosion Severity} = \frac{\text{MEP} \times \text{MRPR (sample dust)}}{\text{MEP} \times \text{MRPR (Pittsburgh coal dust)}} \tag{8.2}$$

The Index of Explosibility is then the
Ignition Sensitivity × Explosion Severity (8.3)

The relative explosion hazard rating is assessed on a basis of the value or the range of value of the three parameters defined in Eqs. (8.1), (8.2), and (8.3) as shown in Table 8.2.[3]

Table 8.2

RELATION BETWEEN HAZARD RATING AND INDEX OF EXPLOSIBILITY[3]

Relative explosion hazard rating	*Ignition sensitivity*	*Explosion severity*	*Index of explosibility*
Weak	<0.2	<0.5	<0.1
Moderate	0.2–1.0	0.5–1.0	0.1–1.0
Strong	1.0–5.0	1.0–2.0	1.0–10
Severe	>5.0	>2.0	>10

Explosion Protection

Ideally it would be most satisfactory to avoid all sources of ignition in order to prevent dust explosions from ever getting started. This laudable aim is not achievable in practice because it is not practicable to preclude all sources of ignition from plant operations. The greatest hazards of dust explosions occur in plants with continuous processing.

Sources of ignition can include: naked flames (the most dangerous source), smoldering matter, hot surfaces (especially horizontal ones), welding and cutting operations (procedural controls can help here), friction and impact (bucket elevators and also low energy sparks from hand tools), electrical sparks (sources such as electric mains, batteries, static electricity... earth), and spontaneous heating (Arrhenious effect). Since the potential ignition sources listed are ubiquitous, it is easy to understand why the objective is to dust proof the plant—operators aim for dust tightness rather than flame proofing. In order to reach an acceptable level of protection against dust explosions, certain design and operational procedures are recommended. Some of these are: minimizing dust cloud formation, where necessary providing equipment containment (e.g., comminution equipment), separating the plant units in order to confine the explosion, providing venting according to publicly accepted standards (the venting duct length is critical, but must not exhaust into occupied or public access area), implementing inerting where necessary (flue gas, N_2), and employing automatic suppression.

In practice, since not all industrial plant and equipment were built yesterday, there is a broad spectrum of ages of plant just as there is a broad spectrum of level of operator knowledge and understanding of the incipient problems of dust explosions. Consequently, dust explosions do occur and their causes are many and varied. Some are itemized in Table 8.3.[4]

Table 8.3

CASE HISTORIES: DUST EXPLOSIONS[4]

Industry	*Source*	*Ignition*	
furniture factory	extraction ducting	match	fluff
Chemical	mixer	mill	
chemical	fluid bed drier	static	dust and ethanol
foodstuff	spray drier	spont heating	
sugar	bucket elevator	friction	
milk	spray drier	spont heat and lub oil	
animal feedstuff	silo	electric lamp	
chemical	pneumatic system	static	

8.3 HEALTH HAZARDS

Lung diseases such as silicosis and various skin ailments have been known to be a product of particulate irritation. There is an increasing technology of particulates which is concerned with physiologically active sprays and the like. These products include chemical and germ warfare agents, insecticides, therapeutic agents, radioactive fallout, and many others.

Inhalation of Particles

The size and density of particles is the major factor in determining whether they are inhaled and what portion of the lung they reach. The respiratory system consists of the nose and mouth leading to the trachea which splits into the left and right bronchial tubes. Each of these divides into successively finer and finer bronchioles eventually terminating in the cavities known as the alveoli (illustrated in Fig. 8.4).

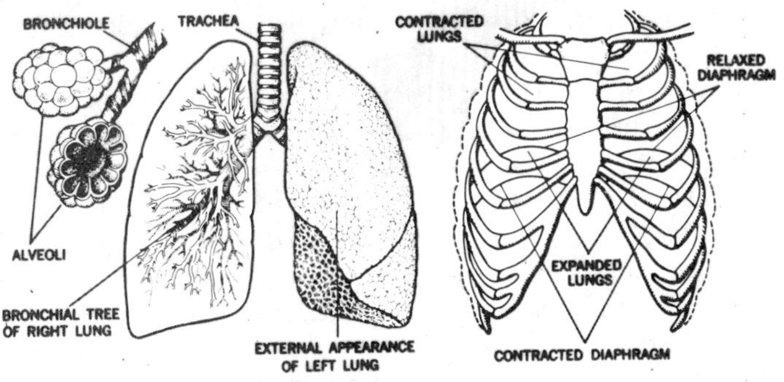

Fig. 8.4 The alveoli are the functional units of the lungs. Oxygen in the air breathed into the lungs is diffused through them into the blood and carbon dioxide carried in the blood from the cells as waste is diffused through them into the lungs and exhaled.[33]

The bronchioles, bronchial tubes, and trachea are covered with fine hairs which are termed cilia. These cilia move back and forth in ripple-like waves. It should be noted that the whole system is very tortuous and that particle deposition in the base of the lung is not easy.

In normal circumstances a breathing man displaces 500 cc of air per breath (the tidal air), and this can increase by 1600 cc in times of great exertion. This extra amount is called the complemental air. A violent expiration will retain 1000 cc of air (residual air) in the lungs but after ordinary expiration there will be 1600 cc of supplemental air left in the lungs. The combined tidal, complemental, and supplemental air is the vital capacity. About 1500 cc corresponds to the anatomical dead space in the lungs. The total internal surface area of both lungs adds up to ca. 300 ft^2.

If a cloud of particles ($0.01\ \mu < d < 10.0\ \mu$) is inhaled, those with diameters in the range ($0.1\ \mu < d < 1.0\ \mu$) have the best chance of being exhaled, because in this size range the behaviors of the particles are optimal for human expiration. The coarse particles tend to impact upon the nasal, tracheal, and bronchial linings. Under the action of the cilia, these deposited particles are swept up into the throat and swallowed. The digestive tract can deal with them more easily than can the lung system. On the other hand, the very fine particles with diameters smaller than 0.1 μ are carried by the air flow into the base of the lung by a mechanism of diffusion.

A calculation using dimensional analysis has been made in a study of the behavior of submicron particles in the lung.[5] It has been suggested that charges on particles may influence their behavior and also that fibrous particles may be retained more compared with spherical particles. There is some evidence that an electric charge on the inhaler's person promotes retention of particles in the lung. Although insoluble particles may reach the alveoli, most of them actually are removed by the cilia, but if the particles are soluble, the cilia may be ineffective in removing them; for example, $PbCO_3$ produces acute Pb poisoning whereas $PbCrO_3$ has little or no ill effect. Ultrafine insoluble particles which reach the alveoli are ingested by large cells called phagocytes which may then transport the particles to the liver, etc. In this way the alveoli are cleansed. Some retention of dust may last for years as evidenced by black sputum from long retired miners.

Effects of Fine Particles

There are three important categories of effects of fine particles: fibrosis-producing dusts, toxic dusts, and irritants.

Fibrosis: This disease is caused by a number of particulates, including quartz (SiO_2), coal, and asbestos, and produces increased collagen in lymph glands adjacent to the alveoli. SiO_2 produces silicosis and the finer particles are the most active ones. The exact reason(s) for the production of the disease is not clear but it may be that the silica acts as a special type of catalyst. Silicosis may be "cured" by Al powder inhalation. This treatment has been and is presently used in control of the disease. Asbestosis is similar to silicosis except that the hydrated Mg silicate is fibrous and that malignant growths may occur more often than in the case of silicosis. Intense functional symptoms precede x-ray indications. Coal dust is another cause of lung disease. Not only does the silica produce silicosis but the combination of coal dust and SiO_2 is said to produce anthrosilicosis–modified silicosis.

Beryllium poisoning is a serious danger to those persons in contact with Be, its alloys, or its compounds. Although people differ in sensitivity to the element, it would be accurate to describe it as highly toxic to the whole system, including the lungs. From the acute form of Be poisoning, which may involve part or all of the pulmonary system, the patient usually recovers. However, the chronic form may be delayed in its onset and is often fatal. It has been found that good ventilation reduces the plant hazard to tolerable proportions,[6] $< 2\ \mu g/m^3$.

Asbestos fibers that may be inhaled are carefully controlled because of their carcinogenic potential. Although the idea that their fibrous shape may be the primary cancer-promoting characteristic has been skeptically received in some quarters, recent Japanese data on the carcinogenic effects of glass fibers may cause some reconsideration of the shape carcinogen hypothesis.

Irritants

Normally innocuous substances (foreign protein) may produce allergic reactions in some people. For example, pollens, and household dust are relatively of large diameter (10-100 microns); hay fever and asthma are therefore largely diseases of the upper respiratory system. Cotton mill workers have an allergic reaction (emphysema, etc.) called byssinosis. Irritants are substances which are normally damaging to lung tissue and an important aerosol is a suspension of sulfuric acid droplets in air. In general it has been found that although small particles rapidly produce irritation, large ones produce effects which wear off more slowly. This is explained by the high rate of solution of the smaller particles–a more rapid process by which its effect quickly wears off.

Toxic Dusts

It is more likely that a toxic dust will poison by inhalation than by swallowing because although the coarser particles are removed by cilia to the digestive tract, the finer (or more soluble) particles enter the blood stream.

The classic poison is Pb poisoning. It is curable once the victim is removed from contact with the metal. Cd oxide poisoning is often fatal but none or few chronic cases occur in U.S.A. More common than any of the above is poisoning from the particles of insecticides. There are three categories:

Nerve Gas Types	1. Organic pyrophosphates $(CH_3O)_4P_2O_3$ 2. Alkyl thiophosphates (e.g., parathion). People have been killed with this substance. It is banned in many countries
Tissue Acting Type	3. Phosphoramides [e.g., ethyl bis (dimethylamido) phosphate]

In addition to the straight effect of the particles, it is possible for an interaction between the particles and a gas to promote an increased reaction or effect. For example, SO_2 is known to deletereously affect cilia action and so permit fine particles to reach and remain in place where they otherwise would not. This interactive effect is termed synergistic.

Radioactive dust presents a special case of particulate damage. Acute damage more often than not kills the cells but the chronic situation may tend to produce partially recovered cells which become malignant. The classic case was with the girls who painted radium on watch dials to make them luminous in the dark. They used to lick the brushes in order to do a precise job. Cancer was rife in these operators. However, it has since been pointed out that the dust hazard may have been partially responsible. In general radioactive poisons are 10^6-10^7 times more dangerous than are chemical poisons.

Therapy usually involves dilation applications (e.g., using phenylephere) and treatments with antibiotics. In general such treatment is equivalent to a slow continuous injection.

Skin

The skin consists of two essential parts—the outer epidermis and the inner corium as shown in Fig. 8.5. The outer layer is the epidermis, which flakes off and contains no blood cells. For bleeding to occur the corium has to be sliced. It appears that the main function of the skin is to protect the body and that this protection is effective against radioactive dust (Japanese fishermen in 1964 near Marshall Island explosion) but not against nerve gases specially developed for their ability to diffuse through the skin.

Air Pollution

This pollution is of two main types: one, known as industrial air pollu-

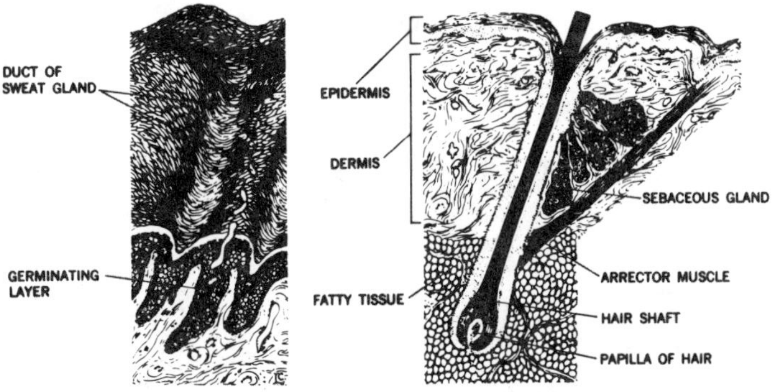

Fig. 8.5 The epidermal structure is clearly seen in the diagram of skin at the soles of the feet, left, where it adapts to the function of the feet and thickens to give necessary additional strength. Diagram right, shows the usual structure of the skin. There are three layers; the protective epidermis; the dermis, which contains blood vessels, nerves, glands, and hair roots; and the fatty cushion of subcutaneous tissue.[33]

tion, refers to single source pollution. It is easy to track down and therefore rectification, although perhaps expensive, may be easy to carry out. The other main type of pollution is smog, of which there are two kinds. One is called photochemical smog produced in automobile-using communities and the other is that produced in coal-burning communities. Strictly speaking smog is a combination of smoke and fog.

1. Photochemical Smog

Various reactions have been proposed for this type of process and two are mentioned here:

$$NO_2 + h\nu = NO + O$$

and

$$O + O_2 + M = O_3 + M$$

$$\text{Low energy} \qquad \text{High energy}$$

$$NO_3 + O_3 = NO_2 + O_2 \qquad (8.4)$$

and peroxy radicals react with oxygen:

$$RO_2 + O_2 = RO + O_3$$

For example, the irradiation of diacetyl in the presence of oxygen is described by

$$CH_3 = C - C - CH_3 + h\nu = 2\,CH_3 - \overset{\overset{\displaystyle O}{\|}}{C}$$

$$2CH_3 - C + 2\,O_2 = 2CH_3 - \overset{\overset{\displaystyle O}{\|}}{C} - OO \qquad (8.5)$$

and

$$CH_3 - C - OO + O_2 = CH_3 - \overset{\overset{\displaystyle O}{\|}}{C} - O + O_3$$

The first set of reactions accounts for 3 pphm O_3 and the second set of reactions accounts for the remainder. During one day the ozone can vary: 2½ pphm at night rising to 40 or 50 pphm at noon.

The ideal conditions for the development of a photochemical smog are little or no wind plus an inversion which minimizes mixing between the warm air above and the cooler air below, as shown in Fig. 8.6.[7] The results of an examination of the contents of such a smog is shown in Table 8.4.[8]

2. Coal Burning Smog

These smogs contain large quantities of aromatic, aliphatic, and cyclic compounds and it is therefore quite likely that they contain carcinogens. When compared with photochemical smog they contain more inorganic gases such as SO_2, NO_2, NO, HF, CO, and NH_3 but a concentration of ozone may be produced although it may be very quickly reacted out of the system. However, in the vicinity of large industrial cities the ozone concentration can be quite high even though the smog is not very intense. Actual figures for a number of U.S. cities and regions for dates shown are given in Tables 8.5 and 8.6. The compounds over the first four cities are mainly photochemical and those over the second four are of the other smog type. The first four have a high percent of benzene soluble and a low percent of sulfate compared with the others. The clean air analysis of the nonurban locations in Table 8.6 shows that these are not free of pollutants.

Fog droplets are allegedly 4.25 μ in diameter with a highly skewed distribution probably due to incorrect assessment of the smaller particle sizes.

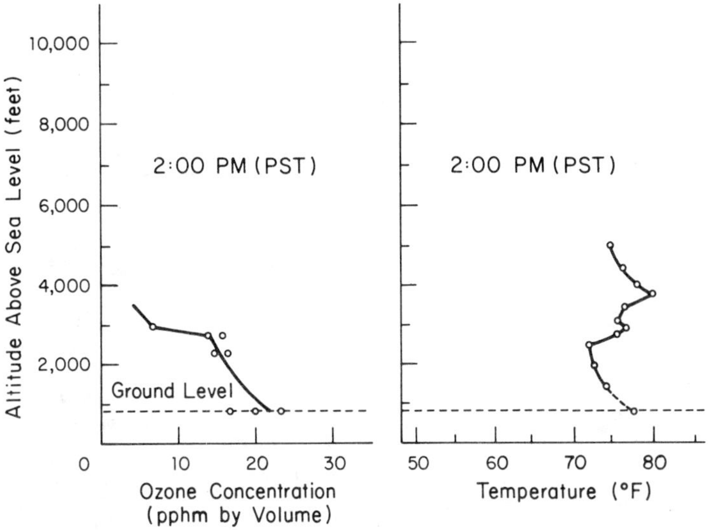

Fig. 8.6 Vertical ozone and temperature profiles over Rose Bowl area, July 17, 1952.[7]

Many actual measurements indicate the presence of carbon agglomerates (soot particles) which form almost immediately upon leaving the chimney, due to secondary bonding forces, metallurgical dusts from 0.001-100 μ, and fly ash 1-300 μ.

Effects of Smog

It would be difficult to name one good effect. Bad effects include poor visibility, health problems, and, not least, the cost of reducing the smog to bearable proportions.

Visibility is decreased in two ways: light intensity reaching an object is reduced and so also the reflected intensity is reduced. Because the smog particles reflect the light in the direction of the observer, the contrast is much reduced. This is discussed more fully elsewhere.[10]

The most serious acute effect on human health was observed in the London Smog of December 5-9, 1952. In this smog more than 3500 persons died. For the most part they were found dead in bed. Because the majority of victims suffered from asthma and emphysema, it must be that their condition was partly a chronic effect of previous smogs. There are many other examples of deaths in cities caused by smogs.

The actual harmful constituents causing the acute effects may be sulfuric

Table 8.4

LOS ANGELES PHOTOCHEMICAL SMOG–ANALYSIS OF PARTICULATE MATTER FROM THE AIR AND ANALYSIS OF SAMPLE COLLECTED FROM WESTINGHOUSE PRECIPITRON[8]

Minerals and other inorganic substances about 60 percent of total	Water-soluble fraction, about 15 percent of total; Water-insoluble fraction, about 45 percent of total	Elements identified by emission spectrograph:	
		Large amount (10 percent plus)	Silicon Aluminum Iron
		Small amount (1–9 percent)	Titanium Calcium
		1 percent	Magnesium Barium Sodium Potassium
		Very small amount (0.1–0.9 percent)	Lead Zinc
		0.1 percent	Vanadium Manganese Nickel
		Trace (0.01–0.1 percent)	Tin Copper Zirconium Strontium
		0.001–0.01 percent	Boron Chromium
		0.001 percent	Bismuth Cobalt
		Substances identified by Chemical means:	PERCENT
		SiO_2	14.3
		Iron and Aluminum	7.8
		Calcium (as Ca)	5.2
		Fluoride (as F)	0.05
		Sulfate (as H_2SO_4)	2.5
		Ammonia	0.70
		Nitrate (as HNO_3)	4.8
		Chloride (as NaCl)	0.26
		Nitrite	0.00
		Sulfide	0.00
		Sodium (as NaCl)	4.6
Organic compounds soluble in organic solvents, about 10 percent; Fibers, pollen, carbon, and highly polymerized organic material, about 15 percent		Mainly hydrocarbons Also small amounts of organic acids (0.27 percent), and aldehydes Peroxides, 0.04 percent (as H_2O_2)	
Water and volatile organic substances (by difference), about 15 percent			

Sample collected March 24 to April 19, 1949

Table 8.5

PARTICULATE CONCENTRATION FOR SELECTED CITIES FOR 1958 – VALUES ARE ARITHMETIC MEANS[9]

Station location	*A* *Suspended Particulate*	*B* *Benzene Soluble Organic Matter*		*C* *Sulfate*		*D* *Nitrate*	
	($\mu g/m^3$)	($\mu g/m^3$)	percent of A	($\mu g/m^3$)	percent of A	($\mu g/m^3$)	percent of A
Los Angeles	213	30.4	21.5	14.8	7.0	8.1	3.8
San Francisco	80	10.6	13.3	6.2	7.7	2.6	3.3
San Diego	93	12.2	13.1	7.7	6.3	4.2	4.5
Denver	110	11.0	10	6.1	5.5	2.3	2.1
New York	164	14.3	8.7	23.0	14.0	2.2	1.3
Pittsburgh	167	13.0	7.8	15.1	9.0	2.6	1.6
Cincinnati	143	13.7	9.6	12.2	8.5	2.6	1.8
Louisville	228	18.0	7.9	20.6	9.1	4.9	2.1

acid droplets in coal smog and ozone in a toxic substance in a photochemical smog. In the latter, the airborne solid particles have little or no irritating effect and formaldehyde is a major cause of acute conditions.

Chronic effects of photochemical smog produce a higher percent of lung cancer in urban areas as compared with nonurban areas. There is some indication that heart disease is also involved. Smokers are more prone to harm than are nonsmokers and therefore it is inferred that some interaction occurs.[11] Many city smogs are carcinogenic in that they have been found to contain cancer-promoting substances such as 3.4 benzpyrene:

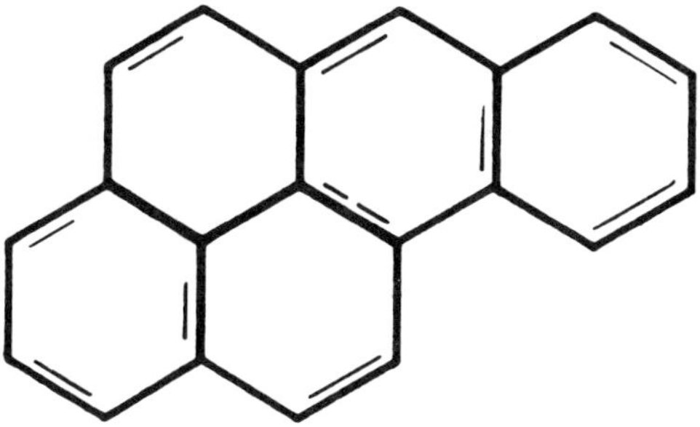

Fig. 8.7 Benzpyrene

Table 8.6

PARTICULATE CONCENTRATIONS FOR NINE NONURBAN STATIONS FOR 1958 – VALUES ARE ARITHMETIC MEANS[9]

Station location	*A* *Suspended Particulate*	*B* *Benzene Soluble Organic Matter*		*C* *Sulfate*		*D* *Nitrate*	
	($\mu g/m^3$)	($\mu g/m^3$)	percent of A	($\mu g/m^3$)	percent of A	($\mu g/m^3$)	percent of A
Acadia Natnl Park, Me.	27	2.5	9.3	5.6	21	0.9	3.3
Baldwin Co., Ala.	27	2.7	10	3.5	12.9	0.9	3.3
Bryce Canyon Park	83	2.2	2.7	1.9	2.3	0.4	0.48
Butte Co., Idaho	23	1.4	6.1	2.2	9.6	0.6	2.6
Cook Co., Minn.	44	2.4	5.5	4.0	9.1	0.5	1.1
Florida Keys	36	2.0	5.6	4.7	13.1	1.2	3.3
Huron Co., Mich.	44	1.6	3.6	6.5	14.8	1.5	3.4
Shannon Co., Mo.	37	2.0	5.4	6.5	17.5	1.8	4.8
Wark Co., N.D.	28	1.9	6.8	3.2	11.4	0.9	3.2

This substance has a melting point of 180°C and it is therefore always present as solid particles. It is more common in coal smogs than in photochemical smogs (10 to 300 or 400 μg per 103 m^3 in coal smogs versus much less in the other). For example in Montgomery, Alabama, the solid particles contained 340 μg per 1000 m^3 of air and the benzene soluble portion was 2000 μg per 1000 m^3. The main cause of the chronic effects must be the long term incorporation of the carcinogen(s) in the system.

Pb is contained in city air at a level of 0.1-1.7 $\mu g/m^3$ with a mean diameter of 0.2 μ. It is fairly soluble but it has been suggested,[7] that it may not be highly concentrated enough for real harm.

Plants are subjected to a wide variety of air pollutant effects. Photochemical smog produces "silver leaf" estimated to cost millions of dollars loss to growers.

Industrial Air Pollution

A major pollutant is fluorine from the fertilizer and the Al industries. In the manufacture of Al the Hall-Heroult process dissolves Al_2O_3 in molten cryolite (Na_3AlF_6) and the F containing particles get into the atmosphere. The Bessemer process and the Open Hearth process are big producers of very fine particulates (see Table 8.7). The electric process also produces very fine particles. In all of the cases cited the major component is iron oxide. The Bessemer process is a particularly serious problem. In the United States and many other countries serious and successful efforts have been made and continue to be made to minimize such furnace effluent. The drift and deposition of smoke from a smoke stack is shown graphically in Fig. 8.8.

Table 8.7

OPEN HEARTH FURNACE FUME

μ	*w/o*	*cum w/o*
1–3	7	7
.5–1	29	36
.15–.5	50	86
$<$.15	14	100

8.4 DESERTS AND SAND MOVEMENT

The quantity of the arable land of the world is more or less static: marginal land may be brought under cultivation but this is countered by the developing urban sprawl. The human population is increasing at a rate which places an ever-growing pressure upon utilization of land for crop and animal husbandry purposes. Although properly watered deserts are fertile, the provision of adequate irrigation is very expensive and little is actually irrigated. Coupled with this, a large proportion of the irrigated land has been destroyed by salinization. Figures for this for a number of countries are given in Table 8.8.[13]

Due to action of the prevailing winds in combination with the over utilization of the land for agricultural purposes the deserts are expanding.

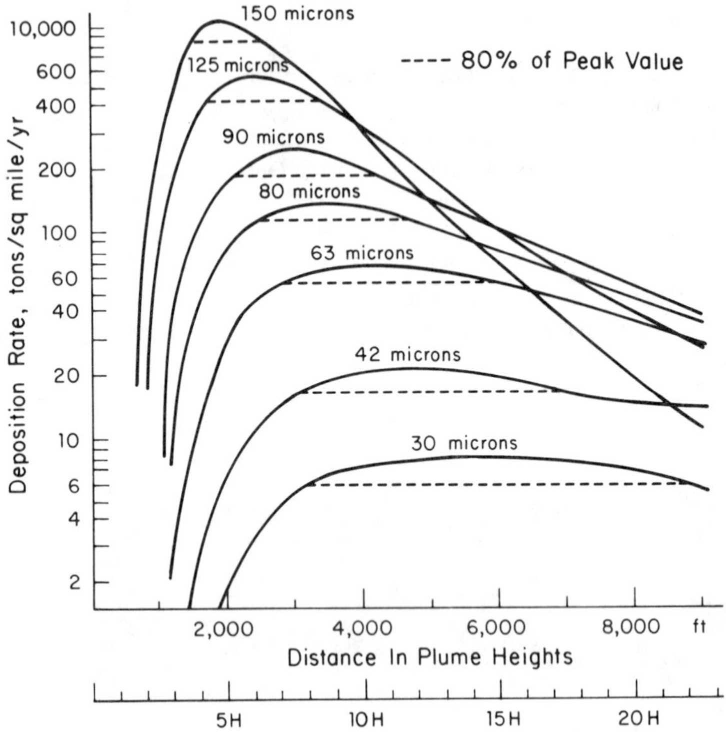

Fig. 8.8 Deposition rates of particles of different diameters.[12] Effective height = 400 ft. Rate of emission of each size of particle = 0.1 ton/h. Wind speed = 20 ft/s. Wind directional frequency, b = 0.2.

The United Nations Environmental Program has estimated that 5 through 7 million hectares (approximately 70,000 square kilometers) of formerly productive land is lost world wide annually. A major part of this is attributed to desertification.[14]

The "characteristic" path[15] of a sand particle is described by Fig. 8.9(a) As shown, the path is completely defined by the initial upward velocity of the particle, by the wind velocity of the particle, and by the wind velocity distribution.

Despite the visual appearance of a sand storm, the bulk of the material flow occurs close to the ground level. As shown in Fig. 8.9(b),[15] the mean height to which the sand particles rise is of the order of 1 cm. This is confirmed by the appearance on upright obstacles of wear caused by sand

Table 8.8

LARGE AREAS OF IRRIGATED AGRICULTURE DESTROYED BY SALINIZATION DUE TO FAULTY IRRIGATION PRACTICES[13]

	Percentage of total irrigated area affected by salinity
Iraq	20–30
Indus Valley	20
India	15
China	20
Syria	25–50
Iran	>50
Peru (coast)	33
Patagonia	20–30

(Reprinted with permission of Heyden & Son, Ltd., London).

saltation. The Bagnold equation for the rate of sand movement relates the flow of a sand of interest to a standard laboratory 0.25 mm sand:[15]

$$q = C\sqrt{\frac{d}{D}}\,\frac{\rho}{g}\,V_*'^3 \tag{8.6}$$

in which

- q is the rate of sand flow (in g/cm width of lane per s);
- d is the diameter of the sand in question;
- D is the grain diameter of the standard 0.25 mm sand;
- C is the empirical coefficient with a value of
  - 1.5 for nearly uniform sand,
  - 1.8 for a naturally graded sand, and
  - 2.8 for a sand with a wide size range;
- ρ is the air density;
- g is the acceleration due to gravity; and
- V_*' is the gradient of the wind that drives the sand. Actually, V_*' is defined by reference 15.

$$V_z = 5.75\,V_*'\log_{10}\frac{z}{K'} + V_t$$

in which

V_z = wind velocity at height z,
K' = ripple height, multiple (fraction) of
V_t = threshold velocity to move sand measured at height K'.

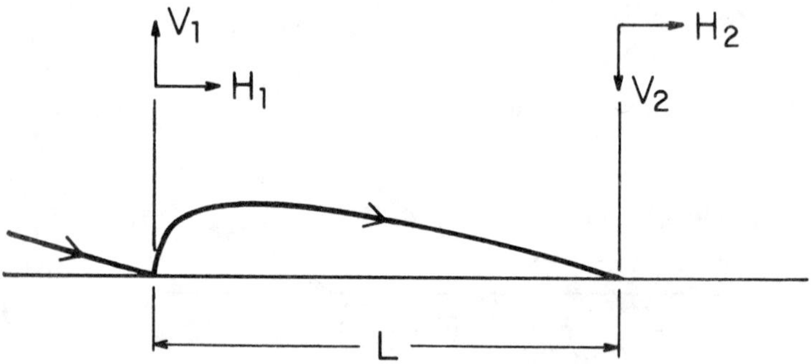

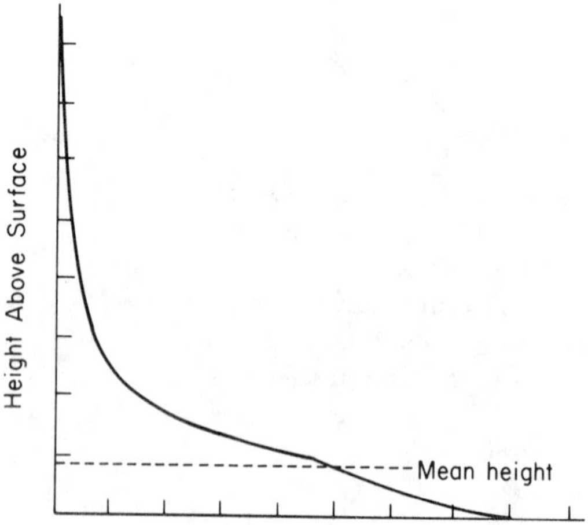

Fig. 8.9(a) "The Characteristic Path"; **(b)** approximate curve showing distribution of sand grains according to height above the surface.[15]

8.5 DUST FLAME PROPAGATION

The total process involves a number of distinct, although not necessarily discrete, steps. These include the preflame reactions, burning velocity, and burning of individual particles. The combustion of individual particles has been studied but it is not possible at the present time to extrapolate the case of the single particle to the case of a cloud of particles. The interactions within the cloud are complex and the identity of the rate controlling process(es) is in doubt in many cases. Consequently, theories often relate to specific systems and are not generalized nor apparently generalizable.[1]

The burning rate of an individual particle is inversely proportional to the radius:

$$dr/dt = -B/r \tag{8.7}$$

Upon integration, one obtains

$$t = Kd^2 \tag{8.8}$$

in which

r is the particle radius,
B is a constant,
t is the time,
d is the initial diameter of the particle, and
K is the burning constant and is of the order of 10^3 for solids.[16]

Making an oversimple assumption that the rate of diffusion of oxygen is the rate controlling factor (this is not always so, see reference 17, for example) it can be shown that the burning constant is proportional to the diffusion coefficient.[7] A comparison between calculated and observed values of k is presented in Table 8.9.[16]

Coal Dust Explosions

The initiation of a coal dust explosion is an easy task to accomplish. Its likelihood is compounded by methane which is always present in mines. Methane is dangerous in concentrations of 2-5 percent and very dangerous above 5 percent. However, in British mines there has been a reduction in the number of firedamp explosions since 1945 with the dust explosions remaining unchanged. Polish mines are much more prone to dust explosions.[18] Some idea of the seriousness of the problem can be obtained from Table 8.10.[18]

The dust originates in various ways, from mining of the coal (50 percent), mechanical preparation of the coal (37 percent) and transportation

Table 8.9

VALUES OF BURNING CONSTANT *K* FOR COKE RESIDUES OF CARBONIZED COAL PARTICLES: COMPARISON OF EXPERIMENTAL AND CALCULATED VALUES[16]

Coal	C (*d.m.f.*)	*Square-Law Index*	*K (Expt.)*	*K (Calc.)*	*Ratio*
1. Stanllyd	93.0	2.02	2125	2720	1.28
2. Five ft.	91.8	1.94	1290	2620	2.03
3. Two ft. Nine	91.2	2.25	1470	2070	1.41
4. Red Vein	89.7	2.09	1475	2275	1.54
5. Garw	88.9	2.01	1410	2030	1.44
6. Silkstone	86.9	2.25	1110	1655	1.49
7. Winter	84.0	2.18	1125	1710	1.52
8. Cowpen	82.7	1.94	1060	1775	1.68
9. High Hazel	81.9	2.14	1450	1725	1.19
10. Lorraine	79.3	2.20	992	1890	1.91
				mean ratio	1.55

of the coal (13 percent). Approximately 5 percent of the dust is retained on the walls, 5 percent on pillars and 5 percent on the road.[18] A conventional method of reducing the hazard is to spread stone dust upon the coal dust layers on the mine surfaces.

A primary variable of importance in coal dust explosions appears to be the percent of volatile matter in the coal particles. In general heat causes pyrolysis of the coal, constituting a chemical-physical change in the nature of the coal. A bituminous coal is termed thermolabile. The percent of volatile matter in coal is determined by heating the coal to 850°C under specified conditions. The processes involved in coal decomposition are complicated. A fast heating of a typical Upper Silesian Coal produces a product that has the following composition:[18] 18.4 percent H_2, 73.4 percent CO, 2.8 percent CO_2, and 4.4 percent CH_4.

The importance of the volatile matter is demonstrated by photographic observation[19] in which it was reported that two distinct stages are noted. In the first stage, comprising a fairly short time, the volatile matter is consumed. In the second stage, which lasts about 20 times as long as the first stage, the remaining solid coke particle is consumed. Various studies of coal particle combustion are listed in Table 8.11.[1]

The explosibility of a given coal dust is measured in a simple manner.

Table 8.10

THE NUMBER OF EXPLOSIONS RESULTING IN FATAL ACCIDENTS INCLUDING THOSE IN WHICH A COAL DUST FACTOR WAS ASCERTAINED [IN BRITISH MINES FROM 1936–1969 (deaths in brackets)] [18]

Years	*Number of explosions*	*Number of explosions excluding minor accidents*	*Number of explosions with the participation of coal dust*
1936–1940	43 (306)	29 (287)	3 (172)
1941–1945	51 (224)	32 (199)	4 (79)
1946–1950	31 (211)	16 (172)	4 (152)
1951–1955	15 (129)	9 (116)	1 (81)
1956–1960	14 (123)	13 (122)	4 (76)
Total for years 1936–1960	154 (993)	99 (896)	16 (560)
1961–1965	8 (68)	8 (68)	3 (25)
1966–1969	1 (2)	1 (2)	0

Table 8.11

REVIEWS OF SINGLE PARTICLE COMBUSTION[1]

Reference	*Fuels considered*	*Type of review*
20	Mainly coal	General description.
16	Coal and allied materials, various liquids.	Mathematical analysis.
21	Coal	Analysis of combustion of volatiles and of carbon.
22	Coal	Comparison of diffusional and chemical rate control.
23	Coal	Comparison of experimental results and analyses.

The limiting value of the total of incombustible solid matter contained in the coal plus the added stone dust (if any) which is just sufficient to prevent the propagation of an explosion is indicated by the symbol S and is termed the explosibility of the coal dust. The explosibility index Z is related to this in the following way:[18]

$$Z = \frac{S}{100 - S} \tag{8.9}$$

The coal dust explosibility is directly related to the percent of volatile matter in the coal as is shown in Fig. 8.10.[18]

Experimental investigations in carefully designed and regulated coal galleries have demonstrated that the explosion flame velocity reaches very high levels indeed. This is shown in Fig. 8.11. Here n is the percent of noncombustibles; all other symbols are as before.[18]

The effect of the degree of fineness of the coal dust is usually taken into account in the relationship between the coal dust explosibility and the specific surface of the dust. In this way it has been observed that there is a more or less linear relationship between Z and F as follows:[18]

$$Z = 0.0010119F + C \tag{8.10}$$

in which

F is the specific surface, and

C is a number that has a value ranging from approximately 0.78 for

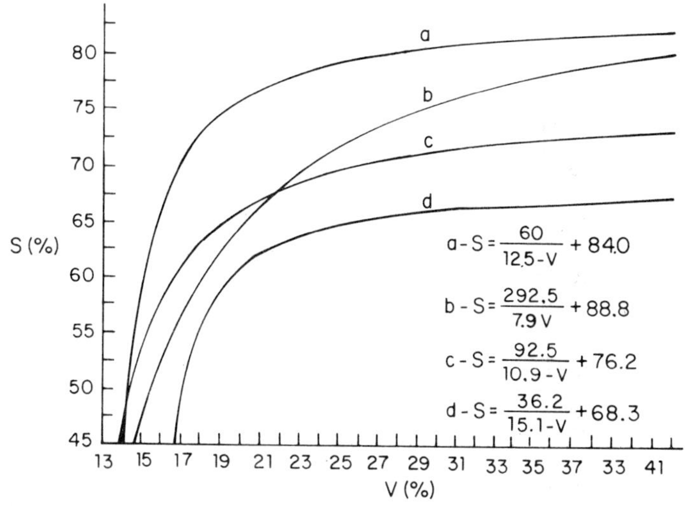

Fig. 8.10 Curves of the dependence of the coal dust explosibility *S* upon the content of volatile matter *V*: I–the d85 dust, initiator 30mPH; II–the d85 dust, initiator 50 m^3 CH_4; III–the d25 dust, initiator 30mPH; IV–the d25 dust, initiator 50 m^3 CH_4.[18]

conditions in which dispersion is easy to 0.25 for conditions in which dispersion is less easy.

Explosion Suppression

Measures that are routinely taken to help prevent dust explosions in coal mines include limiting dust formation, removal of dust, prevention of cloud formation, prevention of ignition, prevention of propagation, and so on.

Stone dust, which may be slate, diatomite, limestone, or gypsum, can be used in the form of stone dust barriers in order to kill an explosion. The dusts usually have perhaps 85 percent of particles sized less than 75 μ and their effectiveness varies depending upon their material substance. For example, in the case of coal dust with a volatile matter of 33.3 percent, the limit percentages of the named dusts required to suppress the explosion are as listed in Table 8.12.[18]

Because the flame of the coal dust explosion is preceded by a blast, the introduction of a ledge or a shelf on top of which is a loose pile of stone dust causes the release of the stone dust due to the agitation caused by the

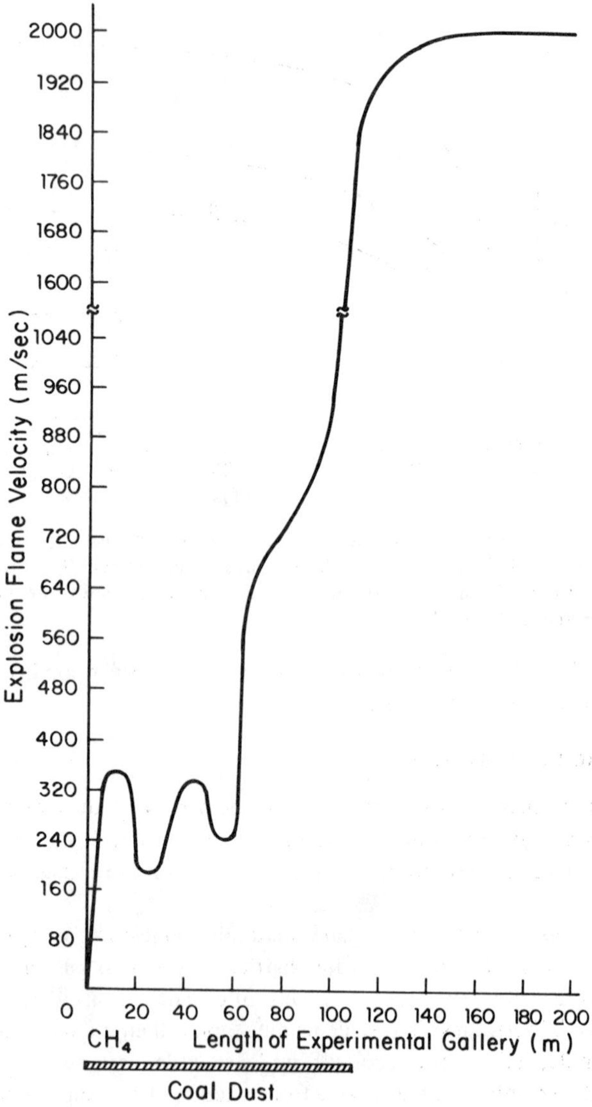

Fig. 8.11 A diagram of the coal dust explosion No. 680; seam 18V; the "powder" dust; $F = 8410$ (14,700) cm^2/g; $V = 17.3$ percent, $n = 28.4$ percent; the easiest dispersion conditions, initiator 27.5 m^3 CH_4.[18]

Table 8.12

THE PERCENT OF DUST REQUIRED TO SUPPRESS EXPLOSION OF COAL DUST WITH 33.3 PERCENT V.M.[18]

Slate	67.5
Diatomite	62.5
Anydrite	60.0
Limestone	57.5
Dolomite	57.5
Gypsum	40.0

pressure wave and ensuing turbulence. The dust particles shield the coal dust particles from radiation and this quenches the reactions occurring. As an example, the insertion of a proper number of adequately designed stone dust barriers at point x in Fig. 8.11 would cause the suppression of the explosion moving along the gallery.[18] Water barriers are also used. In this connection water has a two-fold advantage: the evaporation of the liquid absorbs heat and the water prevents dust cloud formation. However, even wet dusts, if properly initiated, produce an explosion.[18]

8.6 HEALTH HAZARD CASE STUDIES

There are two main areas of concern with respect to health hazards of human subjects. In one type of hazard foreign protein (foreign to the host human being, that is) invades the biological host system and triggers an immunological response. The response is termed an allergic reaction. Asthma is a well-known example of allergic response. Modern health science researchers are indicating that this type of process may be much more important than was hitherto supposed. For example, it is possible for the host system to be tricked into triggering off an immunological response to certain of its own body protein substance, in effect recognizing friend as foe and virtually setting about to destroy itself (lupus disease is an example of this).[24] The interactions between fine particle hazards and foreign protein allergies should be sought out in the literature. Some foreign proteins of importance include animal and vegetable protein. In domestic environments, pets (especially cats) and also house mites, etc., are a steady source of allergic triggers for the unwary. Additional sources include the well-known pollens and and the not so well-known but just as insidious mold spores.[24]

The other type of hazard is that of toxicity of the finely divided material. As has been pointed out, however, it is unwise to consider the substance as dangerous; it is much better to consider that the substance has dangerous properties.[25]

Metal Powders and Their Components

Metal powders and their components contribute to dust pollution problems in industrial plants. Some examples of the dust content of the air in the working spaces in the case of components made from iron powder and nonferrous magnets are shown in Tables 8.13 and 8.14, respectively.

Table 8.13

DUST CONTENT IN WORKING ZONE DURING MANUFACTURE OF SINTERED ARTICLES FROM IRON-BASE POWDER[25]

Operation	*Dust concentration,* mg/m^3		
	minimum	maximum	mean
Screening of iron powders	7.3	12.0	10.2
Screening of graphite powder	10.0	27.0	15.8
Mixing of iron powder with graphite and plasticizer, loading and unloading of ball mills	46.5	72.0	59.5
Pressing of iron-graphite bushings on semiautomatic presses	0	1.3	0.32
Manual proportioning of iron-graphite mixture	4.0	29.5	19.5
Pressing of iron-graphite articles on hydraulic presses (at press operator's position)	0.1	5.0	2.7
Loading of articles into boats with packing powder	33.3	200.0	97.2
Separating of articles from packing material by screening	46.5	133.2	81.5

Clearly, some operations are more dusty and therefore (potentially)

Table 8.14

DUST CONTENT IN AIR OF WORKING ZONE DURING MANUFACTURE OF NONFERROUS MAGNETS (OF ALNICO AND MAGNICO TYPES)[25]

Operation	*Dust concentration*, mg/m^3		
	minimum	maximum	mean
Crushing of ferroaluminum master alloy in jaw crusher	131.0	234.0	193.0
Screening of nonferrous magnet charge on vibrating screen	22.6	252.0	186.0
Loading and unloading of nonferrous magnet charge from ball mill	13.3	65.6	48.8
Pressing of nonferrous magnets (Alnico and others) on semiautomatic presses	0	1.3	0.6
Pouring of nonferrous magnet charge into hopper of semiautomatic presses	57.5	121.0	89.3
Loading of nonferrous magnets into boats and their packing with carbon powder containing iron, nickel, aluminum, and titanium.			
Local ventilation not operating	46.6	153.2	96.2
Local ventilation operating	13.3	41.35	27.3
Unloading of articles from boats (local ventilation operating)	14.5	19.5	16.7
Dry polishing of nonferrous aluminum–nickel–cobalt magnets (Alnico) on flat polisher	64.5	242.0	146.7

more hazardous than others. But note the wide range of variability of the conditions that are reported. Measures of the particle sizes at the various working stages in the powder metallurgical industry are reported in Table 8.15. Note the very high percentages of submicron particles in the powders prior to the compaction stage. These are entirely capable of lung penetration down to the deepest recesses. In this connection, the reader is referred to a paper in which the lung is modelled in such a way that it can be considered as being a series of subsystems each with a different and specific Reynolds number.[26]

Table 8.15

PARTICLE SIZE ANALYSIS OF DUST AT OPERATOR'S POSITIONS DURING MANUFACTURE OF SINTERED ARTICLES, PERCENT[25]

Operation	*Size of dust particles, μ*						
	<1	1–2	2–3	3–4	4–5	5	6–7
Unloading of ferronickel master alloy from mill	34.8	32.4	13.8	11.2	6.2	1.6	–
Screening of ferronickel master alloy on vibrating screen (middle of shop)	41	23	4.8	7	21.2	2.4	0.6
Pressing of aluminum–nickel–cobalt magnets	50.8	14.8	11.2	12.2	1.2	9.2	0.6
Pressing of articles (middle of pressing department)	58	16	4.4	5.8	8.2	5.4	2.2
Loading of ferrobarium charge into mixer	37.6	26.8	18.4	8.2	7.6	1.4	–
Pressing of ferrobarium magnets	29	34.4	15.8	10	6.6	3.4	0.8
Pressing of tungsten–copper articles	18.4	15.6	20.8	7.8	16.6	16.6	4.2
Unloading of copper–graphite mixture from mixers	30.4	30.8	14.2	8	10.8	4	1.8
Milling of ferroaluminum master alloy in muller	24.6	20	10.6	11.8	30	3	–
Proportioning of iron–graphite mixture	16.6	20.6	20.8	29.4	8.2	4.4	
Pressing of ferrobarium magnets	33.4	27.4	22.2	8.2	4.6	2.8	1.4
Screening of copper–graphite mixture of vibrating screens	58	27.8	8.2	4	1.4	0.6	–
Milling of ferroaluminum master alloy in muller	18	24.2	18.6	14.4	15.2	6.4	3.2
Pressing of copper–graphite articles (PV-474 press)	57.8	16.6	8.4	4.8	5.6	4.8	2
Pressing of magnesium–nickel–cobalt magnets (Magnico)	50.8	14.8	11.2	12.2	7.2	3.2	0.6
Pressing of tungsten–copper contacts	34.6	32.6	13.8	11.2	6.2	1.6	–
Pressing of bronze–graphite articles (PV-474 press)	30.4	31.0	14.2	9	10.0	4	1.4
Loading and unloading of ferrobarium charge	18	36	26	12	5	1	2
Loading of nonferrous magnets into boats	15	25	38	10	12	–	–
Unloading of nonferrous magnets after sintering	10	24	36	8	15	2	5
Loading of iron–graphite articles into boats with packing material	13	19	42	7	14	3	2
Unloading bronze–graphite articles from boats	14	28	32	15	9	2	–
Separating from packing material	18	30	24	18	10	–	–

Exposure to metallic-type dusts adversely affects the respiratory system, the cardiovascular and nervous systems, the liver, kidneys, and also the gastrointestinal tract. One complicating factor in relating causes to effects is that the identification of foreign material deposited in the lung tissue (or any other tissue) does not necessarily implicate that deposit as the cause of the observed dysfunction. As an example, the electron micrograph in Fig. 8.12 shows electron-dense localities which when analyzed chemically show the presence of relatively high amounts of Fe, Ni, Co, and Cr such as might be expected from inhalation of welding fumes by the patient during the course of his work (see Fig. 8.13).[27] The inference is that these electron-dense regions represent residues long after much of the original particles have been dissolved and removed.[27] The patient in this case was suffering from lung disease.

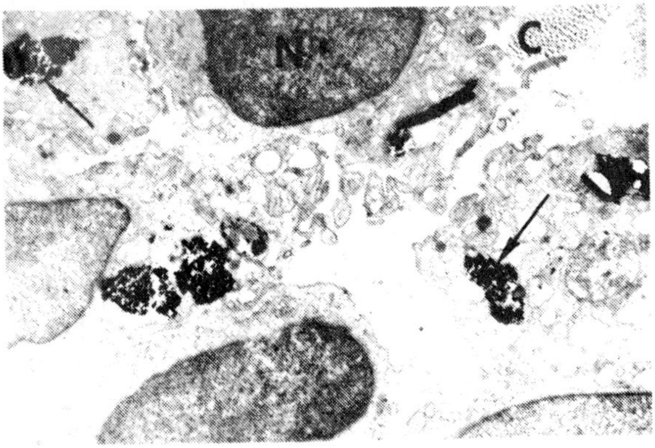

Fig. 8.12 Electron micrograph of a group of alveolar macrophages showing intracellular vacuoles containing electron-dense material (arrows).[27]

It was observed that metal-type powders have less fibrogenic activity than corresponding quartz dust (about one fifth) and that the main observable toxic effects are chronic and acute related to metabolic disorders resulting from the blocking of sulfhydryl, amino, and carboxyl groups of enzymes and other proteins. Also important are the disturbances caused to the biosynthesis of nucleic acids.[25] An overview of the various effects of metal-type powders is given in Fig. 8.14 which shows a scheme of the periodic table. In this are shown: 4) acute poisons, 5) chronic poisons, and 6) pneumoconiotic effect.

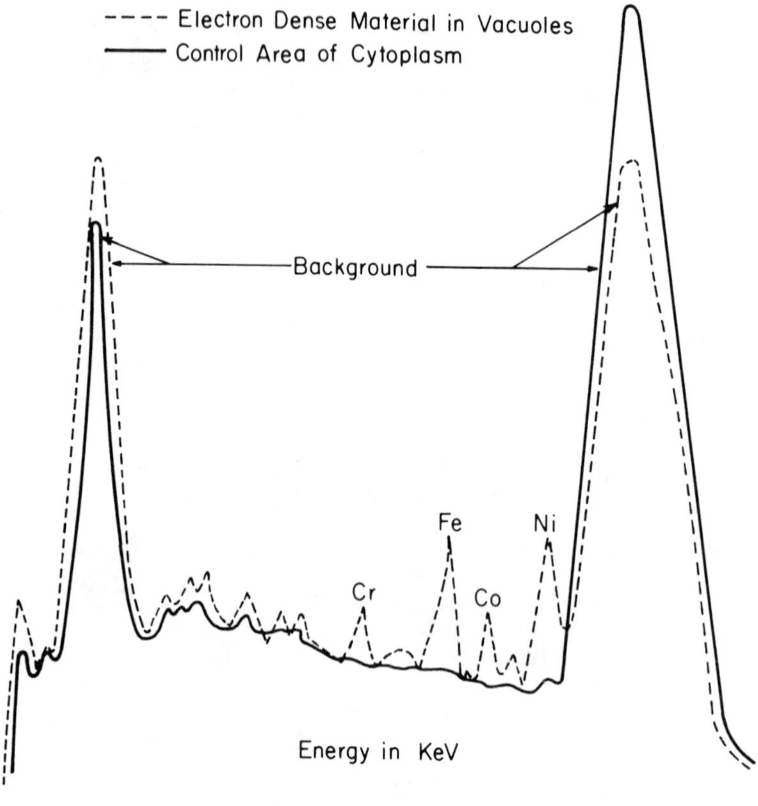

Fig. 8.13 EDXA of thin section of lung showing relative concentrations of cobalt, chromium, iron, and nickel in intracellular dense particles of macrophage. A control analysis of an adjacent area of the cytoplasm is also shown.[27]

Fibers

The extreme tortuosity of the human lung can be matched somewhat by that of a fiber with the result that fibers tend to be retained more readily than a spherical particle of similar mass. An analysis of the mineral fiber content of the human lungs is shown in Table 8.16.[28] These results are an analysis of three different groups of individuals from Pittsburgh, from Charleston, and from the asbestos industry, respectively. One should note

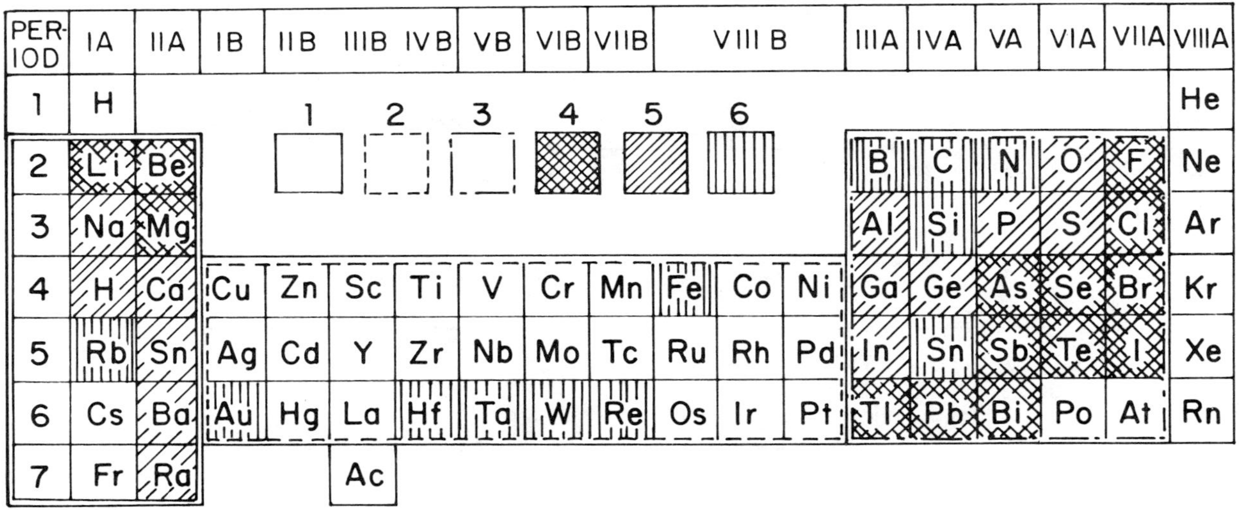

Fig. 8.14 Change of the toxic properties of elements as a function of their electronic structure; 1) s elements; 2) sd transition metals; 3) sp elements; 4) acute poison; 5) chronic poisons; 6) pneumoconiotic effect.[25]

Table 8.16

MINERAL FIBER CONTENT OF HUMAN LUNGS[28]

	Number of cases	*Number of women*	*Age*	*Mineral dust* (mg/g *dry lung*)	$\times 10^3$/g *Dry Lung*		
					Optical-size fibers	*EM-size fibers*	*Fibers less than* 5 μm
Pittsburgh	13	7	67	9.5 (2.6–33.0)	45 (5–170)	1800 (470–6750)	1200 (370–4810)
Charleston	10	8	60	2.8 (0.2–9.3)	7 (0.5–48.8)	1300 (150–6600)	900 (47–4650)
Asbestos workers	7	0	61	19 (5–35)	4000 (22,000–24,000)	150,000 (19,300–340,000)	80,000 (15,700–146,000)

Mineral Fiber Content of Human Lungs

	EM-size fibers/ optically visible fibers	*Percent fibers less than* 5 μm *long*	*Chrysotile* ($\times 10^3$)	*Percent chrysotile of all EM-size fibers*	*Percent chrysotile of fibers less than* 5 μm *long*
Asbestos workers	181 (10–1125)	71	160	7.4	10.5
Pittsburgh	815 (114–3200)	71	130	6.1	9.2
Charleston	325 (14–880)	53			

the percent of Chrysotile, which is the suspected carcinogen.

To illustrate the serious effects on mortality due to exposure to asbestos-containing working environments, the figures for the mortality of insulation workers in Belfast are worth mentioning. One hundred and seventy men were selected as having composed the total set of insulation workers in Belfast in 1940. As of 1966, there had occurred 98 deaths compared with an expected value of 37. The nasty part was that the ratio of observed over expected deaths for cancers of the lower respiratory tract and pleura was 17.6. The graph in Fig. 8.15 summarizes the story.[29]

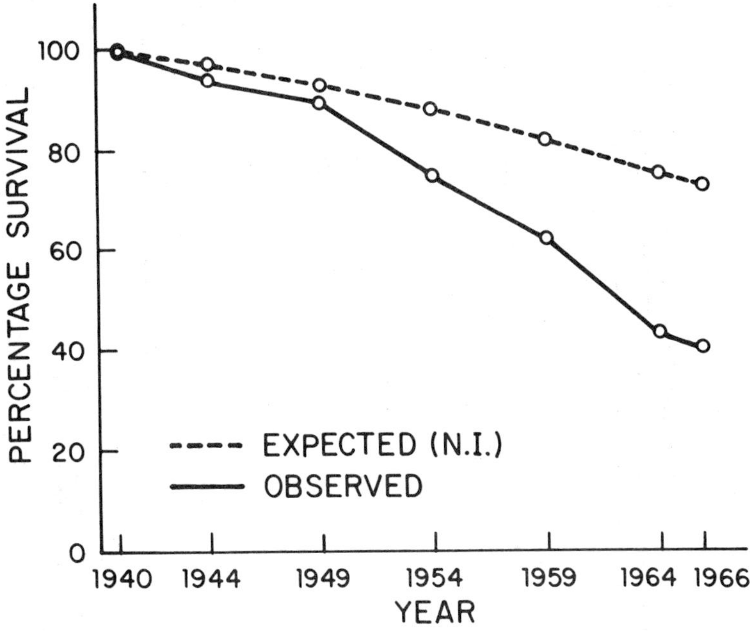

Fig. 8.15 Survival of insulation workers at five-year intervals compared with the male population of Northern Ireland adjusted for age. By the end of 1966 only 41 percent (67 men) survived compared with 73 percent (121 men) expected to survive.[29]

However, not all fibers have to be inorganic to present health hazards. Something as innocuous as wood dust has been shown to be associated with carcinoma of the nasal sinuses[30] and with certain pathological changes in the lungs[31] apart from the allergic problems occurring from time to time. Tables 8.17 and 8.18 show the levels of dust concentration and the median size of particles at five different furniture factories in the High Wycombe area of UK. It was observed in the factories that the workers used the masks supplied for their use only in the operation of the dustiest work tasks.[32]

Table 8.17

CONCENTRATIONS OF AIRBORNE DUST (mg/m^{-3}) MEASURED WITH THE PERSONAL AIR SAMPLER AT FIVE FURNITURE FACTORIES[32]

Factory	*Sample*	*Operation*						
		Band sawing	*Planing*	*Routing*	*Spindle molding*	*Sanding*	*Assembly*	*Turning*
A	Day 1	20.0[1]	2.0	1.8	5.8	2.4	25.5[3]	—
	Day 2	7.3[1]	2.4	3.8	6.3	3.6	2.1	—
B	Day 1	5.0	8.5	8.6	6.5	8.2	3.5	—
	Day 2	7.3	9.1	3.7	8.4	3.2	4.4	—
C	Day 1	1.0	1.8	3.5	1.5	2.0	3.7	—
	Day 2	1.8	2.8	3.3	4.4	2.4	4.5	—
D	Day 1	12.5[1,2]	3.1	94.6[1]	—	25.2[1]	5.9	—
	Day 2	9.2[1]	6.3[1]	8.2[1]	—	22.6	7.6	—
E	Day 1	6.5	10.9	—	3.2	7.9	8.2	4.6
	Day 2	4.1	4.1	—	4.4	12.2	9.8	12.5

[1] Sample contained 'inertials' in addition to fine dust
[2] Includes circular as well as band sawing
[3] This operator was sanding table tops before assembly

Table 8.18

CONCENTRATIONS AND MASS MEDIAN EQUIVALENT DIAMETERS OF AIRBORNE DUST MEASURED WITH THE CASCADE CENTRIPETER[32]

Operation	*No. of measurements*	*Dust concentration* (mg m^{-3}) Range	Mean	*Mass median equivalent diameter* (μm)
Band circular sawing	7	0.8-100	20.1	11.5
Planing	5	1.7- 9.4	3.6	9.2
Routing	4	2.5-11.3	5.5	10.0
Spindle molding	6	2.0-36.3	17.0	10.0
Sanding	8	0.5-34.3	8.0	8.4
Assembly	6	1.3- 5.3	3.4	7.6
Turning	1	–	9.0	11.5
Shaping	1	–	7.2	8.8
Cyclone shed	1	–	15.2	5.6
Near bag filters	1	–	1.4	6.4

REFERENCES

1. Palmer, K. "Dust Explosions Seminar," International Conference on Powder and Bulk Solids Handling, Chicago, May, 1976.
2. Dorsett, H., Jacobson, M., Nagy, W., and Williams, R. P. U.S. Bureau of Mines, RI 5624, 1960.
3. Palmer, K. *Dust Explosions and Fires.* London: Chapman and Hall, 1973.
4. Loss & Prevention Bulletin, 018 Institute of Chemical Engineers, 165–171 Rugby Terrace, Rugby, Warwicks, England.
5. Donaldson, et al, *Am. Ind. Hyg. Ass. J.* 25, 69 (1964).
6. Friedlander, F. J. *Am. Ind. Hyg. Ass. J.* 25, 37 (1964).
7. Cadle, R. *Particle Size.* New York: Van Nostrand Reinhold, 1965.
8. Stanford Research Institute, "The Smog Problem in Los Angeles County." Menlo Park, Calif., 1964.
9. American Industrial Hygiene Association. *Air Pollution Manual Part I.* Detroit, 1960.
10. Van de Hulst. *Light Scattering by Small Particles.* New York: Wiley, 1957.
11. Prindle. *J. Air Pol. Control Assoc.* 9, 12 (1959).
12. Hawkins and Nonhebel. *J. Inst. Fuel* 28, 530 (1955).
13. Schechter, J. "Desertification Processes and the Search for Solutions," *Interdisciplinary Science Reviews* 2, 1 (1977): 36–54.
14. UNEP/GC/30 Review of the environmental situation and of activities relating to the environmental programme, Governing Council, Nairobi, April 12, May 2, 1975.
15. Bagnold, R. A. *Blown Sands and Desert Dunes.* London: Methuen, 1941.
16. Essenhigh, R. H., and Fells, I. "The Physical Chemistry of Aerosols." discussions of the Faraday Society, 30, pp. 208–221, London, 1961.
17. Chamberlain, C. T., and Gray, W. A. *Nature* London, 216 (1967): 1245.
18. Cybulski, W. "Coal Dust Explosion and Their Suppression." translated by Zienkiewicz, A., Warsaw, Poland, TT 73-54001, NTIS, Dept. Commerce, Va., USA, 1975.
19. Godbert, A. L., and Wheeler, R.V. "The Relative Inflammability of Coal Dusts." A Laboratory Study, Safety in Mines Research Board, paper 56, 1929.
20. Brown, K. C., and James, G. J. "Safety in Mines Research Estab-

lishment." Research Report 201, 1962.
21. Field, M. A., Gill, D. W., Morgan, B. B., and Hawksley, P. G. W. "Combustion of Pulverised Coal." BCUR Assoc., Leatherhead, England, 1967.
22. Essenhigh, R. H. *J. Inst. Fuel.* 34 (1961): 239-244.
23. Effenberger, H. *Energietch* 13 (1963): 85-89, 125-129, 162-167, 208-212.
24. Kammermeyer, J., personal communication, November, 1976.
25. Brakhnova, I. T. "Environmental Hazards of Metals." translated from Russian by J. H. Slep, Consultants Bureau, New York, 1975 copyright Plenum Press.
26. Wang, C. S., Shah, M., and Cheng, Y. S. "Dispersion and Deposition of Inhaled Particles in the Human Respiratory Tract." presented at the 8th Annual Meeting of the Fine Particle Society, Chicago, August 1976. See also Shah, M. A., Chen, W. J. R., Whipple, R. T., and Wang, C. S. *Powder Technology* 18 (1977): 53-64.
27. Siegesmund, K. A., Funahashi, A., and Pintar, K. *Arch. Environ. Health* 28 (June, 1974): 345-349.
28. Gross, P., Harley, R. A., Davis, J. M., and Cralley, L. J. *Am. Ind. Hyg. Ass. J.* (March, 1974): 148-151.
29. Elmes, P. C., and Simpson, M. J. C. *Brit. J. Ind. Med.* 28 (1971): 226-236.
30. Acheson, E. D., Cowdell, R. H., Hadfield, E., and Macbeth, R. G. *Brit. Med. J.* 1968 2, pp. 587-596.
31. Michaels, L. *Can. Med. Ass. J.* 1967 96, 1150.
32. Hounam, R. F., and Williams, J. *Brit. Ind. Med.* 31 (1974): 1-9.
33. *The Modern Medical Encyclopedia,* editor-in-chief, Benjamin F. Miller, M.D., 1965, Golden Press, N.Y., pp. 1143, 1205.

AUTHOR INDEX

A

B

C

D

E

F

G

H

I

J

K

L

M

N

O

P

Q

R

S

T

U

V

W

Y

Z

SUBJECT INDEX

E

F

J

K

L

M

Q

R

S

T

W